GLOBAL ANALYSIS AND APPLIED MATHEMATICS

Related Titles from AIP Conference Proceedings

718 Computing Anticipatory Systems: CASYS '03 - Sixth International Conference
Edited by D. M. Dubois, August 2004, 0-7354-0198-5

707 Bayesian Inference and Maximum Entropy Methods in Science and Engineering:
23rd International Workshop on Bayesian Inference and Maximum Entropy Methods in
Science and Engineering
Edited by Gary Erickson and Yuxiang Zhai, May 2004, 0-7354-0182-9

690 The Monte Carlo Method in the Physical Sciences: Celebrating the 50th Anniversary
of the Metropolis Algorithm
Edited by James E. Gubernatis, November 2003, 0-7354-0162-4

676 Experimental Chaos: 7th Experimental Chaos Conference
Edited by V. In, L. Kocarev, T. L. Carroll, B. J. Gluckman, S. Boccaletti, and J. Kurths,
August 2003, 0-7354-0145-4

661 Modeling of Complex Systems: Seventh Granada Lectures
Edited by Pedro L. Garrido and Joaquín Marro, April 2003, 0-7354-0121-7

659 Bayesian Inference and Maximum Entropy Methods in Science and Engineering:
22nd International Workshop on Bayesian Inference and Maximum Entropy Methods in
Science and Engineering
Edited by Christopher J. Williams, March 2003, 0-7354-0119-5

658 Modern Challenges in Statistical Mechanics: Patterns, Noise, and the Interplay of
Nonlinearity and Complexity; Pan American Studies Institute
Edited by V. M. Kenkre and K. Lindenberg, March 2003, 0-7354-0118-7

627 Computing Anticipatory Systems: CASYS 2001 - Fifth International Conference
Edited by D. M. Dubois, September 2002, 0-7354-0081-4

583 Advanced Computing and Analysis Techniques in Physics Research:
VII International Workshop; ACAT 2000
Edited by Pushpalatha C. Bhat and Matthias Kasemann, August 2001, 0-7354-0023-7

511 Unsolved Problems of Noise and Fluctuations: UPoN'99: Second International
Conference
Edited by Derek Abbott and Lazlo B. Kish, March 2000, 1-56396-826-6

To learn more about these titles, or the AIP Conference Proceedings Series, please visit the webpage
http://proceedings/aip.org/proceedings

GLOBAL ANALYSIS AND APPLIED MATHEMATICS

International Workshop on Global Analysis

Ankara, Turkey   15 – 17 April 2004

EDITORS
Kenan Taş
Dumitru Baleanu
Çankaya University, Ankara, Turkey

Demeter Krupka
Olga Krupková
Palacky University, Olomouc, Czech Republic

SPONSORING ORGANIZATION
Çankaya University

Melville, New York, 2004
AIP CONFERENCE PROCEEDINGS ■ VOLUME 729

Editors:

Kenan Taş
Dumitru Baleanu

Çankaya University
Ogretmenler Cad. No. 14
Yuzuncuyil
06530 Ankara
TURKEY

E-mail: kenan@cankaya.edu.tr
         dumitru@cankaya.edu.tr

Demeter Krupka
Olga Krupková

Faculty of Science
Palacky University
Tomkova 40
779 00 Olomouc
CZECH REPUBLIC

E-mail: krupka@inf.upol.cz
         krupkova@inf.upol.cz

L.C. Catalog Card No. 2004112437
ISBN 0-7354-0209-4
ISSN 0094-243X
Printed in the United States of America

CONTENTS

Preface...ix

PLENARY LECTURES

Global Variational Principles: Foundations and Current Problems.............3
    D. Krupka
The Geometry of Variational Equations 19
    O. Krupková
A Compatibility Criterion for Systems of PDEs and Generalized
Lagrange-Charpit Method..39
    B. Kruglikov and V. Lychagin
Motion Coordination Algorithms Resulting from Classical Geometric
Optimization Problems...54
    J. Cortés

GLOBAL ANALYSIS

Zeta-Regularization and Calculus on Infinite Dimensional Spaces.............71
    A. Asada
Fractional Euler-Lagrange Equations for Constrained Systems 84
    T. Avkar and D. Baleanu
Variational Problems in the Geometrized First-Order Jet Framework 91
    V. Balan
Killing-Yano Tensors, Surface Terms and Superintegrable Systems...........99
    D. Baleanu and Ö. Defterli
Compatibility of Non-Generic Supersymmetries and Geometric
Duality for a Subclass of Generalized pp-wave Metrics....................106
    D. Baleanu and S. Başkal
Collineations of $(2+1)$-Dimensional
Friedmann-Robertson-Walker Spacetimes 114
    U. Camci
Symmetries and Supersymmetries of Dirac-type Operators
on Euclidean Taub-NUT Space...124
    I. I. Cotăescu and M. Visinescu
Variational Non-Holonomic Systems in Physics...........................131
    L. Czudková and J. Musilová
A New Natural Hamiltonian System on T^*S^2 Admitting an Integral
of Degree 3 in Momenta 141
    H. R. Dullin and V. S. Matveev
Linear Operators on Generalized Bergman Spaces........................147
    C. de Fabritiis

On the Multi-Component NLS Type Systems and Their Gauge Equivalent: Examples and Reductions 162
    V. S. Gerdjikov and G. G. Grahovski

Discrete Calculus of Variations 170
    G. S. Guseinov

Symmetries and Orbit Theory in 4-Dimensional Lorentz Manifolds 177
    G. S. Hall

Finite Group Invariance and Solution of Jaynes-Cummings Hamiltonian 185
    D. Haydargil and R. Koç

Decomposition of ODEs with an sl_2 Algebra of Symmetries 193
    C. V. Jensen

Dixmier-Douady Sheaves of Groupoids and Brownian Loops 200
    R. Léandre

The Geometry of the Geodesic Equation in the Framework of Jets of Submanifolds 207
    G. Manno

Doubly Warped Product Manifolds and Submanifolds 218
    K. Matsumoto

On Einstein Lagrangian Submanifolds of a Complex Projective Space 225
    Y. Matsuyama

Generalized Connections on Affine Bundles 232
    T. Mestdag

On the Path Integral of Constrained Systems 240
    S. I. Muslih

An Extension of the Mazur-Ulam Theorem 248
    C. P. Niculescu

The Types of Centro-Affine Curves 257
    Ö. Pekşen

Sprays and Cartan Projective Connections 262
    D. J. Saunders

Some Properties of the Cauchy-Type Integral for the Laplace Vector Fields Theory 274
    B. Schneider and M. Shapiro

A Generalization of Lepage Forms in Mechanics 281
    J. Seděnková

On Regularization of Second Order Lagrangians 289
    D. Smetanová

On the Nonholonomic Variational Principle 297
    M. Swaczyna

APPLIED MATHEMATICS

Fredholm Joint Spectrum for Families of Operators 309
    A. Akgül and S. Bayramov

Oscillation Criteria for Second Order Impulsive Delay
Differential Equation 317
    J. Alzabut and A. Zafer
Nonlocal Boundary-Value Problems for PDE: Well-Posedness 325
    A. Ashyralyev
Crossover Behavior between KdV and mKdV Equations in a
Cold Plasma with Negative Ions 332
    D. Grecu, A. Visinescu, and A. S. Cârstea
An Overview of Mean Field Theory in Combinatorial
Optimization Problems. 339
    S. Kasap and T. B. Trafalis
Differential Algebraic Equations in Primal Dual Interior Point
Optimization Methods 347
    S. Kasap and T. B. Trafalis
Nambu-Poisson Dynamics and Its Applications 355
    N. Makhaldiani
Relaxation Phenomena in the (in)activation Gates of the
Voltage-Gated Ion Channels 362
    M. Özer
Stability Analysis of the Steady-State Solution of a Mathematical
Model in Tumor Angiogenesis. 369
    S. Pamuk and A. Gürbüz
The Hirota Method for Reaction-Diffusion Equations with
Three Distinct Roots. 374
    G. Tanoğlu and O. Pashaev
Group Theoretical Treatment of the Jan-Teller Systems:
$T_1 \otimes (e \oplus t_2)$ Coupling 381
    H. Tütüncüler, R. Koç, and B. U. Türkdönmez
Modulational Instability of Some Nonlinear Continuum
and Discrete Systems 389
    A. Visinescu and D. Grecu

Author Index 397

PREFACE

The International Workshop on Global Analysis was held at the Çankaya University in Ankara, Turkey from April 15th through April 17th, 2004. The Workshop took place under the auspices of the President of the Board of Trustees of the Çankaya University, Mr. Sitki Alp.
The members of the Organizing Committee were Dumitru Baleanu, Yurdahan Guler, Demeter Krupka (Vice Chairman) , Olga Krupková (Vice Chairman), Emre Sermutlu and Kenan Taş (Chairman).

More than 90 mathematicians and mathematical physicists from 32 countries participated in the meeting.

The program of the workshop included four plenary lectures which are published in these proceedings.

-Demeter Krupka, "Global Variational Principles: Foundations and Current Problems"

-Olga Krupková, "The Geometry of Variational Equations"

-Valentin Lychagin, "A Compatibility Criterion for Systems of PDEs and Generalized Lagrange-Charpit Method "

-Jorge Cortés Monforte, "Motion Coordination Algorithms Resulting from Classical Geometric Optimization Problems"

Further contributions to the conference were presented in two sections

     A) Global Analysis and Differential Equations – headed by O. Krupková,

     B) Global Variational Calculus, Geometric Optimization – headed by D. Krupka

in the form of lectures, short communications and posters.

These Proceedings contain plenary lectures and selected papers in global analysis and applied mathematics.

The Chairman, and the editors of these Proceedings are grateful to the President of the Board of Trustees of the Çankaya University, Mr.Sitki Alp, and to the Rector of the Çankaya University, Professor Ziya Aktas, for their support of the workshop activities.

We would like to thank all the referees and other colleagues who helped in preparing these Proceedings for publication. Our thanks are also due to all participants for their contributions to the workshop program and to these Proceedings.

The chairman wishes to express his thanks to Ahmet Kabarcik, Tansel Avkar, and Dilara Demirbulak for their successful work on solving several problems regarding the Proceedings. He is also grateful to the colleagues and volunteer students of the Department of Mathematics and Computer Science of Çankaya University for their assistance during the meeting. Special thanks are due to the supporting staff Serafettin Karakoy and to the coordinator of the International Relations Office, Gamze Tanil.

Ankara, July 30, 2004

Kenan Taş

Plenary Lectures

Global Variational Principles: Foundations and Current Problems

D. Krupka

Department of Algebra and Geometry, Palacky University, 779 00 Olomouc, Czech Republic
and
Department of Mathematics, La Trobe University, Melbourne, Victoria 3086, Australia

Abstract. A survey of basic concepts of the theory of unconstrained higher order variational principles in fibered spaces is given, and selected open problems, related to the variational sequence theory, are discussed.
Key words. Fibered manifold, lagrangian, variational principle, Lepage form, Noether symmetry.
PACS. 02.40.Vh, 02.30.Xx.
MSC. 49F05, 58A15, 58A20, 58E99.

1. VARIATIONAL THEORY IN FIBERED SPACES

In this part of the lecture we give a survey of basic concepts of the geometric theory of global, higher order variational functionals in fibered spaces. Our main sources are Garcia [8], Goldschmidt and Sternberg [10], Krupka [17], [19], [23], and Trautman [43], and, for the variational sequence theory, Krupka [25].

1.1. Differential forms on fibered manifolds

Throughout this paper, all manifolds and mappings belong to the category C^∞. Standard symbols of the differential calculus on manifolds are used: d is the exterior derivative, T the tangent functor, ∂_ξ (resp. i_ξ) the Lie derivative (resp. the contraction) by a vector field ξ, and $*$ stands for the pull-back operation. Y is a fibered manifold with base X and projection π, and we denote $n = \dim X$, $n + m = \dim Y$. $J^r Y$, where $r \geq 0$, is the r-jet prolongation of Y, and $\pi^{r,s} : J^r Y \to J^s Y$, $\pi^r : J^r Y \to X$ are the canonical *jet projections*. The points of $J^r Y$ are r-jets $J^r_x \gamma$ of sections γ of Y at $x \in X$; the *r-jet prolongation* of γ is the mapping $x \mapsto J^r \gamma(x) = J^r_x \gamma$. Any *fibered chart* (V, ψ), $\psi = (x^i, y^\sigma)$, on Y induces the *associated charts* (V^r, ψ^r), $\psi^r = (x^i, y^\sigma, y^\sigma_{j_1}, y^\sigma_{j_1 j_2}, \ldots, y^\sigma_{j_1 j_2 \ldots j_r})$, on $J^r Y$, and (U, φ), $\varphi = (x^i)$, on X; here $V^r = (\pi^r)^{-1}(V)$, and $1 \leq i, j_1, j_2, \ldots, j_r \leq n$, $1 \leq \sigma \leq m$. A vector

CP729, *Global Analysis and Applied Mathematics: International Workshop on Global Analysis*,
edited by K. Taş, D. Krupka, O. Krupková, and D. Baleanu
© 2004 American Institute of Physics 0-7354-0209-4/04/$22.00

Ξ at a point $y \in Y$ is said to be π-*vertical*, if $T_y \pi \cdot \Xi = 0$; a differential form ρ on Y is π-*horizontal*, if it vanishes whenever one of its arguments is a π-vertical vector. Clearly, these concepts apply to the canonical jet projections.

For any open set $W \subset Y$, we denote by $\Omega_0^r W$ (resp. $\Omega_k^r W$) the ring of smooth functions (resp. the $\Omega_0^r W$-module of smooth k-forms) on $W^r = (\pi^{r,0})^{-1}(W)$. We also use some submodules of these modules; $\Omega_{k,X}^r W \subset \Omega_k^r W$ (resp. $\Omega_{k,Y}^r W \subset \Omega_k^r W$) are submodules of π^r-horizontal (resp. $\pi^{r,0}$-horizontal) forms. We have a morphism of exterior algebras $h : \Omega_k^r W \to \Omega_{k,X}^{r+1} W$ defined by

$$hf = f\pi^{r+1,r}, \quad hdx^i = dx^i, \quad hdy^\sigma_{j_1 j_2 \dots j_l} = y^\sigma_{j_1 j_2 \dots j_l k} dx^k, \tag{1}$$

where $f : V^r \to \mathbb{R}$ is a function; obviously, $J^r \gamma^* \rho = J^{r+1} \gamma^* h\rho$ for every section γ of Y. We call h the π-*horizontalization*. We say that a form $\rho \in \Omega_k^r W$ is *contact*, if $h\rho = 0$. For any fibered chart (V, ψ), $\psi = (x^i, y^\sigma)$, the 1-forms

$$\omega^\sigma_{j_1 j_2 \dots j_l} = dy^\sigma_{j_1 j_2 \dots j_l} - y^\sigma_{j_1 j_2 \dots j_l p} dx^p, \quad 0 \le l \le r-1, \tag{2}$$

are examples of contact forms. The system of forms $dx^i, \omega^\sigma_{j_1 j_2 \dots j_l}, dy^\sigma_{j_1 j_2 \dots j_r}$ is a basis of linear forms on V^r.

A form $\rho \in \Omega_k^r W$ has a unique decomposition

$$(\pi^{r+1,r})^* \rho = h\rho + p_1\rho + p_2\rho + \dots + p_k\rho, \tag{3}$$

in which $p_k\rho$ contains, in any fibered chart, exactly k exterior factors $\omega^\sigma_{j_1 j_2 \dots j_l}$; transformation properties of these forms guarantee invariance of the decomposition. $p_k\rho$ is the k-*contact component* of ρ. If $k \ge n+1$, then we define $\rho \in \Omega_k^r W$ to be *strongly contact*, if $p_{k-n}\rho = 0$.

1.2. Lagrangians, variational functionals

By a *lagrangian* (of order r) for Y we mean an element λ of the module $\Omega_{n,X}^r W$. In a fibered chart,

$$\lambda = \mathscr{L}\omega_0, \tag{4}$$

where $\omega_0 = dx^1 \wedge dx^2 \wedge \dots \wedge dx^n$; the component $\mathscr{L} : V^r \to \mathbb{R}$ is the Lagrange function. Let Ω be a piece of X, i.e., a compact, n-dimensional submanifold with boundary $\partial\Omega$, and let $\Gamma_\Omega Y$ be the set of sections of Y, defined on Ω. λ gives rise to the *variational functional*

$$\Gamma_\Omega Y \ni \gamma \to \lambda_\Omega(\gamma) = \int_\Omega J^r \gamma^* \lambda \in \mathbb{R}. \tag{5}$$

Our principal aim in this section will be to discuss general properties of (5).

By a π-*projectable* vector field we mean a vector field Ξ on Y such that there exists a vector field ξ on X satisfying

$$T\pi \cdot \Xi = \xi \circ \pi. \tag{6}$$

Let $V \subset Y$ be an open set, $\alpha : V \to Y$ a diffeomorphism, commuting with π, $U = \pi(V)$, and let $\alpha_0 : U \to X$ be the π-projection of α. Setting

$$J^r\alpha(J^r_x\gamma) = J^r_{\alpha_0(x)}\alpha\gamma\alpha_0^{-1} \tag{7}$$

for every $J^r_x\gamma \in V^r$, we obtain the *r-jet prolongation* $J^r\alpha : V^r \to J^rY$. Applying this concept to the flow of a π-projectable vector field Ξ on Y, and differentiating, we obtain the *r-jet prolongation* of Ξ, denoted by $J^r\Xi$. The chart expression of $J^r\Xi$ is given in [23].

Let $U \subset X$ be an open set, let $\gamma : U \to Y$ be a section. Let Ξ be a π-projectable vector field on an open set $W \subset Y$ such that $\gamma(U) \subset W$. If α_t is the flow of Ξ, and $\alpha_{(0)t}$ is its π-projection, then since $\pi\alpha_t = \alpha_{(0)t}\pi$ for all t,

$$\gamma_t = \alpha_t\gamma\alpha_{(0)t}^{-1} \tag{8}$$

is a 1-parameter family of sections of Y, depending smoothly on the parameter t. Sometimes γ_t is called the *variation*, or the *deformation* of γ, induced by Ξ.

Consider the variational functional (5). Choose an element $\gamma \in \Gamma_\Omega Y$ and a π-projectable vector field Ξ on Y, and consider the variation γ_t of γ (8). Since the domain of γ_t contains Ω for all sufficiently small t, we get a real-valued function on a neighborhood $(-\varepsilon, \varepsilon)$ of the origin $0 \in \mathbb{R}$,

$$(-\varepsilon, \varepsilon) \ni t \to \lambda_{\alpha_{(0)t}(\Omega)}(\alpha_t\gamma\alpha_{(0)t}^{-1}) = \int_{\alpha_{(0)t}(\Omega)} J^r(\alpha_t\gamma\alpha_{(0)t}^{-1})^*\lambda \in \mathbb{R}. \tag{9}$$

But $J^r(\alpha_t\gamma\alpha_{(0)t}^{-1})^*\lambda = (J^r\alpha_t \circ J^r\gamma \circ \alpha_{(0)t}^{-1})^*\lambda = (\alpha_{(0)t}^{-1})^*J^r\gamma^*J^r\alpha_t^*\lambda$, so we have

$$\int_{\alpha_{(0)t}(\Omega)} (J^r(\alpha_t\gamma\alpha_{(0)t}^{-1}))^*\lambda = \int_{\alpha_{(0)t}(\Omega)} (\alpha_{(0)t}^{-1})^*J^r\gamma^*J^r\alpha_t^*\lambda = \int_\Omega J^r\gamma^*J^r\alpha_t^*\lambda. \tag{10}$$

Differentiating the function (9) at $t = 0$ we obtain

$$(\partial_{J^r\Xi}\lambda)_\Omega(\gamma) = \int_\Omega J^r\gamma^*\partial_{J^r\Xi}\lambda. \tag{11}$$

The number (11) is the *variation* of the variational functional λ_Ω at γ, *induced* by the vector field Ξ. This formula shows, in particular, that the function $\Gamma_\Omega Y \ni \gamma \to (\partial_{J^r\Xi}\lambda)_\Omega(\gamma) \in \mathbb{R}$ is the variational functional (over Ω) associated with a new lagrangian $\partial_{J^r\Xi}\lambda$. We call this function the *variational derivative*, or the *first variation* of λ_Ω by Ξ.

We say that a section $\gamma \in \Gamma_\Omega Y$ is *stable* with respect to a variation Ξ of γ, if $(\partial_{J^r\Xi}\lambda)_\Omega(\gamma) = 0$. Stable sections with respect to *families* of variations are defined in an obvious way. If γ is stable with respect to all Ξ with support contained in $\pi^{-1}(\Omega)$, we say that γ is an *extremal* of λ_Ω. A section γ which is an extremal of every λ_Ω is said to be an *extremal* (of λ). Formula (11) admits a direct generalization. If Ξ_1 and Ξ_2 are two π-projectable vector fields, then the *second variational derivative*, or the *second variation*, of the variational functional λ_Ω by these vector fields is the mapping $\Gamma_\Omega Y \ni \gamma \to (\partial_{J^r\Xi_1}\partial_{J^r\Xi_2}\lambda)_\Omega(\gamma) \in \mathbb{R}$ defined by the formula

$$(\partial_{J^r\Xi_1}\partial_{J^r\Xi_2}\lambda)_\Omega(\gamma) = \int_\Omega J^r\gamma^*\partial_{J^r\Xi_1}\partial_{J^r\Xi_2}\lambda. \tag{12}$$

It is now obvious how *higher variational derivatives* are defined.

1.3. Lepage equivalents and the first variation formula

Our aim now will be to analyze the structure of the variational derivatives by means of invariant differential-geometric operations. We know that if η is a differential form on a manifold M, and ξ is a vector field on M, then the Lie derivative $\partial_\xi \eta$ decomposes into two terms, $\partial_\xi \eta = i_\xi d\eta + d i_\xi \eta$. We wish to apply this formula to the Lie derivative (11).

Let $\lambda \in \Omega^r_{n,X} W$ be a lagrangian. We say that a form $\rho \in \Omega^s_n W$ is a *Lepage equivalent* of λ, if (a) $h\rho = \lambda$ (up to a canonical jet projection), and (b) $h i_\zeta d\rho = 0$ for every $\pi^{s,0}$-vertical vector field ζ on W^s; condition (b) is equivalent to saying that $p_1 d\rho$ is $\pi^{s+1,0}$-horizontal.

Let $\rho \in \Omega^s_n W$ be a Lepage equivalent of $\lambda \in \Omega^r_{n,X} W$. Condition (a) implies that

$$\int_\Omega J^s \gamma^* \rho = \int_\Omega J^{s+1} \gamma^* h\rho = \int_\Omega J^r \gamma^* \lambda, \tag{13}$$

which means that the variational functional associated with ρ coincides with λ_Ω (5). For any π-projectable vector field Ξ on W, we have

$$J^s \gamma^* \partial_{J^s \Xi} \rho = J^s \gamma^* i_{J^s \Xi} d\rho + d J^s \gamma^* i_{J^s \Xi} \rho. \tag{14}$$

Since (b) is equivalent to saying that $J^s \gamma^* i_\zeta d\rho = 0$ for every ζ and γ, the term $J^s \gamma^* i_{J^s \Xi} d\rho$ in (14) depends (linearly) on $J^s \Xi$ via the vector field Ξ only, but is independent of derivatives of the components of Ξ. Moreover, since $J^s \gamma^* \partial_{J^s \Xi} \rho = J^{s+1} \gamma^* h \partial_{J^s \Xi} \rho = J^{s+1} \gamma^* \partial_{J^{s+1} \Xi} h\rho = J^r \gamma^* \partial_{J^r \Xi} \lambda$, (14) can also be written as $\partial_{J^r \Xi} \lambda = h i_{J^r \Xi} d\rho + h d i_{J^r \Xi} \rho$. Therefore, it is *a priori* clear that for a Lepage form ρ, the decomposition (14) has the properties which are required in the integrand expressions of the classical first variation formula: If $\Omega \subset \pi(W)$ is a piece, then integrating (14) one gets

$$\int_\Omega J^r \gamma^* \partial_{J^r \Xi} \lambda = \int_\Omega J^s \gamma^* i_{J^s \Xi} d\rho + \int_{\partial\Omega} J^s \gamma^* i_{J^s \Xi} \rho. \tag{15}$$

The expression on the right of (14) can be further simplified using the 1-contact component of $d\rho$. Since $J^s \gamma^* i_{J^s \Xi} d\rho = J^{s+1} \gamma^* i_{J^{s+1} \Xi} p_1 d\rho$, we have the following result (the *first variation formula*).

Theorem 1. *For any Lepage equivalent $\rho \in \Omega^s_n W$ of a lagrangian $\lambda \in \Omega^r_{n,X} W$,*

$$\partial_{J^r \Xi} \lambda = i_{J^{s+1} \Xi} p_1 d\rho + h d i_{J^s \Xi} \rho. \tag{16}$$

We now describe further basic properties of Lepage equivalents. The proof of Theorem 2 is based on an analysis of the structure of differential forms on $J^r Y$, Theorem 3 is an easy consequence of definitions, and Theorem 4 can be proved by means of a partition of unity. Denote

$$\omega_i = (-1)^{i-1} dx^1 \wedge dx^2 \wedge \ldots \wedge dx^{i-1} \wedge dx^{i+1} \wedge \ldots \wedge dx^n. \tag{17}$$

Theorem 2. *Let $W \subset Y$ be an open set, and let $\lambda \in \Omega_{n,X}^r W$ be a lagrangian. Assume that in a fibered chart (V, ψ), $\psi = (x^i, y^\sigma)$, on Y such that $V \subset W$, $\lambda = \mathcal{L}\omega_0$. Then a form $\rho \in \Omega_n^s W$ is a Lepage equivalent of λ if and only if $(\pi^{s+1,s})^*\rho$ has an expression*

$$(\pi^{s+1,s})^*\rho = \Theta_\lambda + d\eta + \mu, \qquad (18)$$

where

$$\Theta_\lambda = \mathcal{L}\omega_0 + \sum_{k=0}^{r-1} \left(\sum_{l=0}^{r-1-k} (-1)^l d_{p_1} d_{p_2} \dots d_{p_l} \frac{\partial \mathcal{L}}{\partial y_{j_1 j_2 \dots j_k p_1 p_2 \dots p_l i}^\sigma} \right) \omega_{j_1 j_2 \dots j_k}^\sigma \wedge \omega_i, \qquad (19)$$

η *is a contact $(n-1)$-form, and the order of contactness of μ is ≥ 2.*

The n-form Θ_λ, defined by (19), is the *principal Lepage equivalent* of the lagrangian λ, associated with the fibered chart (V, ψ).

Theorem 3. *Let $W \subset Y$ be an open set, let $\lambda \in \Omega_{n,X}^r W$ be a lagrangian, and let $\rho \in \Omega_n^s W$ be a Lepage equivalent of λ. Then the form $(\pi^{s+1,s})^* d\rho$ has an expression*

$$(\pi^{s+1,s})^* d\rho = E(\lambda) + F, \qquad (20)$$

where $E(\lambda) = p_1 d\rho$, and F is a form whose order of contactness is ≥ 2. If in a fibered chart, $\lambda = \mathcal{L}\omega_0$, then $E(\lambda)$ has the chart expression

$$E(\lambda) = E_\sigma(\mathcal{L})\omega^\sigma \wedge \omega_0, \qquad (21)$$

where

$$E_\sigma(\mathcal{L}) = \sum_{l=0}^{r} (-1)^l d_{p_1} d_{p_2} \dots d_{p_l} \frac{\partial \mathcal{L}}{\partial y_{p_1 p_2 \dots p_l}^\sigma}. \qquad (22)$$

The form $E(\lambda)$ is called the *Euler-Lagrange form* associated with λ. The components $E_\sigma(\mathcal{L})$ (22) are the *Euler-Lagrange expressions*. Obviously, $E(\lambda)$ belongs to $\Omega_{n+1,Y}^{2r} W$. The mapping $\Omega_{n,X}^r W \ni \lambda \to E(\lambda) \in \Omega_{n+1,Y}^{2r} W$, assigning to a lagrangian its Euler-Lagrange form, is called the *Euler-Lagrange mapping*.

The following result guarantees that we can always use the (global) first variation formula (16).

Theorem 4. *Every lagrangian has a Lepage equivalent.*

A section $\gamma \in \Gamma_\Omega Y$ is an extremal of λ if and only if $E(\lambda) \circ J^{2r}\gamma = 0$, or, which is the same, it is a solution of the *Euler-Lagrange equations* $E_\sigma(\mathcal{L}) = 0$. We give the following standard theorem for the record.

Theorem 5. *Let $\lambda \in \Omega_{n,X}^r W$ be a lagrangian, $\rho \in \Omega_n^s W$ be a Lepage equivalent of λ, and γ a section of Y. The following conditions are equivalent:*

(a) *γ is an extremal of λ.*

(b) *For every π-vertical vector field Ξ, $J^{2r}\gamma^* i_{J^{2r}\Xi} d\rho = 0$.*

(c) *For any fibered chart* (V, ψ), $\psi = (x^i, y^\sigma)$, γ *satisfies the system of partial differential equations*

$$E_\sigma(\mathscr{L}) \circ J^{s+1}\gamma = 0, \quad 1 \leq \sigma \leq m. \tag{23}$$

1.4. Invariant variational principles

Let λ be a lagrangian of order r for Y, and let $\alpha : W \to Y$ be an automorphism of Y. We say that α is an *invariant transformation* of λ (resp. $E(\lambda)$), if $J^r\alpha^*\lambda = \lambda$ (resp. $J^s\alpha^*E(\lambda) = E(\lambda)$). A *generator* of invariant transformations of λ (resp. $E(\lambda)$) is a π-projectable vector field on Y whose local one-parameter group consists of invariant transformations of λ (resp. $E(\lambda)$).

Since

$$J^s\alpha^*E(\lambda) = E((J^r\alpha)^*\lambda) \tag{24}$$

for every automorphism α of Y, an invariant transformation of λ is always an invariant transformation of $E(\lambda)$.

Theorem 6. *Let Ξ be a π-projectable vector field on Y.*

(a) Ξ *generates invariant transformations of λ if and only if*

$$\partial_{J^r\Xi}\lambda = 0. \tag{25}$$

(b) Ξ *generates invariant transformations of $E(\lambda)$ if and only if*

$$\partial_{J^s\Xi}E(\lambda) = 0. \tag{26}$$

Equations (24), the *Noether equation*, and (25), the *Noether-Bessel-Hagen equation*, represent relations between λ and the vector field Ξ.

The following is known as the (first) *theorem of Emmy Noether*. Its proof, based on the first variation formula (Theorem 1), is trivial.

Theorem 7. *Let λ be a lagrangian, ρ a Lepage equivalent of λ, let γ be an extremal. Then for every generator Ξ of invariant transformations of λ,*

$$dJ^s\gamma^*i_{J^s\Xi}\rho = 0. \tag{27}$$

1.5. The variational sequence

As we have already seen, the Euler-Lagrange mapping E assigns to an n-form (the lagrangian) an $(n+1)$-form (the Euler-Lagrange form). This leads to an idea to find a proper (cohomological) sequence of forms in which E would be included as one arrow, and to get in this way some *global* characteristics of E. A conclusion, which we discuss below, is the theory of (finite order) variational sequences.

Let $\Omega_{0,c}^r = \{0\}$, and let $\Omega_{k,c}^r$ be the sheaf of *contact k-forms*, if $k \le n$, or the sheaf of *strongly contact k-forms*, if $k > n$, on J^rY. We set

$$\Theta_k^r = \Omega_{k,c}^r + d\Omega_{k-1,c}^r, \tag{28}$$

where $d\Omega_{k-1,c}^r$ is the image sheaf of $\Omega_{k-1,c}^r$ by the exterior derivative d. It can be shown that we get an exact sequence of soft sheaves $0 \to \Theta_1^r \to \Theta_2^r \to \Theta_3^r \to \ldots$, where the morphisms are the exterior derivative, i.e., a subsequence of the *De Rham sequence* $0 \to \mathbb{R} \to \Omega_0^r \to \Omega_1^r \to \Omega_2^r \to \Omega_3^r \to \ldots$. The quotient sequence

$$0 \to \mathbb{R} \to \Omega_0^r \to \Omega_1^r/\Theta_1^r \to \Omega_2^r/\Theta_2^r \to \Omega_3^r/\Theta_3^r \to \ldots \tag{29}$$

which is also exact, is called the *r-th order variational sequence* on Y. We denote the quotient mappings in (29) by $E_k : \Omega_k^r/\Theta_k^r \to \Omega_{k+1}^r/\Theta_{k+1}^r$. The following is a basic property of this sequence.

Theorem 8. *The variational sequence is an acyclic resolution of the constant sheaf $\mathbb{R}$ over Y.*

Denote (29) symbolically as $0 \to \mathbb{R} \to \mathscr{V}^r$. Let $\Gamma(Y, \mathscr{V}^r)$ be the cochain complex $0 \to \Gamma(Y, \mathbb{R}) \to \Gamma(Y, \Omega_0^r) \to \Gamma(Y, \Omega_1^r) \to \Gamma(Y, \Omega_2^r) \to \ldots$ of global sections of (29). We get as a corollary to the abstract De Rham theorem the following identification of the cohomology groups $H^k(\Gamma(Y, \mathscr{V}^r))$ of this complex with the De Rham cohomology groups of the manifold Y

$$H^k(\Gamma(Y, \mathscr{V}^r)) = H^k Y. \tag{30}$$

Now we discuss some consequences of the theory of variational sequences. To understand the meaning of (30), one should compute the classes entering the quotient spaces in (29). Note that the quotient spaces Ω_k^r/Θ_k^r are determined *up to an isomorphism*. Thus, the classes admit various equivalent characterizations. A simple analysis shows that the elements of Ω_n^r/Θ_n^r can be identified, in fibered charts, with some n-forms $\mathscr{L}\omega_0$, i.e., with some *lagrangians* for Y. The elements of $\Omega_{n+1}^r/\Theta_{n+1}^r$ can be identified with $(n+1)$-forms $\varepsilon_\sigma \omega^\sigma \wedge \omega_0$, i.e., with *dynamical*, or *source forms*. More precisely, we have the following result.

Theorem 9. *The sheaf Ω_n^r/Θ_n^r is isomorphic with a subsheaf of the sheaf of lagrangians $\Omega_{n,X}^{r+1}$, $\Omega_{n+1}^r/\Theta_{n+1}^r$, is isomorphic with a subsheaf of the sheaf of dynamical forms $\Omega_{n+1,Y}^{2r+1}$, and the quotient mapping $E_n : \Omega_n^r/\Theta_n^r \to \Omega_{n+1}^r/\Theta_{n+1}^r$ is the Euler-Lagrange mapping.*

Now it is clear what kind of results are described by the variational sequence. Assume that a lagrangian $\lambda = [\rho]$ satisfies $E_n(\lambda) = 0$. Then by exactness of (29), there always exists a class $[\eta]$ such that $E_{n-1}([\eta]) = [\rho] = [d\eta]$. This means that, locally, ρ decomposes as the sum of a closed form and a contact form. Condition

$$E_n(\lambda) = 0 \tag{31}$$

is the *local variational triviality condition*, and may be explicitly expressed with the help of (30). If in addition, $H^n Y = \{0\}$, (30) says that η may be chosen *globally defined* on $J^r Y$. The local variational triviality condition strongly determines the structure of the lagrangians whose Euler-Lagrange forms vanish identically.

Analogously, assume that we have a source form $\varepsilon = [\rho]$ which satisfies the *local variationality condition*

$$E_{n+1}(\varepsilon) = 0. \tag{32}$$

Then there exists a class $[\eta]$ such that $E_n([\eta]) = [\rho] = [d\eta]$. Thus, locally, ρ can be expressed as the sum of a closed form and a strongly contact form. If in addition, $H^{n+1}Y = \{0\}$, (30) guarantees that may be chosen *globally defined* on $J^r Y$. The local variationality condition strongly determines the structure of such source forms which can, at least locally, be treated as the Euler-Lagrange forms of suitable lagrangians.

2. TOPICS: VARIATIONAL SEQUENCES

We discuss further notions, methods and ideas, related to the variational sequence theory. For simplicity, we restrict ourselves to lower orders and dimensions.

2.1. Lepage forms and their generalizations

By a *Lepage form of order s* we mean any form $\rho \in \Omega_n^s W$ such that $p_1 d\rho \in \Omega_{n+1,Y}^{s+1} W$ (Krupka [23]). Thus, the Lepage forms are defined by means of a restriction imposed on their exterior derivative; we have already seen that this restriction implies that $p_1 d\rho$ is exactly the Euler-Lagrange form of the lagrangian $h\rho \in \Omega_{n,X}^{s+1} W$, i.e.,

$$p_1 d\rho = E_{h\rho}. \tag{33}$$

In particular, this formula explains the Euler-Lagrange form as the composite of two geometric operations, acting in spaces of forms, d, and the projection p_1.

The concept of a Lepage form unifies several known examples of forms, used in the first order calculus of variations of multiple integrals, with different properties; the most well-known are the Cartan form, the Poincaré-Cartan form, and the Carathéodory form, and the so called *fundamental form* (see Betounes [3], Crampin and Saunders [4], Dedecker [5], Garcia [8], Goldschmidt and Sternberg [10], Gotay [12], Krupka [18], [19], Olver [35], Saunders [37], Sniatycki [39]). For the second order Lepage form, which can be considered as a direct generalization of the Cartan form, we refer to Krupka [23]. All these forms give the same variational functional, as well as the conservation laws; they lead to different Hamilton equations.

A problem arises, as to whether there exists a construction for forms of higher degrees, satisfying a condition similar to (33). An idea how to introduce such forms comes from the variational sequence theory and from the variational bicomplex theory. For a form $\rho \in \Omega_{n+k}^s W$, the expression on the right of (33) is taken to be the so called *canonical representative* of $d\rho$, constructed by means of the *interior Euler-Lagrange operator* (see Krbek and Musilová [15], Sedenková [41], and the references therein).

10

2.2. The Helmholtz form

In this section, we derive the chart expression of the *Helmholtz form* $E_{n+1}(\varepsilon) = [d\varepsilon]$ for a *third order* dynamical form

$$\varepsilon = \varepsilon_\sigma \omega^\sigma \wedge \omega_0. \tag{34}$$

We want to extract in a canonical manner from the class of the form

$$d\varepsilon = \left(\frac{\partial \varepsilon_\sigma}{\partial y^\nu} \omega^\nu + \frac{\partial \varepsilon_\sigma}{\partial y_j^\nu} \omega_j^\nu + \frac{\partial \varepsilon_\sigma}{\partial y_{jk}^\nu} \omega_{jk}^\nu + \frac{\partial \varepsilon_\sigma}{\partial y_{jkl}^\nu} \omega_{jkl}^\nu \right) \wedge \omega^\sigma \wedge \omega_0 \tag{35}$$

a term belonging to the module Θ_{n+2}^4. A direct method consists in applying proper relations among the elements of a basis of Θ_{n+2}^4, associated with a fibered chart. In particular, the following four formulas are needed:

$$
\begin{aligned}
(\omega_{jkl}^\nu &\wedge \omega^\sigma - \omega_{jkl}^\sigma \wedge \omega^\nu) \wedge \omega_0 \\
&= \tfrac{1}{3} d(\omega_{jk}^\nu \wedge \omega^\sigma \wedge \omega_l - \omega_{jk}^\sigma \wedge \omega^\nu \wedge \omega_l + \omega_{lj}^\nu \wedge \omega^\sigma \wedge \omega_k \\
&\quad - \omega_{lj}^\sigma \wedge \omega^\nu \wedge \omega_k + \omega_{kl}^\nu \wedge \omega^\sigma \wedge \omega_j - \omega_{kl}^\sigma \wedge \omega^\nu \wedge \omega_j) \\
&\quad - \tfrac{1}{6} d(\omega_j^\nu \wedge \omega^\sigma \wedge \omega_k + \omega_l^\nu \wedge \omega^\sigma \wedge \omega_k + \omega_j^\nu \wedge \omega^\sigma \wedge \omega_l \\
&\quad + \omega_k^\nu \wedge \omega_j^\sigma \wedge \omega_l + \omega_k^\nu \wedge \omega_l^\sigma \wedge \omega_j + \omega_l^\nu \wedge \omega_k^\sigma \wedge \omega_j), \\
(\omega_j^\nu &\wedge \omega_k^\sigma - \omega_j^\sigma \wedge \omega_k^\nu) \wedge \omega_0 + (\omega_{jk}^\nu \wedge \omega^\sigma - \omega_{jk}^\sigma \wedge \omega^\nu) \wedge \omega_0 \\
&= \tfrac{1}{2} d(\omega_j^\nu \wedge \omega^\sigma \wedge \omega_k - \omega_j^\sigma \wedge \omega^\nu \wedge \omega_k \\
&\quad + \omega_k^\nu \wedge \omega^\sigma \wedge \omega_j - \omega_k^\sigma \wedge \omega^\nu \wedge \omega_j), \\
(\omega_{jk}^\nu &\wedge \omega^\sigma + \omega_{jk}^\sigma \wedge \omega^\nu) \wedge \omega_0 \\
&= \tfrac{1}{2} d(\omega_j^\nu \wedge \omega^\sigma \wedge \omega_k + \omega_j^\sigma \wedge \omega^\nu \wedge \omega_k \\
&\quad + \omega_k^\nu \wedge \omega^\sigma \wedge \omega_j + \omega^\sigma \wedge \omega^\nu \wedge \omega_j), \\
(\omega_j^\nu &\wedge \omega^\sigma - \omega_j^\sigma \wedge \omega^\nu) \wedge \omega_0 = d(\omega^\nu \wedge \omega^\sigma \wedge \omega_j).
\end{aligned} \tag{36}
$$

Using these formulas, we can get the required Helmholtz form of ε,

$$
\begin{aligned}
E_{n+2}(\varepsilon) &= \frac{1}{2} \left(\frac{\partial \varepsilon_\sigma}{\partial y^\nu} - \frac{\partial \varepsilon_\nu}{\partial y^\sigma} - \frac{1}{2} d_j \left(\frac{\partial \varepsilon_\sigma}{\partial y_j^\nu} - \frac{\partial \varepsilon_\nu}{\partial y_j^\sigma} - \frac{1}{2} d_k d_l \left(\frac{\partial \varepsilon_\sigma}{\partial y_{jkl}^\nu} - \frac{\partial \varepsilon_\nu}{\partial y_{jkl}^\sigma} \right) \right) \right) \\
&\quad \cdot \omega^\nu \wedge \omega^\sigma \wedge \omega_0 \\
&\quad + \frac{1}{2} \left(\frac{\partial \varepsilon_\sigma}{\partial y_j^\nu} + \frac{\partial \varepsilon_\nu}{\partial y_j^\sigma} - d_k \left(\frac{\partial \varepsilon_\sigma}{\partial y_{jk}^\nu} + \frac{\partial \varepsilon_\nu}{\partial y_{jk}^\sigma} \right) \right) \omega_j^\nu \wedge \omega^\sigma \wedge \omega_0 \\
&\quad + \frac{1}{2} \left(\frac{\partial \varepsilon_\sigma}{\partial y_{jk}^\nu} - \frac{\partial \varepsilon_\nu}{\partial y_{jk}^\sigma} - \frac{3}{2} d_l \left(\frac{\partial \varepsilon_\sigma}{\partial y_{jkl}^\nu} - \frac{\partial \varepsilon_\nu}{\partial y_{jkl}^\sigma} \right) \right) \omega_{jk}^\nu \wedge \omega^\sigma \wedge \omega_0 \\
&\quad + \frac{1}{2} \left(\frac{\partial \varepsilon_\sigma}{\partial y_{jkl}^\nu} + \frac{\partial \varepsilon_\nu}{\partial y_{jkl}^\sigma} \right) \omega_{jkl}^\nu \wedge \omega^\sigma \wedge \omega_0.
\end{aligned} \tag{37}
$$

In the well-known sense, the vanishing of the coefficients in (37) are necessary and sufficient conditions for existence of (local) lagrangians for ε (the *Helmholtz conditions*).

11

For general Helmholtz conditions we refer to Anderson and Duchamp [1], and Krupka [19], [21]. The uniqueness of the Helmholtz operator has been proved by Kolar and Vitolo [14].

2.3. The inverse problem: First order field theory

In this section, we give a solution of the equation $E_{n+2}(\tau) = 0$ for the *first order* dynamical forms τ in field theory (Krupková and Haková [13], Krupka [24]). For higher order forms only partial results are known (Krupková [30], [31]).

Assume that we have a dynamical $(n+1)$-form on J^1Y, $\tau = \tau_\sigma \omega^\sigma \wedge \omega_0$. Recall that τ is said to be *variational*, if it coincides with the Euler-Lagrange form of a lagrangian for Y. *Local* variationality is equivalent with the conditions

$$\frac{\partial \tau_\sigma}{\partial y_j^\nu} + \frac{\partial \tau_\nu}{\partial y_j^\sigma} = 0, \qquad \frac{\partial \tau_\sigma}{\partial y^\nu} - \frac{\partial \tau_\nu}{\partial y^\sigma} + d_j \frac{\partial \tau_\nu}{\partial y_j^\sigma} = 0 \tag{38}$$

(see (37)).

One can prove equivalence of the following four conditions: (a) τ is locally variational, (b) there exists a unique $(n+1)$-form α on Y such that $\tau = p_1\alpha$, and $d\alpha = 0$, (c) there exists an n-form η on Y such that $\tau = E_\lambda$, and $\lambda = h\eta$, (4) τ is (globally) variational. In any fibered chart, variationality of τ is equivalent to the existence of functions $A_{\sigma_1\sigma_2...\sigma_k,i_{k+1}i_{k+2}...i_n}$, $0 \leq k \leq n$, defined on V, such that

$$
\begin{aligned}
\tau &= \left(-\sum_{k=1}^n \frac{1}{(k-1)!} \frac{\partial A_{\nu\sigma_2...\sigma_k,i_{k+1}i_{k+2}...i_n}}{\partial x^{i_1}} y_{i_2}^{\sigma_2} y_{i_3}^{\sigma_3} \cdots y_{i_k}^{\sigma_k} \right. \\
&\quad + \left. \sum_{k=1}^n \frac{1}{(k-1)!} \left(-\frac{\partial A_{\nu\sigma_2...\sigma_k,i_{k+1}i_{k+2}...i_n}}{\partial y^{\sigma_1}} + \frac{\partial A_{\sigma_1\sigma_2...\sigma_k,i_{k+1}i_{k+2}...i_n}}{\partial y^\nu} \right) y_{i_1}^{\sigma_1} y_{i_2}^{\sigma_2} \cdots y_{i_k}^{\sigma_k} \right) \\
&\quad \cdot \omega^\nu \wedge dx^{i_1} \wedge dx^{i_2} \wedge \ldots \wedge dx^{i_n}.
\end{aligned}
\tag{39}
$$

If τ is variational, then every lagrangian of the first order for τ has a chart expression $\lambda = \mathscr{L}\omega_0$, where

$$
\begin{aligned}
\mathscr{L} &= \left(A_{i_1i_2...i_n} + \frac{1}{1!}A_{\sigma_1,i_2i_3...i_n}y_{i_1}^{\sigma_1} + \frac{1}{2!}A_{\sigma_1\sigma_2,i_3i_4...i_n}y_{i_1}^{\sigma_1}y_{i_2}^{\sigma_2} \right. \\
&\quad + \ldots + \frac{1}{(n-1)!}A_{\sigma_1\sigma_2...\sigma_{n-1},i_n}y_{i_1}^{\sigma_1}y_{i_2}^{\sigma_2} \cdots y_{i_{n-1}}^{\sigma_{n-1}} \\
&\quad + \left. \frac{1}{n!}A_{\sigma_1\sigma_2...\sigma_n}y_{i_1}^{\sigma_1}y_{i_2}^{\sigma_2} \cdots y_{i_n}^{\sigma_n} \right) \varepsilon^{i_1i_2...i_n},
\end{aligned}
\tag{40}
$$

and the coefficients $A_{i_1i_2...i_n}$, $A_{\sigma_1,i_2i_3...i_n}$, $A_{\sigma_1\sigma_2,i_3i_4...i_n}$, ..., $A_{\sigma_1\sigma_2...\sigma_{n-1},i_n}$, $A_{\sigma_1\sigma_2...\sigma_n}$ depend on x^i, y^σ only. α is given by

$$
\begin{aligned}
\alpha &= \tau_\sigma \omega^\sigma \wedge \omega_0 + \sum_{k=1}^n \frac{1}{k!(k+1)!} \frac{\partial^k \tau_\sigma}{\partial y_{j_1}^{\nu_1} \partial y_{j_2}^{\nu_2} \cdots \partial y_{j_k}^{\nu_k}} \omega^\sigma \wedge \omega^{\nu_1} \\
&\quad \wedge \omega^{\nu_2} \wedge \ldots \wedge \omega^{\nu_k} \wedge \omega_{j_1j_2...j_k},
\end{aligned}
\tag{41}
$$

12

where $\omega_{j_1 j_2 \ldots j_k} = i_{\partial / \partial x^{j_k}} \omega_{j_1 j_2 \ldots j_{k-1}}$.

2.4. Fibered mechanics: Local and global triviality, local and global variationality

In this section, we discuss differences between local and global variational triviality of lagrangians, and local and global variationality of dynamical equations for some fibered manifolds Y over 1-*dimensional* bases X (*higher order fibered mechanics*). Recall that these concepts are defined by the Euler-Lagrange mapping $\Omega^r_{n,X} W \ni \lambda \rightarrow E_1(\lambda) \in \Omega^{2r}_{n+1,Y} W$, where W runs through open subsets of Y.

A lagrangian $\lambda \in \Omega^r_1 Y$ is locally variationally trivial if and only if $E_1(\lambda) = 0$. If λ is locally variationally trivial and $H^1 Y = 0$, then λ is globally variationally trivial. A source form $\varepsilon \in \Omega^s_{n+1,Y} W$ is locally variational if and only if $E_2(\varepsilon) = 0$. If ε is locally variational and $H^2 Y = 0$, then ε is globally variational.

Consider simple examples (Anderson and Duchamp [1], Krupka [25]). Denote by $Q = \mathbb{R}^m, S^m, T, M, K$ the real, m-dimensional Euclidean space, the m-dimensional sphere, the 2-dimensional torus $S^1 \times S^1$, the Möbius strip, and the Klein bottle, respectively. Then $H^0 Q = \mathbb{R}$, and

$$
\begin{aligned}
&H^i \mathbb{R}^m = 0, \quad 1 \le i \le m, \\
&H^1 S^1 = \mathbb{R}, \quad H^i S^m = 0, \quad H^m S^m = \mathbb{R}, \quad m \ge 2, \quad 1 \le i \le m-1, \\
&H^1 T = \mathbb{R} \times \mathbb{R}, \quad H^2 T = \mathbb{R}, \\
&H^1 M = \mathbb{R}, \quad H^2 M = 0, \\
&H^1 K = \mathbb{R}, \quad H^2 K = 0.
\end{aligned}
$$

Since $\dim X = 1$, if we restrict ourselves to *connected* base manifolds, we have essentially two possibilities: (a) $X = \mathbb{R}$, and (b) $X = S^1$.

(a) Let $X = \mathbb{R}$. Assume that $Y = \mathbb{R} \times Q$. Then by the Künneth formula, $H^1(\mathbb{R} \times Q) = H^1 Q$. Thus if $Q = \mathbb{R}^m$, or $Q = S^m$, $m \ge 2$, then local variational triviality always implies global variational triviality. Analogously, $H^2(\mathbb{R} \times Q) = H^2 Q$. Thus if $Q = \mathbb{R}^m$, or $Q = S^m$, $m \ne 2$, or $Q = M$, $Q = K$, local variationality automatically implies global variationality. If Y is a vector bundle over $\mathbb{R}$, then local variational triviality implies global variational triviality, and local variationality implies global variationality.

(b) Let $X = S^1$. Assume that $Y = S^1 \times Q$. Then $H^1(S^1 \times Q) = H^1 Q \times H^0 Q$. Since $H^0 Q$ is always nontrivial, $H^1(S^1 \times Q) \ne 0$, and in this case local variational triviality does not automatically imply global variational triviality. Therefore, one should examine every case independently. Similarly, $H^2(S^1 \times Q) = H^2 Q \times H^1 Q$, and if $Q = \mathbb{R}^m$, or $Q = S^m$, where $m \ge 3$, then local variationality always implies global variationality. If we consider M (resp. K) as a fibered manifold over S^1 with fiber $\mathbb{R}$ (resp. S^1), then in both cases, local variationality implies global variationality. If Y is a vector bundle over S^1, then local variationality implies global variationality.

2.5. Hamilton equations

The theory of *Hamilton equations* for *Hamilton extremals*, presented below (Dedecker [5], Goldschmidt and Sternberg [10], Krupka [20], [22], Krupka and Stepanková [28], has some specific features: (a) It is a theory of *Lagrangian* type, there is no *dual* concept of a Hamiltonian (such a concept arises *locally*), (b) the theory *extends* the Euler-Lagrange theory in the sense that every solution of the associated Euler-Lagrange equations is a Hamilton extremal, (c) no additional assumption, such as existence of a distinguished *time coordinate*, is imposed, and (d) in classical mechanics, where the lagrangian satisfies the standard *regularity condition* (regularity of the Hessian matrix), the theory gives the well-known Hamilton theory. The classical regularity condition has been generalized by Dedecker [5], Krupka and Stepanková [28], and Garcia and Munoz [9] (see also Gotay [11], Krupka and Musilová [27], Shadwick [38], for a wider notion we refer to Krupková [29]-[33]).

We know that every form $\rho \in \Omega_n^r W$ defines a lagrangian $h\rho \in \Omega_{n,X}^{r+1}W$, the corresponding variational functional, and the Euler-Lagrange form $E(h\rho)$ (Sections 1.2, 1.3). ρ also defines another variational functional

$$\Gamma_\Omega J^r W \ni \delta \to \rho_\Omega(\delta) = \int_\Omega \delta^* \rho \in \mathbb{R}, \tag{42}$$

whose domain are sections of $J^r Y$. Note that the sections in $\Gamma_\Omega J^r W$ are not, in general, holonomic, i.e., are not necessarily of the form $\delta = J^r \gamma$.

To be more precise, consider $J^r Y$ fibered over X by the projection π^r. Prolonging $J^r Y$ we get the 1-jet prolongation $J^1 J^r Y$, and the associated horizontalization, denoted $\tilde{h}$, defined by

$$\tilde{h}f = f \circ (\pi^r)^{1,0}, \quad \tilde{h}dx^i = dx^i, \quad \tilde{h}dy^\sigma_{j_1 j_2 \ldots j_p} = y^\sigma_{j_1 j_2 \ldots j_p, k}dx^k, \quad 0 \le p \le r. \tag{43}$$

Note that now the associated coordinates on $J^1 J^r Y$ are denoted by x^i, $y^\sigma_{j_1 j_2 \ldots j_p, k}$, where $0 \le p \le r$. Correspondingly, let $\Lambda = \tilde{h}\rho$ be the lagrangian. Indeed, Λ is a $(\pi^r)^1$-horizontal form on $J^1 J^r Y$. We set

$$H_\rho = E_\Lambda, \tag{44}$$

and call H_ρ the *Hamilton form* of ρ. The corresponding Euler-Lagrange equations are called the *Hamilton equations*, and their solutions δ are called the *Hamilton extremals*.

As an illustration of this general scheme, consider the case $n = 1$, $r = 1$ (*first order fibered mechanics*). Denote by $(t, q^\sigma, \dot{q}^\sigma)$ some induced fibered coordinates on $J^1 Y$, and by $(t, q^\sigma, \dot{q}^\sigma, q_1^\sigma, \dot{q}_1^\sigma)$ the associated coordinates on $J^1 J^1 Y$. Assume that we have a first order lagrangian $\lambda \in \Omega_{1,X}^1$, $\lambda = \mathscr{L}dt$, and consider the *Cartan form* $\Theta = \mathscr{L}dt + (\partial\mathscr{L}/\partial\dot{q}^\sigma)\omega^\sigma$. Then $\lambda = h\Theta = \mathscr{L}dt$ and $\Lambda = \tilde{h}\Theta = \tilde{\mathscr{L}}dt$, where

$$\tilde{\mathscr{L}} = \mathscr{L} + \frac{\partial\mathscr{L}}{\partial\dot{q}^\sigma}(q_1^\sigma - \dot{q}^\sigma), \tag{45}$$

and

$$H_\Theta = \left(\frac{\partial \mathscr{L}}{\partial q^\sigma} + \frac{\partial^2 \mathscr{L}}{\partial q^\sigma \partial \dot{q}^\nu}(q_1^\nu - \dot{q}^\nu) - \frac{d}{dt}\frac{\partial \mathscr{L}}{\partial \dot{q}^\sigma} \right) dq^\sigma \wedge dt$$
$$- \frac{\partial^2 \mathscr{L}}{\partial \dot{q}^\sigma \partial \dot{q}^\nu}(q_1^\nu - \dot{q}^\nu) d\dot{q}^\sigma \wedge dt. \tag{46}$$

If $\mathscr{L}$ is *regular*, i.e., if

$$\det \left(\frac{\partial^2 \mathscr{L}}{\partial \dot{q}^\sigma \partial \dot{q}^\nu} \right) \neq 0, \tag{47}$$

then H_Θ can be computed in the *Legendre coordinates* $p_\sigma = \partial \mathscr{L}/\partial \dot{q}^\sigma$. Denoting $\mathscr{H} = -\mathscr{L} + p_\sigma \dot{q}^\sigma$, we get

$$H_\Theta = -\left(\frac{\partial \mathscr{H}}{\partial q^\sigma} + \frac{dp_\sigma}{dt} \right) dq^\sigma \wedge dt + \left(-\frac{\partial \mathscr{H}}{\partial p_\sigma} + \frac{dq^\sigma}{dt} \right) dp_\sigma \wedge dt. \tag{48}$$

The coefficients in H_Θ are exactly the left-hand sides of the well-known *Hamilton equations*.

It is interesting that an analogous situation arises for the second order *Hilbert lagrangian* of the general relativity theory. Here X is any n-dimensional manifold (*spacetime*), $Y = \mathrm{Met}X$ is the fibered manifold of *regular* tensors of degree $(0,2)$ over X, $\lambda = \mathscr{L}\omega_0$, where $\mathscr{L} = R\sqrt{|\det(g_{ij})|}$, and R is the scalar curvature invariant (considered as a function on $J^2\mathrm{Met}X$). For any coordinates (x^i) on X, we have the associated coordinates (x^i, g_{ij}) on $\mathrm{Met}X$, and $(x^i, g_{jk}, g_{ij,k}, g_{ij,kl})$ on $J^2\mathrm{Met}X$. We can compute the *principal Lepage equivalent* Θ_λ (Section 1.3, Theorem 3). It is easily seen that Θ_λ is a *global n-form* on $J^1\mathrm{Met}X$, defined by

$$\Theta_\lambda = \sqrt{|\det(g_{ij})|} \left(g^{ip}(\Gamma^j_{ip}\Gamma^k_{jk} - \Gamma^j_{ik}\Gamma^k_{jp}) \, \omega_0 \right.$$
$$\left. + (g^{jp}g^{iq} - g^{pq}g^{ij})(dg_{pq,j} + \Gamma^k_{pq}dg_{jk}) \wedge \omega_i \right). \tag{49}$$

The corresponding Hamilton form (48) is a global form on $J^1J^1\mathrm{Met}X$ can be derived by a routine calculation. Since (49) satisfies a *regularity condition*, one may also obtain the corresponding Legendre transformation, and the Hamilton equations (see [28]).

2.6. Variational sequences and bicomplexes

For further information about the properties of variational sequences, we refer to Francaviglia, Palese and Vitolo [7], Krbek, Musilová [15], Krbek, Musilová and Kasparová [16], Krupka [25], Musilová [34], Stefánek [40], Vitolo [47], and to references therein. For comments on relations of the variational sequences to Spencer sequences, see Pommaret [36].

An alternative way of including the Euler-Lagrange mapping into an exact sequence, historically preceding the variational sequence, is the variational bicomplex, an object defined by means of an infinite jet construction (Anderson and Duchamp [1], Dedecker and Tulczyjew [6], Saunders [37], Takens [42], Tulczyjew [44], Vinogradov [45]. The bicomplex has also become a tool to study the geometry of differential equations (Anderson and Pohjanpelto [2], Vinogradov, Krasilschik, Lychagin [46]).

We give a brief comparison of these two theories. A principal difference consists in the fact that the variational sequence stabilizes the order of jets, while the operators, entering the variational bicomplex, increase the order of the elements of the bicomplex. For the calculus of variations it is important that one of the morphisms in the variational sequence is exactly the Euler-Lagrange mapping; on the other hand, the corresponding mapping, entering the bicomplex, is different, because it includes all order jets, i.e., has a wider domain of definition and range. This has a principal consequence for the cohomology: the classes characterizing differences between local and global properties of the mappings in the sequence and the bicomplex have a different meaning. For example, if the condition of (global) variationality $H^{n+1}Y = 0$ is satisfied, then the exactness of the variational sequence on $J^r Y$ guarantees, for a locally variational form, existence of a lagrangian λ of order $r + 1$, and of certain concrete *functional properties*, defined by the *structure* of the sheaf Ω_n^r / Θ_n^r (i.e., $\lambda = h\rho$ for some ρ on $J^r Y$); the bicomplex theory guarantees existence of a global lagrangian (of certain order) only.

Acknowledgments

The author is grateful to the Mathematics Department and the Institute for Advanced Study at La Trobe University, where as an IAS Distinguished Fellow he completed this paper. He also appreciates the support of the Czech Grant Agency (grant 201/03/0512), and the Czech Ministry of Education, Youth, and Sports (grant MSM 153100011).

REFERENCES

1. I. Anderson, T. Duchamp, On the existence of global variational principles, Am. J. Math. 102 (1980), 781-867.
2. I. Anderson, J. Pohjanpelto, Variational principles for differential equations with symmetries and conservation laws, 1. Second order scalar equations, Math. Ann. 299 (1994), 191-222; 2. Polynomial differential equations, Math. Ann. 301 (1995), 627-653.
3. D. Betounes, Extension of the classical Cartan form, Phys. Rev. D 29 (1984), 599-606.
4. M. Crampin, D.J. Saunders, The Hilbert-Caratheodory forms for parametric multiple integral problems in the calculus of variations, Acta Appl. Math. 76 (2003), 37-55.
5. P. Dedecker, On the generalization of symplectic geometry to multiple integrals in the calculus of variations, in: Lecture Notes in Math. 570, Springer, Berlin, 1977, 395-456.
6. P. Dedecker, W.M. Tulczyjew, Spectral sequences and the inverse problem of the calculus of variations, Internat. Colloq., Aix-en-Provence, 1979; in: *Differential Geometric Methods in Mathematical Physics, Lecture Notes in Math.* 836, Springer, Berlin, 1980, 498-503.
7. M. Francaviglia, M. Palese, R. Vitolo, Symmetries in finite order variational sequences, Czech. Math. J. 52 (2002), 197-213.
8. P.L. García, The Poincaré-Cartan invariant in the calculus of variations, Symposia Mathematica 14 (1974), 219-246.
9. P.L. García, J.M. Masqué, Le probléme de la régularité dans le calcul des variations du second ordre, C. R. Acad. Sci. Math. 301 (1985), 639-642.
10. H. Goldschmidt, S. Sternberg, The Hamilton-Cartan formalism in the calculus of variations, Ann. Inst. H. Poincaré 23 (1973), 203-267.

11. M. Gotay, A multisymplectic framework for classical field theory and the calculus of variations, I. Covariant Hamiltonian formalism, in: M. Francaviglia, D.D. Holm, Eds., *Mechanics, Analysis, and Geometry: 200 Years After Lagrange*, North Holland, Amsterdam, 1990, 203-235.
12. M. Gotay, An exterior differential systems approach to the Cartan form, in: Geométrie Symplectique et Physique Mathematique, P. Donato at all., Eds., Birkhauser, Boston, 1991.
13. A. Haková, O. Krupková, Variational first-order partial differential equations, J. Differential. Equations 191 (2003), 67-89.
14. I. Kolar, R. Vitolo, On the Helmholtz operator for Euler morphisms, Math. Proc. Camb. Phil. Soc. 135 (2003), 277-290.
15. M. Krbek, J. Musilová, Representation of the variational sequence, Rep. Math. Phys. 51 (2003), 251-258.
16. M. Krbek, J. Musilová, J. Kasparová, Representation of the variational sequence in field theory, in: *Steps in Differential Geometry*, L. Kozma, P.T. Nagy, and L. Tamássy, Eds., Proc. Colloq. on Diff. Geom., Debrecen, Hungary, July, 2000, Debrecen, Hungary, 147-160.
17. D. Krupka, A geometric theory of ordinary first order variational problems in fibered manifolds, I. Critical sections, J. Math. Anal. Appl. 49 (1975), 180-206; II. Invariance, J. Math. Anal. Appl. 49 (1975) 469-476.
18. D. Krupka, A map associated to the Lepagean forms of the calculus of variations in fibered manifolds, Czech. Math. J. 27 (1977), 114-118.
19. D. Krupka, Lepagean forms in higher order variational theory, in: *Modern Developments in Analytical Mechanics*, Proc. IUTAM-ISIMM Sympos., Turin, June 1982; Academy of Sciences of Turin, 1983, 197-238.
20. D. Krupka, *On the higher order Hamilton theory in fibered spaces*, Sept. 1983; J.E. Purkyne Univ., Brno, 1984, 167-184; ArXiv: math-ph/0203039.
21. D. Krupka, *On the local structure of the Euler-Lagrange mapping of the calculus of variations*, Univerzita Karlova, Prague, 1981, 181-188; ArXiv: math-ph/0203034.
22. D. Krupka, Regular lagrangians and Lepagean forms, in: *Differential Geometry and Its Applications*, Proc. Conf., D. Krupka and A. Svec, Eds., Brno, Czechoslovakia, 1986; D. Reidel, Dorrecht, 1986, 111-148.
23. D. Krupka, *Some Geometric Aspects of the Calculus of Variations in Fibered Manifolds*, Folia Fac. Sci. Nat. UJEP Brunensis 14 (1973); ArXiv: math-ph/0110005.
24. D. Krupka, Variational principles for energy-momentum tensors, Rep. Math. Phys. 49 (2002), 259-268.
25. D. Krupka, Variational sequences in mechanics, Calc. Var. 5 (1997), 557-583.
26. D. Krupka, Variational sequences on finite order jet spaces, in: *Differential Geometry and its Applications*, Proc. Conf., Brno, Czechoslovakia, 1989; World Scientific, Singapore, 1990, 236-254.
27. D. Krupka, J. Musilová, Hamilton extremals in higher order mechanics, Arch. Math. (Brno) 20 (1984), 21-30.
28. D. Krupka, O. Stepankova, On the Hamilton form in second order calculus of variations, Proc. of the Meeting *"Geometry and Physics"*, Florence, October 1982; Pitagora Editrice Bologna, 1983, 85-101.
29. O. Krupková, A geometric setting for higher order Dirac-Bergmann theory of constraints, J. Math. Phys. 35 (1994), 6557-6576.
30. O. Krupková, Hamiltonian field theory, J. Geom. Phys. 43 (2002), 93-132.
31. O. Krupková, Hamiltonian field theory revisited: A geometric approach to regularity, in: *Steps in Differential Geometry*, Proc. Colloq. Diff. Geom., Debrecen, July 2000, L. Kozma, P.T. Nagy, and L. Tamássy, Eds. (University of Debrecen, Debrecen, 2001) 187-207.
32. O. Krupková, Lepagean 2-forms in higher order Hamiltonian mechanics, I. Regularity, II. Inverse problem, Arch. Math. (Brno) 22 (1986), 97-120; 23 (1987), 155-170.
33. O. Krupková, *The Geometry of Ordinary Variational Equations*, Lecture Notes in Math. 1678, Springer, Berlin, 1997.
34. J. Musilová, Variational sequence in higher order mechanics, in: *Differential Geometry and Applications*, Proc. Conf., 1995, Brno, Czechia; Masaryk University, Brno, 1996, 611-624.
35. P. Olver, Equivalence and the Cartan form, Acta Appl. Math. 31 (1993), 99-136.
36. J.F. Pommaret, Spencer sequence and variational sequence, Acta Appl. Math. 41 (1995), 285-296.
37. D. Saunders, *The Geometry of Jet Bundles*, Cambridge Univ. Press, 1989.

17

38. W.F. Shadwick, The Hamiltonian formulation of regular r-th order Lagrangian field theories, Letters in Math. Phys. 6 (1982), 409-416.
39. J. Sniatycki, On the geometric structure of classical field theory in Lagrangian formulation, Proc. Camb. Phil. Soc. 68 (1970) 475-484.
40. J. Stefanek, A representation of the variational sequence in higher order mechanics, in: *Differential Geometry and Applications*, Proc. Conf., 1995, Brno, Czech Republic; Masaryk University, Brno, 1996, 469-478.
41. J. Sedenková, A generalization of Lepage forms in mechanics, in this volume.
42. F. Takens, A global version of the inverse problem of the calculus of variations, J. Differential Geometry 14 (1979), 543-562.
43. A. Trautman, Noether equations and conservation laws, Comm. Math. Phys. 6 (1967) 248-261.
44. W.M. Tulczyjew, The Euler-Lagrange resolution, Internat. Coll. on Diff. Geom. Methods in Math. Phys., Aix-en-Provence, 1979; Lecture Notes in Math. 836, Springer, Berlin, 1980, 22-48.
45. A.M. Vinogradov, A spectral sequence associated with a non-linear differential equation, and the algebro-geometric foundations of Lagrangian field theory with constraints, Soviet Math. Dokl. 19 (1978), 790-794.
46. A.M. Vinogradov, I. S. Krasilschik, and V. V. Lychagin, *Introduction to the Geometry of Non-linear Differential Equations* (Russian), Nauka, Moscow, 1986.
47. R. Vitolo, Finite order Lagrangian bicomplexes, Math. Proc. Cambridge Phil. Soc. 125 (1999), 321-333.

The geometry of variational equations

O. Krupková

Department of Algebra and Geometry, Faculty of Science, Palacký University
Tomkova 40, 779 00 Olomouc, Czech Republic
and
Department of Mathematics, La Trobe University,
Bundoora, Victoria 3086, Australia

Abstract. This lecture brings a survey of topics and results concerning differential equations which arise as equations for extremals of higher-order variational functionals on fibered manifolds.
Key words. Fibered manifold, system of higher-order differential equations, dynamical form, Lepagean form, Lagrangian, inverse problem of the calculus of variations, regular equations, semiregular equations, Euler–Lagrange distribution, Hamiltonian differential system, symmetry, first integral, complete integral, Noether theorem, Liouville theorem.
PACS. 02.40.Vh, 02.30.Xx, 02.30.Hq, 02.30.Jr.
MSC. 34A26, 35G20, 58A15, 70H50, 70S05.

1. INTRODUCTION

The subject of this lecture is the systems of generally higher-order *ordinary differential equations* of the form

$$E_\sigma\left(t,\gamma^\nu,\frac{d\gamma^\nu}{dt},\cdots,\frac{d^s\gamma^\nu}{dt^s}\right)=0, \quad 1\le\sigma\le m, \tag{1.1}$$

where $t\to(\gamma^\nu(t))$, $1\le\nu\le m$, are components of a curve, or *partial differential equations*

$$E_\sigma\left(x^i,\gamma^\nu,D_{j_1}\gamma^\nu,\cdots,D_{j_1\ldots j_s}\gamma^\nu\right)=0, \quad 1\le\sigma\le m, \tag{1.2}$$

for mappings $(x^i)\to(\gamma^\nu(x^i))$, $1\le i\le n$, $1\le\nu\le m$. In the framework of global analysis, equations of this kind are local expressions of differential equations for mappings between smooth manifolds. I will focus on *variational equations*, arising as equations for extremals of variational functionals on fibered manifolds. I will present a geometric setting which enables to study such equations and their solutions by methods of differential geometry and global analysis, topics which appear in the geometric theory of variational equations, and some recent results on their properties and on structure of their solutions. Also I will touch the problem of existence of a Lagrangian to a given system of differential equations, and its local and global aspects.

CP729, *Global Analysis and Applied Mathematics: International Workshop on Global Analysis*,
edited by K. Taş, D. Krupka, O. Krupková, and D. Baleanu

2. VARIATIONAL EQUATIONS IN JET MANIFOLDS

2.1. Structures and calculus

In what follows, we assume all manifolds and maps be smooth, and use standard notations: T and J^r denotes the tangent and the r-jet prolongation functor, respectively, d the exterior derivative, * the pull-back, i_ξ the contraction by a vector field ξ, ∂_ξ the Lie derivative with respect to a vector field ξ, etc. The summation convention is used unless otherwise explicitly stated.

We consider a *fibered manifold* $\pi : Y \to X$ with a base X of dimension n, and total space Y, $\dim Y = m + n$, and its *jet prolongations* $\pi_r : J^r Y \to X$, $r \geq 1$. (For simplicity of notations, we also write $J^0 Y = Y$ and $\pi_0 = \pi$). There naturally arise induced fibered manifolds $\pi_{r,s} : J^r Y \to J^s Y$, where $r > s \geq 0$.

If $\dim X = 1$ (this case refers to *ordinary differential equations*, or *higher-order mechanics*) we denote by (t, q^σ), $1 \leq \sigma \leq m$, local fibered coordinates on Y, and by $(t, q^\sigma, q_1^\sigma, \ldots, q_r^\sigma)$, associated coordinates on $J^r Y$. If $r = 1$ and $r = 2$, we also write $(t, q^\sigma, \dot{q}^\sigma)$ and $(t, q^\sigma, \dot{q}^\sigma, \ddot{q}^\sigma)$ for associated fibered coordinates on $J^1 Y$ and $J^2 Y$, respectively. In the case $\dim X = n > 1$ (*partial differential equations*, or *higher-order field theory*), fibered coordinates on Y are denoted by (x^i, y^σ), where $1 \leq i \leq n$, $1 \leq \sigma \leq m$, and associated coordinates on $J^r Y$ by $(x^i, y^\sigma, y_{j_1}^\sigma, \ldots, y_{j_1 \ldots j_r}^\sigma)$, where $1 \leq j_1 \leq \ldots \leq j_r \leq n$. Next, we denote

$$\omega_0 = dx^1 \wedge \ldots \wedge dx^n, \quad \omega_j = i_{\partial/\partial x^j} \omega_0, \quad \ldots, \quad \omega_{j_1 \ldots j_r} = i_{\partial/\partial x^{j_r}} \omega_{j_1 \ldots j_{r-1}}. \tag{2.1}$$

By a *section* γ of π we mean a mapping $\gamma : U \to Y$, defined on an open subset U of X, such that $\pi \circ \gamma = id_U$. A section of π_r is called *holonomic* if it is the r-jet prolongation of a section of π. In fibered coordinates, components of a section of π take the form (x^i, γ^σ), where the γ^σ's are functions of the x^i's. Similarly, components of a section of π_r read $(x^i, \gamma^\sigma, \gamma_{j_1}^\sigma, \ldots, \gamma_{j_1 \ldots j_r}^\sigma)$. On the other hand, components of a holonomic section of π_r take the form of *derivatives* of the components γ^σ, i.e., $\gamma_{j_1 \ldots j_k}^\sigma = D_{j_1 \ldots j_k} \gamma^\sigma$, $1 \leq k \leq r$.

In fibered manifolds, there are distinguished vector fields and differential forms, adapted to the fibered structure. A vector field ξ on $J^r Y$, $r \geq 0$, is called π_r-*projectable* if there exists a vector field ξ_0 on X such that $T\pi_r . \xi = \xi_0 \circ \pi$, and π_r-*vertical* if it projects onto a zero vector field on X, i.e., $T\pi_r . \xi = 0$. Quite similarly one can define a $\pi_{r,s}$-*projectable* or a $\pi_{r,s}$-*vertical* vector field on $J^r Y$, where $r > s$. A differential k-form η on $J^r Y$ is called π_r-*horizontal* (resp. $\pi_{r,s}$-*horizontal*) if $i_\xi \eta = 0$ for every π_r-vertical (resp. $\pi_{r,s}$-vertical) vector field ξ on $J^r Y$. It is important to notice that π_r-horizontal forms are those which in fibered coordinates contain wedge products of the base differentials dx^i only (of course, with components dependent upon all the fibered coordinates). Similarly, $\pi_{r,0}$-horizontal forms contain wedge products of the total space differentials dx^i's and dy^σ's only, etc. To every k-form η on $J^r Y$ one can assign a unique horizontal k-form on $J^{r+1} Y$, denoted by $h\eta$ and called the *horizontal part of* η. The mapping h is defined to be an R-linear wedge product preserving mapping such that for every function f on $J^r Y$, $hf = f \circ \pi_{r+1,r}$, and

$$hdx^i = dx^i, \quad hdy^\sigma = y_i^\sigma dx^i, \quad hdy_j^\sigma = y_{ji}^\sigma dx^i, \quad \ldots, \quad hdy_{j_1 \ldots j_r}^\sigma = y_{j_1 \ldots j_r i}^\sigma dx^i. \tag{2.2}$$

In particular, $hdf = d_i f dx^i$, where d_i is the i-th *total derivative operator*

$$d_i = \frac{\partial}{\partial x^i} + y_i^\sigma \frac{\partial}{\partial y^\sigma} + y_{ji}^\sigma \frac{\partial}{\partial y_j^\sigma} + \ldots + y_{j_1\ldots j_r i}^\sigma \frac{\partial}{\partial y_{j_1\ldots j_r}^\sigma}. \tag{2.3}$$

For $\dim X = 1$ the above formulas read

$$hdt = dt, \quad hdq^\sigma = q_1^\sigma dt, \quad hdq_1^\sigma = q_2^\sigma dt, \quad \ldots, \quad hdq_r^\sigma = q_{r+1}^\sigma dt, \tag{2.4}$$

and $hdf = \frac{df}{dt} dt$, where

$$\frac{d}{dt} = \frac{\partial}{\partial t} + q_1^\sigma \frac{\partial}{\partial q^\sigma} + q_2^\sigma \frac{\partial}{\partial q_1^\sigma} + \ldots + q_{r+1}^\sigma \frac{\partial}{\partial q_r^\sigma}. \tag{2.5}$$

By definition of h, for any form η of degree $k > n$, $h\eta = 0$.

A k-form η on $J^r Y$ is called *contact* if for every section γ of π, $J^r \gamma^* \eta = 0$. The ideal of contact forms of order r is generated by the following *canonical contact 1-forms*

$$\omega_{j_1\ldots j_k}^\sigma = dy_{j_1\ldots j_k}^\sigma - y_{j_1\ldots j_k i}^\sigma dx^i, \quad 0 \le k \le r - 1, \tag{2.6}$$

and contact 2-forms $d\omega_{j_1\ldots j_{r-1}}^\sigma$. If $\dim X = 1$, we write

$$\omega_k^\sigma = dq_k^\sigma - q_{k+1}^\sigma dt, \quad 0 \le k \le r - 1, \tag{2.7}$$

or, in the "dot notation",

$$\omega^\sigma = dq^\sigma - \dot{q}^\sigma dt, \quad \dot{\omega}^\sigma = d\dot{q}^\sigma - \ddot{q}^\sigma dt, \quad \ddot{\omega}^\sigma = d\ddot{q}^\sigma - \dddot{q}^\sigma dt. \tag{2.8}$$

Working with coordinate expressions of forms in jet bundles, it is worth to use instead of a canonical basis of one forms on $J^r Y$ a *basis adapted to the contact structure*, i.e.,

$$(dx^i, \omega^\sigma, \ldots, \omega_{j_1\ldots j_{r-1}}^\sigma, dy_{j_1\ldots j_r}^\sigma), \quad \text{resp.} \quad (dt, \omega^\sigma, \ldots, \omega_{r-1}^\sigma, dq_r^\sigma). \tag{2.9}$$

A contact k-form η on $J^r Y$ is called *q-contact*, $1 \le q \le k$, if the expression of $\pi_{r+1,r}^* \eta$ in the adapted basis contains in each term exactly q of the canonical contact 1-forms. Throughout the paper we shall often use Krupka's *Theorem on canonical decomposition of forms*, saying that for every k-form η on $J^r Y$ there is a unique decomposition

$$\pi_{r+1,r}^* \eta = h\eta + p_1 \eta + p_2 \eta + \cdots + p_k \eta, \tag{2.10}$$

where $h\eta$ is the horizontal part of η, and $p_q \eta$, $q = 1, 2, \cdots, k$, is a *(uniquely determined)* q-contact form, called the *q-contact part of η*.

Note that a section of π_r is holonomic if and only if it is an integral section of the contact ideal on $J^r Y$. Integral sections of the ideal generated by 2-contact forms on $J^r Y$ are called *quasi-holonomic* (or *Dedecker's sections*).

For more details on jet manifolds and the calculus of horizontal and contact forms we refer to [18], [23], [30].

2.2. Differential equations modelled by dynamical forms

Fibered manifolds and their jet prolongations represent an appropriate background for a global theory of differential equations. Within this framework, differential equations can be represented as invariant (geometric) objects, and their properties (such as compatibility, equivalence, symmetries, transformation properties, integrability, reducibility, etc.) can be investigated by methods of differential geometry and global analysis. Equations and their related geometric structures then can be used to study local or global solutions or families of solutions. An important part of the theory is concerned with formulation of *integration methods*, usually based on symmetries of equations and related objects, or equivalence between different systems of equations.

There are many possibilities, how to globally represent differential equations. Among the most familiar ones, there are *vector fields*, representing systems of first-order ordinary differential equations explicitly solved with respect to the first derivatives, *distributions*, generalizing the latter to systems of explicit first-order partial differential equations, or their higher-order analogs, called *semisprays* (for ODE) and *semispray distributions* (for PDE). To study more complicated systems of differential equations, *exterior differential systems* are used, or the equations are modelled as *submanifolds of the manifolds of jets* (see e.g. [4], [14], [28]). In the context of the calculus of variations, it is convenient to represent differential equations by *dynamical forms* [23]. This setting covers, besides all variational equations, the equations mentioned in Sec. 1. Methods which are developed to investigate variational equations and their solutions can often provide a new insight into the "non-variational" case.

By a *dynamical form* of order $s \geq 1$ on a fibered manifold $\pi : Y \to X$ we mean a differential $(n+1)$-form on J^sY which is 1-contact, and horizontal with respect to the projection onto Y. In fibered coordinates a dynamical form on J^sY reads $E = E_\sigma \omega^\sigma \wedge \omega_0$, where E_σ are functions of order s (if $\dim X = 1$, $\omega_0 = dt$ and E is a 2-form). A *section* γ of π defined on an open set $U \subset X$ is called a *path* of E if

$$E \circ J^s\gamma = 0. \tag{2.11}$$

The above *equation for paths* of a dynamical form E takes in fibered coordinates the form of *a system of m differential equations of order s*,

$$E_\sigma \circ J^s\gamma = 0, \quad 1 \leq \sigma \leq m, \tag{2.12}$$

where $m = \dim Y - \dim X$ is the fiber dimension. These, however, are the ODE (1.1) and PDE (1.2) if $\dim X = 1$ and $\dim X = n > 1$, respectively. In this way, *equations for paths of a dynamical form on a fibered manifold* (with the base dimension n and fiber dimension m) *can be regarded as a global characterization of differential equations (1.1) or (1.2) for graphs of mappings* $R^n \to R^m$, $n \geq 1$.

It should be pointed out that dynamical forms represent quite a wide class of differential equations. For example, for $\dim X = 1$, as apparent from (1.1), this class includes not only vector fields and semisprays, but also those first and higher-order ODE which are "not solvable" with respect to the highest derivatives, i.e. which *do not* have an equivalent "normal form"

$$\frac{d^s \gamma^\sigma}{dt^s} = F^\sigma\left(t, \gamma^\nu, \frac{d\gamma^\nu}{dt}, \ldots, \frac{d^{s-1}\gamma^\nu}{dt^{s-1}}\right), \quad 1 \le \sigma \le m. \tag{2.13}$$

2.3. Variational equations

Let $r \ge 1$. A horizontal n-form λ on $J^r Y$ (where $n = \dim X$) is called a *Lagrangian of order r*. To every Lagrangian λ one assigns a family of n-forms, called *Lepagean equivalents* of the Lagrangian. For example, Lepagean equivalents of a first-order Lagrangian $\lambda = L\omega_0$ take the form $\rho = \theta_\lambda + \mu$, where

$$\theta_\lambda = L\omega_0 + \frac{\partial L}{\partial y_j^\sigma} \omega^\sigma \wedge \omega_j \tag{2.14}$$

is the famous *Poincaré–Cartan form*, and μ is an arbitrary at least 2-contact form. It should be pointed out that among Lepagean equivalents of a first-order Lagrangian there is also another important n-form,

$$\rho_\lambda = L\omega_0 + \sum_{k=1}^n \frac{1}{(k!)^2} \frac{\partial^k L}{\partial y_{j_1}^{\sigma_1} \cdots \partial y_{j_k}^{\sigma_k}} \omega^{\sigma_1} \wedge \ldots \wedge \omega^{\sigma_k} \wedge \omega_{j_1 \cdots j_k}, \tag{2.15}$$

called *Krupka form* (see [16], [3]). For $\dim X = 1$ the Lepagean equivalent of $\lambda = Ldt$ is *unique*, and coincides with the Cartan form.

Lepagean equivalents of Lagrangians have the following properties: *If ρ is a Lepagean equivalent of λ then:*
- $h\rho = \lambda$.
- *The action functions of ρ and λ are the same.*
- *The paths of the dynamical form*

$$E_\lambda = p_1 d\rho \tag{2.16}$$

are extremals of the Lagrangian λ; therefore E_λ is called the *Euler–Lagrange form* of λ, and its components are called *Euler–Lagrange expressions*. It holds $E_\lambda = E_\sigma(L)\omega^\sigma \wedge dt$, where

$$E_\sigma(L) = \frac{\partial L}{\partial q^\sigma} - \sum_{l=1}^r (-1)^{l-1} \frac{d^l}{dt^l} \frac{\partial L}{\partial q_l^\sigma}, \quad 1 \le \sigma \le m, \tag{2.17}$$

if $\dim X = 1$, and $E_\lambda = E_\sigma(L)\omega^\sigma \wedge \omega_0$, where

$$E_\sigma(L) = \frac{\partial L}{\partial y^\sigma} - \sum_{l=1}^r (-1)^{l-1} d_{p_1} d_{p_2} \ldots d_{p_l} \frac{\partial L}{\partial y_{p_1 p_2 \cdots p_l}^\sigma}, \quad 1 \le \sigma \le m, \tag{2.18}$$

if $\dim X = n > 1$.

• *Equations for extremals of λ (Euler–Lagrange equations) take one of the following equivalent intrinsic forms:*

$$J^{2r-1}\gamma^* i_\xi d\rho = 0 \quad \text{for every vertical vector field } \xi \text{ on } J^{2r-1}Y.$$

$$E_\lambda \circ J^{2r}\gamma = 0. \tag{2.19}$$

The first equation comes from the first variation formula for the Lagrangian λ. The second equation simply reflects the fact that the Euler–Lagrange form is a dynamical form. In fibered coordinates Euler–Lagrange equations take the form of a system of m ordinary (resp. partial) differential equations of order at most $2r$ for the components γ^ν, $1 \leq \gamma \leq m$, of sections γ of π,

$$\left(\frac{\partial L}{\partial q^\sigma} - \sum_{l=1}^{r} (-1)^{l-1} \frac{d^l}{dt^l} \frac{\partial L}{\partial q_l^\sigma} \right) \circ J^{2r}\gamma = 0, \quad 1 \leq \sigma \leq m, \tag{2.20}$$

respectively,

$$\left(\frac{\partial L}{\partial y^\sigma} - \sum_{l=1}^{r} (-1)^{l-1} d_{p_1} d_{p_2} \ldots d_{p_l} \frac{\partial L}{\partial y_{p_1 p_2 \ldots p_l}^\sigma} \right) \circ J^{2r}\gamma = 0, \quad 1 \leq \sigma \leq m. \tag{2.21}$$

Lagrangians, giving rise to the same Euler–Lagrange form (i.e. having the same Euler–Lagrange expressions), are called *equivalent*. It is known that Lagrangians λ_1 and λ_2 (of possibly different orders) are equivalent if and only if (up to projection) $\lambda_1 - \lambda_2 = hd\eta$ for a $(n-1)$-form η. In mechanics, η is a function, f, therefore the above formula gives $L_1 - L_2 = df/dt$. The difference of two equivalent Lagrangians is a Lagrangian with the identically zero Euler–Lagrange form; it is called a *trivial Lagrangian*.

For more details on the variational principle, higher-order Lepagean n-forms, Euler–Lagrange form, the first variation formula, Euler–Lagrange equations, and related topics, we refer to Krupka [15], [18]-[20] (see also [23]).

2.4. Locally variational forms and the inverse problem of the calculus of variations

We have seen that variational equations arise as equations for paths of certain dynamical forms. However, *how to recognize whether a given dynamical form comes from a Lagrangian or not?* Since our underlying structures are topologically non-trivial, this question needs a subtle investigation: it really may happen that a dynamical form possesses Lagrangians, but only *local* ones, which cannot be "put together" to define a global Lagrangian. Therefore one must distinguish not only between "variationality and non-variationality", but also between "local and global variationality".

Let E be a dynamical form on $J^s Y$. E is called *locally variational* if to every point in $J^s Y$ there exists a neighbourhood U, an integer $r \leq s$, and a Lagrangian λ defined on $\pi_{s,r}(U)$ such that $E|_U = E_\lambda$. E is called *globally variational* if λ is defined on $J^r Y$, i.e., if E is locally variational and has a global Lagrangian.

In view of these definitions, *variationality of a system of differential equations* can be introduced. To a system of m s-order differential equations for graphs of mappings from R^n to R^m one can associate a dynamical form E of order s in such a way that the "left-hand sides" of the equations become the components of E. More precisely, one sets $E = E_\sigma \omega^\sigma \wedge dt$ (resp. $E = E_\sigma \omega^\sigma \wedge \omega_0$), where the functions E_σ, $1 \le \sigma \le m$, are defined as follows: For a system of ODE (1.1),

$$E_\sigma \circ J^s \gamma = E_\sigma \left(t, \gamma^\nu, \frac{d\gamma^\nu}{dt}, \cdots, \frac{d^s \gamma^\nu}{dt^s} \right), \tag{2.22}$$

and for a system of PDE (1.2),

$$E_\sigma \circ J^s \gamma = E_\sigma \left(x^i, \gamma^\nu, D_{j_1} \gamma^\nu, \cdots, D_{j_1 \dots j_s} \gamma^\nu \right). \tag{2.23}$$

Now, *a system of differential equations* ((1.1), resp. (1.2)) is called *variational* if the associated dynamical form E is locally variational.

Necessary and sufficient conditions for a system of differential equations be variational, were originally found by Helmholtz for the case of second-order ODE [12]. A generalization to ODE of an arbitrary order is due to Vanderbauwhede [38]. For the case of higher-order PDE the conditions were obtained independently by Anderson and Duchamp [1], and Krupka [17], [18]. To a variational system of differential equations of order $s \ge 1$ a Lagrangian can be computed directly by using the Tonti–Vainberg formula [35], [37]. The results can be summarized as follows:

Theorem 1.

(1) *A system of s-order PDE (1.2) is variational if and only if*

$$\frac{\partial E_\sigma}{\partial y^\nu_{j_1 j_2 \dots j_l}} - (-1)^l \frac{\partial E_\nu}{\partial y^\sigma_{j_1 j_2 \dots j_l}} - \sum_{k=l+1}^{s} (-1)^k \binom{k}{l} d_{j_{l+1}} d_{j_{l+2}} \dots d_{j_k} \frac{\partial E_\nu}{\partial y^\sigma_{j_1 j_2 \dots j_k}} = 0 \tag{2.24}$$

for all σ, ν, and $0 \le l \le s$. A (local) Lagrangian then takes the form

$$L = y^\sigma \int_0^1 E_\sigma (x^i, uy^\nu, uy^\nu_j, \dots, uy^\nu_{j_1 \dots j_s}) du. \tag{2.25}$$

(2) *A system of s-order ODE (1.1) is variational if and only if*

$$\frac{\partial E_\sigma}{\partial q^\nu_l} - (-1)^l \frac{\partial E_\nu}{\partial q^\sigma_l} - \sum_{k=l+1}^{s} (-1)^k \binom{k}{l} \frac{d^{k-l}}{dt^{k-l}} \frac{\partial E_\nu}{\partial q^\sigma_k} = 0 \tag{2.26}$$

for all σ, ν, and $0 \le l \le s$. A (local) Lagrangian then takes the form

$$L = q^\sigma \int_0^1 E_\sigma (t, uq^\nu, uq^\nu_1, \dots, uq^\nu_s) du. \tag{2.27}$$

Notice that the above Tonti–Vainberg formula gives Lagrangians of order s (i.e. of the same order as the equations). There arises a question if these Lagrangians can be *reduced to a lower order*, or even to *the minimal possible order*, which is $s/2$ for an even s, and $(s+1)/2$ for an odd s, respectively. (By a "reduced Lagrangian" we mean an equivalent Lagrangian of order $r < s$). As proved in [21] and [38], *every variational system of ordinary differential equations has local Lagrangians of the minimal possible order*, and an explicit formula for minimal-order Lagrangians was obtained. In the case of partial differential equations a complete answer to the problem of local reducibility is yet not known.

Examples. We shall write down the variationality conditions for some lower order differential equations explicitly.

• **First-order ODE.** Necessary and sufficient conditions for variationality read

$$\frac{\partial E_\sigma}{\partial \dot{q}^v} + \frac{\partial E_v}{\partial \dot{q}^\sigma} = 0, \qquad \frac{\partial E_\sigma}{\partial q^v} - \frac{\partial E_v}{\partial q^\sigma} + \frac{d}{dt}\frac{\partial E_v}{\partial \dot{q}^\sigma} = 0, \qquad 1 \le \sigma, v \le m. \tag{2.28}$$

Equivalently, the "left-hand sides" E_σ must be affine in the first derivatives, i.e. $E_\sigma = A_\sigma + B_{\sigma v}\dot{q}^v$, where the matrix $B_{\sigma v}$ is *skewsymmetric*, and

$$\frac{\partial A_\sigma}{\partial q^v} - \frac{\partial A_v}{\partial q^\sigma} + \frac{\partial B_{v\sigma}}{\partial t} = 0, \qquad \frac{\partial B_{\sigma\rho}}{\partial q^v} + \frac{\partial B_{v\sigma}}{\partial q^\rho} + \frac{\partial B_{\rho v}}{\partial q^\sigma} = 0. \tag{2.29}$$

The latter conditions simply mean that the 2-form

$$\alpha = A_\sigma dq^\sigma \wedge dt + \tfrac{1}{2}B_{\sigma v}dq^\sigma \wedge dq^v \tag{2.30}$$

associated with the equations is *closed* (i.e. $d\alpha = 0$). Next, notice that for $m = 1$ (one equation) the variationality conditions are reduced to $B = 0$, meaning that none single first-order ODE is variational.

• **Second-order ODE (classical Helmholtz conditions).** For second-order ODE conditions (2.26) become

$$\begin{aligned}
\frac{\partial E_\sigma}{\partial \ddot{q}^v} &- \frac{\partial E_v}{\partial \ddot{q}^\sigma} = 0, \qquad \frac{\partial E_\sigma}{\partial \dot{q}^v} + \frac{\partial E_v}{\partial \dot{q}^\sigma} - 2\frac{d}{dt}\frac{\partial E_v}{\partial \ddot{q}^\sigma} = 0, \\
\frac{\partial E_\sigma}{\partial q^v} &- \frac{\partial E_v}{\partial q^\sigma} + \frac{d}{dt}\frac{\partial E_v}{\partial \dot{q}^\sigma} - \frac{d^2}{dt^2}\frac{\partial E_v}{\partial \ddot{q}^\sigma} = 0.
\end{aligned} \tag{2.31}$$

Equivalently, the E_σ are affine in the second derivatives, i.e. $E_\sigma = A_\sigma + B_{\sigma v}\ddot{q}^v$, the matrix B is *symmetric*, and

$$\begin{aligned}
\frac{\partial B_{\sigma v}}{\partial \dot{q}^\rho} &= \frac{\partial B_{\sigma\rho}}{\partial \dot{q}^v}, \qquad \frac{\partial A_\sigma}{\partial \dot{q}^v} + \frac{\partial A_v}{\partial \dot{q}^\sigma} = 2\frac{\bar{d}B_{\sigma v}}{dt}, \\
\frac{\partial A_\sigma}{\partial q^v} &- \frac{\partial A_v}{\partial q^\sigma} = \frac{1}{2}\frac{\bar{d}}{dt}\left(\frac{\partial A_\sigma}{\partial \dot{q}^v} - \frac{\partial A_v}{\partial \dot{q}^\sigma}\right),
\end{aligned} \tag{2.32}$$

where the "cut" total derivative operator $\bar{d}/dt$ is defined by $\bar{d}/dt = \partial/\partial t + \dot{q}^\rho \partial/\partial q^\rho$. Also these conditions mean that a certain 2-form (on J^1Y) is closed, namely,

$$\alpha = A_\sigma \omega^\sigma \wedge dt + \frac{1}{4}\left(\frac{\partial A_\sigma}{\partial \dot{q}^v} - \frac{\partial A_v}{\partial \dot{q}^\sigma}\right)\omega^\sigma \wedge \omega^v + B_{\sigma v}\omega^\sigma \wedge d\dot{q}^v. \tag{2.33}$$

- **First-order PDE.** Variationality conditions (2.24) take a simple form

$$\frac{\partial E_\sigma}{\partial y_j^\nu} + \frac{\partial E_\nu}{\partial y_j^\sigma} = 0, \quad \frac{\partial E_\sigma}{\partial y^\nu} - \frac{\partial E_\nu}{\partial y^\sigma} + d_i \frac{\partial E_\nu}{\partial y_i^\sigma} = 0, \quad 1 \leq \sigma, \nu \leq m, \ 1 \leq j \leq n. \quad (2.34)$$

It can be shown that these conditions are equivalent with the requirement that the E_σ's are *polynomials in the first derivatives*, i.e.,

$$E_\sigma = B_{\sigma \nu_1 \cdots \nu_n}^{j_1 \cdots j_n} \frac{\partial y^{\nu_1}}{\partial x^{j_1}} \cdots \frac{\partial y^{\nu_n}}{\partial x^{j_n}} + \ldots + B_{\sigma \nu_1 \nu_2}^{j_1 j_2} \frac{\partial y^{\nu_1}}{\partial x^{j_1}} \frac{\partial y^{\nu_2}}{\partial x^{j_2}} + B_{\sigma \nu_1}^{j_1} \frac{\partial y^{\nu_1}}{\partial x^{j_1}} + A_\sigma, \quad (2.35)$$

where the coefficients are functions of (x^i, y^p), completely antisymmetric in the upper and lower indices, and the $(n+1)$ form

$$
\begin{aligned}
\alpha &= E_\sigma \omega^\sigma \wedge \omega_0 + \sum_{k=1}^{n} \frac{1}{k!(k+1)!} \frac{\partial^k E_\sigma}{\partial y_{j_1}^{\nu_1} \cdots y_{j_k}^{\nu_k}} \omega^\sigma \wedge \omega^{\nu_1} \wedge \ldots \wedge \omega^{\nu_k} \wedge \omega_{j_1 \cdots j_k} \\
&= A_\sigma dy^\sigma \wedge \omega_0 + \sum_{k=1}^{n} \frac{1}{(k+1)!} B_{\sigma \nu_1 \cdots \nu_k}^{j_1 \cdots j_k} dy^\sigma \wedge dy^{\nu_1} \wedge \ldots \wedge dy^{\nu_k} \wedge \omega_{j_1 \cdots j_k},
\end{aligned}
$$
(2.36)

is closed [11] (in fact, it is the exterior derivative of the Krupka form ρ_λ (2.15)). From (2.35) and the formula for the Tonti–Vainberg Lagrangian (2.25) we can see that (first-order) Lagrangians for first-order variational PDE are polynomials in the first derivatives. Notice that similarly as in the ODE case, conditions (2.34) for one equation ($m = 1$) mean that E does not depend upon y_j, $1 \leq j \leq n$, i.e., none single first-order PDE is variational.
- **Second-order PDE.** Variationality conditions take the form

$$
\begin{aligned}
\frac{\partial E_\sigma}{\partial y_{ij}^\nu} - \frac{\partial E_\nu}{\partial y_{ij}^\sigma} &= 0, \quad \frac{\partial E_\sigma}{\partial y_j^\nu} + \frac{\partial E_\nu}{\partial y_j^\sigma} - 2d_i \frac{\partial E_\nu}{\partial y_{ij}^\sigma} = 0, \\
\frac{\partial E_\sigma}{\partial y^\nu} - \frac{\partial E_\nu}{\partial y^\sigma} &+ d_i \frac{\partial E_\nu}{\partial y_i^\sigma} - d_i d_j \frac{\partial E_\nu}{\partial y_{ij}^\sigma} = 0.
\end{aligned}
$$
(2.37)

Similarly as above, these conditions are equivalent with the existence of a certain closed $(n+1)$-form, which, however, is no more unique [25].

One could hope to find a solution of the inverse problem as contained in Theorem 1 by a direct approach, i.e. to find obstructions for integrability of the overdetermined system of equations $E_\sigma(L) = E_\sigma$, or to solve it with respect to L (here E_σ are the "left-hand sides" of given equations and $E_\sigma(L)$ are Euler–Lagrange expressions of an unknown Lagrangian). This requires complicated considerations within the theory of formal integrability of differential equations with yet almost no chance to obtain explicit results in the non-analytical case (cf. [9], [10]). On the other hand, as seen in the above examples, existence of Lagrangians for differential equations means that these equations can be related with closed forms. Indeed, a geometric setting enables one to discover and explore *a deep connection between the Euler–Lagrange operator in the calculus of variations and the exterior derivative operator in differential geometry*. Within this context, variationality conditions turn out to be "closedness conditions", the (rather mysterious) Tonti–Vainberg formula appears simply due to Poincaré Lemma, and global existence results follow in fact from De Rham Theorem. To illustrate this situation more precisely, let us recall the following assertions ([18], [21], [11], [13]).

Theorem 2. *A dynamical form E on J^sY is locally variational if and only if in a neighbourhood of every point in J^sY there exists an at least 2-contact form F such that $d(E+F) = 0$. If $\dim X = 1$, F is unique (hence global), and $\alpha = E + F$ is projectable onto $J^{s-1}Y$. If $\dim X = n$ and $s = 1$, there is a unique at least 2-contact form F such that $\alpha = E + F$ is projectable onto Y.*

The relation of locally variational forms with closed $(n+1)$-forms gives the concept of a *Lepagean equivalent of a locally variational form E* as a closed $(n+1)$-form α such that $p_1\alpha = E$ [21], [25]. In the above examples, the forms (2.30), (2.33), and (2.36) are the corresponding (global and unique) Lepagean equivalents. Since a Lepagean equivalent of a locally variational form E locally equals to $d\rho$ where ρ is a Lepagean equivalent of a Lagrangian λ for E, one gets $\lambda = h\rho = h(A\alpha) = AE$, where A is the homotopy operator adapted to the contact structure, appearing in the Poincaré Lemma; explicitly, this is precisely the Tonti–Vainberg formula. However, Lepagean $(n+1)$-forms, as closed global counterparts of locally variational forms, are useful not only in the study of the inverse variational problem. They play an even more important role in the *geometric theory of variational equations*, providing a tool for investigating properties and structure of solutions of variational equations. Questions of this kind will be discussed in the next section.

As already mentioned, existence of local Lagrangians does not mean that a global Lagrangian should exist. A similar fact concerns trivial Lagrangians. Obstructions for global variationality, or global triviality, are given by *topology of Y*, and can be discovered within the theory of *variational bicomplexes* ([1], [6], [34], [36], [39]), or *variational sequences* [19] (in the following theorem, $H^k(Y)$ denotes, as usual, the k-th cohomology group of the manifold Y, i.e. the quotient of closed modulo exact k-forms on Y):

Theorem 3. *If $H^{n+1}(Y) = \{0\}$ then every locally variational form on J^sY is globally variational. If $H^n(Y) = \{0\}$ then every trivial Lagrangian is globally trivial.*

Being a quotient sequence of De Rham sequence, the variational sequence reflects the deep relation between the Euler–Lagrange mapping and the exterior derivative. Moreover, it helps to introduce new objects to the calculus of variations and the theory of differential equations, which have not been known before (e.g. the Helmholtz $(n+1)$-form [19], or a general concept of Lepagean k-form for $k \geq n$ [20], [31]).

3. DIFFERENTIAL SYSTEMS FOR VARIATIONAL EQUATIONS

3.1. The geometric meaning and classification of variational ODE

First, let us turn to ordinary differential equations. Analyzing the variationality conditions (2.26), one finds that the equations must be *affine with respect to the highest derivatives*, i.e., we can write

$$E_\sigma = A_\sigma(t, q^\rho, \dot{q}^\rho) + B_{\sigma\nu}(t, q^\rho, \dot{q}^\rho) q_s^\nu, \quad 1 \le \sigma \le m. \qquad (3.1)$$

From Sec. 2.4 we already know that there is a *one-to-one correspondence between locally variational forms of order s and Lepagean 2-forms of order s − 1*, and that every locally variational form of order s has local Lagrangians of order $s/2$ (resp. $(s+1)/2$) if s is even (resp. odd). This means that in a neighbourhood of every point in J^sY, the Lepagean equivalent α of a locally variational form E takes the form $\alpha = d\theta_\lambda$, where λ is a local minimal-order Lagrangian for E. Now, instead of working with dynamical forms one can *equivalently* work with Lepagean 2-forms, and explore that they are *closed* forms.

Due to (2.19), we can consider variational equations of order s in the following form:

$$J^{s-1}\gamma^* i_\xi \alpha = 0, \quad \text{for every } \pi_{s-1}\text{-vertical vector field } \xi \text{ on } J^{s-1}Y, \qquad (3.2)$$

where α is the corresponding Lepagean 2-form (recall, $\alpha = p_1 E$ and $d\alpha = 0$). In this formulation, variational ODE represented by a dynamical form E are equations for *holonomic* integral sections of a distribution

$$\Delta_E = annih\{i_\xi \alpha \,|\, \xi \text{ runs over } \pi_{s-1}\text{-vertical vector fields on } J^{s-1}Y\}, \qquad (3.3)$$

and called the *Euler–Lagrange distribution*. Equations for integral sections of Δ_E, i.e.

$$\delta^* i_\xi \alpha = 0, \quad \text{for every } \pi_{s-1}\text{-vertical vector field } \xi \text{ on } J^{s-1}Y, \qquad (3.4)$$

are called *Hamilton equations*, their solutions are called *Hamilton extremals*. Besides that we have the *characteristic distribution* of the 2-form α,

$$\mathscr{D}_E = annih\{i_\xi \alpha \,|\, \xi \text{ runs over all vector fields on } J^{s-1}Y\} \qquad (3.5)$$

(or equivalently, spanned by vector fields ζ on $J^{s-1}Y$ such that $i_\zeta \alpha = 0$). The above distributions have the following properties:

• $\mathscr{D}_E$ is a subdistribution of Δ_E, and all its holonomic integral sections correspond to extremals. However, extremals (more precisely, the $(s-1)$-prolongations of extremals) are only a part of solutions, since there may be integral sections which are not of the form of prolongations of sections of π.

• The rank of $\mathscr{D}_E$ or Δ_E need not be constant. In such a case, a distribution annihilated by smooth 1-forms cannot be generated by local *continuous* vector fields.

• It holds $rank\,\alpha = corank\,\mathscr{D}_E$. Moreover, if $rank\,\alpha$ is constant, it is *even*.

• The distributions need not be "horizontal": at some points they may be integrable but posses no integral *sections*. It can be proved that Δ_E *is horizontal at each point if and only if* Δ_E *and* $\mathscr{D}_E$ *coincide*.

Analyzing the properties of the two related distributions, we obtain the following *geometric classification of variational ODE*, and can learn the structure of extremals and Hamilton extremals:

Regular equations. E is called *regular* if $rank\,\Delta_E = 1$. In this most simple case both the distributions Δ_E and $\mathscr{D}_E$ coincide, and the equations (3.2) and (3.4) are *equivalent*.

29

Regularity simply means that equations (1.1) are solvable with respect to the highest derivatives (i.e., have an equivalent normal form (2.13)). The *regularity condition* reads

$$det\left(\frac{\partial E_\sigma}{\partial q_s^\nu}\right) \neq 0, \qquad (3.6)$$

or with help of a Lagrangian of the *minimal possible order* r_0 (recall, $r_0 = s/2$ if s is even, or $r_0 = (s+1)/2$ if s is odd):

$$det\left(\frac{\partial^2 L}{\partial q_{r_0}^\sigma \partial q_{r_0}^\nu}\right) \neq 0, \quad \text{and} \quad det\left(\frac{\partial^2 L}{\partial q_{r_0}^\sigma \partial q_{r_0-1}^\nu} - \frac{\partial^2 L}{\partial q_{r_0}^\nu \partial q_{r_0-1}^\sigma}\right) \neq 0, \qquad (3.7)$$

for *even-order* and *odd-order variational equations*, respectively. Note that the geometric definition of regularity based on the Δ_E gives a more precise concept of regularity than that of a "regular Lagrangian" commonly used in the calculus of variations: our regularity is a *property equations* (not of a particular Lagrangian), generalizes the concept also to *odd-order* equations, and provides *new regularity conditions for Lagrangians* (cf. (3.7) for the odd-order case; moreover, (3.6) can be rewritten for *any* Lagrangian for E, not only a minimal-order one, and the arising conditions can look quite differently (see [23]). The distribution (3.3) (resp. (3.5)) for regular equations is locally spanned by *one smooth vector field* (a local solution of the equation $i_\zeta \alpha = 0$), the semispray

$$\zeta = \frac{\partial}{\partial t} + \sum_{j=0}^{s-2} q_{j+1}^\sigma \frac{\partial}{\partial q_j^\sigma} + F^\sigma \frac{\partial}{\partial q_{s-1}^\sigma}, \quad \text{where} \quad F^\sigma = B^{\sigma\nu} A_\nu. \qquad (3.8)$$

(The above notations correspond to (2.13) and (3.1), $(B^{\sigma\nu})$ is the inverse matrix to (3.6)). This means that the dynamics is *completely integrable*, i.e. through every point of the "evolution space" $J^{s-1}Y$ there passes a unique maximal solution—prolonged extremal (locally, graph of a curve). In terms of physics, the dynamics obey the Newtonian determinism: *a solution is completely determined by initial conditions* (to fix a point in $J^{s-1}Y$ means to fix an initial position—a point in Y, and velocities up to the order $s-1$). We can also say that *the structure of global (prolonged) solutions of regular equations* (1.1) *is given by a one-dimensional foliation in the evolution space* J^1Y.

Semiregular equations. The second step in the classification concerns equations for which $\Delta_E = \mathscr{D}_E$, and $rank\,\Delta_E$ is constant: such equations are called *semiregular*. Since $\mathscr{D}_E$ is a characteristic distribution of a closed 2-form, $corank\,\Delta_E = 2p$, and the above definition guarantees that the distributions are *completely integrable and "horizontal"* (the latter means that they admit integral mappings in the form of sections). Equations (3.2) and (3.4) are not equivalent (some Hamilton extremals are not prolongations of extremals). Maximal integral manifolds define a foliation of $J^{s-1}Y$ with leaves of dimension $ms + 1 - 2p$, which is generally greater than 1. Through every point in the evolution space $J^{s-1}Y$ there passes a unique leaf; any section lying in a leaf is a Hamilton extremal, and conversely, every Hamilton extremal (hence every prolonged extremal) lies in a leaf. Roughly speaking, the dynamics obeys a "partial determinism": initial conditions do not determine a solution completely, one can have more solutions through one point, however, all within a leaf. For obtaining a more subtle dynamical picture describing prolonged extremals in the evolution space, one must apply the *geometric constraint algorithm* (see below).

Weakly regular, and strongly semiregular equations. Variational ODE are called *weakly regular* if, for some k, the Euler–Lagrange distribution Δ_E has a subdistribution Σ locally spanned by k semisprays. In this case equations (3.2) and (3.4) are not equivalent, however, all integral sections of Σ are prolonged extremals, which means that among Hamilton extremals, solutions of (3.2) can be easily specified. It can be proved that for weakly regular equations

- $\mathscr{D}_E = \Delta_E$ (the rank need not be constant),
- $\operatorname{rank} B = \operatorname{rank}(B|A) = \text{const.}$ (where in the notations of (3.1), $B = (B_{\sigma\nu})$, and $(B|A)$ contains the additional column (A_σ)),
- if $k > 1$, the semispray subdistribution Σ is not completely integrable (defines no foliation).

If weakly regular equations are semiregular, they are called *strongly semiregular*. In this case it is possible to further specify the "partially deterministic" picture of the semiregular dynamics: Hamilton extremals proceed within the leaves of Δ_E, prolonged extremals are specified among Hamilton extremals as being integral sections of the subdistribution Σ. Although there is no subfoliation defined by Σ, the prolonged extremals are "forced" to proceed within the leaves defined by the Euler–Lagrange distribution.

Equations with internal constraints. This terminology reflects the work of Dirac, who studied dynamics of second-order variational equations arising from singular Lagrangians [7]. Internal constraints on the motion can be classified with help of geometric properties of the Euler–Lagrange and the characteristic distribution, and specified by means of the so-called *geometric constraint algorithm*, a procedure, how to find a local and global dynamical picture for non-regular variational equations. Typically this means to specify the sets of points in the evolution space $J^{s-1}Y$ where passes no solution, i.e., to exclude initial conditions which are not admitted (in particular, this concerns all points where $\Delta_E \neq \mathscr{D}_E$), and to find the structure of Hamilton extremals and extremals near the remaining points.

A more detailed exposition of this topic including concrete examples can be found in [23], for a generalization to not necessarily variational ODE we refer to [24].

3.2. Hamiltonian differential systems for variational PDE

To study variational partial differential equations we can use locally variational forms and their related Lepagean $(n+1)$-forms (i.e. such that $\alpha = p_1 E$ and $d\alpha = 0$) similarly as in the case of ODE. However, as mentioned in Sec. 2.4, we have to take into account that *for PDE of order $s \geq 2$ the Lepagean equivalent is no more unique*, and generally is *not projectable onto $J^{s-1}Y$*. While variational equations of order s still have the form (3.2) (where $s-1$ should be replaced by the order of α, say r, and α is any Lepagean equivalent of E)—clearly, the equations are the same for any choice of α, the associated *Hamilton equations* (3.4) depend upon the choice of the Lepagean equivalent α of E. Another significant difference appears in the geometric interpretation of the equations. We have no more a distribution, now, (3.4) are *equations for integral sections of an* *exterior differential system* $\mathcal{H}_\alpha$, generated by a system of *n-forms* $i_\xi \alpha$, where ξ runs over π_r-vertical vector fields on $J^r Y$, and the variational equations themselves are equations for *holonomic* integral sections of $\mathcal{H}_\alpha$. We call $\mathcal{H}_\alpha$ *Hamiltonian differential system* of α.

Denote by $\hat{\alpha}$ the at most 2-contact part of α (it is called the *principal part*), and consider the corresponding Hamiltonian differential system $\mathcal{H}_{\hat{\alpha}}$. Note that if α is of order r, $\hat{\alpha}$, and $\mathcal{H}_{\hat{\alpha}}$ are of order $r+1$. Apparently $\mathcal{H}_{\hat{\alpha}}$ and $\mathcal{H}_\alpha$ have the same holonomic integral sections—the prolonged extremals.

In this way, one gets to given variational equations, represented by a dynamical form E, a family of associated Hamiltonian systems. Their properties and solutions may be different, however, in all cases (prolonged) extremals form a subset in the set of solutions. Of course, one would be interested to choose a Hamiltonian system in such a way that its solution set would be as much close to the set of extremals as possible. This idea leads to a geometrical concept of regularity of variational PDEs as follows, similar to the case of ODEs discussed in the preceding section.

Let E be a locally variational form, α its Lepagean equivalent on $J^r Y$, $r \geq 1$, r_0 the minimal possible order of Lagrangians for E. The form α is called *regular* if the system of local generators of $\mathcal{H}_{\hat{\alpha}}$ contains all the *n-forms* $\omega^\sigma \wedge \omega_i$, $\omega^\sigma_{(j_1} \wedge \omega_{i)}, \ldots, \omega^\sigma_{(j_1 \ldots j_{r_0-1}} \wedge \omega_{i)}$, where (..) means symmetrization in the indicated indices. By this definition, every solution of the corresponding principal system $\mathcal{H}_{\hat{\alpha}}$ is *holonomic up to the order* r_0. It can be shown that *for a regular* α, *quasi-holonomic integral sections of the Hamiltonian differential system of* $\pi^*_{r+1,r} \alpha$ *are up to the order* r_0 *prolongations of extremals.* We say that α is *strongly regular* if *all* Hamilton extremals are up to the order r_0 prolongations of extremals. For the most frequent case of second-order variational equations coming from first-order Lagrangians regularity means that extremals are in *bijective* correspondence with quasi-holonomic Hamilton extremals, and strong regularity means that Hamilton equations of α and the corresponding variational PDE are *equivalent*.

The above mentioned relations between extremals and integral sections of regular and strongly regular Hamiltonian differential systems can be used for a deeper investigation of solutions of variational PDE (e.g. classification, symmetries, equivalence properties, integration methods). The above as well as some other results concerning Hamiltonian systems of first and higher-order PDE can be found in [25], [26], [11] (see also [32]). However, a comprehensive theory similar to that known for variational ODE is yet far from being complete.

3.3. Symmetries and first integrals

In differential geometry one has a standard definition of a symmetry of a differential form η on a manifold M as a vector field ξ on M such that $\partial_\xi \eta = 0$. This can be applied directly to a Lagrangian or its Euler–Lagrange form: this is the origin of definitions of symmetries of Lagrange functions and Euler–Lagrange equations which appear (usually in a coordinate form) in the calculus of variations.

Let λ be a Lagrangian of order r, E_λ its Euler–Lagrange form. A π-projectable vector field ξ on Y is called a *symmetry of* λ if $\partial_{J^r \xi} \lambda = 0$, and a *generalized symmetry of* λ if it is a symmetry of the Euler–Lagrange form of λ, i.e. if $\partial_{J^{2r-1}\xi}E_\lambda = 0$. One has the relation $\partial_{J^{2r-1}\xi}E_\lambda = E_{\partial_{J^r\xi}\lambda}$, which means that *every symmetry of λ is its generalized symmetry*, and conversely, *if ξ is a generalized symmetry of λ then there exists an $(n-1)$-form η such that $\partial_{J^r\xi}\lambda = hd\eta$*. Now, from the first variation formula one easily gets the famous Noether Theorem and it generalizations [29], [15]):

Theorem 4. *Let λ be a Lagrangian on J^rY, and ρ its Lepagean equivalent. Let a π-projectable vector field ξ on Y be a symmetry of λ. If γ is an extremal then the $(n-1)$-form $\eta = J^{2r-1}\gamma^* i_{J^{2r-1}\xi}\rho$ is closed.*

The *conservation law* corresponding to variational *partial* differential equations, contained in Noether Theorem, takes the form $\operatorname{div} F = 0$, where $F = (F^1, \ldots, F^n)$ are components of η. For *ordinary* differential equations η is a *function*, f, hence the conservation law reads $f = \operatorname{const}$.

Functions constant along solutions of ordinary differential equations can be used *to find solutions* of the equations. In the sequel I will briefly discuss *applications of the theory of symmetries in the theory of ordinary differential equations, and related integration methods*. For a more detail and advanced exposition the reader can consult [23].

The above definition of symmetry of a differential form can be applied directly to a *locally variational form*, its *Lepagean equivalent*, or a particular *Lagrangian*, and its *Cartan form*. Also, one can apply to the *Euler–Lagrange distribution*, or to the *characteristic distribution* the standard definition of symmetry of a distribution: ξ is called a *symmetry of a distribution* Δ if Δ is invariant under the flow ϕ_u of ξ (i.e., for all values of the parameter u and every point x, $T\phi_u(\Delta_x) \subset \Delta_{\phi_u(x)}$). We shall also say that a vector field ξ is a *symmetry of the equations* $E \circ J^s\gamma = 0$ if it is a symmetry of the characteristic distribution $\mathscr{D}_E$. In the theory of symmetries of ODE the most important role is played by vector fields on the evolution space $J^{s-1}Y$, where the Lepagean equivalent α of E, and both the related distributions Δ_E and $\mathscr{D}_E$ are defined: symmetries on $J^{s-1}Y$ are called *dynamical symmetries*. Due to the prolongation structure, one has a particular kind of dynamical symmetries, the so-called *point symmetries* which are of the form of prolongations of projectable vector fields defined on Y (as we have seen, they appear in the Noether Theorem). The following result can be proved (see [23]):

Theorem 5.

(1) *Every point symmetry of a Lagrangian is a point symmetry of its Euler–Lagrange form.*

(2) *Point symmetries of a Lagrangian λ and of its Lepagean equivalent θ_λ coincide.*

(3) *Point symmetries of a locally variational form E and of its Lepagean equivalent α coincide.*

(4) *If α has a constant rank then every its dynamical symmetry is a dynamical symmetry of the characteristic distribution $\mathscr{D}_E$, hence it is a dynamical symmetry of the corresponding equations. The same assertion holds for point symmetries.*

Dynamical symmetries of Lepagean 2-forms give rise to *first integrals* of the corresponding equations. Indeed, if ξ is a dynamical symmetry of α, we have $0 = \partial_\xi \alpha =$

$i_\xi d\alpha + di_\xi\alpha \Rightarrow di_\xi\alpha = 0$, since α is closed, hence locally $i_\xi\alpha = df$ for a function f. By definition of $\mathscr{D}_E$, the 1-form $i_\xi\alpha = df$ belongs to the annihilator of $\mathscr{D}_E$, meaning that f is a first integral of the distribution $\mathscr{D}_E$; in other words, f *is constant along solutions of the corresponding equations* $E \circ J^s\gamma = 0$. By Frobenius Theorem, first integrals of completely integrable distributions can be used to construct integral manifolds. In our case, when $rank\,\alpha = corank\,\mathscr{D}_E = 2p$, any family of $2p$ independent first integrals $f^1, \ldots f^{2p}$ of $\mathscr{D}_E$ can be completed to a local chart on $J^{s-1}Y$, adapted to the distribution $\mathscr{D}_E$. If $(f^1, \ldots f^{2p}, x^J)$, where $1 \leq J \leq \dim J^{s-1}Y - 2p$ is such a chart then maximal integral manifolds of the distribution $\mathscr{D}_E$ are locally given by the equations $f^1 = c^1, \ldots, f^{2p} = c^{2p}$, where the c's are constants. This means that the problem of local integration of the corresponding equations is solved if a system of $2p$ *independent first integrals of* $\mathscr{D}_E$ (defined on an open subset of $J^{s-1}Y$) is found.

3.4. Geometric integration methods for semiregular ODE

The above mentioned relations between symmetries and first integrals can be used to develop *integration methods for variational ODE*, i.e. procedures and formulas for constructing a corresponding *complete family of exact solutions*. In the classical calculus of variations, such methods are developed for second-order ODE, coming from *regular first-order Lagrangians*, and are based on the Liouville theorem, Hamilton–Jacobi equation, or canonical transformations. In global analysis, there is known the *Lie method of integration*, an algorithm for constructing local foliations related with completely integrable distributions (see e.g. [28]). Having in mind the geometric classification of variational ODE (Sec. 3.1) we can see that general methods are applicable also to *semiregular variational ODE of any order*. It turns out, however, that the existence of Lagrangians means that for these equations alternative integration methods can be developed, which in many cases are more simple. In fact, analyzing the geometric content of the above mentioned integration methods for equations of classical mechanics, it is possible to find their generalization to *semiregular* equations both of *even and odd order*. In the final part of my lecture, I will mention basic ideas of the *generalized Liouville integration method*, since it is based on a nice geometric interpretation of some classical variational concepts. The exposition is concise, and the interested reader is suggested to consult [22], [23].

Assume that E is semiregular. Recall that this means that the distributions Δ_E and $\mathscr{D}_E$ coincide, have a constant corank equal to an *even* number, say $2p$, and are *completely integrable*. Consequently, Frobenius theorem guarantees the *existence* of a global foliation of $J^{s-1}Y$, whose leaves are $2p$-codimensional integral manifolds of the distribution $\mathscr{D}_E$. To find the (local) foliation explicitly, one must find $2p$ independent first integrals of $\mathscr{D}_E$. The essence of the Liouville integration method is the following: due to *variationality*, it is sufficient to find only p appropriate symmetries (and their corresponding first integrals), the remaining p first integrals then can be *computed* using concrete integral formulas.

Let us start with a basic definition. A *complete integral* of α is defined to be a completely integrable distribution $\mathscr{I}$ of corank p on an open set $U \subset J^{s-1}Y$, such that

α belongs to the differential ideal generated by $\mathscr{I}$ (i.e. $\alpha = 0$ on integral manifolds of $\mathscr{I}$). Complete integrals have the following properties:

• If $\mathscr{I}$ is a complete integral of α then the distribution $\mathscr{D}_E$ is its subdistribution. Consequently, every leaf of the foliation of $\mathscr{I}$ is foliated by the leaves of $\mathscr{D}_E$.

• If (x^J, a^K) is a chart on $J^{s-1}Y$ adapted to $\mathscr{I}$, i.e. if $\mathscr{I}$ is annihilated by the forms da^K, $1 \leq K \leq p$, then $a^1, a^2, \ldots, a^p$ are independent first integrals of $\mathscr{D}_E$.

• If $a^1, a^2, \ldots, a^p$ are independent first integrals of $\mathscr{D}_E$ satisfying the conditions

$$\{a^K, a^L\} \equiv i_{\xi_K} da^L = i_{\xi_K} i_{\xi_L} \alpha = 0, \tag{3.9}$$

where ξ_K is a symmetry corresponding to the first integral a^K, then the distribution $\mathscr{I}$, annihilated by $da^1, \ldots, da^K$, is a complete integral of α.

The bracket of first integrals defined in (3.8) is called *Poisson bracket*. We also say that first integrals $a^1, \ldots a^p$ of $\mathscr{D}_E$ satisfying condition (3.8) are *in involution*.

The vague concept of "appropriate symmetries" used above concerns nothing but that of a complete integral. Indeed, the **Generalized Liouville Theorem** says that *if for a semiregular locally variational form E a complete integral is known* (i.e. if one has a system of *p* independent first integrals in involution), *then remaining p first integrals can be computed "by quadratures"* (meaning that for their explicit computation integral formulas are used). The corresponding formulas follow quite simply by using the definition of the complete integral, and the Poincaré Lemma (see [23]).

Hence, the integration procedure based on the Liouville Theorem can be started, if a complete integral is known. To this purpose, it is sufficient to find *p* first integrals satisfying condition (3.8): this, however, can be done for example with help of the Noether Theorem and/or by computing some symmetries of α. Another possibility is to solve a (generalized) Hamilton–Jacobi equation (a complete integral appearing in this equation represented by a function *S* is in fact a complete integral in the sense of the above geometric definition, possessing an additional regularity property) [22], [23].

A remark on variational integrators. The Liouville integration method discussed above applies to variational equations, however, in fact, the existence of a Lagrangian is not so essential. It turns out that it is sufficient to have for given equations a *local closed 2-form of a constant rank* such that solutions are integral sections of its *characteristic distribution* [22], [23]. Such a form, however, is easily obtained if the equations are variational: simply one takes the corresponding Lepagean 2-form. On the other hand, for non-variational equations one does not know how to find such a form (unless a complete system of independent first integrals is known [22]) which makes the Liouville Theorem practically not applicable. This fact, and also some other nice properties of variational equations can serve as an motivation to extend the inverse problem of the calculus of variations to ask the following question: *Given a system of differential equations, is it possible to find a Lagrangian such that solutions of these equations would be extremals of the Lagrangian?* Unfortunately, yet nobody has been able to find a solution of this problem in such a general formulation. Even a more simple version of it, *the problem of existence of variational integrators for second-order ordinary differential equations* has not yet been completely solved. The latter problem is as follows: *Given a system of* *second-order ODE in normal form*, i.e.

$$\ddot{q}^\sigma = F^\sigma(t, q^\nu, \dot{q}^\nu), \quad 1 \le \nu \le m, \tag{3.10}$$

does there exist functions $B_{\sigma\rho}(t, q^\nu, \dot{q}^\nu)$ such that the matrix $(B_{\sigma\rho})$ is regular, and the equations $B_{\sigma\rho}(\ddot{q}^\rho - F^\rho) = 0$ are variational? (in the sense that there exists a Lagrangian L such that the functions $E_\sigma = B_{\sigma\rho}(\ddot{q}^\rho - F^\rho)$ are Euler–Lagrange expressions of L). Functions $B_{\sigma\rho}$ are then called *variational integrators* for (3.10). Since (3.10) are equations for integral sections of a semispray ("second-order vector field")

$$\xi = \frac{\partial}{\partial t} + \dot{q}^\sigma \frac{\partial}{\partial q^\sigma} + F^\sigma \frac{\partial}{\partial \dot{q}^\sigma}, \tag{3.11}$$

one also speaks about a problem *under what conditions a second-order vector field is variational*. A straightforward approach to answer the question is to apply Helmholtz conditions to the functions $E_\sigma = B_{\sigma\rho}(\ddot{q}^\rho - F^\rho)$, i.e. put in (2.32) $A_\sigma = -B_{\sigma\rho}F^\rho$. In this way one obtains *equations for an unknown variational integrator*, and the answer depends on the existence and number of solutions of these partial differential equations.

The problem of variational integrators was first set and solved for a single second-order ordinary differential equation by Sonin [33], who proved that *every* such equation has a variational integrator. A generalization to a system of 2 equations was later successfully treated by Douglas [8], who discovered that there exist equations with *no* variational integrator, and provided a complete classification of equations with respect to the existence and corresponding forms of variational integrators, which in this case are local (2×2)-matrices on J^1Y (see also [5] for a geometric analysis of Douglas' results). A similar discussion in case of more equations seems to be too complicated, and no final results have been achieved yet. On the other hand there is a lot of papers dealing with different aspects of this problem, where interesting and deep results can be found (e.g. relations with closed forms, symmetries, metrizability of linear connections, Finsler geometry and its generalizations, variational forces in mechanics, etc.); for more details the reader can consult e.g. [2], [9], [23], [27], and references therein.

ACKNOWLEDGMENTS

Research supported by grants GAČR 201/03/0512 of the Czech Grant Agency, and MSM 153100011 of the Czech Ministry of Education, Youth and Sports. The author also highly appreciates support and hospitality of the Mathematics Department and the Institute for Advanced Study at La Trobe University, Australia, where this paper was completed.

REFERENCES

1. I. Anderson and T. Duchamp, *On the existence of global variational principles*, Am. J. Math. 102 (1980) 781–867.

2. I. Anderson and G. Thompson, *The Inverse Problem of the Calculus of Variations for Ordinary Differential Equations*, Memoirs of the AMS 98, No. 473 (1992) 110 pp.
3. D.E. Betounes, *Extension of the classical Cartan form*, Phys. Rev. D 29 (1984) 599–606.
4. R.L. Bryant, S.S. Chern, R.B. Gardner, H.L. Goldschmidt and P.A. Griffiths, *Exterior Differential Systems*, Springer, New York, 1991.
5. M. Crampin, W. Sarlet, E. Martínez, G.B. Byrnes and G.E. Prince, *Towards a geometrical understanding of Douglas's solution of the inverse problem of the calculus of variations*, Inverse Problems 10 (1994) 245–260.
6. P. Dedecker and W.M. Tulczyjew, *Spectral sequences and the inverse problem of the calculus of variations*, Internat. Coll. on Diff. Geom. Methods in Math. Phys., Aix-en-Provence, 1979, in: Lecture Notes in Math. 836 (Springer, Berlin, 1980) 498–503.
7. P.A.M. Dirac, *Generalized Hamiltonian dynamics*, Canad. J. Math. II (1950), 129–148.
8. J. Douglas, *Solution of the inverse problem of the calculus of variations*, Trans. Amer. Math. Soc. 50 (1941) 71–128.
9. J. Grifone and Z. Muzsnay, *Variational Principles for Second-order Differential Equations*, World Scientific, Singapore, 2000.
10. J. Grifone, J. Muñoz Masqué and L.M. Pozo Coronado, *Variational first-order quasilinear equations*, in: Steps in Differential Geometry, Proc. Colloq. Diff. Geom., Debrecen, July 2000, L. Kozma, P.T. Nagy and L. Tamássy, eds. (University of Debrecen, Debrecen, 2001) 131–138.
11. A. Haková and O. Krupková, *Variational first-order partial differential equations*, J. Differential Equations 191 (2003) 67–89.
12. H. Helmholtz, *Ueber die physikalische Bedeutung des Prinzips der kleinsten Wirkung*, J. für die reine u. angewandte Math. 100 (1887), 137–166.
13. L. Klapka, *Euler–Lagrange expressions and closed two–forms in higher order mechanics*, in: Geometrical Methods in Physics, Proc. Conf. on Diff. Geom. and Appl. Vol. 2, Nové Město na Moravě, Sept. 1983, D. Krupka, ed. (J. E. Purkyně Univ. Brno, Czechoslovakia, 1984) 149–153.
14. I.S. Krasilschik, V.V. Lychagin and A.M. Vinogradov, *Geometry of Jet Spaces and Differential Equations*, Gordon and Breach, 1986.
15. D. Krupka, *A geometric theory of ordinary first order variational problems in fibered manifolds. I. Critical sections, II. Invariance*, J. Math. Anal. Appl. 49 (1975) 180–206; 469–476.
16. D. Krupka, *A map associated to the Lepagean forms of the calculus of variations in fibered manifolds*, Czechoslovak Math. J. 27 (1977) 114–118.
17. D. Krupka, *On the local structure of the Euler-Lagrange mapping of the calculus of variations*, Univerzita Karlova, Prague, 1981, 181–188; ArXiv: math-ph/0203034.
18. D. Krupka, *Lepagean forms in higher order variational theory*, in: Modern Developments in Analytical Mechanics I: Geometrical Dynamics, Proc. IUTAM-ISIMM Symposium, Torino, Italy 1982, S. Benenti, M. Francaviglia and A. Lichnerowicz, eds. (Accad. delle Scienze di Torino, Torino, 1983) 197–238.
19. D. Krupka, *Variational sequences on finite order jet spaces*, in: Differential Geometry and Its Applications, Proc. Conf., Brno, Czechoslovakia, 1989, J. Janyška and D. Krupka eds. (World Scientific, Singapore, 1990) 236–254.
20. D. Krupka, *Global variational principles: Foundations and current problems*, in this volume.
21. O. Krupková, *Lepagean 2-forms in higher order Hamiltonian mechanics, I. Regularity, II. Inverse problem*, Arch. Math. (Brno) 22 (1986) 97–120; 23 (1987) 155–170.
22. O. Krupková, *Liouville and Jacobi theorems for vector distributions*, in: Differential Geometry and Its Applications, Proc. Conf., Opava, August 1992, O. Kowalski and D. Krupka, eds. (Mathematical Publications, Vol 1, Silesian University in Opava, Opava, Czechoslovakia, 1993) 75–88.
23. O. Krupková, *The Geometry of Ordinary Variational Equations*, Lecture Notes in Mathematics 1678, Springer, Berlin, 1997.
24. O. Krupková, *Differential systems in higher-order mechanics*, in: Proceedings of the Seminar on Differential Geometry, D. Krupka, ed. (Mathematical Publications, Vol. 2, Silesian University in Opava, Opava, 2000) 87–130.
25. O. Krupková, *Hamiltonian field theory*, J. Geom. Phys. 43 (2002) 93–132.
26. O. Krupková, *Hamiltonian field theory revisited: A geometric approach to regularity*, in: Steps in Differential Geometry, Proc. Colloq. Diff. Geom., Debrecen, July 2000, L. Kozma, P.T. Nagy and L. Tamássy, eds. (University of Debrecen, Debrecen, 2001) 187–207.

27. O. Krupková, *Variational metric structures*, Publ. Math. Debrecen 62 (2003) 461–495.
28. V. Lychagin, *Lectures on Geometry of Differential Equations, Part I, Part II*, Consiglio Nazionale delle Ricerche (G.N.F.M.), Roma, 1992.
29. E. Noether, *Invariante Variationsprobleme*, Nachr. kgl. Ges. Wiss. Göttingen, Math. Phys. Kl., 1918, 235–257.
30. D.J. Saunders, *The Geometry of Jet Bundles*, London Math. Soc. Lecture Notes Series 142, Cambridge Univ. Press, Cambridge, 1989.
31. J. Šeděnková, *A generalization of Lepage forms in mechanics*, in this volume.
32. D. Smetanová, *On regularization of second order Lagrangians*, in this volume.
33. N.Y. Sonin, *On the definition of maximal and minimal properties*, Warsaw Univ. Izvestiya, No. 1–2 (1886), 1–68 (in Russian).
34. F. Takens, *A global version of the inverse problem of the calculus of variations*, J. Diff. Geom. 14 (1979) 543–562.
35. E. Tonti, *Variational formulation of nonlinear differential equations I, II*, Bull. Acad. Roy. Belg. Cl. Sci. 55 (1969) 137–165, 262–278.
36. W.M. Tulczyjew, *The Euler-Lagrange resolution*, Internat. Coll. on Diff. Geom. Methods in Math. Phys., Aix-en-Provence 1979, in: Lecture Notes in Math. 836 (Springer, Berlin, 1980) 22–48.
37. M.M. Vainberg, *Variational methods in the theory of nonlinear operators*, GITL, Moscow, 1959 (in Russian).
38. A.L. Vanderbauwhede, *Potential operators and variational principles*, Hadronic J. 2 (1979), 620–641.
39. A.M. Vinogradov, *The $\mathscr{C}$-spectral sequence, Lagrangian formalism, and conservation laws. I. The linear theory. II. The nonlinear theory*, J. Math. Anal. Appl. 100 (1984) 1–40, 41–129.

A compatibility criterion for systems of PDEs and generalized Lagrange-Charpit method

Boris Kruglikov[*] and Valentin Lychagin[*]

[*]*Institute of Mathematics and Statistics, University of Tromsø, Tromsø 90-37, Norway*

Abstract. In this paper we give a general compatibility theorem for overdetermined systems of scalar partial differential equations of complete intersection type in terms of generalized Mayer brackets. As an applications we propose a generalization of the classical Lagrange-Charpit method for integration of a single scalar PDE.
Key words. Jacobi and Mayer brackets, Spencer δ-cohomology, Weyl tensor, integrals, characteristics, compatibility of PDEs.
PACS. 02.30.Jr, 45.10.Na, 02.40.Yy.
MSC. 35N10, 58A20, 58H10, 35A30.

1. INTRODUCTION

In this paper we give an efficient compatibility criterion for a certain class of overdetermined systems $\mathscr{E} \subset J^k M$ of PDEs. Namely we consider the systems with the characteristic symbolic ideal $I(g) \subset STM$ being *complete intersection*. This means that it can be generated by elements $f_1, \ldots, f_r$ (one can think of them as of homogeneous polynomials), the only relations between which are the Koszul relations: $f_j f_i = f_i f_j$.

We also call such systems of differential equations *complete intersections*. By the above arguments they can be represented by (in general non-linear) PDEs $F_1 = 0, \ldots, F_r = 0$ with the symbols f_i of differential operators F_i satisfying the above requirements.

Compatibility (or formal integrability) of the system $\mathscr{E}$ is an algebraic problem, which is studied via the Spencer theory: the obstruction to integrability belongs to the second Spencer δ-cohomology group $H^{*,2}(\mathscr{E})$ (see [16, 3] or the main text). Moreover, the precise obstruction is a kind of curvature (structural) tensor and its component $W_l(\mathscr{E}) \in H^{l-1,2}$ corresponding to l-jets is called the Weyl tensor of order l ([13, 9]).

Leaving the precise definitions of the basic notions to the main text let us formulate the two main results already now.

Theorem A. *Let $\mathscr{E}$ be a system of scalar differential equations on a manifold M, which is of complete intersection type. Let its generating PDEs (without Koszul relations on the symbolic level) have orders $k_1, \ldots, k_r$ (clearly $r \leq n = \dim M$). Then the only non-zero*

CP729, *Global Analysis and Applied Mathematics: International Workshop on Global Analysis,*
edited by K. Taş, D. Krupka, O. Krupková, and D. Baleanu
© 2004 American Institute of Physics 0-7354-0209-4/04/$22.00

Spencer δ-cohomologies are:

$$H^{*,0}(\mathscr{E}) = \mathbb{R}^1, \ H^{*,1}(\mathscr{E}) = \mathbb{R}^r, \ H^{*,2}(\mathscr{E}) = \mathbb{R}^{r(r-1)/2}, \dots, H^{*,r}(\mathscr{E}) = \mathbb{R}^1.$$

More precisely, the generators of $H^{*,s}(\mathscr{E}) = \oplus_k H^{k,s}(\mathscr{E}) = \mathbb{R}^{\binom{r}{s}}$ *correspond to* $k = k_{i_1} + \cdots + k_{i_s} - s$ *for various choices of* $1 \le i_1 < \cdots < i_s \le r$.

In order to evaluate the Weyl tensors $W_l(\mathscr{E})$ we introduce the special brackets. Mayer bracket $[F, G]$ for a pair of differential operators generalizes the classical Mayer bracket known in the first order case ([10]). Its restriction to the equation prolongation $[,]_{\mathscr{E}}$ is defined canonically, whence the bracket for functions on $J^\infty M$, see details below. Note that for linear operators F, G the bracket $[F, G]_{\mathscr{E}}$ is the usual (reduced) commutator.

Theorem B. *For a complete intersection scalar system $\mathscr{E}$ the only non-zero Weyl tensors $W_k = W^{(i,j)}(\mathscr{E}) \in H^{k-1,2}(\mathscr{E})$ occur at orders $k = k_i + k_j - 1$ for various choices of $1 \le i < j \le r$. Let the system $\mathscr{E}$ be represented in the form*

$$\begin{cases} F_1\left(x, u(x), \dots, \frac{\partial^{|\varsigma|} u}{\partial x^\varsigma}\right) = 0, & |\varsigma| \le k_1, \\ \qquad \dots\dots\dots \\ F_r\left(x, u(x), \dots, \frac{\partial^{|\tau|} u}{\partial x^\tau}\right) = 0, & |\tau| \le k_r. \end{cases} \tag{1}$$

Then the Weyl tensors express via the restricted Mayer brackets as follows:

$$W^{(i,j)}(\mathscr{E}) = [F_i, F_j]\big|_{\mathscr{E}_{k_i+k_j-1}} \cdot \Omega^{(i,j)},$$

where the classes $\Omega^{(i,j)} = [\omega^{(i,j)}] \in H^{k_i+k_j-2,2}(\mathscr{E})$ *form a basis of* $H^{*,2}(\mathscr{E})$. *Moreover, they can be represented locally via the forms* $\omega^{(i,j)} = \sum_{\alpha,\beta} e^{(i,j)}_{\alpha,\beta} \otimes dx^\alpha \wedge dx^\beta$, *where $e^{(i,j)}_{\alpha,\beta}$ are some sections of the symbol bundle g associated to $\mathscr{E}$, which depend only on pair-wise resultants $R(f_i, f_j)$ of the symbols $f_i = \sigma(F_i)$, $f_j = \sigma(F_j)$ of the generating differential operators.*

Notice that the condition of complete intersection can be reformulated as follows: The characteristic varieties $\mathrm{Char}^{\mathbb{C}}(F_i) \subset P^{\mathbb{C}}T^* M$ are jointly transversal, i.e. the codimension of $\cap_{i=1}^r \mathrm{Char}^{\mathbb{C}}(F_i)$ is exactly r.

Corollary. *System $\mathscr{E}$ is formally integrable if and only if all the pair-wise Mayer brackets vanish due to the system:*

$$[F_i, F_j]\big|_{\mathscr{E}_{k_i+k_j-1}} = 0 \quad \forall \ 1 \le i < j \le r.$$

If in addition $r = n$, then the system is locally smoothly integrable.

Particular cases of the above theorem were proved in [10, 11].

Using the above formula for compatibility we generalize the known (in the first order) Lagrange-Charpit method for integration of partial differential equations: One imposes on the system an overdetermination of a special kind.

This resembles the other known method of differential constraints. The difference is however that in the former case we require the new system to be compatible, while in the latter it needs to be only solvable, so that compatibility conditions are to be imposed on the system additionally.

We dot not assume any restriction on the order of the equations in the system and check compatibility via our criterion.

2. BASICS FROM THE DIFFERENTIAL EQUATIONS THEORY

2.1. Jet bundles and prolongations

Let M be a smooth manifold and $J^k M$ the corresponding jet-bundle. Recall that its fiber $J^k_x M$ over a point $x \in M$ consists of classes $[f]^k_x$ of functions $f \in C^\infty_{loc}(M)$ by the following equivalence relation: $f_1 \sim f_2$ iff $f_1(z) - f_2(z) = o(z^k)$, where z is a local coordinate centered at x. The equivalence class is called the k-jet $x_k = [f]^k_x = j_k f(x)$.

The jet space has a bundle structure given by the natural projection $\pi_k : J^k M \to M$. There are also natural bundle morphisms $\pi_{k,l} : J^k M \to J^l M$. Any smooth function $f \in C^\infty_{loc}(M)$ induces a section $j_k f : x \mapsto [f]^k_x$ of the bundle π_k.

The fiber $F(x_{k-1}) = \pi^{-1}_{k,k-1}(x_{k-1})$ has a natural affine structure associated to the vector space $S^k \tau_x^*$, where $\tau_x = T_x M$. We write $F(x_{k-1}) \simeq S^k \tau_x^*$.

An order k scalar differential equation (or system) is represented as a submanifold $\mathcal{E} \subset J^k M$ ([8]). The ith prolongation of the equation $\mathcal{E}$ is defined by the formula

$$\mathcal{E}^{(i)} = \{ x_{k+i} = [f]^{k+i}_x \in J^{k+i} M \, | \, j_k f(M) \text{ is tangent to } \mathcal{E} \text{ at } x_k \text{ with order} \geq i \}.$$

In order to cover the case of several equations of different orders we modify the usual definition. By a differential equation/system of (maximal) order k we mean a sequence $\mathcal{E} = \{\mathcal{E}_l\}_{-1 \leq l \leq k}$ of submanifolds $\mathcal{E}_l \subset J^l(\pi)$ with $\mathcal{E}_{-1} = M$, $\mathcal{E}_0 = J^0 M = M \times \mathbb{R}$ and such that for all $0 < l \leq k$ the following conditions hold:

1. $\pi^{\mathcal{E}}_{l,l-1} : \mathcal{E}_l \to \mathcal{E}_{l-1}$ are smooth fiber bundles.

2. The first prolongations $\mathcal{E}^{(1)}_{l-1}$ are smooth subbundles of π_l and $\mathcal{E}_l \subset \mathcal{E}^{(1)}_{l-1}$.

Such a system $\mathcal{E}$ is called *formally integrable* (compatible) if $\pi_{k+i,k+i-1} : \mathcal{E}^{(i)}_k \to \mathcal{E}^{(i-1)}_k$ are smooth bundle projections for all $i > 0$. It is called *solvable* if the fiber $\mathcal{E}_\infty = \mathcal{E}^{(\infty)}_k$ over every $x \in M$ is non-empty.

Due to Cartan-Kuranishi theorem on prolongations there exists a (minimal) number l_1 such that $\mathcal{E}^{(1)}_{l-1} = \mathcal{E}_l$ for all $l \geq l_1$. Each number l, where the previous equality fails is called an *order* of the system. The codimension of $\mathcal{E}_l$ in $\mathcal{E}^{(1)}_{l-1}$ is called the multiplicity $m(l)$ of the order. Thus Cartan-Kuranishi theorem can be reformulated as finiteness of the set of orders and multiplicities. In addition the theorem says that the problems of formal integrability and solvability can be resolved in a finite number of steps.

41

Denote the set of all orders of the system by $\mathrm{ord}(\mathcal{E})$. Their totality counted with multiplicity is called *formal codimension*. This number $r = \mathrm{codim}(\mathcal{E})$ of involved PDEs is an important invariant of the system. If $r > 1$ the system is overdetermined, while in the case $r = 1$ we have a single (determined) equation.

2.2. Cartan and metasymplectic structures

Fix a point $x_k \in J^k M$. Then the tangent planes to jet-sections span a subspace

$$\mathcal{C}(x_k) = \langle \cup (j_k f)_* \tau_x \mid [f]_x^k = x_k \rangle,$$

which is called the *Cartan subspace* ([8]). The corresponding distribution $\mathcal{C}(x_k) \subset T_{x_k} J^k M$ on $J^k M$ is called the *Cartan distribution*.

In $J^1 M$ this distribution is nothing else but the standard contact structure. For general k we can describe $\mathcal{C}(x_k)$ in local coordinates as follows.

Every (local) coordinate system (x^i) on M induces coordinates (x^i, p_σ) on $J^k M$ (multiindex σ has length $|\sigma| \leq k$), where $p_\sigma\left([f]_x^k\right) = \dfrac{\partial^{|\sigma|} f}{\partial x^\sigma}(x)$. Then the differential forms (Cartan forms)

$$\omega_\sigma = dp_\sigma - \sum \omega_{\sigma+1_i} dx^i$$

span the annulator of the distribution $\mathcal{C}$, i.e. $\mathcal{C} = \mathrm{Ker}\{\omega_\sigma\}_{0 \leq |\sigma| < k}$.

To describe a basis of sections of $\mathcal{C}$ we recall ([8]) the *operator of total derivative* $\mathcal{D} : C^\infty_{\mathrm{loc}}(J^k M) \to \Omega^1(M) \otimes_{C^\infty_{\mathrm{loc}}(M)} C^\infty_{\mathrm{loc}}(J^{k+1} M)$. To define $\mathcal{D}$ we note that every function on $J^k M$ is a differential operator of order k. Composing it with a vector field $X \in \mathcal{D}(M)$ we get a differential operator of order $k+1$ producing the needed operator $\mathcal{D}_X : C^\infty_{\mathrm{loc}}(J^k M) \to C^\infty_{\mathrm{loc}}(J^{k+1} M)$.

In coordinates $\mathcal{D}_i = \mathcal{D}_{\partial_{x^i}}$ is given by infinite series

$$\mathcal{D}_i = \partial_{x^i} + \sum p_{\sigma+1_i} \partial_{p_\sigma}. \tag{2}$$

If in the above sum we restrict $|\sigma| < k$ we get vector fields $\mathcal{D}_i^{(k)}$ on $J^k M$. In terms of them the Cartan distribution on $J^k M$ is given by

$$\mathcal{C} = \langle \mathcal{D}_i^{(k)}, \partial_{p_\sigma} \rangle_{1 \leq i \leq n, |\sigma| = k} \tag{3}$$

Recall that for any distribution Π its curvature $\Xi_\Pi \in \Lambda^2 \Pi^* \otimes v$ is the 2-form on Π with values in the normal bundle $v = TM/\Pi$, defined as $\Xi_\Pi(X, Y) = [X, Y] \mod \Pi$ (although the formula uses vector fields, the tensor Ξ_Π depends only on their values at the considered point). We will however consider a smaller space of values — the linear span of the $\mathrm{Im}\,\Xi_\Pi \subset v$ — and denote it by v.

The curvature of the Cartan distribution is called *metasymplectic structure*. In this case $v \simeq S^{k-1} \tau^* (= F(x_{k-2}))$ and so the metasymplectic structure can be viewed as the following 2-form

$$\Omega_k = \Omega_{x_k} \in \Lambda^2(\mathcal{C}^*(x_k)) \otimes S^{k-1} \tau_x^*.$$

Thus for every $\lambda \in S^{k-1}\tau_x$ the evaluation $\Omega_k(\lambda) = \langle \Omega_{x_k}, \lambda \rangle$ is an ordinary 2-form on the Cartan space $\mathscr{C}(x_k)$. In particular for $k = 1$ we have the standard symplectic form Ω_1 on $\mathscr{C}(x_1)$.

Fix some $f \in C_{\mathrm{loc}}^{\infty}(M)$ with $[f]_x^k = x_k$. Denote $H_f(x_k) = (j_k f)_* \tau_x$ (this subspace does not depend on a particular choice of f, but only on $x_{k+1} = [f]_x^{k+1}$; it is often denoted by $L(x_{k+1})$). Then we have a decomposition

$$\mathscr{C}(x_k) = H_f(x_k) \oplus F(x_{k-1}) \simeq \tau_x \oplus S^k \tau_x^*. \tag{4}$$

Note that $\Omega_k|_H = 0$ for this choice of $H = H_f$ because $j_k f(M)$ is an integral manifold of the Cartan distribution. On the other hand it is obvious that $\Omega_k|_F = 0$.

So to calculate Ω_k it is enough to know the value on the bivector $X \wedge \theta$ with $\theta \in S^k \tau_x^*$, $X \in H(x_k) \simeq \tau_x$. This value can be expressed in terms of the Spencer operator $\delta : S^k \tau^* \to \tau^* \otimes S^{k-1} \tau^*$ as follows (cf. [13]):

$$\Omega_k(X, \theta) = \delta_X \theta \in S^{k-1} \tau^*, \tag{5}$$

where $\delta_X = i_X \circ \delta$ is the differentiation along X. The introduced structure does not depend on the point x_{k+1} determining the decomposition because the subspace $H(x_k) = L(x_{k+1})$ is Ω_k-isotropic.

Let us write the metasymplectic structure is coordinates. The commutators of the basic fields (3) are:

$$[\mathscr{D}_i, \mathscr{D}_j] = 0, \ [\partial_{p_\sigma}, \partial_{p_\gamma}] = 0 \ \text{and} \ [\partial_{p_\sigma}, \mathscr{D}_i] = \partial_{p_\theta} \ \text{if } \sigma = \theta + 1_i \ \text{and} \ 0 \ \text{otherwise.}$$

So we can take $\nu \simeq \langle \partial_{p_\theta} \rangle_{|\theta|=k-1}$ and the above relations represent the metasymplectic structure $\Omega_k = \Xi_\mathscr{C}$ on $J^k M$. Another form of it is

$$\Omega_k = \sum_{|\sigma|=k-1} d\omega_\sigma \otimes \partial_{p_\sigma}. \tag{6}$$

If $\mathscr{E}$ is a PDEs system represented by a submanifold $\mathscr{E}_k \subset J^k M$ then the restriction of the Cartan distribution on $\mathscr{E}_k$

$$\mathscr{C}_{\mathscr{E}_k}(x_k) = \mathscr{C}(x_k) \cap T_{x_k}(\mathscr{E}_k)$$

is called Cartan distribution on $\mathscr{E}_k$. The restriction of the metasymplectic structure to $\mathscr{C}_{\mathscr{E}_k}$ will be denoted by $\Omega_{\mathscr{E}_k}$.

2.3. Spencer cohomology and Weyl tensor

The *symbol* of differential equation $\mathscr{E}$ at a point $x_k \in \mathscr{E}_k$ is the vertical tangent space to $\mathscr{E}_k$:

$$g_k(x_k) = T_{x_k}(F(x_{k-1})) \cap T_{x_k}(\mathscr{E}) \subset S^k \tau_x^*.$$

Note that the metasymplectic structure of equation $\mathscr{E}$ takes values in symbols:

$$\Omega_{\mathscr{E}_k} \in \Lambda^2 \left(\mathscr{C}_{\mathscr{E}_k}^*(x_k) \right) \otimes g_{k-1}(x_{k-1}).$$

43

The symbol of prolongations $\mathscr{E}_k^{(1)}$ at the point $x_{k+1} \in F(x_k)$ is a subspace

$$g_k^{(1)}(x_k) = \{\theta \in S^{k+1}\tau_x^* \mid \delta\theta \in \tau_x^* \otimes g_k(x_k)\},$$

where $\delta : \Lambda^i \tau_x^* \otimes S^{j+1}\tau_x^* \to \Lambda^{i+1}\tau_x^* \otimes S^j\tau_x^*$ is the Spencer δ-operator (one can think of δ as of de Rham operator on forms with polynomial coefficients). The space $g_k^{(1)}$ is called the algebraic prolongation of g_k.

Thus any sequence $\{x_l \in \mathscr{E}_l \mid \pi_{l,l-1}(x_l) = x_{l-1}\}$ determines the maps $\delta : g_l(x_l) \to g_{l-1}(x_{l-1}) \otimes \tau_x^*$ and therefore the sequence of symbols $\{g_l(x_l)\}$ defines the Spencer δ-complex

$$0 \to g_k(x_k) \xrightarrow{\delta} g_{k-1}(x_{k-1}) \otimes \tau_x^* \xrightarrow{\delta} \cdots \xrightarrow{\delta} g_{k-n}(x_{k-n}) \otimes \Lambda^n \tau_x^* \to 0,$$

where $n = \dim M$.

The cohomology group at the term $g_i(x_i) \otimes \Lambda^j \tau_x^*$ is called the *Spencer δ-cohomology group* at the point x_k and shall be denoted by $H^{i,j}(\mathscr{E}; x_k)$.

Though the system $\mathscr{E}$ of maximal order k determines only $\{g_l(x_k)\}_{l \le k}$, we can study higher cohomology as well by setting $g_l(x_k) = g_k^{(l-k)}(x_k)$ for $l > k$.

We define *regular system of PDEs* of maximal order k as a submanifold $\mathscr{E} = \mathscr{E}_k \subset J^k M$ filtered by $\mathscr{E}_l$ and such that the symbolic system and the Spencer cohomology form graded bundles over it. Thus we often omit reference to the point.

A subspace $H \subset T_{x_k}\mathscr{E}_k$ is called *horizontal* if $(\pi_k)_* : H \to T_x M$ is an isomorphism.

Definition 1. *A horizontal distribution $H(x_k)$ on $\mathscr{E}_k$ is called a* Cartan connection *if $H(x_k) \subset \mathscr{C}_{\mathscr{E}_k}(x_k)$.*

One can prove that near every point x_k there exists a Cartan connection H. Any such connection determines the splitting similar to (4):

$$\mathscr{C}_{\mathscr{E}_k}(x_k) = H(x_k) \oplus g_k(x_k). \tag{7}$$

Definition 2. *The tensor $\Omega_H(x_k) = \Omega_{x_k}|_{H(x_k)}$ is called the* curvature *of the Cartan connection H at the point $x_k \in \mathscr{E}_k$.*

This curvature due to isomorphism $(\pi_k)_* : H(x_k) \simeq \tau_x$ can be viewed as a 2-form

$$\Omega_H(x_k) \in \Lambda^2(\tau_x^*) \otimes g_{k-1}(x_{k-1}).$$

Note that calculation of Ω_k in decomposition (7) is different of that in a *special* decomposition (4), where $H(x_k) = L(x_{k+1})$. However (5) produces a way to calculate Ω_H.

The curvature form of the Cartan connection enjoys the following properties: It is δ-closed, $\Omega_H \in \Lambda^2(\tau^*) \otimes g_{k-1}(x_{k-1})$. In addition if H' is another Cartan connection then $\Omega_{H'} = \Omega_H + \delta\sigma$, where $\sigma \in \tau_x^* \otimes g_k(x_k)$ represents H' in decomposition (7): $H' = \text{graph}\{\sigma : H \to g_k\}$.

Thus we get a canonical cohomology class:

Definition 3. *The* Weyl tensor *of differential equation $\mathscr{E}$ at a point $x_k \in \mathscr{E}_k$ is the δ-cohomology class*

$$W_k(\mathscr{E}; x_k) = \Omega_H \bmod \delta(\tau_x^* \otimes g_k(x_k)) \in H^{k-1,2}(\mathscr{E}; x_k).$$

For geometric structures represented as PDEs systems this tensor coincides with the classical structural function.

Theorem 1 ([13]). *Let $\pi_{k,k-1} : \mathscr{E}_k \to \mathscr{E}_{k-1}$ be a smooth bundle. Then differential equation $\mathscr{E}_k \subset J^k M$ admits the first prolongation $\mathscr{E}_k^{(1)}$ at a point x_k if and only if $W_k(\mathscr{E}_k; x_k) = 0$.*

So if $W_k(\mathscr{E}) = 0$ we consider the Weyl tensor $W_{k+1}(\mathscr{E}) \in H^{k,2}(\mathscr{E})$ for the equation $\mathscr{E}^{(1)}$, study the equation $W_{k+1}(\mathscr{E}) = 0$ etc. Due to Poincaré δ-lemma after some number t of prolongations the second Spencer δ-cohomologies vanish, $H^{k-1+i,2}(\mathscr{E}) = 0$, $i \geq t$. Therefore the number of conditions $W_l = 0$ is finite.

A system of different orders should be investigated for formal integrability successively by the maximal order k. If some prolongation $\mathscr{E}_k^{(1)}$ is not regular, its projections $\{\pi_{k+1,l}(\mathscr{E}_{k+1})\}_{l \leq k}$, form a new system of maximal order k. Taking the regular part one continues with prolongations. The process stops in a finite number of steps (by the Cartan-Kuranishi theorem).

Remark 1. *If $g_k(x_k) = 0$ the equation $W_k(\mathscr{E}) = 0$ is exactly the Frobenius theorem for the horizontal (in this case) distribution $\mathscr{C}_{\mathscr{E}}$. More generally if $\mathscr{E}$ is a regular equation of finite type, i.e. $g_k^{(i)} = 0$ for some i, then the conditions $W_k = 0, \ldots, W_{k+i-1} = 0$ guarantee the existence of local smooth solutions.*

2.4. Characteristic variety

Consider the symbolic system $\{g_l(x_k) \subset S^l \tau^*\}$. Let $g^*(x_k) = \oplus g_l^*$ be its $\mathbb{R}$-dual. It bears the structure of an $S\tau$-module given by

$$(v \cdot \varkappa)p = \varkappa(\delta_v p), \; v \in S\tau, \; \varkappa \in g^*, \; p \in g.$$

Call g^* the *symbolic module*. It is Noetherian and the Spencer cohomology of g dualizes to the Koszul homology of g^*.

Define the *characteristic ideal* by $I(g) = \operatorname{ann}(g^*) \subset S\tau$. It can be also described as follows.

Let $\delta_u : S^{l+1}\tau^* \to S^l\tau^*$ be the differentiation along $u \in \tau$ as above. Extend it to symmetric multi-vectors by the formula $\delta_v = \delta_{u_1} \cdots \delta_{u_k} : S^{k+l}\tau^* \to S^l\tau^*$ on decomposable $v = u_1 \cdots u_k \in S^k\tau$.

Define $I_k^{(l)} = \operatorname{ann} g_{k+l} \subset S^k\tau$ via the pairing $S^k\tau \otimes S^{k+l}\tau^* \to S^l\tau^*$ for $l \geq 0$: $(v, p) \mapsto \delta_v p$. Let $I^{(l)} = \oplus_k I_k^{(l)} \subset S\tau$ be the corresponding ideal. Then (by Noetherian property the intersection is actually finite):

Proposition 2 ([11]). *The characteristic ideal satisfies $I(g) = \cap_{l=0}^\infty I^{(l)}$.*

Define the *characteristic variety* as the set of $v \in T^* \setminus \{0\}$ such that for every k there exists a $w \in N \setminus \{0\}$ with $v^k \otimes w \in g_k$. This is a conical affine variety. We projectivize its complexification and denote the result by $\operatorname{Char}(g) \subset P^{\mathbb{C}}T^*$. Since only complex characteristics will be used, we omit the $\mathbb{C}$-superscript. Also for the characteristic variety we will denote by (co)dim its complex (co)dimension.

Another description of this variety is given via the characteristic ideal $I(g) = \oplus I_k$:

Proposition 3 ([16]). $\mathrm{Char}(g) = \{p \in P^{\mathbb{C}}T^* \mid f(p^k) = 0 \forall f \in I_k, \forall k\}$.

A symbolic system g is called an *(algebraic) complete intersection* if the algebra $S\tau/I(g)$ is such, i.e. if the ideal $I(g)$ is generated by $r = \mathrm{codim}\,\mathrm{Char}(g)$ elements (we mean codimension at the non-singular stratum).

For a system $\mathscr{E}$ of PDEs we denote by $\mathrm{Char}(\mathscr{E})$ the characteristic variety of the corresponding symbolic system.

Definition 4. *Call a regular PDEs system $\mathscr{E} \subset J^k M$ a complete intersection if the corresponding symbolic system g is such at each point $x_k \in \mathscr{E}_k$.*

A scalar system $\mathscr{E}$ is a complete intersection if it can be represented by differential equations $F_1 = 0, \ldots, F_r = 0$ and the characteristic varieties, given by the symbols $f_i = 0$, $1 \leq i \leq r$, are jointly transversal.

Consider an example. The system $\{u_{xx} = 0, u_{yy} = 0, u_{zz} = 0\}$ is a complete intersection, while $\{u_{xy} = 0, u_{yz} = 0, u_{zy} = 0\}$ is not.

3. COMPATIBILITY RESULT

3.1. Mayer and Jacobi brackets

Define the higher total derivative operators $\mathscr{D}_\sigma : C^\infty(J^l M) \to C^\infty(J^{l+|\sigma|} M)$ by the formula $\mathscr{D}_\sigma = \mathscr{D}_1^{i_1} \ldots \mathscr{D}_n^{i_n}$, where $\mathscr{D}_i$ is the total derivative (2) by x^i and $\sigma = (i_1, \ldots, i_n)$ is a multiindex. Denote $F_\sigma = \partial_{p_\sigma}(F)$.

The *Jacobi brackets* of $F \in C^\infty(J^k M)$ and $G \in C^\infty(J^l M)$ is the following function:

$$\{F, G\} = \sum_\sigma \mathscr{D}_\sigma(F) G_\sigma - \sum_\tau \mathscr{D}_\tau(G) F_\tau \in C^\infty(J^{k+l} M).$$

This bracket is canonical (independent of coordinates [8]), its $(k+l)$-symbol vanishes and so $\{F, G\}$ is a scalar (non-linear) differential operator of order $k+l-1$. The same concerns the following bracket (shrunk summation), which is however not canonical:

$$[F, G] = \sum_{|\sigma|=l} \mathscr{D}_\sigma(F) G_\sigma - \sum_{|\tau|=k} \mathscr{D}_\tau(G) F_\tau \in C^\infty(J^{k+l-1} M).$$

But the difference between the brackets belongs to the $(k+l-1)$-st order ideal generated by F and G:

$$[F, G] - \{F, G\} \in \mathscr{I}^{(k,l)}(F, G) = \left\langle \mathscr{D}_\sigma(F), \mathscr{D}_\tau(G) \right\rangle_{0 \leq |\tau| < k, 0 \leq |\sigma| < l}.$$

Definition 5. *The Mayer brackets of functions $F \in C^\infty(J^k M)$ and $G \in C^\infty(J^l M)$ is the restriction $[F, G]_{\mathscr{E}}$ of any of the above brackets to the prolongation $\mathscr{E}_{k+l-1}$ (that always exists!) of the system $\mathscr{E} = \{F = G = 0\}$.*

Now this bracket is canonical ([10]), but by abuse of notations its representative $[F, G]$ will be also called the Mayer bracket.

3.2. Proof of theorem A

Consider at first the case when the codimension of the system equals the dimension of the base: $r = n$. This is precisely the case, when the characteristic variety is empty: $\mathrm{Char}(\mathscr{E}) = 0$.

As we have noted before, the Spencer cohomology of $\mathscr{E}$ (or g) is $\mathbb{R}$-dual to the Koszul homology of the symbolic module g^*, which in the scalar case becomes the algebra $S\tau/I(g)$.

For the complete intersection $S\tau/I(g)$ the Koszul homology are known. They are given by the following characterization due to Tate and Assmus ([1]): a module g^* is a complete intersection iff $H_i(g^*) = \Lambda^i H_1(g^*)$, i.e. $H_*(g^*)$ is the exterior algebra generated by the first homology group.

But $H_1(g^*) \simeq H^{*,1}(g)$ has the rank equal to the formal codimension $\mathrm{codim}(\mathscr{E})$, because the first Spencer cohomology counts the number of relations ([11]). Thus in the considered case the statement is proved.

In the general case $r \leq n$ it follows from the reduction theorem proved in [11], which we cite in the simplified for our purposes form.

Consider the scalar symbolic system $g = \{g_l \subset S^l \tau^*\}$ and let $V^* \subset \tau^*$ be a subspace. Then we can define another scalar symbolic system $\tilde{g} = \{g_l \cap S^l V^*\} \subset SV^*$. It is called the V^*-reduction of g.

Theorem 4 ([11]). *Let g be a symbolic system of complete intersection type and the subspace $V^* \subset \tau^*$ be transversal to the characteristic variety of g: $\mathrm{codim}(\mathrm{Char}(g) \cap P^{\mathbb{C}}V^*) = r =: \mathrm{codim}\,\mathrm{Char}(g)$. For instance this is so if V^* is a non-characteristic subspace of dimension r. Then the Spencer cohomology of the system g and of its V^*-reduction $\tilde{g}$ are isomorphic:*

$$H^{i,j}(g) \simeq H^{i,j}(\tilde{g}).$$

Remark 2. *The theorem was proved for more general Cohen-Macaulay modules (not necessary scalar case) and complete intersections are Cohen-Macaulay.*

By the Noether normalization lemma ([15]) we can always choose a transversal to $\mathrm{Char}(g)$ subspace V^* of dimension r, which intersects it only at zero and so is non-characteristic and transversal simultaneously. The corresponding V^*-reduction $\tilde{g}$ has the same Spencer cohomology and is a complete intersection of finite type. The result follows.

3.3. Reduction and transversality

In this section we wish to discuss the reduction procedure used in the previous theorem. Since only symbolic systems are concerned we use the notation g, not $\mathscr{E}$.

Definition 6. *We call a symbolic system $\{g_k\}$ on τ reductive from dimension n to dimension $m < n$ if there is a subspace $W \subset \tau$ of codimension m and a system $\{\tilde{g}_k\}$ (V^*-reduction) on the quotient $V = \tau/W$ with order multiplicities $\tilde{m}(r)$ (zero for $r \notin \mathrm{ord}(\tilde{g})$) such that*

$$g_k \cap S^k V^* = \tilde{g}_k \text{ and } \dim H^{k-1,1}(g) = \tilde{m}(k).$$

The second condition means $g_k = g_{k-1}^{(1)}$ until k is an order of $\tilde{g}$, in which case the quotient $g_{k-1}^{(1)}/g_k$ has dimension equal to the multiplicity $\tilde{m}(k)$.

In the statement below we do not assume g to be a complete intersection.

Proposition 5. *If the system is reductive, then for a generic $V^* = \operatorname{ann}(W)$ of $\dim = r$ and any $k > 0$ the subspace g_k is transversal to $g_{k-1}^{(1)} \cap S^k V^*$ in $g_{k-1}^{(1)}$.*

Proof. Choose V^* to be non-characteristic, which means that $L = P^{\mathbb{C}} V^*$ does not intersect $X = \operatorname{Char}(g)$. By the Noether normalization lemma ([15]) such subspaces L are generic. Let $f_1, \ldots, f_m$ be homogeneous (complex) polynomials defining the characteristic variety X.

Consider at first the case $\deg f_1 = \cdots = \deg f_m = k$. The condition we need to prove is that $g_k = \{f_1 = \cdots = f_m = 0\}$ is transversal to $S^k V^*$ in $S^k \tau^*$. If it is not the case, then intersection $g_k \cap S^k V^*$ has codimension less than m and so there is a non-trivial linear relation $\lambda^1 f_1|_V + \cdots + \lambda^m f_m|_V = 0$. Since $m - 1$ curves of degree k on $\mathbb{C}P^{m-1}$ have a common point (generically k^{m-1} points), the subspace L intersects X that contradicts our assumption.

Consider the general case. Let homogeneous generators of the ideal I of the equation g be $f_1^j, \ldots, f_{m_j}^j$ of degrees k_j $(1 \le j \le s)$ and $k_1 < \cdots < k_s$. By a modification of the above argument we obtain a flag $V_1^* \subset \ldots V_s^* = V^*$ of subspaces of dimensions $m_1, m_1 + m_2, \ldots, m_1 + \cdots + m_s = m$ such that the functions $\{f_i^j\}_{i=1,\ldots,m_j}^{j=1,\ldots,l}$ have no common zeros on V_l^*.

Now we prove the claim $g_k + g_{k-1}^{(1)} \cap S^k V^* = g_{k-1}^{(1)}$ using induction by k and starting from obvious $k = 0$. If the number $k \ne m_j$ the statement holds because $g_k = g_{k-1}^{(1)}$. So we should study only the cases $k = m_l$. We are going to prove by induction a more general transversality:

$$g_{m_l} + g_{m_l-1}^{(1)} \cap S^{m_l} V_l^* = g_{m_l-1}^{(1)}.$$

If this fails, then equations $f_1^l = \cdots = f_{m_l}^l = 0$ are dependent on $g_{m_l-1}^{(1)} \cap S^{m_l} V_l^*$ and define variety of codimension $< m_l$.

This in turn means that we can exclude one of the functions $\{f_i^j|_{V_l}\}_{i=1,\ldots,m_j}^{j=1,\ldots,l}$ that have degree k_l (for example $f_1^l = \sum_{t=2}^{m_l} \lambda^t f_t^l + \sum_{j=1}^{l-1} \sum_{i=1}^{m_j} p_j^i f_i^j$ for some numbers λ^t and polynomials p_j^i) and the resulting set will have the same zeros. Since the number of functions in the resulting set is $m_1 + \cdots + m_l - 1 = \dim V_l - 1$ there is a zero. This contradicts our assumptions about V_j^*. $\square$

Note that the statement is equivalent to the surjectivity of the map $\delta_w : g_k \to \delta(g_{k-1}^{(1)}) \subset g_{k-1}$ for all nonzero $w \in W$. Actually in [11] we proved more, namely the surjectivity of $\delta_w : g_k \to g_{k-1}$ for $w \in W \setminus \{0\}$. But maybe this weaker condition is in fact equivalent to the reducibility.

Remark 3. *It is interesting to compare this with a criteria of involutivity ([5]) for the first order PDEs systems: $g_1 \subset T^* \otimes N$ is involutive iff there exists a filtration*

$\{0\} = W_0 \subset \dots W_i \subset \dots W_n = T$ with $\dim W_i = i$ such that for $g[i] = g_1 \cap \operatorname{ann}(W_i) \otimes N$ the map $\delta_w : g[i]^{(1)} \to g[i]$ is epimorphic for some $w \in W_{i+1} \setminus W_i$.

3.4. Proof of theorem B

We give here only a sketch. A detailed exposition will appear elsewhere.

Again we restrict to the case $r = n$, the general situation is to be treated via the reduction theorem (using a transversal non-characteristic subspace $V^* \subset \tau^*$) and the ideas developed in the proof of the case $r = 2$, n arbitrary, from [11].

Let $k_i \in \operatorname{ord}(\mathscr{E})$ be orders of the operators F_i, generating the system (1). Recall that the Weyl tensor $W_{k_i+k_j-1}(\mathscr{E})$ of a system $\mathscr{E}$ is the δ-cohomology class of the metasymplectic structure restriction $(\Omega_{k_i+k_j-1})\big|_H$, where $H = \langle \nabla_1, \dots, \nabla_n \rangle$ is a horizontal subspace generated by

$$\nabla_t = \mathscr{D}_t^{(k_i+k_j-1)} + \sum\nolimits_{|\sigma|=k_i+k_j-1} a_t^\sigma \partial_{p_\sigma}, \quad 1 \le t \le n,$$

and $\mathscr{D}_i^{(s)}$ is the total derivative restricted to $J^s(M)$. However the specified space H is not unique and has $\dim g_{k_i+k_j-1}$-parametric freedom.

The coefficients a_t^σ can be found from the following condition: $H \subset T\mathscr{E}_{k_i+k_j-1}$ is equivalent to the linear algebraic system

$$\sum_{|\tau|=k_s} a_t^{\sigma+\tau} \cdot (F_s)_\tau = -\mathscr{D}_{\sigma+1_t}^{(k_i+k_j-1)} F_s, \quad |\sigma| = k_i + k_j - 1 - k_s, \; 1 \le s,t \le n. \quad (8)$$

This system is underdetermined until $k = 1$ or $n = 2$, in which case it is determined (the first case is classically known and the second was considered in details in [10, 11]). In fact, for a fixed t it consists of $\binom{2k+n-2}{n-1}$ unknowns a_t^σ (they are symmetric in multiindices σ) and $n \cdot \binom{k+n-1}{n-1}$ equations.

Whenever the symbols of $F_1, \dots, F_n$ are independent (form complete intersection as in the assumptions of the theorem), linear system (8) has full rank and so is compatible. Each solution determines a subspace $H \subset \mathscr{C}(x_{k_i+k_j-1})$.

The metasymplectic structure has in coordinates form (6). Therefore for such a horizontal space H we have:

$$(\Omega_{k_i+k_j-1})\big|_H = \sum_{|\sigma|=k_i+k_j-1} \sum_{\alpha,\beta} d\omega_\sigma(\nabla_\alpha, \nabla_\beta) \partial_{p_\sigma} \otimes dx^\alpha \wedge dx^\beta. \quad (9)$$

Let us evaluate the coefficients:

$$d\omega_\sigma(\nabla_\alpha, \nabla_\beta) = a_\beta^{\sigma+1_\alpha} - a_\alpha^{\sigma+1_\beta}.$$

It turns out that they do not depend on a particular choice of a solution a_t^σ of (8) and substitution of this value into formula (9) for the curvature shows that the Weyl tensors $W_{k_i+k_j-1}$ are proportional to the Mayer brackets as indicated in the statement.

49

3.5. Remark on the space of solutions

Let the system $\mathcal{E} = \{F_1 = 0, \ldots, F_n = 0\}$ satisfy the hypotheses of theorem B. Denote by $\mathcal{R}_{\mathcal{E}}$ the space of (germs of) its solutions. It is finite-dimensional if and only if $r = n$. Let $\mathrm{ord}(\mathcal{E}) = \{k_1, \ldots, k_r\}$.

Proposition 6. *If $r = n$, then $\mathcal{R}_{\mathcal{E}}$ is smooth and* $\dim \mathcal{R}_{\mathcal{E}} = k_1 \cdots k_n$.

Proof. Since the considered case is of finite type, Frobenius theorem applied to $\mathcal{E}_{\infty} \simeq \mathcal{E}_{k_1 + \cdots + k_n - n + 1}$ implies smoothness. For the dimension we have: $\dim \mathcal{R}_{\mathcal{E}} = \dim \mathcal{E}_{\infty} - n = \sum_0^{\infty} \dim g_k$.

We can deform the symbolic system g preserving the complete intersection requirement to the product of 1-dimensional systems $g^{[i]} \subset SV_{[i]}^1$ (i.e. ODEs, $T = \oplus V_{[i]}^1$), $1 \le i \le n$, of orders $k_1, \ldots, k_n$. But for an ODE of order k_i the space of solutions depends on k_i parameters. The dimension formula follows. $\qquad\square$

4. GENERALIZATION OF THE LAGRANGE-CHARPIT METHOD

4.1. Auxiliary integrals

Consider a compatible (i.e. formally integrable) system $\mathcal{E}$ of PDEs.

Definition 7. *Call a system $\tilde{\mathcal{E}}$ an* auxiliary integral *(or a set of integrals) for the system $\mathcal{E}$ if the joint system $\mathcal{E} \cap \tilde{\mathcal{E}}$ is also compatible.*

It was noted in [11] that classical objects, such as point symmetries, contact symmetries and intermediate integrals ([4, 6, 8, 12]) as well as higher symmetries are partial cases of this notion. Moreover, some of the newly introduced generalized symmetries are also auxiliary integrals.

Consider the particular case of one single scalar PDE $\mathcal{E} = \{F = 0\}$. We wish to add to it a set of PDEs $\tilde{\mathcal{E}} = \{F_2 = 0 \ldots, F_r = 0\}$. Then by the corollary of theorem B $\tilde{\mathcal{E}}$ is the set of auxiliary integrals iff $[F_i, F_j]_{\mathcal{E} \cap \tilde{\mathcal{E}}} = 0$ (where we denoted $F_1 = F$). Equivalently we can write

$$\{F_i, F_j\} = \sum_l \lambda_{i,j}^l \circ F_l, \tag{10}$$

where $\lambda_{i,j}^l$ are some differential operators of orders $\mathrm{ord}(\lambda_{i,j}^l) \le k_i + k_j - k_l - 1$.

Traditional methods of solving PDEs are based on usage of a certain kind of auxiliary integral ([4, 5, 17]). The Lagrange-Charpit method consists of finding an overdetermination of a special form for a given PDE to solve it.

Given a single PDE $F = 0$ we will search for the set of auxiliary integrals $F_2 = 0, \ldots, F_r = 0$ with $r = n$. Moreover we will suppose that the symbols have jointly transversal characteristic varieties, so that the total system is a complete intersection. In this case our compatibility criterion is applicable. Moreover, the system is of finite type and thus has a finite dimensional space of solutions.

They can be found as follows. Whenever the integrability conditions are fulfilled, the system has zero symbol at order $k = k_1 + \cdots + k_n - n + 1$. Thus $\mathscr{E}_k$ is equipped with the integrable horizontal distribution $\mathscr{C}_{\mathscr{E}_k}$. It can be integrated to a foliation formed by the jet-extension of solutions. The latter step means solution of a system of ODEs, which is convenient to solve via a symmetry algebra by the method of S.Lie ([2]).

Thus we obtain the solution space with parametrized solutions. We call this procedure the *generalized Lagrange-Charpit method*.

Let us notice that even on symbolic level the proposed method imposes certain obstructions. Actually, consider a scalar PDE $F = 0$. Denote the symbol of an operator H by $\sigma(H)$. If $F_2, \ldots, F_n$ is a set of auxiliary integrals for F, then we have an identity for the symbols:

$$\{\sigma(F_i), \sigma(F_i)\} = \sum_l \lambda_l^{i,j} \cdot \sigma(F_l), \tag{11}$$

where $\{,\}$ is the standard Poisson bracket on T^*M. Thus the ideal $\langle \sigma(F_i) \,|\, 1 \le i \le n \rangle$ is Poisson.

So at the beginning we search for an extension of $\sigma(F)$ to a Poisson ideal, solving symbolic equation (11). By a parameter count, generically there are obstructions for such an extension. Then we try to extend the symbols $\sigma(F)$ to differential operators and to satisfy the next approximation to the system (10). Thus we adjust symbols of order $k_i - 1$ of F_i etc.

4.2. Lagrange complete integrals and variation of constants

By results of the previous section the generalized Lagrange-Charpit method, i.e. an overdetermination of a single PDE $F = 0$ by auxiliary integrals $F_2, \ldots, F_n$, is equivalent to finding a k-parametric family of solutions to $F_1 = 0$, with $k = \dim \mathscr{R}_{\mathscr{E}}$ being found via proposition 6.

Another choice is to use a complete overdetermination on the level of $m = \mathrm{ord}(F)$ jets, i.e. to impose $\binom{n+m-1}{m} - 1$ additional auxiliary integrals, so that the m-th symbol of the joint system vanishes. Let's restrict for simplicity to this case, when $k = \binom{n+m}{m} - 1$. The former case is treated quite similarly.

Definition 8. *Call a k-parametric family of solutions $u = V(x_1, \ldots, x_n; a_1, \ldots, a_k)$ a complete integral if the following non-degeneracy condition is fulfilled: The lift of these solutions to an appropriate jet-space form a submanifold of dimension $n + k$.*

Thus a complete integral corresponds to a parametrization (chart) of a domain in the solutions space of $\{F = 0\}$. Notice that for the case of first order equations, when $k = n$, this coincides with the classical Lagrange complete integral.

Knowing a complete integral (which is the case, when we applied our generalization of the Lagrange-Charpit method) one can fix constants a_i and obtain a solutions. But in fact one can more: Take some functions $a_i(x)$ instead of them and obtain an infinite-dimensional family of solutions. This method of *variation of constants* extends the classical relation between complete and general integrals ([6, 7, 17]). The functions are not arbitrary, but depend on a choice of the complete integral.

Theorem 7. *The functions $a_i(x)$ give a solution upon substitution into the complete integral $u = V(x;a)$ if and only if*

$$
\begin{bmatrix}
V_{a_1} & \cdots & V_{a_k} \\
\cdots & \cdots & \cdots \\
V_{x_\sigma a_1} & \cdots & V_{x_\sigma a_k} \\
\cdots & \cdots & \cdots
\end{bmatrix}
\cdot
\begin{bmatrix}
\frac{\partial a_1}{\partial x_1} & \cdots & \frac{\partial a_1}{\partial x_n} \\
\vdots & \ddots & \vdots \\
\frac{\partial a_k}{\partial x_1} & \cdots & \frac{\partial a_k}{\partial x_n}
\end{bmatrix}
= 0, \qquad |\sigma| < k. \tag{12}
$$

This system is determined (NB: don't be mislead by a mere equations count) and thus is formally integrable. In particular, if the integral $V(x;a)$ is analytic it possesses an analytic solution (with arbitrariness in initial conditions as in the Cauchy problem). The same concerns the elliptic case: If the systems is elliptic and regular it is solvable.

Proof. The family keeps parametrizing solutions upon substitution of $a_i(x)$ iff the Cartan forms ω_σ vanish on its lift. This means that the equations $p_\sigma = \partial_{x^\sigma} V$ preserve their form after the substitution $a_i = a_i(x)$. This condition is equivalent to the following determined system of PDEs:

$$
\sum_i V_{a_i} \frac{\partial a_i}{\partial x_s} = 0,
$$

$$
\sum_i \left(V_{x_s a_i} \frac{\partial a_i}{\partial x_t} + V_{x_t a_i} \frac{\partial a_i}{\partial x_s} \right) + \sum_{i,j} V_{a_i a_j} \frac{\partial a_i}{\partial x_s} \frac{\partial a_j}{\partial x_t} + \sum_i V_{a_i} \frac{\partial^2 a_i}{\partial x_s \partial x_t} = 0,
$$

$$
\cdots\cdots
$$

First order in V equations are exactly those from (12). If we differentiate them by x^t and subtract from the equations having the second order in V we obtain the next equations from (12) etc. $\qquad\square$

Thus choosing different overdeterminations (auxiliary integrals) we cover the whole space of solutions to $F = 0$.

Remark 4. *The vector-rows of the first matrix in (12) are orthogonal in $\mathbb{R}^k$ to the vectors-columns of the second matrix. If we additionally assume they are linearly independent a.e., we get the following restriction: $\binom{n+m-1}{m-1} + n \le \binom{n+m}{m} - 1$. This inequality is strict for $n > 2$, $m \ne 1$ or $n \ne 1$, $m > 2$. If $n = m = 2$ we have equality and for $m = 1$ the inequality fails to hold. In this latter case $k = n$ and whenever $\nabla V \ne 0$ we get $\det \|\frac{\partial a_i}{\partial x_j}\| = 0$, i.e. there are functional relations between a_i: $\Phi_s(a_1,\ldots,a_n) = 0$, $1 \le s \le r$. This is an essential difference between the first and higher order cases.*

4.3. Examples

1. For a PDE $F(u_{xx}, u_{xy}, u_{yy}) = 0$ we have the following complete integral

$$
V(x_1, x_2; a_1, a_2, a_3, a_4, a_5) = a_1 + a_2 x_1 + a_3 x_2 + a_4 x_1^2 + a_5 x_1 x_2 + \tilde{a}_{45} x_2^2,
$$

where the last constant is found from the equation $F(2a_4, a_5, 2\tilde{a}_{45}) = 0$. For instance, if we study the Laplace equation $\Delta u = 0$, then $\tilde{a}_{45} = -a_4$. Variation of constants for this choice of V yields all polynomial solutions (and others too).

2. Let (M^n, g) be a Riemannian manifold and $F = \Delta_g$ the corresponding Beltrami-Laplace equation. If the metric g is geodesically equivalent to another metric $\bar{g}$, non-proportional to g at least at one point, then by a result of Matveev and Topalov ([14]) there are commuting independent second order differential operators $F_2, \ldots, F_n$. They allow to integrate the Laplace equation and moreover the equations $\Delta_g u = \lambda u$ for eigenvalues completely using our approach.

3. If a PDE $F = 0$ is linear and has constant coefficients along some subspace W, we can exclude them via the Fourier transform. Alternatively we can impose auxiliary integrals $G_\sigma = p_\sigma$ with the multi-index corresponding to the W-direction: They will definitely commute with F. Thus we need to establish a smaller set of auxiliary integrals to integrate F. For instance, if the coefficient depend only on one variable, it is reducible to an ODE and so is integrable.

REFERENCES

1. W. Bruns, J. Herzog, *"Cohen-Macaulay rings"*, Cambridge University Press, Cambridge, U.K. (1993)
2. S. V. Duzhin, V. V. Lychagin, *"Symmetries of Distributions and Quadrature of Ordinary Differential Equations"*, Acta Appl. Math., **24** (1991), 29–51.
3. H. Goldschmidt, *"Integrability criteria for systems of nonlinear partial differential equations"*, J. Diff. Geom., **1**(3) (1967), 269–307.
4. E. Goursat, *"Lecons sur l'intégration des équations aux dérivées partielles du premier ordere"*, Hermann, Paris (1891)
5. V. Guillemin, S. Sternberg, *"An algebraic model of transitive differential geometry"*, Bull. A.M.S., **70** (1964), 16–47.
6. N. M. Gunter, *"Integration of PDEs of the first order"*, ONTI (Russian) Leningrad-Moscow (1934).
7. E. Kamke, *"Differentialgleichungen. Lösungsmethoden und lösunden. II"* (Germain), Leipzig (1959).
8. I. S. Krasilschik, V. V. Lychagin, A. M. Vinogradov, *"Geometry of jet spaces and differential equations"*, Gordon and Breach (1986).
9. B. S. Kruglikov, V. V. Lychagin, *"On equivalence of differential equations"*, Acta et Comment. Univ. Tartuensis Math. 3 (1999), 7–29.
10. B. S. Kruglikov, V. V. Lychagin, *"Mayer brackets and solvability of PDEs – I"*, Diff. Geom. and its Appl. **17** (2002), 251–272.
11. B. S. Kruglikov, V. V. Lychagin, *"Mayer brackets and solvability of PDEs – II"*, Trans. A.M.S. (2004), to appear.
12. S. Lie, F. Engel, *"Theorie der Transformationsgruppen"*, vol. II *"Begründungstransformationen"*, Leipzig, Teubner (1888-1893).
13. V. V. Lychagin, *"Homogeneous geometric structures and homogeneous differential equations"*, in A. M. S. Transl., *"The interplay between differential geometry and differential equations"*, V. Lychagin Eds., ser. 2, **167** (1995), 143–164.
14. V. S. Matveev, P. J. Topalov *"Quantum integrability of Beltrami-Laplace operator as geodesic equivalence"*, Math. Z. **238** (2001), 833–866.
15. D. Mumford, *"Algebraic geometry I. Complex projective varieties"*, Springer-Verlag (1976).
16. D. C. Spencer, *"Overdetermined systems of linear partial differential equations"*, Bull. Amer. Math. Soc., **75** (1969), 179–239.
17. V. V. Stepanov, *"Lectures on differential equations"*, PhisMatGis, Moscow (1959).

Motion coordination algorithms resulting from classical geometric optimization problems

Jorge Cortés

Coordinated Science Laboratory
University of Illinois at Urbana-Champaign
1308 West Main Street, Urbana, Illinois 61801
`http://motion.csl.uiuc.edu/~jorge`
E-mail: `jcortes@uiuc.edu`

Abstract. This paper introduces various geometric optimization problems and explores their relationship with motion coordination algorithms for networks of mobile agents. For each problem, the objective is the optimization of an appropriate multi-center function encoding the sensing task to be achieved by the mobile network in a dynamic environment. We present five different scenarios: the expected value scenario, the expected value scenario with limited range interactions, the area scenario, the worst-case scenario and the non-interference scenario. We carefully analyze the smoothness properties and gradient information of the multi-center functions. Based on this investigation, we propose distributed motion coordination algorithms specifically tailored for each scenario. The multi-center functions play the role of network aggregate cost functions certifying the validity of the coordination algorithms. Various numerical simulations illustrate the results.
Key words. Coverage optimization, motion coordination algorithms, geometric optimization, non-smooth analysis, proximity graphs, Voronoi partitions.
PACS. 02.30.Yy, 45.10.Db, 89.70.+c.
MSC. 37N35, 68W15, 93D20, 49J52, 05B40.

1. INTRODUCTION

Consider the following scenario: let $(p_1, \ldots, p_n)$ denote the location of n mobile agents in a convex polygonal environment Q. Assume that the agents have the ability to sense its immediate environment, communicate with other agents, process the received data, and plan its own motions accordingly. Furthermore, assume that certain events of interest are taking place in the environment Q that the mobile network has to take care of. How should the mobile network plan its motion in order to optimize the coverage of the environment? More precisely, we aim at designing motion coordination algorithms that, implemented over each mobile agent, will achieve the objective of optimally positioning the agents in the environment with respect to the desired sensing task.

Since the initial works from the graphics and ecology communities on distributed coordination on swarms and flocking [20, 24], the design of coordination algorithms has been studied extensively by the behavioral control and the control theory communities (see [2, 3, 10, 15, 17, 21–23] and references therein). In this paper, we describe an

CP729, *Global Analysis and Applied Mathematics: International Workshop on Global Analysis*,
edited by K. Taş, D. Krupka, O. Krupková, and D. Baleanu
© 2004 American Institute of Physics 0-7354-0209-4/04/$22.00

innovative technical approach that relies on non-smooth distributed descent algorithms and on aggregate utility functions that encode optimal coverage and sensing policies. We introduce various notions of quality-of-service provided by an adaptive mobile network in a dynamic environment. We refer to these notions as multi-center functions. We set up five different geometric optimization problems: the expected value scenario, the expected value scenario with limited range interactions, the area scenario, the worst-case scenario and the non-interference scenario. We carefully investigate the smoothness properties and the (generalized) gradient information of the multi-center functions. Building on this analysis, we design coordination algorithms implementable by a mobile network with sensing, communication and motion capabilities that optimize the multi-center functions.

When designing motion coordination algorithms, we take into careful consideration all constraints present on the mobile network. In particular, the coordination algorithms should be: (i) adaptive, in order to provide the network with the ability to address changing environments, sensing task, and network topology (due to agents departures, arrivals, or failures); (ii) distributed, in the sense that the behavior of each agent depends only on the location of its neighbors. They should typically not require a fixed-topology communication graph, i.e., allow for the neighborhood relationships to change as the network evolves. The advantages of distributed algorithms are scalability and robustness; (iii) verifiable asymptotically correct, i.e., guarantee monotonic optimization of the cost function encoding the sensing task. Asymptotically, the evolution of the mobile network should be guaranteed to converge to the critical points of the optimal sensor coverage problem. The importance of formal verification proofs increases with the dimension and complexity of vehicle networks; and (iv) amenable to asynchronous implementation, meaning that the algorithms should be implementable in a network composed of agents evolving at different speeds, with different computation and communication capabilities. In such a case, no global synchronization is required and convergence properties are preserved even if information about neighboring vehicles propagates with some delay.

The organization of the paper is as follows. In Section 2, we introduce some basic concepts from computational geometry [5, 19] and present the model of synchronous mobile network considered throughout the paper. Section 3 introduces various geometric optimization problems in which service sites are spatially allocated to fulfill a specified request [1, 6, 12, 13]. We propose different notions of quality-of-service to measure the network performance, depending on the specific sensing task, the capabilities of the agents and the information about the distribution of events taking place in the environment. We state the differentiable properties of these functions and the expressions of their (generalized) gradients via nonsmooth analysis [8]. We show that their critical points are *center Voronoi configurations*. Finally, Section 4 presents distributed motion coordination algorithms for each of the scenarios introduced in Section 3. Using the stability tools in [4, 14, 25], we propose novel strategies based on the nonsmooth generalized gradients of the aggregate cost functions. We investigate their asymptotic behavior and show that the algorithms are guaranteed to continuously improve the corresponding network performance measure. A detailed analysis of all results presented in this paper can be found in [9–11].

2. PRELIMINARIES

Let $\|\cdot\|$ denote the Euclidean distance function on $\mathbb{R}^N$, $N \in \mathbb{N}$, and let $v \cdot w$ denote the scalar product of $v, w \in \mathbb{R}^N$. Let versus(v) denote the unit vector in the direction of $0 \neq v \in \mathbb{R}^N$, i.e., versus$(v) = v/\|v\|$. Given $S \subset \mathbb{R}^N$, co(S) and int S denote its convex hull and interior set, respectively. Let $1_S : \mathbb{R}^N \to \{0,1\}$ be the indicator function defined by $1_S(q) = 1$ if $q \in S$, and $1_S(q) = 0$ if $q \notin S$. If S is convex, let proj$_S : \mathbb{R}^N \to S$ denote the orthogonal projection onto S and let $D_S : \mathbb{R}^N \to \mathbb{R}$ denote the distance function to S. For $p \in \mathbb{R}^N$ and $r \in \mathbb{R}_+ = [0, +\infty)$, let $B_r(p) = \{q \in \mathbb{R}^N \mid \|q - p\| \leq r\}$ denote the closed ball in $\mathbb{R}^N$ centered at p of radius r. Let $n_{B_r(p)}(q)$ denote the unit outward normal to $B_r(p)$ at $q \in \partial B_r(p)$.

Let Q be a simple convex polygon in $\mathbb{R}^2$. We denote by Ed$(Q) = \{e_1, \ldots, e_M\}$ and Ve$(Q) = \{v_1, \ldots, v_L\}$ the set of edges and vertices of Q, respectively. If $e \in$ Ed(Q), we let n_e denote the unit normal to e pointing toward int(Q). The *diameter of Q* is defined as diam$(Q) = \max_{q,p \in Q} \|q - p\|$. Let $P = (p_1, \ldots, p_n) \in Q^n \subset (\mathbb{R}^2)^n$ denote the location of n generators in Q.

2.1. Voronoi partitions and proximity graphs

In this section, we review the notion of Voronoi partition generated by sets of points on the Euclidean plane; we refer the reader to [5, 19] for comprehensive treatments. Next, we shall present some relevant concepts on proximity graph functions. This notion is an extension of the notion of proximity graph as explained in the survey article [16].

A *covering* of $\mathbb{R}^2$ is a collection of subsets of $\mathbb{R}^2$ whose union is $\mathbb{R}^2$; a *partition* of $\mathbb{R}^2$ is a covering whose subsets have disjoint interiors. Let $\mathscr{P}$ be a set of n distinct points $\{p_1, \ldots, p_n\}$ in $\mathbb{R}^2$. The *Voronoi partition of $\mathbb{R}^2$* generated by $\mathscr{P}$ with respect to the Euclidean norm is the collection of sets $\{V_i(\mathscr{P})\}_{i \in \{1, \ldots, n\}}$ defined by

$$V_i(\mathscr{P}) = \{q \in \mathbb{R}^2 \mid \|q - p_i\| \leq \|q - p_j\|, \text{ for all } p_j \in \mathscr{P}\}.$$

It is customary and convenient to refer to $V_i(\mathscr{P})$ as V_i. The boundary of each set V_i is the union of a finite number of segments and rays. A vertex $v \in$ Ve$(V_i(P))$ is *nondegenerate* if it is determined by exactly three elements. Otherwise it is *degenerate*. The configuration P is *nondegenerate* if all its vertices are nondegenerate, otherwise it is *degenerate*.

Let Σ_n be the set of permutations of n elements. A map $f : X^n \to 2^{X \times X}$ is Σ_n-*equivariant* if for all $(x_1, \ldots, x_n) \in X^n$ and $\sigma \in \Sigma_n$, $(x_i, x_j) \in f(x_1, \ldots, x_n)$ implies $(x_{\sigma(i)}, x_{\sigma(j)}) \in f(x_{\sigma(1)}, \ldots, x_{\sigma(n)})$. A *proximity graph function* associates to a set of n distinct points $\mathscr{P} = \{p_1, \ldots, p_n\}$ in $\mathbb{R}^2$ a graph with vertex set $\mathscr{P}$ and edge set $\mathscr{E}(p_1, \ldots, p_n)$, where $\mathscr{E} : (\mathbb{R}^2)^n \to 2^{\mathbb{R}^2 \times \mathbb{R}^2}$ is a Σ_n-equivariant map with the property that $\mathscr{E}(p_1, \ldots, p_n) \subseteq \mathscr{P}^2 = \{p_1, \ldots, p_n\}^2 = \{p_1, \ldots, p_n\} \times \{p_1, \ldots, p_n\}$. Note that, since the map $\mathscr{E}$ is Σ_n-equivariant, the value of $\mathscr{E}(p_1, \ldots, p_n)$ is independent of the ordering of the elements $(p_1, \ldots, p_n)$, and therefore we will write it as $\{p_1, \ldots, p_n\} = \mathscr{P} \mapsto \mathscr{E}(\mathscr{P})$, and refer to it as the *proximity edge function* corresponding to the proximity graph function $\mathscr{P} \mapsto \mathscr{G}(\mathscr{P})$. To each proximity graph function $\mathscr{G}$, one can associate the *set of neigh-*

bors map $\mathcal{N}_{\mathcal{G}} \colon \mathbb{R}^2 \times 2^{\mathbb{R}^2} \to 2^{\mathbb{R}^2}$, defined by $\mathcal{N}_{\mathcal{G}}(p,\mathcal{P}) = \{q \in \mathcal{P} \mid (p,q) \in \mathcal{E}_{\mathcal{G}}(\mathcal{P})\}$. We will often write $\mathcal{N}_{\mathcal{G},p}(\mathcal{P})$ to denote $\mathcal{N}_{\mathcal{G}}(p,\mathcal{P})$. For $r \in \mathbb{R}_+$, we have the following proximity graph functions:

(i) the *Delaunay graph* $\mathcal{P} \mapsto \mathcal{G}_{\mathrm{D}}(\mathcal{P}) = (\mathcal{P}, \mathcal{E}_{\mathrm{D}}(\mathcal{P}))$ with edge set

$$\mathcal{E}_{\mathrm{D}}(\mathcal{P}) = \{(p_i, p_j) \in \mathcal{P}^2 \setminus \mathrm{diag}(\mathcal{P}^2) \mid V_i(\mathcal{P}) \cap V_j(\mathcal{P}) \neq \emptyset\};$$

(ii) the *r-disk graph* $\mathcal{P} \mapsto \mathcal{G}_{\mathrm{disk}}(\mathcal{P},r) = (\mathcal{P}, \mathcal{E}_{\mathrm{disk}}(\mathcal{P},r))$ with edge set

$$\mathcal{E}_{\mathrm{disk}}(\mathcal{P},r) = \{(p_i, p_j) \in \mathcal{P}^2 \setminus \mathrm{diag}(\mathcal{P}^2) \mid \|p_i - p_j\| \leq r\};$$

(iii) the *r-limited Delaunay (or, limited-range Delaunay) graph* $\mathcal{P} \mapsto \mathcal{G}_{\mathrm{LD}}(\mathcal{P},r) = (\mathcal{P}, \mathcal{E}_{\mathrm{LD}}(\mathcal{P},r))$ consists of the edges $(p_i, p_j) \in \mathcal{P}^2 \setminus \mathrm{diag}(\mathcal{P}^2)$ with the property

$$\left(V_i(\mathcal{P}) \cap B_{\frac{r}{2}}(p_i)\right) \cap \left(V_j(\mathcal{P}) \cap B_{\frac{r}{2}}(p_j)\right) \neq \emptyset. \tag{1}$$

There are other important proximity graphs such as the Gabriel graph or the Euclidean Minimum Spanning Tree. We refer to [11, 16] for further details.

Note that in the previous definitions we have emphasized the fact that the points $\{p_1, \ldots, p_n\}$ are distinct. Occasionally though, we will consider ordered sets of possibly coincident points. In this case, it is useful to adopt the following notation: given a tuple $(p_1, \ldots, p_n) \in (\mathbb{R}^2)^n$, we let $\{p_1, \ldots, p_n\}$, or equivalently $\mathcal{P}$, denote the associated point set that only contains the corresponding distinct points. The cardinality of $\mathcal{P} = \{p_1, \ldots, p_n\}$ is less than or equal to n. More precisely, if $\mathcal{S}$ denotes the set

$$\mathcal{S} = \{(p_1, \ldots, p_n) \in (\mathbb{R}^2)^n \mid p_i = p_j \text{ for some } i, j \in \{1, \ldots, n\}, i \neq j\}, \tag{2}$$

then $\#\mathcal{P} < n$ if $(p_1, \ldots, p_n) \in \mathcal{S}$ and $\#\mathcal{P} = n$ if $(p_1, \ldots, p_n) \notin \mathcal{S}$. The *Voronoi covering* $\mathcal{V}(p_1, \ldots, p_n) = \{V_i(p_1, \ldots, p_n)\}_{i \in \{1, \ldots, n\}}$ *generated by the tuple* $(p_1, \ldots, p_n)$ is defined by assigning to each point p_i its corresponding Voronoi cell in the Voronoi partition generated by $\mathcal{P}$. Note that coincident points in the tuple $(p_1, \ldots, p_n)$ have the same Voronoi cell.

We are now in a position to discuss distributed control laws and algorithms in formal terms. Let $\mathcal{G}$ be a proximity graph function and let Y be a set. A map $f \colon (\mathbb{R}^2)^n \to Y^n$ is *spatially distributed over* $\mathcal{G}$ if there exist maps $\tilde{f}_i \colon \mathbb{R}^2 \times 2^{(\mathbb{R}^2)^n} \to Y$, $i \in \{1, \ldots, n\}$, with the property that for all $(p_1, \ldots, p_n) \in (\mathbb{R}^2)^n$

$$f_i(p_1, \ldots, p_n) = \tilde{f}_i(p_i, \{p_j \in \mathbb{R}^2 \mid p_j \in \mathcal{N}_{\mathcal{G},p_i}(\mathcal{P})\}),$$

where f_i denotes the *i*th-component of f. A vector field X on $(\mathbb{R}^2)^n$ is *spatially distributed over* $\mathcal{G}$ if its associated map $X \colon (\mathbb{R}^2)^n \to (\mathbb{R}^2)^n$, where the canonical identification between the tangent space of $(\mathbb{R}^2)^n$ and $(\mathbb{R}^2)^n$ itself is understood, is spatially distributed in the above sense. In other words, to compute the *i*th component of a spatially-distributed function or vector field at $(p_1, \ldots, p_n)$, it is only required the knowledge of the vertex p_i and the neighboring vertexes in the proximity graph $\mathcal{G}(\{p_1, \ldots, p_n\})$.

One can prove that r-limited Delaunay graph is spatially distributed over $\mathscr{G}_{\mathrm{disk}}(r)$. More precisely, for $r \in \mathbb{R}_+$, the map $\mathscr{N}_{\mathscr{G}_{\mathrm{LD}}(\cdot,r)} \colon (\mathbb{R}^2)^n \to \left[2^{(\mathbb{R}^2)^n} \right]^n$, defined by

$$(p_1, \ldots, p_n) \mapsto \left(\mathscr{N}_{\mathscr{G}_{\mathrm{LD}}(r),p_1}(\mathscr{P}), \ldots, \mathscr{N}_{\mathscr{G}_{\mathrm{LD}}(r),p_n}(\mathscr{P}) \right),$$

is spatially distributed over $\mathscr{G}_{\mathrm{disk}}(r)$. Loosely speaking, this proposition states that the r-limited Delaunay graph $\mathscr{G}_{\mathrm{LD}}$ can be computed in a spatially localized way: each agent needs to know only the location of all other agents in a disk of radius r.

2.2. Modelling a network of mobile agents

Here, we introduce the notions of *mobile agent* and of *network of mobile agents*. Let n be the number of agents in the network. Each agent has the following sensing, computation, communication, and motion control capabilities. The ith agent has a processor with the ability of allocating continuous and discrete states and performing operations on them. The ith agent occupies a location $p_i \in Q \subset \mathbb{R}^2$ and it is capable of moving in space, at any time $t \in \mathbb{R}_+$ according to a first order continuous dynamics of the form

$$\dot{p}_i(t) = u_i. \tag{3}$$

Here, the control u_i takes values in a bounded subset of $\mathbb{R}^2$. The processor has access to the agent's location p_i and determines the *control* u_i. The processor of the ith agent is capable of transmitting information to any other agent within a closed disk of radius $r_i \in \mathbb{R}_+$. We will consider two different sensing and communication models: in the first one, we assume that the communication radius r_i is a quantity controllable by the ith processor; in the second one, we assume that the communication radius r_i is a fixed quantity and equal for all agents, $r_i = r \in \mathbb{R}_+$, $i \in \{1, \ldots, n\}$. In both models, the communication bandwidth is assumed to be limited. Throughout the paper, we shall specify the concrete communication model used.

Equivalently, we shall consider groups of mobile agents without communication capabilities, but instead capable of measuring the relative position of each other agent within a closed disk of radius $r_i \in \mathbb{R}_+$. We assume that all communication between agents and all sensing of agents locations are accurate. Note that this network model is synchronous. We refer the reader to [10, 17, 18] for various asynchronous network models.

3. MULTI-CENTER FUNCTIONS AS NETWORK PERFORMANCE MEASURES

In this section, we begin by introducing the precise notions of quality-of-service provided by the mobile network in a dynamic environment. These notions will be later the criteria to judge the asymptotic correctness and performance of the motion coordination algorithms. We end the section by characterizing the smoothness properties of the multi-center functions.

Expected value scenario

Assume that a certain density function $\phi : Q \to \mathbb{R}_+$ is known, describing the probability distribution in Q of the events of interest. Assume further that the mobile agents can control and tune their communication radius (respectively their sensing radius). In such a case, the network tries to minimize the expected distance from any event in the environment to one of the mobile agents of the network (since, because of noise and loss of resolution, the closest agent will be the one which is able to take the best possible measurement of that event or can reach its location more quickly). Accordingly, we set up the following geometric optimization problem

$$\text{minimize}_{p_1,\ldots,p_n \in Q} \ \left\{ \mathscr{H}_C(p_1,\ldots,p_n) = \int_Q \min_{i \in \{1,\ldots,n\}} \|q - p_i\|^2 \phi(q) dq \right\}. \qquad (4)$$

This problem is referred to as the *p*-median problem in [12]. On $Q^n \setminus \mathscr{S}$, $\mathscr{H}_C$ reads

$$\mathscr{H}_C(P) = \sum_{i=1}^{n} \int_{V_i} \|q - p_i\|^2 \phi(q) dq.$$

Given a polytope W in $\mathbb{R}^N$, its centroid, CM_W, is the center of mass of W with respect to the density function ϕ, i.e.,

$$\text{CM}_W = \frac{1}{\text{M}_W} \int_W q\phi(q) dq, \quad \text{M}_W = \int_W \phi(q) dq.$$

Centroidal Voronoi configurations satisfy $p_i = \text{CM}_{V_i(P)}$ for all $i \in \{1,\ldots,n\}$.

We can slightly modify the previous scenario to set up a different geometric optimization problem. If, under the same hypothesis, the mobile agents have a fixed (and common) communication radius $r \in \mathbb{R}_+$ (respectively sensing radius), then the problem turns out to be harder. The objective is still to solve the optimization in equation (4), but now the agents can only communicate to or sense other agents and events up to a fixed distance. We refer to it as the *expected value scenario with range-limited interactions*. To deal with this situation, we introduce the multi-center function

$$\mathscr{H}_{\frac{r}{2}}(p_1,\ldots,p_n) = \int_Q \max_{i \in \{1,\ldots,n\}} \left(\|q - p_i\|^2 \, 1_{B_{\frac{r}{2}}(p_i)}(q) + \text{diam}(Q)^2 \cdot 1_{Q \setminus B_{\frac{r}{2}}(p_i)}(q) \right) \phi(q) dq,$$

where we assume $r \leq 2\,\text{diam}(Q)$ (since otherwise $Q \subset B_{\frac{r}{2}}(p_i)$ for all $i \in \{1,\ldots,n\}$, and the setting would be the same as in the expected value scenario). The factor $1/2$ multiplying the radius comes out of technical reasons. There are two arguments to explain why this function is important to deal with this scenario. The first one is given by the following constant-factor approximation of the value of $\mathscr{H}_C$ (cf. [11]),

$$\mathscr{H}_{\frac{r}{2}}(P) \geq \mathscr{H}_C(P) \geq \beta \, \mathscr{H}_{\frac{r}{2}}(P) > 0, \qquad (5)$$

for all $P \in Q^n$, and for $\beta = \left(\frac{r}{2\,\text{diam}(Q)} \right)^2 \in [0,1]$. That is, the optimization of $\mathscr{H}_{\frac{r}{2}}$ is equivalent, to the extent determined by equation (5), to the optimization of $\mathscr{H}_C$. The second reason is related with the gradient of $\mathscr{H}_{\frac{r}{2}}$. We postpone its exposition to Section 4.

Area scenario

An alternative optimization problem is the area scenario. Assuming that the mobile agents have a fixed finite communication (respectively, sensing) radius $r \in \mathbb{R}_+$, the mobile network should maximize the amount of area of the environment covered. If a distribution density function $\phi : Q \to \mathbb{R}_+$ is known, then the area can be weighted accordingly. The following geometric optimization problem describes this scenario:

$$\text{maximize}_{p_1,\ldots,p_n \in Q} \left\{ \mathcal{H}_{\text{area}}(p_1,\ldots,p_n) = \int_Q \left(\max_i 1_{B_{\frac{r}{2}}(p_i)}(q) \right) \phi(q) dq \right\}.$$

Worst-case scenario

Assume now that no information is available about the distribution of events taking place in the environment. Assume further that the mobile agents can control and tune their communication radius (respectively their sensing radius). Since no information is available, it seems reasonable to consider the *worst-case scenario*, that is, that the event of interest will occur at the furthest-away point from the network in the environment. Therefore, the network tries to minimize the largest possible distance from any point in the domain to one of the agent locations,

$$\text{minimize}_{p_1,\ldots,p_n \in Q} \left\{ \mathcal{H}_{\text{DC}}(p_1,\ldots,p_n) = \max_{q \in Q} \left(\min_{i \in \{1,\ldots,n\}} \|q - p_i\| \right) \right\}.$$

This problem is referred to as the *p-center problem* in [12, 26]. In terms of the Voronoi partition, the function $\mathcal{H}_{\text{DC}}$ admits the following alternative expression

$$\mathcal{H}_{\text{DC}}(P) = \max_{i \in \{1,\ldots,n\}} \left\{ \max_{q \in V_i} \|q - p_i\| \right\}.$$

It is conjectured in [26] that the *p-center problem* can be restated as a disk-covering problem: how to cover a region with disks of minimum radius, which reads

$$\min\{R \mid \cup_{i \in \{1,\ldots,n\}} B(p_i, R) \supseteq Q\}.$$

In Theorem 3.2 we provide a positive answer to this question. Given a polytope W in $\mathbb{R}^N$, its circumcenter, CC_W, is the center of the minimum-radius sphere that contains W. We say that P is a *circumcenter Voronoi configuration* if $p_i = \text{CC}_{V_i(P)}$, for all $i \in \{1,\ldots,n\}$.

Non-interference scenario

The objective here is to maximize the coverage of the domain in such a way that the various sensing radius do not overlap or leave the environment (because of interference). In this situation, the network tries to solve the following geometric optimization problem

$$\text{maximize}_{p_1,\ldots,p_n \in Q} \left\{ \mathcal{H}_{\text{SP}}(p_1,\ldots,p_n) = \min_{i \neq j \in \{1,\ldots,n\}} \left(\tfrac{1}{2}\|p_i - p_j\|, \text{D}(p_i, \partial Q) \right) \right\},$$

60

so that each agent can fit a circular sensing region as large as possible within the environment and without overlapping with the regions belonging to other agents. In terms of the Voronoi partition, the function $\mathcal{H}_{SP}$ admits the following alternative expression

$$\mathcal{H}_{SP}(P) = \min_{i \in \{1,\dots,n\}} \left\{ \min_{q \notin \mathrm{int}V_i} \|q - p_i\| \right\}.$$

A similar conjecture to the one presented above is that this problem can be restated as a sphere-packing problem: how to maximize the coverage of a region with non-overlapping disks (contained in the region) of minimum radius. The problem reads:

$$\max\{R \mid \cup_{i \in \{1,\dots,n\}} B_2(p_i, R) \subseteq Q, \ \mathrm{int}(B(p_i, R)) \cap \mathrm{int}(B(p_j, R)) = \emptyset\}.$$

In Theorem 3.2 we provide a positive answer to this question. Given a polytope W in $\mathbb{R}^N$, its incenter set, IC_W, is the set of the centers of maximum-radius spheres contained in W. We say that $P \in Q^n$ is an *incenter Voronoi configuration* if $p_i \in \mathrm{IC}_{V_i(P)}$, for all $i \in \{1,\dots,n\}$. If P is an incenter Voronoi configuration, and each Voronoi region $V_i(P)$ has a unique incenter, $\mathrm{IC}_{V_i(P)} = \{p_i\}$, then P is a *generic incenter Voronoi configuration*.

Smoothness analysis of the multi-center functions

The following discussion gathers the results concerning the smoothness properties of the multi-center functions introduced above. The proof of the following result is based on a generalized statement of the Conservation-Of-Mass Lemma. For further details, the reader is referred to [11] and references therein.

Theorem 3.1 *The multi-center functions* $\mathcal{H}_C$, $\mathcal{H}_{\frac{r}{2}}$ *and* $\mathcal{H}_{area}$ *are globally Lipschitz on* Q^n, *and continuously differentiable on* $Q^n \setminus \mathscr{S}$, *where for each* $i \in \{1,\dots,n\}$

$$\frac{\partial \mathcal{H}_C}{\partial p_i}(P) = \int_{V_i} \frac{\partial}{\partial p_i} \|q - p_i\|^2 \phi(q) dq = 2\mathrm{M}_{V_i(P)}(p_i - \mathrm{CM}_{V_i(P)}),$$

$$\frac{\partial \mathcal{H}_{\frac{r}{2}}}{\partial p_i}(P) = 2\mathrm{M}_{V_i(P) \cap B_{\frac{r}{2}}(p_i)}(p_i - \mathrm{CM}_{V_i(P) \cap B_{\frac{r}{2}}(p_i)})$$

$$+ \left(\left(\frac{r}{2}\right)^2 - \mathrm{diam}(Q)^2\right) \sum_{k=1}^{M_i(r)} \int_{\mathrm{arc}_{i,k}(r)} n_{B_{\frac{r}{2}}(p_i)} \phi,$$

$$\frac{\partial \mathcal{H}_{area}}{\partial p_i}(P) = \sum_{k=1}^{M_i(r)} \int_{\mathrm{arc}_{i,k}(r)} n_{B_{\frac{r}{2}}(p_i)} \phi.$$

with $\mathrm{arc}_{i,k}(r), k \in \{1,\dots,M_i(r)\}$ *the arcs in the boundary of* $V_i(P) \cap B_{\frac{r}{2}}(p_i)$. *As a consequence, the critical points of* $\mathcal{H}_C$ *are centroidal Voronoi configurations.*

Concerning the smoothness properties of the multi-center functions $\mathcal{H}_{DC}$ and $\mathcal{H}_{SP}$, let us consider the following alternative expressions. Let

$$\mathcal{H}_{DC}(P) = \max_{i \in \{1,\dots,n\}} G_i(P), \quad \mathcal{H}_{SP}(P) = \min_{i \in \{1,\dots,n\}} F_i(P),$$

61

where $G_i(P) = \max_{q \in V_i(P)} \|q - p_i\|$ and $F_i(P) = \min_{q \notin \text{int} V_i(P)} \|q - p_i\|$. For a convex polygon $W \subset \mathbb{R}^2$, define the functions

$$\text{lg}_W(p) = \max\{\|q - p\| \mid q \in W\} = \max\{\|v - p\| \mid v \in \text{Ve}(W)\},$$
$$\text{sm}_W(p) = \min\{\|q - p\| \mid q \notin \text{int}(W)\} = \min\{\text{D}_e(p) \mid e \in \text{Ed}(W)\}.$$

These functions are locally Lipschitz and regular (the reader is referred to [8] for a comprehensive treatment of nonsmooth analysis), and their generalized gradients are given by $\partial \text{lg}_Q(p) = \text{co}\{\text{versus}(p - v) \mid v \in \text{Ve}(Q), \text{lg}_Q(p) = \|p - v\|\}$ and $\partial \text{sm}_Q(p) = \text{co}\{n_e \mid e \in \text{Ed}(Q), \text{sm}_Q(p) = \text{D}_e(p)\}$. Note that $G_i(P) = \text{lg}_{V_i(P)}(p_i)$ and $F_i(P) = \text{sm}_{V_i(P)}(p_i)$. Despite the slight abuse of notation, it is convenient to let $\partial \text{lg}_{V_i(P)}(p_i)$ denote $\partial \text{lg}_V(p_i)|_{V=V_i(P)}$, and let $\partial \text{sm}_{V_i(P)}(p_i)$ denote $\partial \text{sm}_V(p_i)|_{V=V_i(P)}$, i.e., holding fixed the Voronoi cell V_i.

The properties of the functions G_i and F_i are strongly affected by the dependence on the Voronoi partition $\mathcal{V}(P)$. These properties can be fully characterized (the interested reader is referred to [9] for a detailed discussion): indeed, both $G_i, -F_i : Q^n \to \mathbb{R}$ are locally Lipschitz and regular, and their generalized gradients can be described in a precise way by means of a careful analysis of the vertexes and edges where their values are attained, and of the degenerate/nondegenerate character of the Voronoi partition. In the sake of brevity, here we will only highlight the fact that the knowledge of the generalized gradients of G_i and F_i is key to describe the generalized gradients of the functions $\mathcal{H}_{DC}$ and $\mathcal{H}_{SP}$, as we do next.

Theorem 3.2 *The multi-center functions $\mathcal{H}_{DC}, -\mathcal{H}_{SP} : Q^n \to \mathbb{R}$ are locally Lipschitz and regular. Their generalized gradients can be expressed as*

$$\partial \mathcal{H}_{DC}(P) = \text{co}\{\partial G_i(P) \mid i \text{ such that } G_i(P) = \mathcal{H}_{DC}(P)\},$$
$$\partial \mathcal{H}_{SP}(P) = \text{co}\{\partial F_i(P) \mid i \text{ such that } F_i(P) = \mathcal{H}_{SP}(P)\}.$$

Moreover,

(i) *if $P \in Q^n$ is nondegenerate and $0 \in \text{int} \partial \mathcal{H}_{DC}(P)$, then P is a strict local minimum of $\mathcal{H}_{DC}$, all generators verify $G_i(P) = \mathcal{H}_{DC}(P)$ and P is a circumcenter Voronoi configuration;*

(ii) *if $P \in Q^n$ and $0 \in \text{int} \partial \mathcal{H}_{SP}(P)$, then P is a strict local maximum of $\mathcal{H}_{SP}$, all generators verify $F_i(P) = \mathcal{H}_{SP}(P)$ and P is a generic incenter Voronoi configuration.*

Remark 3.3 Theorem 3.2(i) and (ii) provide the interpretation of the multi-center problems given at the beginning of this section: since all generators are active, they share the same radius.

4. MOTION COORDINATION ALGORITHMS

In this section, we develop continuous-time implementations of the gradient ascent for the multi-center functions introduced in the previous section. Recall that the agents' location obeys a first order dynamical behavior, as described in equation (3).

We start by considering the multi-center functions $\mathcal{H}_C$, $\mathcal{H}_{\frac{r}{2}}$ and $\mathcal{H}_{area}$. Building on the result of Theorem 3.1, pick one of the multi-center functions as an aggregate objective cost to be optimized and impose that the location p_i follows its gradient flow. In more precise terms, we set up the following control laws defined over the set $Q^n \setminus \mathcal{S}$

$$u_i = -\frac{\partial \mathcal{H}_C}{\partial p_i}(P) = 2 M_{V_i(P)}(\text{CM}_{V_i(P)} - p_i), \tag{6a}$$

$$u_i = -\frac{\partial \mathcal{H}_{\frac{r}{2}}}{\partial p_i}(P) = 2 M_{V_i(P) \cap B_{\frac{r}{2}}(p_i)}(\text{CM}_{V_i(P) \cap B_{\frac{r}{2}}(p_i)} - p_i)$$

$$+ \left(\text{diam}(Q)^2 - \left(\frac{r}{2}\right)^2 \right) \sum_{k=1}^{M_i(r)} \int_{\text{arc}_{i,k}(r)} n_{B_{\frac{r}{2}}(p_i)} \phi, \tag{6b}$$

$$u_i = \frac{\partial \mathcal{H}_{area}}{\partial p_i}(P) = \sum_{k=1}^{M_i(r)} \int_{\text{arc}_{i,k}(r)} n_{B_{\frac{r}{2}}(p_i)}, \tag{6c}$$

where we assume that the partition $\mathcal{V}(P) = \{V_1, \dots, V_n\}$ is continuously updated. One can prove the following result.

Proposition 4.1 *Consider the gradient dynamical system on $Q^n \setminus \mathcal{S}$ defined by equation (6) for each of the multi-center functions $\mathcal{H}_C$, $\mathcal{H}_{\frac{r}{2}}$ and $\mathcal{H}_{area}$. Then, we have*

(i) The gradient dynamical system (6a) is spatially distributed over the Delaunay graph $\mathcal{G}_D$, and the gradient dynamical systems (6b) and (6c) are spatially distributed over the r-limited Delaunay graph $\mathcal{G}_{LD}(r)$.

(ii) For each of the flows in (6), the agents' location evolution starting at $P_0 \in Q^n \setminus \mathcal{S}$ remains in $Q^n \setminus \mathcal{S}$ and converges asymptotically to the set of critical points of the corresponding aggregate objective function. Assuming this set is finite, the agents' location converges to a single critical point.

Remark 4.2 Note that the gradient ascent is not guaranteed to find the global optimum. For example, in the vector quantization literature, it is known that for "bimodal" distribution density functions, the solution to the gradient flow reaches local maxima where the number of agents allocated to the two region of maxima are not optimally partitioned.

We are now in a position to recover the discussion about the expected value scenario with range-limited interactions that we began in Section 3. There, we introduced the function $\mathcal{H}_{\frac{r}{2}}$ to deal with this situation. One of the reasons that we gave was the constant-factor approximation (5) of the value of $\mathcal{H}_C$. The other reason is given by Proposition 4.1(i). The gradient flow of $\mathcal{H}_C$ is spatially distributed over the Delaunay graph $\mathcal{G}_D$, and, therefore, not generally implementable over a network of mobile agents with a fixed (and common) communication radius $r \in \mathbb{R}_+$ (respectively sensing radius). On the other hand, the gradient flow of $\mathcal{H}_{\frac{r}{2}}$ is spatially distributed over the r-limited Delaunay graph $\mathcal{G}_{LD}(r)$, and therefore can be implemented over networks of mobile agents with fixed radius r.

Let us consider now the worst-case scenario and the non-interference scenario. Consider the (signed) generalized gradient flow for the multi-center functions $\mathcal{H}_{DC}$ and $\mathcal{H}_{SP}$,

$$\dot{P} = -\text{Ln}(\partial \mathcal{H}_{DC}(P)), \quad \dot{P} = \text{Ln}(\partial \mathcal{H}_{SP}(P)),$$

63

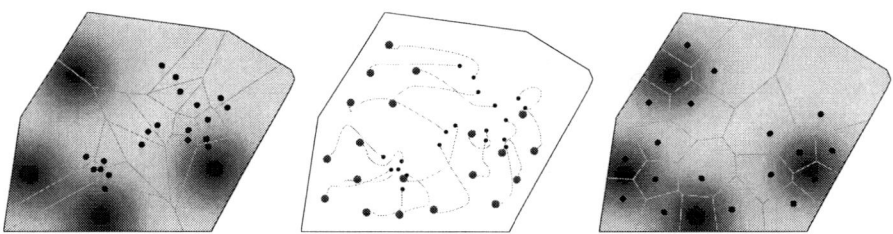

FIGURE 1. Expected value scenario: continuous-time algorithm (6a) for 20 mobile agents in the convex polygonal environment defined by the vertexes of coordinates $\{(0,0),(2.125,0),(2.9325,1.5),(2.975,1.6),(2.9325,1.7),(2.295,2.1),(0.85,2.3),(0.17,1.2)\}$ (in meters). The density ˙ function ϕ is the sum of four Gaussian functions of the form $11\exp(6(-(x-x_{\text{center}})^2-(y-y_{\text{center}})^2))$ and is represented by means of its contour plot. The centers $(x_{\text{center}},y_{\text{center}})$ of the Gaussians are given by $(2.15,.75)$, $(1.,.25)$, $(.725,1.75)$ and $(.25,.7)$, respectively. The left (respectively, right) figure illustrates the initial (respectively, final) locations and Voronoi partition. The central figure illustrates the gradient descent flow. After 13 seconds, the value of the multi-center function is approximately .515.

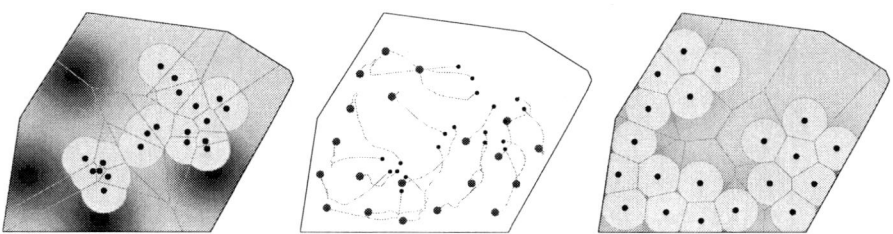

FIGURE 2. Expected value scenario with range-limited interactions: continuous-time algorithm (6b) for 20 mobile agents in the same convex polygonal environment and with the same density function ϕ as in Figure 1. Each agent operates with a finite sensing/communication radius equal to $r = .47$. The left (respectively, right) figure illustrates the initial (respectively, final) locations and Voronoi partition. The central figure illustrates the gradient ascent flow. For each agent i, the intersection $V_i \cap B_{\frac{r}{2}}(p_i)$ is plotted in light gray. After 13 seconds, the value of the multi-center function is approximately 4.794. From the constant-factor approximation (5), we compute $\beta \approx 0.00484$, where P_{final} denotes the final configuration in Figure 2. The absolute error is guaranteed to be less than or equal to $(1-\beta)\mathscr{H}_{\frac{r}{2}}(P_{\text{final}}) \approx 4.77$. In order to compare the performance of this execution with the performance of the algorithm in the expected value scenario (cf. Figure 1), we compute the percentage error in the value of the multi-center function $\mathscr{H}_{\text{C}}$ at their final configurations. This percentage error is approximately equal to 3.277%. As expected, we verified in simulations that the percentage error of the performance of the limited-range implementation improves with higher values of the ratio $r/\operatorname{diam}(Q)$.

where $\operatorname{Ln}: 2^{\mathbb{R}^N} \to \mathbb{R}$ is the map that associates to each convex set $S \subset \mathbb{R}^N$ its least-norm element, $\operatorname{Ln}(S) = \operatorname{proj}_S(0)$. Alternatively, we may write for each $i \in \{1,\ldots,n\}$,

$$\dot{p}_i = -\pi_i(\operatorname{Ln}(\partial\mathscr{H}_{\text{DC}})(p_1,\ldots,p_n)), \tag{7a}$$

$$\dot{p}_i = \pi_i(\operatorname{Ln}(\partial\mathscr{H}_{\text{SP}})(p_1,\ldots,p_n)), \tag{7b}$$

where $\pi_i : (\mathbb{R}^2)^n \to \mathbb{R}^2$ is the canonical projection onto the ith factor. Note that the vector fields (7a) and (7b) are discontinuous, and therefore, we understand its solution

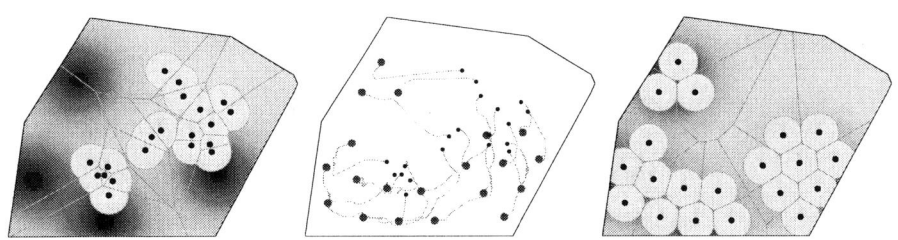

FIGURE 3. Area scenario: continuous-time algorithm (6c) for 20 mobile agents in the same convex polygonal environment and with the same density function ϕ as in Figure 1. Each agent operates with a finite sensing/communication radius equal to $r = .4$. The left (respectively, right) figure illustrates the initial (respectively, final) locations and Voronoi partition. The central figure illustrates the gradient ascent flow. For each agent i, the intersection $V_i \cap B_{\frac{r}{2}}(p_i)$ is plotted in light gray. After 36 seconds, the value of the multi-center function is approximately 14.141.

in the Filippov sense [14]. One needs to first compute the generalized gradients at P, $\partial \mathcal{H}_{DC}(P)$ and $\partial \mathcal{H}_{SP}(P)$, then compute the least-norm element, and finally project to each of the n components. Note that the least-norm element of convex sets can be computed efficiently, see [7], however closed-form expressions are not available in general. One can also see that the compact set Q^n is strongly invariant for both vector fields $- \mathrm{Ln}(\partial \mathcal{H}_{DC})$ and $\mathrm{Ln}(\partial \mathcal{H}_{SP})$ (cf. [9]).

Proposition 4.3 *For the dynamical system* (7a) *(respectively* (7b)*), the generators' location* $P = (p_1, \ldots, p_n)$ *converges asymptotically to the set of critical points of* $\mathcal{H}_{DC}$ *(respectively,* $\mathcal{H}_{SP}$*).*

The gradient dynamical systems (7a) and (7b) enjoy the convergence guarantees stated in Proposition 4.3, but their implementation is not spatially distributed over the Delaunay graph $\mathcal{G}_D$ because of two reasons. First, the values of all functions G_i (respectively F_i) need to be compared in order to determine which generators are active. Second, the least-norm element of the generalized gradients depends on the relative position of the active generators with respect to each other and to the environment. This is the reason why in what follows we propose a distributed implementation of the previous gradient dynamical systems and explore their relation with behavior-based rules.

Consider the following variations of the gradient dynamical systems in equation (7),

$$\dot{p}_i = - \mathrm{Ln}(\partial \, \mathrm{lg}_{V_i(P)})(P), \tag{8a}$$

$$\dot{p}_i = \mathrm{Ln}(\partial \, \mathrm{sm}_{V_i(P)})(P), \tag{8b}$$

for $i \in \{1, \ldots, n\}$. Note that the systems (8a) and (8b) are spatially distributed over the Delaunay graph $\mathcal{G}_D$, since $\mathrm{Ln}(\partial \, \mathrm{lg}_{V_i(P)})(P)$ is determined only by the position of p_i and of its Voronoi neighbors $\mathcal{N}_{\mathcal{G}_D, p_i}(P)$, and $\mathrm{Ln}(\partial \, \mathrm{sm}_{V_i(P)})(P)$ is determined only by the position of p_i and its nearest neighbors (which, in particular, must be Voronoi neighbors). As for the previous dynamical systems, note that these vector fields are discontinuous, and therefore, we understand its solutions in the Filippov sense. One can see that the compact set Q^n is strongly invariant for both vector fields. Moreover, for $P \in Q^n$, the solutions of the dynamical systems (8a) and (8b) starting at P are unique. The following statement summarizes the results concerning these dynamical systems (for the notion of weakly invariant set, we refer to [4, 14]).

Proposition 4.4 *Consider the dynamical systems on Q^n defined by equations (8a) and (8b) for the multi-center functions $\mathcal{H}_{DC}$ and $\mathcal{H}_{SP}$. Then, we have*

(i) *Both dynamical systems are spatially distributed over the Delaunay graph $\mathcal{G}_D$.*

(ii) *For the dynamical system (8a) (resp. the dynamical system (8b)), the generators' location $P = (p_1, \ldots, p_n)$ converges asymptotically to the largest weakly invariant set contained in the closure of $A_{DC}(Q) = \{P \in Q^n \mid i \text{ such that } G_i(P) = \mathcal{H}_{DC}(P) \implies p_i = \mathrm{CC}_{V_i}\}$ (resp. the largest weakly invariant set contained in the closure of $A_{SP}(Q) = \{P \in Q^n \mid i \text{ such that } F_i(P) = \mathcal{H}_{SP}(P) \implies p_i \in \mathrm{IC}_{V_i}\}$).*

Remarks 4.5 (Relation with behavior-based robotics) The dynamical systems (8a) and (8b) have an interesting connection with basic interaction laws in behavior-based robotics.

Move toward the furthest-away vertex: Consider the distributed gradient control law in the disk-covering setting (8a). For the ith generator, if the maximum of $\lg_{V_i(P)}$ is attained at a single vertex v of its Voronoi cell V_i, then (at fixed $V_i(P)$) $\lg_{V_i(P)}$ is differentiable at that configuration, and its derivative corresponds to versus$(p_i - v)$. Therefore, the control law (8a) corresponds to the behavior "move toward the furthest vertex in own Voronoi cell." If there are two or more vertexes of V_i where the value $\lg_{V_i(P)}(p_i)$ is attained, then (8a) provides an average behavior by computing the least-norm element in the convex hull of all versus$(p_i - v)$ such that $\|p_i - v\| = \lg_{V_i(P)}(p_i)$.

Move away from the nearest neighbor Consider the distributed gradient control law in the sphere-packing setting (8b). For the ith generator, if the minimum of $\mathrm{sm}_{V_i(P)}$ is attained at a single edge e, then (at fixed $V_i(P)$) $\mathrm{sm}_{V_i(P)}$ is differentiable at that configuration, and its derivative is n_e. The control law (8b) corresponds to the behavior "move away from the nearest neighbor" (where a neighbor can also be the boundary of the environment). If there are two or more edges where the value $\mathrm{sm}_{V_i(P)}(p_i)$ is attained, then (8b) provides an average behavior in an analogous manner as before.

One could also consider other distributed dynamical systems that also optimize the multi-center functions $\mathcal{H}_{DC}$ and $\mathcal{H}_{SP}$, based on the idea of geometric centering. Roughly speaking, each agent moves toward the circumcenter of its own Voronoi region, for the worst-case scenario, and toward the incenter of its own Voronoi region, for the non-interference scenario. These strategies are the counterparts of the "move toward the centroid" law in equation (6a) for the expected value scenario. They also enjoy similar convergence properties to those of the dynamical systems (8a) and (8b). A detailed discussion can be found in [9].

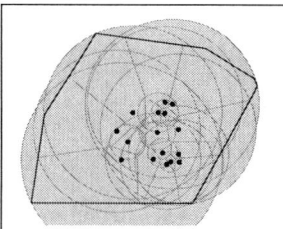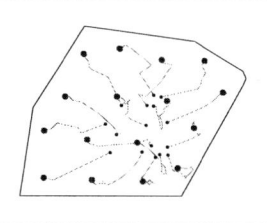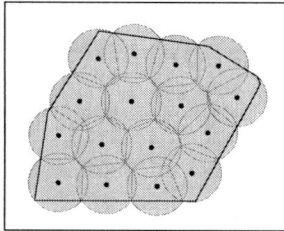

FIGURE 4. Worst-case scenario: "move-toward-the-furthest-away-vertex" algorithm for 16 mobile agents in a convex polygonal environment determined by the vertexes with coordinates $\{(0,0),(2.5,0),(3.45,1.5),(3.5,1.6),(3.45,1.7),(2.7,2.1),(1.,2.4),(.2,1.2)\}$ (in meters). The left (respectively, right) figure illustrates the initial (respectively, final) locations and Voronoi partition. The central figure illustrates the network evolution. After 2 sec., the value of the multi-center function $\mathscr{H}_{\text{DC}}$ is approximately .39504 m.

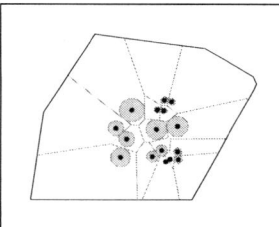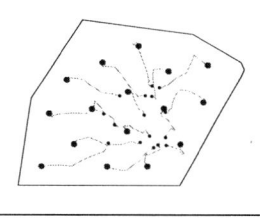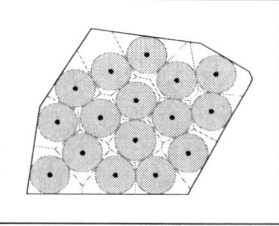

FIGURE 5. Non-interference scenario: "move-away-from-closest-neighbor" algorithm for 16 mobile agents in the same convex polygonal environment as in Figure 4. The left (respectively, right) figure illustrates the initial (respectively, final) locations and Voronoi partition. The central figure illustrates the network evolution. After 2 sec., the value of multi-center function $\mathscr{H}_{\text{SP}}$ is approximately .26347 m.

ACKNOWLEDGMENTS

This material is based upon work done in joyful collaboration with Francesco Bullo and Sonia Martínez. The author thanks the organizers of the International Workshop on Global Analysis for the warm hospitality throughout the meeting and the opportunity to present this work. The partial support by NSF SENSORS Award IIS-0330008 is gratefully acknowledged.

REFERENCES

1. Agarwal, P. K., and Sharir, M., *ACM Computing Surveys*, **30**, (1998), 412-458.
2. Ando, H., Oasa, Y., Suzuki, I., and Yamashita, M., *IEEE Transactions on Robotics and Automation*, **15**, (1999), 818-828.
3. Arkin, R. C., *Behavior-Based Robotics*, Cambridge University Press, Cambridge, UK, 1998, ISBN 0262011654.
4. Bacciotti, A., and Ceragioli, F., *ESAIM. Control, Optimisation & Calculus of Variations*, **4**, (1999), 361-376.

5. de Berg, M., van Kreveld, M., and Overmars, M., *Computational Geometry: Algorithms and Applications*, Springer Verlag, New York, NY, 1997, ISBN 354061270X.
6. Boltyanski, V., Martini, H., and Soltan, V., *Geometric methods and optimization problems*, vol. 4 of *Combinatorial optimization*, Kluwer Academic Publishers, Dordrecht, The Netherlands, 1999, ISBN 0792354540.
7. Boyd, S., and Vandenberghe, L., *Convex Optimization*, Cambridge University Press, Cambridge, UK, 2004, ISBN 0521833787.
8. Clarke, F. H., *Optimization and Nonsmooth Analysis*, Canadian Mathematical Society Series of Monographs and Advanced Texts, John Wiley & Sons, 1983, ISBN 0-471-87504-X.
9. Cortés, J., and Bullo, F., *SIAM Journal on Control and Optimization* (2003), to appear.
10. Cortés, J., Martínez, S., Karatas, T., and Bullo, F., *IEEE Transactions on Robotics and Automation*, **20**, (2004), 243–255.
11. Cortés, J., Martínez, S., and Bullo, F., *ESAIM. Control, Optimisation & Calculus of Variations* (2004), submitted.
12. Drezner, Z., editor, *Facility Location: A Survey of Applications and Methods*, Springer Series in Operations Research, Springer Verlag, New York, NY, 1995, ISBN 0-387-94545-8.
13. Du, Q., Faber, V., and Gunzburger, M., *SIAM Review*, **41**, (1999), 637-676.
14. Filippov, A. F., *Differential Equations with Discontinuous Righthand Sides*, vol. 18 of *Mathematics and Its Applications*, Kluwer Academic Publishers, Dordrecht, The Netherlands, 1988.
15. Jadbabaie, A., Lin, J., and Morse, A. S., *IEEE Transactions on Automatic Control*, **48**, (2003), 988-1001.
16. Jaromczyk, J. W., and Toussaint, G. T., *Proceedings of the IEEE*, **80**, (1992), 1502-1517.
17. Lin, J., Morse, A. S., and Anderson, B. D. O., "The multi-agent rendezvous problem: an extended summary," in *Proceedings of the 2003 Block Island Workshop on Cooperative Control*, edited by N. E. Leonard, S. Morse, and V. Kumar, Lecture Notes in Control and Information Sciences, Springer Verlag, New York, NY, 2004, to appear.
18. Lynch, N. A., *Distributed Algorithms*, Morgan Kaufmann Publishers, San Mateo, CA, 1997, ISBN 1558603484.
19. Okabe, A., Boots, B., Sugihara, K., and Chiu, S. N., *Spatial Tessellations: Concepts and Applications of Voronoi Diagrams*, Wiley Series in Probability and Statistics, John Wiley & Sons, New York, NY, 2000, second edn., ISBN 0471986356.
20. Okubo, A., *Advances in Biophysics*, **22**, (1986), 1-94.
21. Olfati-Saber, R., and Murray, R. M., *IEEE Transactions on Automatic Control* (2003), to appear.
22. Ögren, P., Fiorelli, E., and Leonard, N. E., *IEEE Transactions on Automatic Control* (2004), to appear.
23. Passino, K. M., *Biomimicry for Optimization, Control, and Automation*, Springer Verlag, New York, NY, 2004, in print.
24. Reynolds, C. W., *Computer Graphics*, **21**, (1987), 25-34.
25. Shevitz, D., and Paden, B., *IEEE Transactions on Automatic Control*, **39**, (1994), 1910-1914.
26. Suzuki, A., and Drezner, Z., *Location Science*, **4**, (1996), 69-82.

Global Analysis

Zeta-regularization and calculus on infinite dimensional spaces

Akira Asada

Sinsyu University, 3-6-21 Nogami Takarazuka, 665-0022 Japan

Abstract. ζ-regularization of the Jacobian of scaling transformation of an infinite dimensional integral is derived. It simplifies mathematical justification of the appearance of the Ray-Singer determinant in the Gaussian path integral given in our previous paper "Regularized Calculus: An application of zeta regularization to infinite dimensional Geometry and Analysis, Int.J.Geo, Meth. Mod. Phys., 1(2004)".
Key words. Spectral ζ-function, Ray-Singer determinant, Regularized infinite product, Gaussian path integral.
PACS. 02.40.Vh.
MSC. 58J52, 81S40, 46L57.

1. INTRODUCTION

Let H be a Hilbert space with the complete orthonormal basis $\{e_n\}$. $\{e_n\}$ defines a coordinate $x \to (x_1, x_2, \ldots)$, $x = \sum_n x_n e_n$. Since H is an infinite dimensional space, calculus on H often diverges. For example, a smooth function f on an open set of H is totally differentiable, if and only if $\sum_n |\frac{\partial f}{\partial x_n}|^2 < \infty$. Another example is the Gaussian path integral $\int e^{-\pi(x,Dx)} \mathscr{D}x$, where D is a positive elliptic operator. Assuming D has the spectral decomposition of the form $Df = \sum_n \lambda_n (f, e_n) e_n$, formally this integral is calculated as follows;

$$\lim_{n \to \infty} \int_{-\infty}^{\infty} e^{-\pi \lambda_1 x_1^2} dx_1 \cdots \int_{-\infty}^{\infty} e^{-\pi \lambda_n x_n^2} dx_n$$

$$= \lim_{n \to \infty} \frac{1}{\sqrt{\lambda_1 \cdots \lambda_n}}.$$

Formally, $\lim_{n \to \infty} \lambda_1 \cdots \lambda_n$ is the determinant of D. Hawking proposed to use the Ray-Singer determinant of D as the regularized value of this limit [8]. Here, the Ray-Singer determinant $det.D$ of D is defined by

$$det.D = e^{-\zeta'(D,0)}, \quad \zeta(D,s) = \sum_n \lambda_n^{-s},$$

CP729, *Global Analysis and Applied Mathematics: International Workshop on Global Analysis*,
edited by K. Taş, D. Krupka, O. Krupková, and D. Baleanu
© 2004 American Institute of Physics 0-7354-0209-4/04/$22.00

[10, 12]. If D is defined on a compact manifold, $\zeta(D,s)$ allows analytic continuation to whole complex plane and holomorphic at $s = 0$ [4, 7]. Hence $det.D$ exists. It is a non zero positive real number [1]. By definition, we have

$$det.D = \prod_{n=1}^{\infty} \lambda_n^{\lambda_n^{-s}}|_{s=0}.$$

Here $|_{s=0}$ means analytic continuation to $s = 0$.

Following this expression of the Ray-Singer determinant, we introduced the regularized infinite product $: \prod_n x_n :$ by

$$: \prod_{n=1}^{\infty} x_n := \prod_{n=1}^{\infty} x_n^{\lambda_n^{-s}}|_{s=0}.$$

It was shown $: \prod_n x_n :$ is linear in each variable [2]. So we expect

$$\lim_{N \to \infty} \frac{\partial^N}{\partial x_1 \cdots \partial x_N} : \prod_{n=1}^{\infty} x_n := 1.$$

In this paper, we show this is true, in weak sense. The proof based on the equality

$$\int_0^{\infty} \cdots \int_0^{\infty} \left(\frac{\partial^n}{\partial x_1 \cdots \partial x_n} x_1^{\mu_1^s} \cdots x_n^{\mu_n^s}\right) e^{-x_1 - \cdots - x_n} dx_1 \cdots dx_n$$
$$= \Gamma(1 + \mu_1^s) \cdots \Gamma(1 + \mu_n^s), \quad \mu_n = \lambda_n^{-1}.$$

Since

$$\sum_{n=1}^{\infty} \log \Gamma(1 + \mu_n^s) = -\gamma \zeta(D,s) + \sum_{n=2}^{\infty} (-1)^n \frac{\zeta(n)}{n} \zeta(D, ns),$$

$\prod_n \Gamma(1 + \mu_n^s)$ is continued analytically on the right half plane with singularities on the real axis which are accumulated to 0. By Abel's continuity Theorem, if s tends to 0 along a piecewise smooth curve in the right half plane which is not tangent to the real and imaginary axes, then

$$\lim_{s \to 0} \prod_{n=1}^{\infty} \Gamma(1 + \mu_n^s) = 1.$$

Similar discussion is possible taking $\exp(-\|x\|^2)$ as the testing function. If $\lim_{N \to \infty} \int_0^{\infty} \cdots \int_0^{\infty} f(x_1, \ldots) dx_1 \cdots dx_N$ exists, we define regularization of $\lim_{N \to \infty} \int_0^{\infty} \cdots \int_0^{\infty} f(a_1 x_1, \ldots) dx_1 \cdots dx_N, a_1 > 0, a_2 > 0, \ldots$ by

$$\int_{\mathbb{R}_+^{\infty}} f(a_1 x_1, \ldots) : d^{\infty} x :$$
$$= \lim_{N \to \infty} \int_0^{\infty} \cdots \int_0^{\infty} \left(\frac{\partial^N}{\partial x_1 \cdots \partial x_N} x_1^{\mu_1^s} \cdots x_N^{\mu_N^s}\right) \times$$
$$\times f(a_1 x_1, \ldots) dx_1 \cdots dx_N|_{s=0}.$$

If : $\prod_n a_n$: exists, we have

$$\int_{\mathbb{R}_*^\infty} f(a_1 x_1, \ldots) : d^\infty x :$$

$$= \lim_{N \to \infty} \int_0^\infty \cdots \int_0^\infty (: \prod_{n=1}^\infty a_n :)^{-1} f(x_1, \ldots) dx_1 \cdots dx_N.$$

In other word, the Jacobian of the scaling transformation $T : T x_n = a_n x_n$ is regularized as : $\prod_n a_n$:. Regularization of the Jacobian of a scaling transformation for integrals on $\mathbb{R}^\infty$ is similarly defined. Then, since : $\prod_n \lambda_n := det.D$, we obtain

$$\int_{\mathbb{R}^\infty} e^{-\pi(x,Dx)} : d^\infty x := \frac{1}{\sqrt{det.D}}.$$

This regularization procedure can be interpreted by using fractional calculus, which was presented in [3]. We also note that above regularization of the Jacobian of T is reinterpreted by using Paycha's regularized trace [11]. Because to define regularized determinant of a linear operator T by

$$det.T = e^{tr(D^{-s}S)}|_{s=0}, \quad T = e^S,$$

we have

$$: \prod_{n=1}^\infty a_n := det.T, \quad T x_n = a_n x_n.$$

This paper is organized as follows: The Ray-Singer determinant, Hilbert space equipped with a Schatten class operator and regularized infinite product are reviewed in Sec.2-Sec.4. Hilbert space equipped with a Schatten class operator is closely related to Connes' spectral triple [5] in spirit and necessary to the study of regularized infinite product. Weak derivation of regularized infinite product is discussed in Sec.5. Especially, we show $\exp(-\sum_n x_n^2)$ can be taken as a testing function in the calculation of weak derivation. Then in Sec.6, regularized Jacobian of a scaling transformation is derived. It gives a mathematical justification of appearance of the Ray-Singer determinant of Gaussian path integral.

2. THE RAY-SINGER DETERMINANT

Let D be a non-degenerate elliptic pseudo differential operator defined on a smooth manifold X. We assume D has the Agmon angle θ, that is there is a ray $re^{i\theta}; r \geq 0$, such that spectres of D does not meet this ray. Then the ζ-function $\zeta_\theta(D, s)$ of D with respect to θ is defined by

$$\zeta_\theta(D, s) = tr(D_\theta^{-s}), \quad D_\theta^s = \int_\gamma \lambda_\theta^s (\lambda I - D)^{-1} d\lambda. \tag{1}$$

Here, $\lambda_\theta^s = |\lambda| \exp(is \arg \lambda), \theta < \arg \lambda < \theta + 2\pi$, and γ encloses the spectrum of D [10]. If D is selfadjoint,

$$\zeta_\theta(D,s) = \sum_{\lambda_n > 0} \lambda_n^{-s} + \sum_{\lambda_n < 0} (-1)^s |\lambda_n|^{-s}.$$

Hence there are two-kinds of ζ-functions in this case, according to $0 < \theta < \pi$ and $\pi < \theta < 2\pi$. In the first case, $(-1)^s = e^{-\pi i s}$ and in the second case, $(-1)^s = e^{\pi i s}$. The corresponding ζ-functions are denoted by $\zeta_-(D,s)$ and $\zeta_+(D,s)$, respectively [1].

It is known $\zeta_\theta(D,s)$ and $\eta(D,s)$ are continued meromorphically on whole complex plane with possible poles of order 1 at $s = d/k, (d-1)/k, \ldots$, where $d = dim.X$ and k is the order of D [13]. Moreover, if D is selfadjoint and X is compact, $\zeta_\pm(D,s)$ and $\eta(D,s)$ are holomorphic at $s = 0$ [4, 7]. In this case, $\zeta(D,s)$ takes real value on the real axis, and residues at poles are also real numbers. Therefore, the following definition is possible.

Definition 1. *We define the determinant $det_\theta.D$ of D (with respect to θ) by*

$$det_\theta.D = e^{-\zeta_\theta'(D,0)}, \quad \zeta_\theta'(D,s) = \frac{d}{ds} \zeta_\theta(D,s). \tag{2}$$

$det_\theta.D$ is said to be the Ray-Singer determinant or simply, determinant of D with respect to the Agmon angle θ. In general, $det_\theta.D$ depends on the choice of θ. Precisely, we have

$$det_\theta.D = \prod_{n=1}^{\infty} \lambda_n^{\lambda_n^{-s}} |_{s=0}, \tag{3}$$

where the argument of λ_n is selected to be $\theta < \arg \lambda_n < \theta + 2\pi$. As for the uniqueness of the determinant of a selfadjoint D, the following Theorem is proved in [1].

Theorem 1 *Let D be a nondegenerate selfadjoint elliptic pseudo differential operator on a compact manifold X, then the followings are equivalent.*

1. *$det.D$ is a real number.*
2. *$det.D$ is unique.*
3. *v_- is an integer.*

Here v_- is defined by

$$v_- = \frac{v - \eta(D,0)}{2}, \quad v = \zeta(D^2, 0). \tag{4}$$

If D is positive, $det_\theta.D$ is uniquely defined non-zero positive number. Hence we denote $det.D$ instead of $det_\theta.D$ for a positive elliptic operator D. $\zeta_\theta(D,s)$ also uniquely defined. So we denote it by $\zeta(D,s)$. $v = \zeta(D,0)$ in this case.

Formally, v counts the number of eigenvalues of D. Hence we say v *the regularized dimension of H*, the Hilbert space spanned by the eigenfunctions of D.

$det.D_1 D_2 \neq det.D_1 \cdot det.D_2$ in general. But we have

$$det.D^m = (det.D)^m, \quad det.(tD) = t^v det.D, \, t > 0. \tag{5}$$

In the rest, we denote d the location of the first pole of $\zeta(D, s)$. By definition, $\sum_n \lambda_n^{-s}$ converges if $\Re s > d$, and $\sum_n \lambda_n^{-d}$ diverges.

Note. Let G be the Green operator of D. The ζ-function $\zeta(G, s)$ is defined by $tr(G^s)$. Since $G^s = D^{-s}$, we have

$$\zeta(G, s) = \zeta(D, s), \quad det.G = e^{\zeta'(G,0)}.$$

3. HILBERT SPACE EQUIPPED WITH A SCHATTEN CLASS OPERATOR

Let H be a (real) Hilbert space, G a positive Schatten class operator (cf.[14]) on H, such that its ζ-function $\zeta(G, s) = tr(G^s)$ allows analytic continuation to $s = 0$ and holomorphic at $s = 0$. In the rest, we consider the pair $\{H, G\}$.

Example. Let X be a compact Riemannian manifold, E a vector bundle over X, D a nondegenerate selfadjoint elliptic differential operator acting on the sections of E. Then taking $H = L^2(X, E)$, G the Green operator of D, we obtain such pair $\{H, G\}$.

Definition 2. *We say* $v = \zeta(G, 0)$ *the regularized dimension of* H.

The location d of the first pole of $\zeta(G, s)$ is another numerical invariant of the pair $\{H, G\}$. We also define the regularized determinant $det.G$ of G by $\exp(\zeta'(G, 0))$. If G is the Green operator of an elliptic operator D, then we have

$$det.G = (det.D)^{-1}.$$

Let μ_n be an eigenvalue of G with the normalized eigenvector e_n:$Ge_n = \mu_n e_n$. Similar to (3), we obtain

$$det.G = \prod_{n=1}^{\infty} \mu_n^{\mu_n^s}|_{s=0}.$$

We arrange the eigenvalues of G to be $\mu_1 \geq \mu_2 \geq \cdots > 0$. The complete orthonormal basis of H is fixed to be $\{e_1, e_2, \ldots\}$. We also introduce the Sobolev k-norm $\|x\|_k$ of $x \in H$ by

$$\|x\|_k = \|G^{-k}x\|, \quad provided \ G^{-k}x \in H. \tag{6}$$

Note. In general, $\| \ \|_k$ is different from ordinary Sobolev k-norm. But if G is the Green operator of a 1-st order operator, $\| \ \|_k$ is an equivalent norm of ordinary Sobolev k-norm.

The Sobolev space obtained by this norm and H is denoted by W^k. $H = W^0$ by this notation. By definition, the complete orthonormal basis of W^k is given by $\{e_{1,k}, e_{2,k}, \ldots\}$,

75

where $e_{n,k} = \mu_n^k e_n$. Since $x \in W^k$ is uniquely written $x = \sum_n x_n e_{n,k}$, $\sum_n |x_n|^2 < \infty$, we consider $(x_1, x_2, \ldots)$, to be the coordinate of $x \in W^k$. Alternatively, we can take $(x_{1,k}, x_{2,k}, \ldots)$, $x_{n,k} = x_n \mu_n^k$, as a coordinate of $x \in W^k$.

By definition, $W^k \subset W^l$, $k > l$, as sets. We set

$$W^{k-0} = \bigcap_{l<k} W^l, \quad W^{k+0} = \bigcup_{l>k} W^l.$$

If $k = 0$, we denote H^- and H^+ instead of W^{0-0} and W^{0+0}. W^{k-0} and W^{k+0} are dual each other. But we do not use W^{k+0} in the rest. Let

$$e_{\infty,k} = \sum_{n=1}^{\infty} \mu_n^{d/2} e_{n,k}, \quad e_\infty = e_{\infty,0} = \sum_{n=1}^{\infty} \mu_n^{d/2} e_n. \tag{7}$$

Then $e_{\infty,k} \in W^{k-0}$, but $e_{\infty,k} \notin W^k$. We set

$$W^{k-0}(finite) \;=\; \{\sum_{n=1}^{\infty} x_n e_{n,k} \in W^{k-0} |\lim_{n\to\infty} \mu_n^{-d/2} x_n \ exists\}, \tag{8}$$

$$W^{k-0}(0) \;=\; \{\sum_{n=1}^{\infty} x_n e_{n,k} \in W^{k-0} |\lim_{n\to\infty} \mu_n^{-d/2} x_n = 0\}. \tag{9}$$

Coordinates of these spaces are similarly defined as the coordinate of W^k. As vector spaces, we have

$$W^{k-0}(finite) = W^{k-0} \oplus \mathbb{R} e_{\infty,k}. \tag{10}$$

The topology of $W^{k-0}(0)$ is the subspace topology of W^{k-0}, while the topology of $W^{k-0}(finite)$ is the product space topology of $W^{k-0}(0)$ and $\mathbb{R}$.

Note. The line $\mathbb{R} e_{\infty,k}$ in (10) can be understood as the fibre of the determinant bundle of $W^{k-0}(0)$.

By (10), $x \in W^{k-0}(finite)$ is written uniquely

$$x = x_0 + t e_{\infty,k}, \quad x_0 \in W^{k-0}(0), \quad \lim_{n\to\infty} \mu_n^{-d/2} x_n = t, \tag{11}$$

where $x = \sum_n x_n e_{n,k}$.

Fixing a coordinate of $W^{k-0}(finite)$, we denote

$$x \gg 0 \quad if \ x = (x_1, x_2, \ldots), x_n > 0 \ for \ all \ n.$$

If $x \gg 0$, we define cubes $Q(x)$ and $Q(x, +)$ in $W^{k-0}(finite)$ by

$$Q(x) \;=\; \{(y_1, y_2, \ldots) \in W^{k-0}(finite)| |y_n| \leq x_n\}, \tag{12}$$

$$Q(x, +) \;=\; \{(y_1, y_2, \ldots) \in W^{k-0}(finite)| 0 \leq y_n \leq x_n\}. \tag{13}$$

By the definitions of $W^{k-0}(finite)$ and $Q(x)$, we have

$$\bigcup_{t>0} Q(tx) = W^{k-0}(finite). \tag{14}$$

76

4. REGULARIZED INFINITE PRODUCT

By (11), if $x = \sum_n x_n e_{n,k} \in W^{k-0}(finite)$, we have

$$x_n = x_{0,n} + t\mu_n^{d/2}, \quad \sum_{n=1}^{\infty} x_{0,n} e_{n,k} \in W^{k-0}(0).$$

We assume $t \neq 0$. We rewrite $x_n = t\mu_n^{d/2}(1 + \mu_n^{-d/2}(x_{0,n}/t))$. Then

$$\prod_{n=1}^{\infty} x_n^{\mu_n^s} = \prod_{n=1}^{\infty} \left(t^{\mu_n^s} \mu_n^{\mu_n^s d/2} (1 + \mu_n^{-d/2}\frac{x_{0,n}}{t})^{\mu_n^s} \right)$$

$$= t^{\zeta(G,s)} \left(\prod_{n=1}^{\infty} \mu_n^{\mu_n^s} \right)^{d/2} \prod_{n=1}^{\infty} (1 + \mu_n^{-d/2}\frac{x_{0,n}}{t})^{\mu_n^s}.$$

Hence we have

$$\prod_{n=1}^{\infty} x_n^{\mu_n^s}|_{s=0} = t^{\nu}(det.G)^{d/2} \prod_{n=1}^{\infty} (1 + \mu_n^{-d/2}\frac{x_{0,n}}{t})^{\mu_n^s}|_{s=0}.$$

Similarly, we get

$$\prod_{n=1}^{\infty} x_{n,k}^{\mu_n^s}|_{s=0} = t^{\nu}(det.G)^{k+d/2} \prod_{n=1}^{\infty} (1 + \mu_n^{-d/2}\frac{x_{0,n}}{t})^{\mu_n^s}|_{s=0}.$$

Definition 3. *We define the regularized infinite product* $: \prod_n x_n :$ *by*

$$: \prod_{n=1}^{\infty} x_n := \prod_{n=1}^{\infty} x_n^{\mu_n^s}|_{s=0}. \tag{15}$$

In this definition, $x_1, x_2, \ldots$ need not be the coordinates of an element of $H^-(finite)$, etc.. It is known [2]

Theorem 2. *The followings are hold.*

1. $: \prod_n x_n :$ *is linear in each variable.*
2. $: \prod_n x_n :$ *is positive if all x_n are positive.*
3. $: \prod_n x_n :$ *is a real number if all x_n are real and ν is an integer.*
4. $| : \prod_n x_n : |$ *is equal to* $: \prod_n |x_n| :.$

Note. Let x be $(x_1, x_2, \ldots)$ and T_x be the linear operator defined by $T_x e_n = x_n e_n$, $S_x = \log T_x : T_x = \exp S_x$. Then we get

$$: \prod_{n=1}^{\infty} x_n := e^{tr(\zeta(G,s)S_x)}|_{s=0}. \tag{16}$$

77

Hence regularized infinite product is related to Paycha's ζ-regularized trace [11].

By Theorem 2, we may regard

$$: \prod_{n \notin \{i_1, \dots, i_m\}} x_n := \frac{\partial^m}{\partial x_{i_1} \cdots \partial x_{i_m}} : \prod_{n=1}^{\infty} x_n : . \tag{17}$$

But in ordinary sense, we can not calculate $\lim_{N \to \infty} \dfrac{\partial^N}{\partial x_1 \cdots \partial x_N} : \prod_n x_n :$. To calculate this limit, we used fractional calculus [2].

Fractional integral of order α of a function on $\mathbb{R}_+$, the space of positive real numbers, is defined by

$$I_0^{\alpha} f(x) = \frac{1}{\Gamma(\alpha)} \int_0^x (x-t)^{\alpha-1} f(t) dt. \tag{18}$$

Fractional derivative $\dfrac{d^{\alpha}}{dx^{\alpha}} f$ of order α is defined by $I_0^{1-\alpha} f'(x)$. There are several alternative definitions of fractional order derivations and integrals started from 0. But in any case, we have

$$\frac{d^{\alpha}}{dx^{\alpha}} x^n = \frac{n!}{\Gamma(n-\alpha+1)} x^{n-\alpha}.$$

By using fractional derivation, we define regularized infinite order derivation $\dfrac{\partial^{\infty}}{\prod_n \partial x_n}$ and regularized infinite dimensional integral on $Q(x,+)$, $x \in W^{k-0}(finite)$; $\displaystyle\int_{Q(x,+)} f(x) : d^{\infty} x :$ by

$$\frac{\partial^{\infty}}{\prod_{n=1}^{\infty} \partial x_n} f = \prod_{n=1}^{\infty} \frac{1}{\Gamma(1+\mu_n^s)} \frac{\partial^{\mu_n^s}}{\partial x_n^{\mu_n^s}} f|_{s=0}. \tag{19}$$

$$\int_{Q(x,+)} f(x) : d^{\infty} x : = \prod_{n=0}^{\infty} \Gamma(1+\mu_n^s) I_0^{\mu_n^s} f(x)|_{s=0}. \tag{20}$$

Then we have [2, 3]

$$\frac{\partial^{\infty}}{\prod_{n=1}^{\infty} \partial x_n} : \prod_{n=1}^{\infty} x_n : = 1. \tag{21}$$

$$\int_{Q(x,+)} 1 : d^{\infty} x : = : \prod_{n=1}^{\infty} x_n : . \tag{22}$$

These calculi are called regularized calculus [3]. In [3], they were used to justify appearance of the Ray-Singer determinant in the calculus of Gaussian path integral. In this paper, we give an alternative justification which is same in spirit with the regularized calculus.

5. WEAK DERIVATIVE OF REGULARIZED INFINITE PRODUCT

In this section, we show the equality

$$\lim_{N \to \infty} \frac{\partial^N}{\partial x_1 \cdots \partial x_N} : \prod_{n=1}^{\infty} x_n := 1, \tag{23}$$

in some weak senses. I thank Prof. Léandre who suggested to treat regularized calculus from the point of view of distribution [9].

In this section, we assume G is the Green operator of an elliptic operator on a compact manifold.

Example 1. Since we have

$$\int_0^{\infty} \cdots \int_0^{\infty} \left(\frac{\partial^N}{\partial x_1 \cdots \partial x_N} x_1^{\mu_1^s} \cdots x_N^{\mu_N^s} \right) e^{-x_1 - \cdots - x_N} dx_1 \ldots dx_N$$

$$= \int_0^{\infty} \cdots \int_0^{\infty} x_1^{\mu_1^s} \cdots x_N^{\mu_N^s} e^{-x_1 - \cdots - x_N} dx_1 \cdots dx_N$$

$$= \Gamma(1 + \mu_1^s) \cdots \Gamma(1 + \mu_N^s),$$

we obtain

$$\lim_{N \to \infty} \int_0^{\infty} \cdots \int_0^{\infty} \left(\frac{\partial^N}{\partial x_1 \cdots \partial x_N} x_1^{\mu_1^s} \cdots x_N^{\mu_N^s} \right) \times$$

$$\times e^{-x_1 - \cdots - x_N} dx_1 \cdots dx_N = \prod_{n=1}^{\infty} \Gamma(1 + \mu_n^s).$$

Since $\log \Gamma(1 + x) = -\gamma x + \sum_{m \geq 2} (-1)^m (\zeta(m)/m) x^m$, $\zeta(m) = \sum_n n^{-m}$, where γ is the Euler constant [6], we have

$$\log \left(\prod_{n=1}^{\infty} \Gamma(1 + \mu_n^s) \right) = -\gamma \zeta(G, s) + \sum_{m=2}^{\infty} (-1)^m \frac{\zeta(m)}{m} \zeta(G, ms). \tag{24}$$

Therefore $\prod_n \Gamma(1 + \mu_n^s)$ allows analytic continuation to the right half plane $\{z \in \mathbb{C} | \Re z > 0\}$ with singularities on the real axis, which accumulate to 0. Since $\Gamma(2) = 1$, if $\gamma(t)$ is a piecewise smooth curve in the right half plane such that $\gamma(0) = 0$, does not meet the singularities of $\prod_n \Gamma(1 + \mu_n^s)$, and not tangent to the real and imaginary axes, then by the Abel's continuity Theorem, we have

$$\lim_{t \to 0} \prod_{n=1}^{\infty} \Gamma(1 + \mu_n^s) = 1, \quad s = \gamma(t).$$

Let $C_b^1(\mathbb{R}_+^n)$ be the space of bounded C^1-class functions on $\mathbb{R}_+^n = \{x \in \mathbb{R}^n | x \geq 0\}$. If $f \in C_b^1(\mathbb{R}_+^n) \cap L^1(\mathbb{R}_+^n)$, we have

$$\lim_{s \to 0} \int_0^\infty \cdots \int_0^\infty \left(\frac{\partial^n}{\partial x_1 \cdots \partial x_n} x_1^{\mu_1^s} \cdots x_n^{\mu_n^s}\right) \times$$
$$\times f(x_1, \ldots, x_n) dx_1 \cdots dx_n = \int_0^\infty \cdots \int_0^\infty f(x_1, \ldots, x_n) dx_1 \cdots dx_n.$$

We denote the space generated by $f \exp(-\sum_{m>n} x_m)$, $f \in C_b^1(\mathbb{R}_+^n) \cap L^1(\mathbb{R}_+^n)$, $n = 1, 2, \ldots$, by $L_b^{1,1}(\mathbb{R}_+^\infty)$. Taking completion, this space is considered to be a Sobolev 1-class L^1-type space. Then above discussions show (23) is hold as an element of the dual space of $L_b^{1,1}(\mathbb{R}_+^\infty)$.

Note. $\exp(-\sum_n x_n)$ is defined if $\sum_n |x_n| < \infty$. So if $H = L^2(X)$, this function is not defined on $H = l^2(\{e_1, e_2, \ldots\})$, but defined on $l^1(\{e_1, e_2, \ldots\})$.

Example 2. Since we have

$$\int_{-\infty}^\infty \left(\frac{d}{dx} |x|^{\mu^s}\right) e^{-\pi x^2} dx = \pi^{-\mu^s/2} \Gamma(1 + \mu^s/2),$$

we obtain

$$\lim_{N \to \infty} \int_{-\infty}^\infty \cdots \int_{-\infty}^\infty \left(\frac{\partial^N}{\partial x_1 \cdots \partial x_N} x_1^{\mu_1^s} \cdots x_N^{\mu_N^s}\right) \times$$
$$\times e^{-\pi(x_1^2 + \cdots + x_N^2)} dx_1 \ldots dx_N = \prod_{n=1}^\infty \frac{\Gamma(1 + \mu_n^s/2)}{\pi^{\mu_n^s/2}}.$$

Since

$$\log\left(\prod_{n=1}^\infty \frac{\Gamma(1 + \mu_n^s/2)}{\pi^{\mu_n^s/2}}\right)$$
$$= -\frac{\log \pi + \gamma}{2} \zeta(G, s) + \sum_{m=2}^\infty (-1)^m \frac{\zeta(m)}{m 2^m} \zeta(G, ms),$$

$\prod_n(\Gamma(1 + \mu_n^s/2) \pi^{-\mu_n^s/2})$ allows analytic continuation on the right half plane with singularities on the real axis which accumulate to 0. Hence taking piecewise smooth $\gamma(t)$, $\gamma(0) = 0$, similar to Example 1, we have

$$\lim_{t \to 0} \prod_{n=1}^\infty \frac{\Gamma(1 + \mu_n^s/2)}{\pi^{\mu_n^s/2}} = 1, \quad s = \gamma(t).$$

Hence to define $L_b^{2,1}(\mathbb{R}^\infty)$ to be the Sobolev 1-type L^2-space generated by $f \exp(-\pi(\sum_{m>n} x_m^2))$, $f \in C_b^1(\mathbb{R}^n) \cap L^2(\mathbb{R}^n)$, $n = 1, 2, \ldots$,

$$\lim_{N \to \infty} \frac{\partial^N}{\partial x_1 \cdots \partial x_N} : \prod_{n=1}^\infty |x_n| := 1, \tag{25}$$

is hold as the element of the dual space of $L_b^{2,1}(\mathbb{R}^\infty)$. Here $C_b^1(\mathbb{R}^n)$ means the space of C^1-class bounded functions on $\mathbb{R}^n$. Note that $\exp(-\pi(x_1^2 + x_2^2 + \cdots))$ is defined on H.

Note. Let c_p be selected to $\displaystyle\int_{-\infty}^{\infty} e^{-c_p|x|^p} dx = 1$. By using $\exp(-c_p \sum_n |x_n|^p)$, we can construct an L^p-type function space on $\mathbb{R}^\infty$ on which (25) is hold as the element of the dual space of this space.

6. REGULARIZED JACOBIAN OF A SCALING TRANSFORMATION

Let T_a: $a = (a_1, a_2, \ldots)$ be a nondegenerate scaling transformation

$$T_a e_n = a_n e_n, \quad a_n \neq 0, \ n = 1, 2, \ldots.$$

Note. T_a need not be defined on H. For example, if $a_n = \mu_n^{-1} = \lambda_n$, then regarding W^1 to be a subspace of H, T_a is defined on W^1, not on H.

For a function f on H, $T_a^* f$ means $f(T_a x)$. We assume f is C^1-class and

$$\lim_{n \to \infty} \int_{-\infty}^{\infty} f \, dx_1 \cdots dx_n \quad exits. \tag{26}$$

Since

$$\int_{-\infty}^{\infty} cx^{c-1} f(ax) dx = \int_{-\infty}^{\infty} |a|^{-c} cy^{c-1} f(y) dy, \quad y = ax,$$

we obtain

$$\lim_{n \to \infty} \int_{-\infty}^{\infty} \cdots \int_{-\infty}^{\infty} \left(\frac{\partial^n}{\partial x_1 \cdots \partial x_n} |x_1|^{\mu_1^s} \cdots |x_n|^{\mu_n^s} \right) T_a^* f \, dx_1 \cdots dx_n$$

$$= \lim_{n \to \infty} \int_{-\infty}^{\infty} \cdots \int_{-\infty}^{\infty} |a_1|^{-\mu_1^s} \cdots |a_n|^{-\mu_n^s} \times$$

$$\times \left(\frac{\partial^n}{\partial y_1 \cdots \partial y_n} |y_1|^{\mu_1^s} \cdots |y_n|^{\mu_n^s} \right) f(y) dy_1 \cdots dy_n, \quad y_n = a_n x_n,$$

if $f \in L_b^{2,1}(\mathbb{R}^\infty)$. Hence we obtain

Theorem 3. *If $f \in L_b^{2,1}(\mathbb{R}^\infty)$ and $: \prod_n a_n :$ exists, then*

$$\lim_{n \to \infty} \int_{-\infty}^{\infty} \cdots \int_{-\infty}^{\infty} \left(\frac{\partial^n}{\partial x_1 \cdots \partial x_n} |x_1|^{\mu_1^s} \cdots |x_n|^{\mu_n^s} \right) T_a^* f \, dx_1 \cdots dx_n|_{s=0}$$

$$= \lim_{n \to \infty} \int_{-\infty}^{\infty} \cdots \int_{-\infty}^{\infty} | : \prod_{n=1}^{\infty} a_n : |^{-1} f \, dx_1 \cdots dx_n. \tag{27}$$

Corollary. *Let G be the Green operator of D, a positive elliptic operator, then*

$$\lim_{n\to\infty}\int_{\mathbb{R}^n}\Big(\frac{\partial^n}{\partial x_1\cdots\partial x_n}|x_1|^{\mu_1^s}\cdots|x_n|^{\mu_n^s}\Big)e^{-\pi(x,Dx)}d^nx\big|_{s=0}=\frac{1}{\sqrt{\det.D}}. \tag{28}$$

Here, the analytic continuation is taken along a piecewise smooth path in the right half plane which is not tangent to real and imaginary axes.

We may claim (28) is a mathematical justification of the formula

$$\int_H e^{-\pi(x,Dx)}\mathscr{D}x=\frac{1}{\sqrt{\det.D}}.$$

Definition 4. *We denote*

$$\lim_{n\to\infty}\int_{\mathbb{R}^n}\Big(\frac{\partial^n}{\partial x_1^n\cdots\partial x_n}|x_1|^{\mu_1^s}\cdots|x_n|^{\mu_n^s}\Big)f(x)d^nx\big|_{s=0}=\int_{\mathbb{R}^\infty}f(x):d^\infty x:. \tag{29}$$

Note. (16) shows : $\prod_n a_n$: can be interpreted as the regularized determinant of T_a. Hence (27) gives a regularization of the Jacobian of the coordinate transformation T_a.

Similarly, if $a\gg 0$, $f\in L_b^{1,1}(\mathbb{R}_+^\infty)$ and : $\prod_n a_n$: exists, then we have

$$\lim_{N\to\infty}\int_0^\infty\cdots\int_0^\infty\Big(\frac{\partial^N}{\partial x_1\cdots\partial x_N}x_1^{\mu_1^s}\cdots x_N^{\mu_N^s}\Big)T_a^*f dx_1\cdots dx_N\big|_{s=0}$$
$$=\lim_{N\to\infty}\int_0^\infty\cdots\int_0^\infty(:\prod_{n=1}^\infty a_n:)^{-1}f dx_1\cdots dx_N. \tag{30}$$

By (30), we can use the notation $\int_{\mathbb{R}_+^\infty}f(x):d^\infty x:$, which is defined by the same way as (29).

Since we have $T_a(Q(c))=Q(a\cdot c)$, $T_a(Q(c,+))=Q(a\cdot c,+)$, if $a\gg 0$, $c\gg 0$ and $c\in H^-(finite)$, $a\cdot c\in H^-(finite)$, where $a\cdot c=(a_1c_1,a_2c_2,\ldots)$, the following formulae hold if the integrals are defined.

$$\lim_{N\to\infty}\int_{-c_1}^{c_1}\cdots\int_{-c_N}^{c_N}\Big(\frac{\partial^n}{\partial x_1\cdots\partial x_N}|x_1|^{\mu_1^s}\cdots|x_N|^{\mu_N^s}\Big)T_a^*f dx_1\cdots dx_N\big|_{s=0}$$
$$=\lim_{N\to\infty}\int_{-a_1c_1}^{a_1c_1}\cdots\int_{-a_Nc_N}^{a_Nc_N}(:\prod_{n=1}^\infty a_n:)^{-1}f dx_1\cdots dx_N, \tag{31}$$

$$\lim_{N\to\infty}\int_0^{c_1}\cdots\int_0^{c_N}\Big(\frac{\partial^N}{\partial x_1\cdots\partial x_N}x_1^{\mu_1^s}\cdots x_N^{\mu_N^s}\Big)T_a^*f dx_1\cdots dx_N\big|_{s=0}$$
$$=\lim_{N\to\infty}\int_0^{a_1c_1}\cdots\int_0^{a_Nc_N}(:\prod_{n=1}^\infty a_n:)^{-1}f dx_1\cdots dx_N. \tag{32}$$

82

We may regard these integrals to be the integrals on the domains $Q(c)$, $Q(a \cdot c)$, $Q(c, +)$ and $Q(a \cdot c, +)$, respectively. We denote these integrals by $\int_{Q(a \cdot c)} f(x) : d^\infty x :$ and $\int_{Q(a \cdot c, +)} f(x) : d^\infty x :$, respectively.

Note. The function $\exp(-(x, Dx))$ is defined on $W^{1/2}$. But (31) and (32) suggest the path integral should be taken on $W^{1/2-0}(finite)$.

REFERENCES

1. Asada,A.: Renarks on the zeta-regularized determinant of differential operators, Proc. Conf. Moshé Flato II, 25-36. Kluwer, 2000.
2. Asada,A.: Regularized product of infinitely many independent variables on a Hilbert space and regularization of infinite dimensional indefinite integral via fractional calculus, Review Bull. Cal. Math. Soc.,9(2001), 69-82.
3. Asada,A.: Regularized Calculus; An application of zeta-regularization to infinite dimensional geometry and analysis, Int. Journ. Geo. Meth. in Modern Phys. 1(2004), 107-157.
4. Atiyah,M.F. Patodi,V.K. Singer,I.M.: Spectral asymmetry and Riemannian geometry, I, II, III, Math. Proc. Cambridge Phil. Soc., 77(1975),43-69, 78(1975),405-432, 79(1976),71-99.
5. Connes,A.: Geometry from the spectral point of view, Lett. Math. Phys.,34(1995), 203-238.
6. Erdélyi,A. *et al.*: Higher Transcendental Functions, I. New York, 1953.
7. Gilkey,P.R.: The residue of the global η function at the origin, Adv. in Math., 40(1981), 290-307.
8. Hawking,S.W.: Zeta function regularization of path integrals in curved space time, Commun. Math. Phys.,55(1977), 133-148.
9. Léandre,R.: Theory of distribution in the sense of Connes-Hida and Feynman path integral on a manifold, Inf. Dim. Anal., Quant. Prob. and related Topics, 6(2003), 505-517.
10. Okikiolu,K.: The Campbell-Hausdorff theorem for elliptic operaators and a related trace formula, The multiplicative anomaly for determinants of elliptic operators, Duke Math. Journ.79(1995), 687-722, 723-750.
11. Paycha,S.: Renormalization trace as a looking glass into infinite dimensional geometry, Inf. Dim. Anal. Quant. Prob. and related Topics, 4(2001), 221-266.
12. Ray,D. Singer,I.M.: R-torsion and Laplacians on Riemannian manifolds, Adv. Math.7(1971), 145-210.
13. Seeley,R.T.: Complex powers of an elliptic operator, Proc. Symp. Pure Math. 10, 288-307, Providence, 1967.
14. Simon,B.: Trace Ideals and Their Applications, Cambridge, 1979.

Fractional Euler-Lagrange Equations for Constrained Systems

Tansel Avkar[1*] and Dumitru Baleanu[*†]

*Department of Mathematics and Computer Science, Faculty of Arts and Sciences, Cankaya University, 06530, Ankara, Turkey
†On leave of absence from Institute of Space Sciences, R 76900 Magurele-Bucharest, Romania, e-mails: baleanu@venus.nipne.ro, dumitru@cankaya.edu.tr

Abstract. The fractional calculus is the name for the theory of integrals and derivatives of arbitrary order, which generalize the notions of n-fold integration and integer-order differentiation.
Differential equations of fractional order appear in certain applied problems and in theoretical researches. In this paper, the Euler-Lagrange equations of the Lagrangians linear in velocities were derived using the fractional calculus. Two examples of constrained systems possessing a gauge invariance are investigated in details, the explicit solutions of Euler-Lagrange equations are obtained, and the recovery of the classical results is discussed.
Key words. Riemann-Liouville fractional derivative, constrained systems, fractional Euler-Lagrange equations.
PACS. 11.10.Ef.
MSC. 26A33, 70H45.

1. INTRODUCTION

The mathematical idea of fractional calculus, which goes back to the seventeenth century, has been investigated by many mathematicians for years [24, 25, 26]. The topic of fractional calculus has found applications in various fields of science, engineering and finance [3, 4, 8, 15, 21, 22, 24, 25, 26, 28, 31, 34, 35].

One of the problems encountered in this field is to find what kind of fractional derivative is to be used instead of the integer order derivative for a given problem [11, 14, 24, 25, 26]. Depending on the specific physical situation different authors have applied different derivatives [3, 15].

For the past two decades, the fractional generalization of the standard diffusion equation, which is gaining a great importance in economics, has been considered by several authors [10, 23, 27, 32]. Feynman and Hibbs used a Gaussian probability distribution in the space of all possible paths, for a quantum mechanical particle, to derive the Schrödinger equation [9]. Later, Laskin constructed space fractional quantum mechanics using Feynman's path integral approach, the difference being the use of Lévy distributions instead of Gaussian distributions for the set of possible paths [18, 19, 20].

Nonconservative Lagrangian and Hamiltonian mechanics were investigated by Riewe

---

[1] E-mail: avkar@cankaya.edu.tr

CP729, *Global Analysis and Applied Mathematics: International Workshop on Global Analysis*, edited by K. Taş, D. Krupka, O. Krupková, and D. Baleanu
© 2004 American Institute of Physics 0-7354-0209-4/04/$22.00

within fractional calculus [29, 30]. Besides, in Lagrangian and Hamiltonian fractional sequential mechanics, the models with symmetric fractional derivative were studied [16, 17].

Recently, extensions of the simplest fractional variational problem and of the Lagrange fractional variational problem have been obtained by Agrawal [1, 2]. A natural and interesting generalization of Agrawal's approach is to apply the fractional calculus to the constrained systems [6, 12, 13, 33].

2. RIEMANN - LIOUVILLE FRACTIONAL DERIVATIVES

In this section the definitions of the right and left RL derivatives as well as their basic properties are briefly presented.

RL fractional derivatives [1, 26] are defined as follows

$$
_a\mathbf{D}_t^\alpha f(t) = \frac{1}{\Gamma(n-\alpha)} \left(\frac{d}{dt} \right)^n \int_a^t (t-\tau)^{n-\alpha-1} f(\tau) d\tau , \tag{1}
$$

$$
_t\mathbf{D}_b^\alpha f(t) = \frac{1}{\Gamma(n-\alpha)} \left(-\frac{d}{dt} \right)^n \int_t^b (\tau-t)^{n-\alpha-1} f(\tau) d\tau, \tag{2}
$$

where the order α fulfills $n-1 \leq \alpha < n$ and Γ represents the Euler's gamma function. The first derivative is called the RL left fractional derivative and in (2) the expression of the RL right fractional is presented. If α is any integer, the relations to the usual derivative are obtained as follows

$$
_a\mathbf{D}_t^\alpha f(t) = \left(\frac{d}{dt} \right)^\alpha , \quad _t\mathbf{D}_b^\alpha f(t) = \left(-\frac{d}{dt} \right)^\alpha . \tag{3}
$$

Under the assumptions that $f(t)$ is continuous and $p \geq q \geq 0$, the most general property of RL fractional derivatives can be written as

$$
_a\mathbf{D}_t^p \left(_a\mathbf{D}_t^{-q} f(t) \right) = _a\mathbf{D}_t^{p-q} f(t). \tag{4}
$$

For $p > 0$ and $t > a$ we obtain

$$
_a\mathbf{D}_t^p \left(_a\mathbf{D}_t^{-p} f(t) \right) = f(t), \tag{5}
$$

which means that the RL fractional differentiation operator is a left inverse to the RL fractional integration operator of the same order. The relation (5) is called the fundamental property of the RL fractional derivative. In addition, the fractional derivative of a constant is not zero and the RL fractional derivative of the power function $(t-a)^v$ is given by

$$
_aD_t^p (t-a)^v = \frac{\Gamma(v+1)}{\Gamma(-p+v+1)} (t-a)^{v-p}, \tag{6}
$$

where $v > -1$. The normal derivatives $\dfrac{d^n}{dt^n}$ and $_aD_t^p$ commute only if $f^{(j)}(a) = 0$, $j = 0, 1, \ldots, n-1$ is fulfilled and two RL fractional derivative operators $_aD_t^p$ and $_aD_t^q$ commute only if

$$\left[_a\mathbf{D}_t^{p-j} f(t) \right]_{t=a} = 0, \qquad j = 1, \cdots, m \tag{7}$$

and

$$\left[_a\mathbf{D}_t^{q-j} f(t) \right]_{t=a} = 0, \qquad j = 1, \cdots, n. \tag{8}$$

3. FRACTIONAL EULER-LAGRANGE EQUATIONS FOR LAGRANGIANS WITH LINEAR VELOCITIES

Let $J[q^1, \ldots, q^n]$ be a functional of the form

$$\int_a^b L\left(t, q^1, \ldots, q^n, {}_a\mathbf{D}_t^\alpha q^1, \ldots, {}_a\mathbf{D}_t^\alpha q^n, {}_t\mathbf{D}_b^\beta q^1, \ldots, {}_t\mathbf{D}_b^\beta q^n \right) dt, \tag{9}$$

defined on the set of functions $q^i(t)$, $i = 1, \ldots, n$, which have continuous left RL fractional derivative of order α and right RL fractional derivative of order β in $[a, b]$ and satisfy the boundary conditions $q^i(a) = q_a^i$ and $q^i(b) = q_b^i$. In [1], a necessary condition for $J[q^1, \ldots, q^n]$ to have an extremum for given functions $q^i(t)$, $i = 1, \ldots, n$ was found to be Euler-Lagrange equations

$$\frac{\partial L}{\partial q^j} + {}_t\mathbf{D}_b^\alpha \frac{\partial L}{\partial {}_a\mathbf{D}_t^\alpha q^j} + {}_a\mathbf{D}_t^\beta \frac{\partial L}{\partial {}_t\mathbf{D}_b^\beta q^j} = 0, \qquad j = 1, \ldots, n. \tag{10}$$

Let us consider the following Lagrangian

$$L = a_j\left(q^i\right)\dot{q}^j - V\left(q^i\right), \tag{11}$$

where $a_j\left(q^i\right)$ and $V\left(q^i\right)$ are functions of their arguments.

The first step is to construct the corresponding fractional generalization of (11). The fractional Lagrangian is not unique. In other words, there are several possibilities to replace the time derivative with fractional derivatives. The requirement is to obtain the same Lagrangian expression if the order α is 1. Having in mind the above considerations, for $0 < \alpha \leq 1$, we propose two fractional Lagrangians. The first one is as follows

$$L' = a_j\left(q^i\right){}_a\mathbf{D}_t^\alpha q^j - V\left(q^i\right). \tag{12}$$

Using (10) and (12), the corresponding Euler-Lagrange equations emerge as

$$\frac{\partial L'}{\partial q^k} + {}_t\mathbf{D}_b^\alpha a_k\left(q^i\right) = 0, \tag{13}$$

or, explicitly,

$$\frac{\partial a_j\left(q^i\right)}{\partial q^k}\,_a\mathbf{D}_t^\alpha q^j - \frac{\partial V\left(q^i\right)}{\partial q^k} + {}_t\mathbf{D}_b^\alpha a_k\left(q^i\right) = 0 . \tag{14}$$

The second fractional Lagrangian is given by

$$L' = -a_j\left(q^i\right){}_t\mathbf{D}_b^\alpha q^j - V\left(q^i\right) . \tag{15}$$

Using (10) and (15), the corresponding Euler-Lagrange equations become [5]

$$\frac{\partial a_j\left(q^i\right)}{\partial q^k}\,_t\mathbf{D}_b^\alpha q^j + \frac{\partial V\left(q^i\right)}{\partial q^k} + {}_a\mathbf{D}_t^\alpha a_k\left(q^i\right) = 0 . \tag{16}$$

4. EXAMPLES

Example 1. Consider the following gauge invariant Lagrangian

$$L = \dot{q}^1 q^2 - \dot{q}^2 q^1 - (q^1 - q^2)q^3 . \tag{17}$$

We propose the corresponding fractional Lagrangian as

$$L' = -\left({}_t\mathbf{D}_b^\alpha q^1\right) q^2 + \left({}_t\mathbf{D}_b^\alpha q^2\right) q^1 - (q^1 - q^2)q^3 . \tag{18}$$

Using (16), the Euler-Lagrange equations corresponding to (18) have the form

$$\frac{\partial L'}{\partial q^3} = 0 , \tag{19}$$

$$\frac{\partial L'}{\partial q^1} + {}_a\mathbf{D}_t^\alpha \frac{\partial L'}{\partial \left({}_t\mathbf{D}_b^\alpha q^1\right)} = 0 , \tag{20}$$

$$\frac{\partial L'}{\partial q^2} + {}_a\mathbf{D}_t^\alpha \frac{\partial L'}{\partial \left({}_t\mathbf{D}_b^\alpha q^2\right)} = 0 . \tag{21}$$

After some algebraic manipulations, the above system becomes

$$q^1 = q^2 , \quad -q^3 + {}_t\mathbf{D}_b^\alpha q^2 - {}_a\mathbf{D}_t^\alpha q^2 = 0 , \quad q^3 - {}_t\mathbf{D}_b^\alpha q^1 + {}_a\mathbf{D}_t^\alpha q^1 = 0 . \tag{22}$$

The solution of (22) is given by

$$q^2 = q^1 , \tag{23}$$
$$q^3 = (-{}_a\mathbf{D}_t^\alpha + {}_t\mathbf{D}_b^\alpha)q^1 . \tag{24}$$

The classical solutions are recovered if $\alpha \to 1$.

Example 2. Let us consider the second Lagrangian [7], which is given by

$$L = \frac{1}{2}\left[\left(\dot{q}^2 - e^{q^1}\right)^2 + \left(\dot{q}^3 - e^{q^2}\right)^2\right] . \tag{25}$$

87

First, we propose the fractional generalization of (25) as follows

$$L = \frac{1}{2}\left[\left({}_a\mathbf{D}_t^\alpha q^2 - e^{q^1}\right)^2 + \left({}_a\mathbf{D}_t^\alpha q^3 - e^{q^2}\right)^2\right].$$
(26)

Using (14), the Euler-Lagrange equations corresponding to (26) are given as follows

$$-\left({}_a\mathbf{D}_t^\alpha q^2 - e^{q^1}\right)e^{q^1} = 0,$$
(27)

$${}_t\mathbf{D}_b^\alpha\left({}_a\mathbf{D}_t^\alpha q^3 - e^{q^2}\right) = 0,$$
(28)

$$-\left({}_a\mathbf{D}_t^\alpha q^3 - e^{q^2}\right)e^{q^2} + {}_t\mathbf{D}_b^\alpha\left({}_a\mathbf{D}_t^\alpha q^2 - e^{q^1}\right) = 0.$$
(29)

From (27), we have

$${}_a\mathbf{D}_t^\alpha q^2 = e^{q^1},$$
(30)

which yields the solution of q^2 as

$$q^2(t) = C_1(t-a)^{\alpha-1} + \frac{1}{\Gamma(\alpha)}\int_a^t (t-\tau)^{\alpha-1}e^{q^1(\tau)}d\tau.$$
(31)

Using (29) and (30), we obtain

$${}_a\mathbf{D}_t^\alpha q^3 = e^{q^2}.$$
(32)

Finally, the equation (32) together with (31) give the solution for q^3 as follows

$$q^3(t) = C_2(t-a)^{\alpha-1}$$
$$+ \frac{1}{\Gamma(\alpha)}\int_a^t (t-\tau)^{\alpha-1}\exp\left[C_1(\tau-a)^{\alpha-1} + \frac{1}{\Gamma(\alpha)}\int_a^\tau (\tau-\eta)^{\alpha-1}e^{q^1(\eta)}d\eta\right]d\tau.$$
(33)

Here C_1 and C_2 are arbitrary constants. If $\alpha \to 1$, $a \to 0$, $b \to 1$, then the standard solutions

$$q^1(t) = q^1(t),$$
(34)

$$q^2(t) = \tilde{c}_1 + \int e^{q^1(\tau)}d\tau,$$
(35)

$$q^3(t) = \tilde{c}_2 + \int e^{\left[\tilde{c}_1 + \int e^{q^1(\eta)}d\eta\right]}d\tau$$
(36)

are recovered.

Secondly, we propose the fractional generalization of (25) as

$$L = \frac{1}{2}\left[\left({}_t\mathbf{D}_b^\alpha q^2 + e^{q^1}\right)^2 + \left({}_t\mathbf{D}_b^\alpha q^3 + e^{q^2}\right)^2\right].$$
(37)

Using (16), the Euler-Lagrange equations of (37) are given by

$$\left({}_t\mathbf{D}_b^\alpha q^2 + e^{q^1}\right)e^{q^1} = 0 , \tag{38}$$

$$_a\mathbf{D}_t^\alpha\left({}_t\mathbf{D}_b^\alpha q^3 + e^{q^2}\right) = 0 , \tag{39}$$

$$\left({}_t\mathbf{D}_b^\alpha q^3 + e^{q^2}\right)e^{q^2} + {}_a\mathbf{D}_t^\alpha\left({}_t\mathbf{D}_b^\alpha q^2 + e^{q^1}\right) = 0 . \tag{40}$$

From (38), the solution of q^2 has the form

$$q^2(t) = C_1'(b-t)^{\alpha-1} - \frac{1}{\Gamma(\alpha)}\int_t^b (\tau-t)^{\alpha-1}e^{q^1(\tau)}d\tau . \tag{41}$$

Then, using the equations (38), (39) and (40), we obtain

$$_t\mathbf{D}_b^\alpha q^3 = -e^{q^2} , \tag{42}$$

which gives the solution of q^3 as

$$\begin{aligned}
q^3(t) = {} & C_2'(b-t)^{\alpha-1} \\
& - \frac{1}{\Gamma(\alpha)}\int_t^b (\tau-t)^{\alpha-1}\exp\left[C_1'(b-\tau)^{\alpha-1} - \frac{1}{\Gamma(\alpha)}\int_\tau^b (\eta-\tau)^{\alpha-1}e^{q^1(\eta)}d\eta\right]d\tau,
\end{aligned} \tag{43}$$

where C_1' and C_2' are arbitrary constants. If $\alpha \to 1$, $a \to 0$, $b \to 1$, then the standard solutions are again recovered.

ACKNOWLEDGMENTS

One of the authors (D.B.) would like to thank O. Agrawal, M. Naber and J. A. Tenreiro Machado for important references and interesting discussions.

REFERENCES

1. Agrawal O. P., *Formulation of Euler-Lagrange equations for fractional variational problems*, J. Math. Anal. Appl., vol. **272**, (2002), 368-379.
2. Agrawal O. P., *Lagrangian and Lagrange equation of motion for fractionally damped systems*, Transactions of the ASME, J. Appl. Mech., vol. **68**, no. 2, (2001), 339-341.
3. Agrawal O. P., *Solution for a Fractional Diffusion-Wave Equation Defined in a Bounded Domain*, Nonlinear Dynamics, vol. **29**, (2002), 145-155.
4. Anh V. V., and Nguyen C. N., *Semimartingale representation of fractional Riesz-Bassel motion*, Finance Stoch., vol. **5**, no. 1, (2001), 83-101.

5. Baleanu D., and Avkar T., *Lagrangians with linear velocities within Riemann-Liouville fractional derivatives*, accepted for publication in Il Nuovo Cimento B, arXiv:math-ph/0405012.
6. Baleanu D., and Güler Y., *A general treatment of singular Lagrangians with linear velocities*, Nuovo Cimento B, vol. **115**, no. 3, (2000), 319-324.
7. Banerjee R., *The commutativity principle and lagrangian symmetries*, arXiv:hep-th/0001087.
8. Engheta N., *On fractional paradigm and intermediate zones in electromagnetism: I - Planar observation*, Microw. Opt. Techn. Let., vol. **22**, no. 4, (1999), 236-241.
9. Feynman R. P., and Hibbs A. R., *Quantum Mechanics and Path Integrals*, McGrawHill, Inc., (1965).
10. Fujita Y., *Integrodifferential equation which interpolates the heat equation and the wave equation*, Osaka J. Math., vol. **27**, (1990), 309-321.
11. Gorenflo R., and Mainardi F., *Fractional calculus: Integral and differential equations of fractional order*, in A. Carpinteri and F. Mainardi (Eds.), Fractals and Fractional Calculus in Continuum Mechanics, Springer-Verlag, Wien, (1997).
12. Hanson A., Reggȩ T., and Teitelboim C., *Constrained Hamiltonian Systems*, Academia Nationale dei Lincei, Rome, (1976).
13. Henneaux M., and Teitelboim C., *Quantization of Gauge Systems*, Princeton University Press, (1992).
14. Hilfer R., *Experimental evidence for fractional time evolution in glass forming materials*, Chem. Phys., vol. **284**, no. (1-2), (2002), 399-408.
15. Hilfer R. (Ed.), *Applications of Fractional Calculus in Physics*, Word Scientific Publishing Company, Singapore, New Jersey, London and Hong Kong, (2000).
16. Klimek M., *Fractional sequaential mechanics - models with symmetric fractional derivative*, Czech. Journ. Phys., vol. **51**, no. 12, (2001), 1348-1354.
17. Klimek M., *Lagrangian and Hamiltonian fractional sequential mechanics*, Czech. Journ. Phys., vol. **52**, no. 11, (2002), 1247-1253.
18. Laskin N., *Fractional quantum mechanics*, Phys. Rev. E, vol. **62**, **3**, (2000), 3135-3145.
19. Laskin N., *Fractional quantum mechanics and Lévy path integrals*, Phys. Lett. A, vol. **268**, (2000), 298-305.
20. Laskin N., *Fractional Schrödinger equation*, Phys. Rev. E, vol. **66**, art. no. 056108, (2002), 1-7.
21. Mainardi F., *Fractional calculus: Some basic problems in continuum and statistical mechanics*, in A. Carpinteri and F. Mainardi (Eds.), Fractals and Fractional Calculus in Continuum Mechanics, Springer-Verlag, Wien, (1997).
22. Mainardi F., and Pagnini G., *The Wright functions as solutions of the time-fractional diffusion equation*, Appl. Math. and Comp., vol. **141**, no. 1, (2003), 51-62.
23. Mainardi F., and Tomirotti M., *On a special function arising in the time fractional diffusion-wave equation*, in P. Rusev, I. Dimovski and V. Kiryakova (Eds.), Transform Methods and Special Functions, Sofia, (1994), Science Culture Technology, Singapore, (1995).
24. Miller K. S., and Ross B., *An introduction to the Fractional Integrals and Derivatives-Theory and Applications*, Gordon and Breach, Longhorne, PA, (1993).
25. Oldham K. B., and Spanier J., *The Fractional Calculus*, Academic Press, New York, (1974).
26. Podlubny I., *Fractional Differential Equations*, Academic Press, London, (1999).
27. Prüss J., *Evolutionary Integral Equations and Applications*, Birkhäuser, Basel, (1993).
28. Rida S. Z., and AMA El-Sayed, *Fractional calculus and generalized Rodrigues formula*, Appl. Math. and Comp., vol. **147**, no. 1, (2004), 29-43.
29. Riewe F., *Mechanics with fractional derivatives*, Phys. Rev. E, vol. **55**, (1997), 3582-3592.
30. Riewe F., *Nonconservative Lagrangian and Hamiltonian mechanics*, Phys. Rev. E, vol. **53**, (1996), 1890-1899.
31. Rosu H. C., Madueno L., and Socorro J., *Transform of Riccati equation of constant coefficients through fractional procedure*, J. Phys. A, vol. **36**, (2003), 1087-1094.
32. Schneider W. R., and Wyss W., *Fractional diffusion and wave equations*, J. Math. Phys., vol. **30**, (1989), 134-144.
33. Sundermeyer K., *Constrained Dynamics*, Lecture Notes in Physics, vol. **169**, Springer-Verlag, New-York, (1982).
34. Tenreiro Machado J. A., *A probabilistic interpretation of the fractional-order differentiation*, Fract. Calc. Appl. Anal., vol. **6**, no. 1, (2003), 73-80.
35. Tenreiro Machado J. A., *Discrete-time fractional-order controllers*, Fract. Calc. Appl. Anal., vol. **4**, no. 1, (2001), 47-66.

Variational problems in the geometrized first-order jet framework

Vladimir Balan

University Politehnica of Bucharest, Department Mathematics I
Splaiul Independenței 313, RO-060042 Bucharest, Romania
E-mail: vbalan@mathem.pub.ro

Abstract. In the framework of geometrized jet spaces of first order endowed with a Lagrange structure, is discussed the existence of Lagrangian canonic nonlinear connections. The Euler-Lagrange equations for certain Kronecker-type Lagrangian cases are derived, and extensions of the known results are provided.
Key words. Lagrangian, jet space, nonlinear connection, Finsler structure.
PACS. 02.30.Xx, 02.30.Jr, 02.40.Pc.
MSC. 58A20, 49J10, 49K10, 53B50.

1. BASIC STRUCTURES ON $J^1(T,M)$

The concepts of non-holonomic (or iterated) and semi-holonomic jets, first introduced by Ehresmann ([12, 13]), are commonly used in differential geometry; on the other side, variational problems and Lagrangian structures in fibered manifolds have been intensively studied ([14]-[21], [35]). During the last decades, significant efforts were devoted to the study of certain *basic geometric structures* (Riemannian, Finsler, Lagrange and Generalized Lagrange) on the tangent bundle $\xi_T = (TM, \pi, M)$ of a real differentiable manifold M endowed with a nonlinear connection (see e.g., [25, 26, 27, 11, 10]). The relations among these structures can be briefly summarized in brief as

$$\{\mathcal{R}^n\} \subset \{\mathcal{F}^n\} \subset \{\mathcal{L}^n\} \subset \{\mathcal{GL}^n\}. \tag{1}$$

The tangent bundle approach was further extended to the so-called "osculating" spaces of higher order $\xi_{O,k} = (Osc^{(k)}(M) = J^k(\mathbb{R}, M), \pi_k, M)$, $k \geq 1$, the tangent case being recollected as $\xi_T = \xi_{O,1}$ (Acad. R.Miron and collaborators [24, 25, 22]). The structures (1) provide naturally a framework (called the distinguished, or $d-$ geometrized approach), in which the main geometric object, is a nonlinear connection in the corresponding bundle, provided by the basic structure. This permits to consider and study the distinguished $(d-)$tensor fields (see Def.1.1).

As well, in parallel there was derived a dual extension of the tangent case, modelled on the bundles $\xi_O^* = (T^{*k}M \equiv T^{k-1}M \times T^*M, \pi_k^*, M)$ (R.Miron, [28, 23]).

CP729, *Global Analysis and Applied Mathematics: International Workshop on Global Analysis,*
edited by K. Taş, D. Krupka, O. Krupková, and D. Baleanu
© 2004 American Institute of Physics 0-7354-0209-4/04/$22.00

The present framework (named further $d-$jet framework) is developed on the fibration of jets $\xi = (J^1(T,M), \pi, T \times M)$, (dim $T = m \geq 1$) and has its origins in the works of G.S.Asanov and S.F.Ponomarenko ([1], [2], [3], [4]), being further developed by M.Neagu ([30]-[33]) and by other authors ([5]-[8],[7]-[8], [9]). The $d-$jet framework provides an alternative to the osculator extensions $\xi_{O,k}$, while still generalizing the structures (1) of the tangent case ξ_T, which is recaptured as being the $m = 1$ autonomous flat subcase of ξ.

Consider $\xi = (E = J^1(T,M), \pi, T \times M)$, the first order jet bundle of mappings $\varphi : T \to M$, where T and M are $\mathcal{C}^\infty$ real differentiable manifolds with dim $T = m$, dim $M = n$ respectively. The local coordinates in E will be denoted by

$$(t,x,y) = (t^\alpha, x^i, y^A)_{(\alpha,i,A) \in I_*} \equiv (y^\mu)_{\mu \in I},$$

where we consider the sets of indices

$$I = I_h \cup I_v, \ I_h = I_{h_1} \cup I_{h_2}, \ I_v = \overline{m+n+1, m+n+mn}$$
$$I_{h_1} = \overline{1,m}, \quad I_{h_2} = \overline{m+1, m+n}, \ I_* = I_{h_1} \times I_{h_2} \times I_v.$$

Throughout the paper, the indices will implicitly take values as follows:

$$\alpha, \beta, \ldots \in I_{h_1}; \ i, j, \ldots \in I_{h_2}; \ A, B, \ldots \in I_v; \ \lambda, \mu, \ldots \in I.$$

When appropriate, for any index $A = m+n+n(i-m-1)+\alpha$, we shall identify $A \equiv \binom{i}{\alpha}$ and $y^A \equiv x^{\binom{i}{\alpha}} = \frac{\partial x^i}{\partial t^\alpha}$.

Definition 2.1. A tensor field on E is said to be distinguished tensor field (briefly, $d-$tensor field), if its components change linearly in terms of the Jacobian matrices $[J(t'(t))]$ and $[J(x'(x))]$, as consequence of a coordinate change on the total space $E = J^1(T,M)$ (for details, see [32, 33]).

We can introduce on $J^1(T,M)$ in a natural way the similar to (1) geometric structures, as follows.

Definition 2.2. a) A *generalized metric* on ξ is a (0,2) nondegenerate symmetric vertical $d-$t.f. $\tilde{g}$ of constant signature. The pair $\mathcal{JGL}^n = (\xi, \tilde{g})$ is called *jet-generalized Lagrange space*.

b) A *Lagrangian* on ξ is a smooth mapping $L : E \to \mathbb{R}$ such that the vertical $d-$t.f.

$$\tilde{g}_{AB} = 2^{-1} \dot{\partial}^2_{AB} L \tag{2}$$

is everywhere non-degenerate and of constant signature. The pair $\mathcal{JL}^n = (\xi, L)$ is called *jet-Lagrange space*.

c) A *Finsler function* (Finsler "norm", or "metric") on ξ is a mapping $F : E \to \mathbb{R}$ which obeys the following properties:
i) F is continuous on E and C^∞ outside the null section of ξ;
ii) $F(t,x,y) > 0$ for $y \neq 0$;
iii) F is positively homogeneous of first order on the fibers of ξ, i.e.,

$$F(t,x,\lambda y) = |\lambda| F(t,x,y), \forall \lambda \in \mathbb{R};$$

iv) The (0,2) vertical $d-$t.f. locally given by $\tilde{g}_{AB} = 2^{-1} \dot{\partial}^2_{AB} F^2$ is everywhere non-degenerate and of constant signature. The pair $\mathcal{JF}^n = (\xi, F)$ is called *jet-Finsler space*.

d) A *jet-(pseudo)Riemannian space* $\mathcal{JR}^n = (\xi, \tilde{g})$ is a jet-Generalized Lagrange space with $\tilde{g}$ vertical (pseudo)Riemannian metric dependent only on the coordinates of $T \times M$.

Remarks. The following obvious inclusions take place

$$\{\mathcal{JR}^n\} \subset \{\mathcal{JF}^n\} \subset \{\mathcal{JL}^n\} \subset \{\mathcal{JGL}^n\}. \tag{3}$$

We note that the first studies on the structures $\mathcal{JGL}^n$ and $\mathcal{JL}^n$ were performed by Neagu ([32],[34]).

2. LAGRANGIANS AND NONLINEAR CONNECTIONS ON $\mathcal{JL}^N$

We consider on ξ the Lagrangian of the form

$$L(t,x,y) = \tilde{g}_{AB}(t,x,y)y^A y^B + b_A(t,x,y)y^A + c(t,x,y), \tag{4}$$

with $\tilde{g}_{AB}$ nondegenerate tensor field, $b_A(t,x,y)$ a 1-form on E and $c(t,x,y)$ a scalar function on E. This Lagrangian represents a natural extension of the $J^1(T,M)-$Lagrangian of the model for relativity and electromagnetism considered in ([31]).

The associated Euler-Lagrange equations for L produce a spray which only under certain restrictive conditions, provides further a *non-linear connection* $N = \{N_\mu^A\}_{\mu \in I_h, A \in I_v}$ on E which produces a corresponding splitting $TE = HE \oplus VE$, where $VE = Ker\ \pi_*$ ([33, 27]). As well, N determines the local adapted basis of $\mathfrak{X}(E)$

$$\mathcal{B} = \{\delta_\alpha, \delta_i, \delta_A\}_{(\alpha,i,A)\in I_*} \equiv \{\delta_\mu\}_{\mu \in I}, \tag{5}$$

where we denote briefly $\partial_\alpha = \frac{\partial}{\partial t^\alpha}, \partial_i = \frac{\partial}{\partial x^i}$ and

$$\delta_\alpha = \partial_\alpha - N_\alpha^A \delta_A, \quad \delta_i = \partial_i - N_i^A \delta_A, \quad \delta_A = \dot{\partial}_A = \frac{\partial}{\partial y^A}. \tag{6}$$

The dual basis of $\mathcal{B}$ in (5) is $\mathcal{B}^* = \{\delta^\alpha, \delta^i, \delta^A\}_{(\alpha,i,A)\in I_*} \equiv \{\delta^\mu\}_{\mu \in I}$, where

$$\delta^\alpha = dt^\alpha, \quad \delta^i = dx^i, \quad \delta^A \equiv \delta y^A = dy^A + N_\alpha^A dt^\alpha + N_i^A dx^i. \tag{7}$$

A nontrivial open problem is related to the existence of Lagrangian-produced non-linear connections in the general Kronecker case ([32]). To this goal, we state the following result.

Theorem 2.1. *Assuming that $h_{\alpha\beta}$ and g_{ij} are non-degenerate tensor fields, the Euler-Lagrange equations for the Kronecker case*

$$\tilde{g}_{AB} \equiv \tilde{g}_{\binom{i}{\alpha}\binom{j}{\beta}} = h^{\alpha\beta}(t,x,y)g_{ij}(t,x,y), \tag{8}$$

are

$$E^i(L) \equiv h^{\alpha\beta} y^{\binom{i}{\alpha\beta}} + 2G^i + 2E^i_* + 2E^i_{**} = 0, \tag{9}$$

where

$$2G^i = \left|{}^{\;i}_{jk}\right| y^{\binom{i}{\alpha}} y^{\binom{j}{\beta}} + \left|{}^{\;\alpha}_{\alpha\beta}\right| h^{\beta\gamma} y^{\binom{i}{\gamma}} +$$

$$+ \tfrac{1}{2} g^{ik} (b_{\binom{k}{\gamma},j} y^{\binom{j}{\gamma}} - c_{,k}) + \tfrac{1}{2} g^{ik} (2h^{\alpha\beta} g_{kj,\alpha} y^{\binom{j}{\beta}} + b_{\binom{k}{\alpha}} \left|{}^{\;\gamma}_{\gamma\alpha}\right|),$$

$$2E^i_* = -\tfrac{1}{2} g^{ik} (h^{\alpha\beta}_{\;\;,k} Y_{\alpha\beta} + b_{A,k} y^A) - \tfrac{1}{4} g^{ik} H_k (Y + b_A y^A + c) +$$

$$+ \tfrac{1}{2} g^{ik} \left\{ [g_{lm,\binom{k}{\gamma}} h^{\alpha\beta} - g_{lm} h^{\alpha\rho} h^{\beta\sigma} h_{\sigma\rho,\binom{k}{\gamma}}] y^{\binom{l}{\alpha}} y^{\binom{m}{\beta}} \right.$$

$$+ b_{\binom{k}{\gamma},\gamma} + b_{\binom{k}{\gamma},\binom{m}{\sigma}} y^{\binom{m}{\sigma\gamma}} + b_{B,\binom{k}{\gamma}} y^B + c_{,\binom{k}{\gamma}} \left. \right\}_{\perp\gamma} +$$

$$+ g^{ik} (g_{jk,\binom{m}{\sigma}} y^{\binom{m}{\sigma\gamma}} h^{\gamma\rho} y^{\binom{j}{\rho}} + g_{jk} y^{\binom{j}{\sigma}} h^{\gamma\sigma}_{\;\;\perp\gamma}) +$$

$$+ \tfrac{1}{4} g^{ik} H_\gamma (b_{A,\binom{k}{\gamma}} y^A + c_{,\binom{k}{\gamma}} + \tilde{g}_{AB,\binom{k}{\gamma}} y^A y^B),$$

$$2E^i_{**} = \tfrac{1}{2\sqrt{h}} g^{ik} (\sqrt{h})_{\perp\gamma} (2g_{\binom{k}{\gamma}B} y^B + g_{AB,\binom{k}{\gamma}} y^A y^B + R_{,\binom{k}{\gamma}}).$$

with $u_{,\gamma} = \partial_\gamma u$, $u_{\perp\gamma} = \partial_\gamma u + \partial_m u \cdot y^{\binom{m}{\gamma}} + \partial_{\binom{m}{\sigma}} u \cdot y^{\binom{m}{\gamma\sigma}}$, $y^{\binom{i}{\alpha\beta}} = \partial_\beta y^{\binom{i}{\alpha}}$, *and*

$$Y_{\alpha\beta} = g_{lm} y^{\binom{l}{\alpha}} y^{\binom{m}{\beta}}, Y = \tilde{g}_{AB} y^A y^B,$$

$$H_\mu = h^{\alpha\beta} h_{\alpha\beta,\mu}, R_A = b_{B,A} y^B + b_A + c_{,A}.$$

Proof. Tedious computation using the Hilbert-Palatini variational principle for the Lagrangian density $\mathcal{L} = L\sqrt{|h|}$ and the relations

$$\frac{\partial h^{\alpha\beta}}{\partial w} = -h^{\alpha\delta} h^{\beta\varepsilon} \partial_w h_{\delta\varepsilon}, \quad \frac{\partial \sqrt{|h|}}{\partial w} = \frac{\sqrt{|h|}}{2} h^{\alpha\beta} \partial_w h_{\alpha\beta}, \quad h^{\delta\varepsilon} \partial_\alpha h_{\delta\varepsilon} = 2 \left|{}^{\;\beta}_{\alpha\beta}\right|,$$

where $|h| = \det(h_{\alpha\beta})_{\alpha\beta=\overline{1,m}}$, provide the stated equations. ∎

Assuming that the fields b and c in the Lagrangian (4) do not depend on y, we further describe several notable cases regarding the derivation of the nonlinear connection from the equations provided above.

I. In the case when $h_{\alpha\beta}$ is a metric tensor on T and

$$\tilde{g}_{AB} \equiv \tilde{g}_{\binom{i}{\alpha}\binom{j}{\beta}} = h^{\alpha\beta}(t) g_{ij}(t,x,y),$$

the theorem leads to the known results derived in [33].

II. In the AFL (almost Finsler Lagrange) jet subcase, in which $g_{ij}(t,x,y)$ is $0-$homogeneous in y, leads in the $m=1$ autonomous case to the AFL case described in [29].

We emphasize the following three subcases of the AFL jet case.

II.1. The ARL (almost Riemann Lagrange) jet case, where $h_{\alpha\beta}$ is a metric tensor on T and

$$\tilde{g}_{AB} \equiv \tilde{g}_{\binom{i}{\alpha}\binom{j}{\beta}} = h^{\alpha\beta}(t)g_{ij}(t,x). \tag{10}$$

In this particular case we have $E^i_* = 0, 2E^i_{**} = -\left|\begin{matrix}\alpha\\\alpha\beta\end{matrix}\right| h^{\beta\gamma} y^{(i)}_{(\gamma)}$ in the Theorem 2.1, which permits to determine the spray and the nonlinear connection from [33], [32]. Namely, the Lagrangian L in (4) provides in this case the canonical nonlinear connection $N = (N^{(i)}_{\beta}{}^{(\alpha)}, N^{(i)}_{j}{}^{(\alpha)})$ with the $h-$coefficients $N^{(i)}_{j}{}^{(\alpha)} = h^{\alpha\gamma}\partial_{\binom{j}{\gamma}}G^i$, given by

$$N^{(i)}_{\beta}{}^{(\alpha)} = -\left|\begin{matrix}\gamma\\\alpha\beta\end{matrix}\right| y^{(i)}_{(\gamma)}, \quad N^{(i)}_{j}{}^{(\alpha)} = \left|\begin{matrix}i\\jk\end{matrix}\right| y^{(k)}_{(\alpha)} + \frac{1}{4}g^{ik}(2\partial_\alpha g_{jk} + h_{\alpha\beta}b_{\binom{k}{\beta}j}), \tag{11}$$

where we denoted the h_2-curl of b by $b_{\binom{k}{\beta}j} = \delta_j b_{\binom{k}{\beta}} - \delta_k b_{\binom{j}{\beta}}$. The Lagrangian L is in this case a Kronecker $h^{\alpha\beta} - h-$regular Lagrangian and produces the vertical metric tensor field $\tilde{g}_{AB}$ by

$$\tilde{g}_{AB} = 2^{-1}\dot{\partial}^2_{AB}L. \tag{12}$$

II.2. More particular, in the ARLS (almost Riemann Lagrange separated) jet case, the ARL case in which g_{ij} is a metric tensor field on M, the theorem leads to the spray and the non-linear connection known in literature (Neagu, [31]). In this case g, h and b_A determine the nonlinear connection $N = (N^{(i)}_{\beta}{}^{(\alpha)}, N^{(i)}_{j}{}^{(\alpha)})$ of coefficients

$$N^{(i)}_{\beta}{}^{(\alpha)} = -\left|\begin{matrix}\gamma\\\alpha\beta\end{matrix}\right| y^{(i)}_{(\gamma)}, \quad N^{(i)}_{j}{}^{(\alpha)} = \left|\begin{matrix}i\\jk\end{matrix}\right| y^{(k)}_{(\alpha)} + \frac{1}{4}g^{ik} \cdot h_{\alpha\beta}b_{\binom{k}{\beta}j}. \tag{13}$$

II.3. In the ALML (almost locally Minkowski Lagrange) jet case, where in a sub-atlas $\mathcal{A}$ of E one has

$$\tilde{g}_{AB} \equiv \tilde{g}_{\binom{i}{\alpha}\binom{j}{\beta}} = h^{\alpha\beta}(t,x)g_{ij}(y), \tag{14}$$

one derives that $T \times M$ is necessarily $\mathcal{A}-$locally affine. This extends the case $J^1(\mathbb{R},M) \equiv TM$ considered in [29].

We remark that for $b \equiv 0$ the derived nonlinear connection depends on h and g only, and hence denoting by h^* the dual of h, one easily concludes that for the $\mathcal{JGL}^n$ structure on $J^1(T,M)$ given by a vertical metric $\tilde{g} = h^*(t) \otimes g(t,x)$, such a connection always exists.

The complete characterization of the class of non-linear connections produced by certain geometric objects (metrics or Lagrangians) on $J^1(T,M)$ represents still an open problem (see [32, 31], [5]). However, there are certain cases where due to a specific form of the vertical metric on ξ, the problem is tractable. To this aim, the splittability of the Lagrangian-derived vertical metric field $\tilde{g}$ (the reducibility of L) plays a major role.

Definition 2.1 A Lagrangian L on E is called Kronecker-reducible, if the $d-$tensor field $\tilde{g}_{AB} \equiv 2^{-1}\dot{\partial}^2_{AB}L$ is non-degenerate and has the form

$$\tilde{g}_{\binom{i}{\alpha}\binom{j}{\beta}} = h^{\alpha\beta}(t,x,y)g_{ij}(t,x,y), \tag{15}$$

where h and g are $d-$tensor fields of constant signature on E.

Regarding the splittability, for $m,n \geq 2$ holds the known result

Theorem 2.2. (Neagu, [32]) *If L is a Kronecker reducible Lagrangian on E, and $\tilde{g}$ is provided by (12), then the following statements are equivalent:*
a) L is $h-$normal, i.e., $\tilde{g}$ is of the form

$$\tilde{g}_{\binom{i}{\alpha}\binom{j}{\beta}} = h^{\alpha\beta}(t)g_{ij}(t,x,y); \tag{16}$$

b) The field g_{ij} does not depend on y as well, and L is of the form (4).

A natural extension of this result is the following

Theorem 2.3. *If L is a Kronecker reducible Lagrangian on E and $\tilde{g}$ is provided by (12), then the following statements are equivalent:*
a) one of the two factors in (15) is independent on y;
b) both the fields $h_{\alpha\beta}$ and g_{ij} are independent on y, and L is of the form (4).

Proof. Remark first that the nontrivial implication a) $\Rightarrow$ b) in Theorem 2.2 can be proved easier in the following way: we apply $\dot{\partial}_C$ to the equality (12), and applying the Schwarz theorem, we infer

$$\dot{\partial}^3_{ABC}L = \dot{\partial}^3_{CAB}L \Leftrightarrow \dot{\partial}_C g_{AB} = \dot{\partial}_B g_{CA} \Leftrightarrow \dot{\partial}_{\binom{k}{\gamma}}g_{ij}\cdot h^{\alpha\beta} = \dot{\partial}_{\binom{j}{\beta}}g_{ik}\cdot h^{\alpha\gamma},$$

for $A \equiv \binom{i}{\alpha}, B \equiv \binom{j}{\beta}, C \equiv \binom{k}{\gamma}$. Transvection with $h_{\alpha\beta}$, interchange $j \leftrightarrow k$ and subtraction yield

$$(m-1)(\dot{\partial}_{\binom{k}{\gamma}}g_{ij} - \dot{\partial}_{\binom{j}{\gamma}}g_{jk}) = 0,$$

which for $m \geq 2$ infers $\dot{\partial}_{\binom{k}{\gamma}}g_{ij} = 0, \forall i,j,k \in I_{h_2}, \forall \alpha \in I_{h_1}$, and hence g is $y-$independent. As for the proof of the Theorem, b) $\Rightarrow$ a) is straightforward; for the converse let (15) hold true, with one of the independent on y; then $y-$derivation yields

$$\dot{\partial}^3_{\binom{i}{\alpha}\binom{j}{\beta}\binom{k}{\gamma}}L = \dot{\partial}\binom{k}{\gamma}g_{ij}\cdot h^{\alpha\beta} + g_{ij}\dot{\partial}_{\binom{k}{\gamma}}h^{\alpha\beta}; \tag{17}$$

using $\dfrac{\partial h^{\alpha\beta}}{\partial\Phi} = -h^{\alpha\sigma}h^{\beta\rho}\partial_\Phi h_{\sigma\rho}$ and the Schwarz relation we infer

$$h^{\alpha\beta}\dot{\partial}_{kiak\gamma}g_{ij} - g_{ij}h^{\alpha\sigma}h^{\beta\rho}\partial_{\binom{k}{\gamma}}h_{\sigma\rho} = h^{\alpha\gamma}\dot{\partial}_{kiaj\beta}g_{ik} - g_{ik}h^{\alpha\sigma}h^{\beta\rho}\partial_{\binom{j}{\beta}}h_{\sigma\rho}. \qquad (18)$$

But using g is independent on y, (18) becomes $g_{i[j}h^{\alpha\sigma}h^{\beta\rho}\dot{\partial}_{\binom{k]}{\gamma}}h_{\sigma\rho} = 0$, which leads via appropriate h−transvection to $g_{i[j}\dot{\partial}_{\binom{k]}{\gamma}}h^{\alpha\beta} = 0$. Then g^{ij} transvection lead to $(n-1)\partial_{\binom{k}{\gamma}}h^{\alpha\beta} = 0$, and for $n \geq 2$ this yields $\partial_{\binom{k}{\gamma}}h^{\alpha\beta} = 0, \forall k \in I_{h_2}, \forall \alpha,\beta,\gamma \in I_{h_1}$, whence h results y−independent.

Similarly, for h independent on y, (18) becomes $h^{\alpha\beta}\dot{\partial}_{\binom{k]}{\gamma}}g_{ij]} = 0$, which by a procedure described at the opening remark, leads to g independent on y. ∎

As well, in the case of jet-Finslerian structures $\mathcal{JF}^n$ where the considered Lagrangian is 2−homogeneous in y, we derive the following

Corollary 2.1. *Let L be 2-homogeneous in y Kronecker-splittable Lagrangian and h, g related to L via (12) and (15). Then the following relations hold true:*

a) $y^{\binom{k}{\tau}}g^{il}\dot{\partial}_{\binom{l}{\tau}}g_{ik} = y^{\binom{l}{\tau}}h^{\gamma\rho}\dot{\partial}_{\binom{l}{\tau}}h_{\tau\rho};$

b) $my^{\binom{j}{\gamma}}g^{il}\dot{\partial}_{\binom{l}{\gamma}}g_{ij} + ny^{\binom{j}{\gamma}}h^{\gamma\rho}\dot{\partial}_{\binom{j}{\tau}}h_{\tau\rho}.$

Proof. Transvection of (18) with $y^{\binom{j}{\gamma}}h_{\alpha\tau}g^{il}$ and homogeneity of L followed by transvection $\beta\#\tau, l\#j$ leads to a); transvection with $y^{\binom{j}{\gamma}}h_{\alpha\tau}g^{il}$ and $l\#k, \beta\#\tau$ yield b). ∎

Acknowledgments

The author thanks to the organizers of *IWGA 2004* for their hospitality and for the excellent conditions provided during the Conference. Special thanks go to Prof.Dr. D.Krupka for his constructive useful suggestions. The present work was partially supported by the Grant CNCSIS MEN 1478/2004.

REFERENCES

1. G.S.Asanov, *Finsler Geometry, Relativity and Gauge Geometry*, D.Reidel, Dordrecht, 1985.
2. G.S.Asanov, *Jet extension of Finslerian gauge approach derivation of generalized Einstein-Yang-Mills equations*, Seminarul de Mecanica, No. 16, Univ. of Timisoara, 1988, 1-47.
3. G.S.Asanov, S.F.Ponomarenko, *Finslerian Fiberings over Space-Time, Associated Gauge Fields and Connections* (Russian), Nauka Eds., Acad.Sci.Mold.Rep., 1989.
4. G.S.Asanov, S.F.Ponomarenko, *Finslerian jet and higher-order gauge approach stochastic space-time. Distribution function*, Seminarul de Mecanica, No. 21, Univ. of Timisoara, 1989, 1-51.

5. V. Balan, *Lagrangian structures in geometrized jet framework*, Proc. of The Conf. of Applied Diff. Geom. – General Relativity, June 27 - July 1 2002 Thessaloniki, BJGA Proceedings, 2004, to appear.
6. V. Balan, *Lorentz-type equations in first-order jet spaces endowed with nonlinear connection*, Proceedings of The First French-Romanian Colloquium of Numerical Physics, October 30-31, 2000, Bucharest, Romania, Geometry Balkan Press 2002, 105-114.
7. V. Balan, *Notable curves in geometrized $J^1(T,M)$ jet framework*, BJGA 8 (2003), 2, 1-10.
8. V.Balan, *Synge-Beil and Riemann-Jacobi jet structures with applications to physics*, IJMMS, Hindawi Publ., 27 (2003), 1693-1702.
9. V. Balan, K. Trencevski, *Stationary Lorentz curves in geometrized jet spaces. Numerical simulation and analytic solutions*, AGG, Palm Harbor, 20 (2003), 2, 197-210.
10. D. Bao, S.-S. Chern, Z. Shen, *An Introduction to Riemann-Finsler Geometry*, Springer-Verlag, 2000.
11. A. Bejancu, Finsler Geometry and Applications, E.Horwood, 1990.
12. Ch. Ehresmann, *Extension du calcul des jets aux jets non holonomes*, C.R. Acad Sci, Paris 239, 1762-1764 (1954).
13. Ch. Ehresmann, *Oeuvres completes* (vol. I and II), Amiens, 1984.
14. P.L. Garcia, *The Poincare-Cartan invariant in the calculus of variations*, Symposia Math. 14 (1974), 219-246.
15. I.M. Gelfand, S. Fomin, *Calculus of variations*, New Jersey, 1967.
16. H. Goldschmidt, S. Sternberg, *The Hamilton-Cartan formalism in the calculus of variations*, Ann. Inst. H. Poincare 23 (1973), 203-267.
17. D. Krupka, *A geometric theory of ordinary first order variational problems in fibered manifolds*, I., II, J. Math. Anal. Appl. 49 (1975), 180-206, 469-476.
18. D. Krupka, *Anti-holonomic jets and the Lie bracket*, Archivum Mathematicum (Brno), 34 (1998), 311-319.
19. D. Krupka, *Geometry of Lagrangean structures* (English), Arch. Math., Brno 22, 159-173; 211-228 (1986).
20. D. Krupka, *Lagrange theory in fibered manifolds*, Rep. Math. Phys. 2 (1971), 121-133.
21. D. Krupka, *Some Geometric Aspects of Variational Problems in Fibered Manifolds*, Folia Fac. Sci. Nat. Univ. Purk. Brunensis, Physica, XIV, Brno, Czechoslovakia,1973, 65.
22. R. Miron, *The Geometry of Higher Order Finsler Spaces*, Hadronic Press, 1998.
23. R. Miron, *The Geometry of Higher-Order Hamilton Spaces*, Kluwer, 2003.
24. R. Miron, *The Geometry of Higher Order Lagrange Spaces. Applications in Mathematics and Physics*, Kluwer, FTPH, 1997.
25. R. Miron, *The Geometry of Lagrange Spaces: Theory and Applications*, Kluwer, 1994.
26. R. Miron, M. Anastasiei, *The Geometry of Vector Bundles. Theory and Applications*, Kluwer, 1994.
27. R. Miron, M.Anastasiei, *Vector Bundles. Lagrange Spaces. Applications to Relativity*, Geometry Balkan Press, 1996.
28. R. Miron, D. Hrimiuc, H. Shimada, S.V. Sabău, *The Geometry of Hamilton and Lagrange Spaces*, Kluwer, FTPh. 198, 2001.
29. R. Miron, R.Roşca, M.Anastasiei, K.Buchner, *New aspects in Lagrangian relativity*, Found. of Phys. Lett., v. 5 (1992), n.2, 141-171.
30. M. Neagu, *Canonical Nonlinear connections on jet bundles of first order*, arXiv:math.DG/0111163 v1, 14 Nov 2001.
31. M. Neagu, *The geometry of autonomous metrical multi-time Lagrange space of electrodynamics*, IJMMS, Hindawi Publ., 29.1 (2002), 7-15; http://xxx.lanl.gov/abs/math.DG/0010091, (2000).
32. M. Neagu, *Generalized metrical multi-time Lagrangian geometry of physical fields*, Forum Mathematicum, 15 (2003), 1, 63-92; http://www.degruyter.de/journals/forum/2003/pdf/15_63.pdf; http://xxx.lanl.gov/abs/math.DG/0011003, 2000.
33. M. Neagu, C. Udrişte, *The geometry of metrical multi-time Lagrange spaces*, http://xxx.lanl.gov/abs/math.DG/0009071, 2000.
34. M. Neagu, C. Udrişte, *From PDE Systems and Metrics to Geometric Multi-Time Field Theories*, Sem. Mec. 79, West Univ. of Timisoara, 2001, http://xxx.lanl.gov/abs/math.DG/0009071, 2000.
35. D. Saunders, The Geometry of Jet Bundles, Cambridge Univ. Press, 1989.

Killing-Yano tensors, surface terms and superintegrable systems

Dumitru Baleanu*† and Özlem Defterli*

*Department of Mathematics and Computer Science, Faculty of Arts and Sciences, Çankaya
University, 06530, Ankara, Turkey
†On leave of absence from Institute of Space Sciences, P.O BOX, MG-23, R 76900
Magurele-Bucharest, Romania,
e-mails: baleanu@venus.nipne.ro, dumitru@cankaya.edu.tr

Abstract. Killing-Yano and Killing tensors are investigated corresponding to a set of two
dimensional superintegrable systems. A suitable surface term is added to the corresponding free
Lagrangian describing the motion of a particle on a 2-sphere of unit radius and we analyze the
symmetries of the obtained geometries.
Key words. Killing-Yano tensors, Killing tensors, superintegrability, surface term.
PACS. 04.20.-q, 02.0.-k.
MSC. 70H06.

1. INTRODUCTION

The Killing-Yano (KY) tensors [33, 18, 11] are connected to the supersymmetric
classical and quantum mechanics on curved manifolds and they play an important role in
theories with spin [2]-[8]. The KY tensor is an antisymmetric tensor $f_{\mu\nu} = -f_{\nu\mu}$ which
satisfies the equation

$$f_{\mu\nu;\lambda} + f_{\mu\lambda;\nu} = 0. \tag{1}$$

As it is known, a Killing tensor $k_{\mu\nu}$ generates a constant of motion quadratic in the
four-momentum p_μ as follows

$$L = \frac{1}{2} k^{\mu\nu} p_\mu p_\nu. \tag{2}$$

If $f_{\mu\eta}$ satisfies Eq. (1), then it generates a Killing tensor as

$$k_{\mu\nu} = f_{\mu\lambda} f_\nu^\lambda. \tag{3}$$

Another method to obtain a Killing tensor [39, 27] is to solve its corresponding
equations

$$k_{\nu\lambda;\mu} + k_{\lambda\mu;\nu} + k_{\mu\nu;\lambda} = 0, \tag{4}$$

CP729, *Global Analysis and Applied Mathematics: International Workshop on Global Analysis*,
edited by K. Taş, D. Krupka, O. Krupková, and D. Baleanu
© 2004 American Institute of Physics 0-7354-0209-4/04/$22.00

where $k_{\mu\nu}$ is a symmetric tensor.

The superintegrability in two dimensions represents today one of the important topics in applied mathematics. During the last decade extensive studies were performed about systems with second-order integrals of motion, either in Euclidean space or in spaces with nonzero constant curvature. Highly symmetric systems are often integrable, and in some special cases, superintegrable and exactly solvable [43, 31].

Superintegrable systems on the two dimensional Euclidean spaces have been classified in [15], and also extended to the two dimensional spheres [21]. Recently, several authors have investigated the classifications of superintegrable systems for these two dimensional Riemannian spaces [36, 29, 30].

The main aim of this paper is to investigate the symmetries of a two dimensional superintegrable system and to analyze the importance of the surface terms in generating four dimensional geometries possessing non-generic symmetries.

2. KILLING AND KILLING-YANO TENSORS FOR TWO-DIMENSIONAL SUPERINTEGRABLE SYSTEMS

A D-dimensional Hamiltonian system H is said to be Liouville integrable [1, 19] if there exists a system of D functionally independent functions in involution. The system is superintegrable [43, 15] if a further m integrals $\{Y_1,\ldots,Y_m, 1 \leq m \leq n-1\}$ exist such that the set of constants $\{I_1 = H, I_2,\ldots,I_n, Y_1,\ldots,Y_m\}$ are functionally independent. The additional integrals have vanishing Poisson bracket with H, but not necessarily with each other or with the I_i's [17]-[34].

Assuming the existence of a second order Killing tensor, Darboux and Koenigs classified the two-dimensional Riemann spaces which admits extra constants of motion [32]. These are called Darboux spaces denoted by D_1, D_2, D_3 and D_4 respectively [32].

The infinitesimal distances of the corresponding spaces D_1 and D_4 are

$$ds^2 = (x+y)dxdy \tag{5}$$

and

$$ds^2 = (ae^{-\frac{x+y}{2}} + be^{-x-y})dxdy, \tag{6}$$

respectively.

Using (5) we identify the metric as [31]

$$g_{ij}^{(1)} = \begin{pmatrix} 0 & x+y \\ x+y & 0 \end{pmatrix}. \tag{7}$$

If we change the coordinates as $(x,y) \to (u,v)$, where $x = u+iv$, $y = u-iv$ and $i^2 = -1$, then the metric (7) becomes

$$g_{ij}^{(1)} = \begin{pmatrix} 2u & 0 \\ 0 & 2u \end{pmatrix}. \tag{8}$$

100

Solving (3), the components of the Killing tensor for the metric (8) were calculated as

$$k_{11} = \frac{u}{2}(C_1 v^2 + 2C_2 v + 2C_3),$$

$$k_{12} = -u^2(C_1 v + C_2),$$

$$k_{22} = \frac{u}{2}(C_1 v^2 + 2C_2 v + 4C_1 u^2 + 2C_3 + 2C_4 u), \qquad (9)$$

where C_1, C_2, C_3 and C_4 are constants.

Solving (1) the non-zero component of the KY tensor is given as

$$f_{12} = C_1 u, \qquad (10)$$

where C_1 is a constant.

The second metric has the following form

$$g_{ij}^{(2)} = \begin{pmatrix} ae^{-u} + be^{-2u} & 0 \\ 0 & ae^{-u} + be^{-2u} \end{pmatrix}, \qquad (11)$$

and the components of its corresponding Killing tensors are found to be

$$k_{11} = \frac{(C_1' + C_2' \sin(v) + C_3' \cos(v))(a + be^{-u})^2}{(ae^u + b)},$$

$$\begin{aligned} k_{12} = &-[(a^3 e^{-u} C_2' + 2a^2 (e^{-u})^2 C_2' b + a(e^{-u})^3 C_2' b^2 \\ &+ be^{-2u} C_2' a^2 + 2b^2 e^{-2u} C_2' ae^{-u} + b^3 e^{-2u} C_2' (e^{-u})^2) \cos(v) \\ &+ (-a^3 e^{-u} C_3' - 2a^2 (e^{-u})^2 C_3' b - a(e^{-u})^3 C_3' b^2 - be^{-2u} C_3' a^2 \\ &- 2b^2 e^{-2u} C_3' ae^{-u} - b^3 e^{-2u} C_3' (e^{-u})^2) \sin(v)]/((ae^u + b)ae^{-u}) \end{aligned} \qquad (12)$$

$$\begin{aligned} k_{22} = &[(-2ae^{-3u} b^5 C_2' - 9a^2 e^{-2u} b^4 C_2' - 6ba^5 e^u C_2' - 16a^3 e^{-u} b^3 C_2' \\ &- 14a^4 b^2 C_2' - a^6 e^{2u} C_2') \sin(v) + (-6ba^5 e^u C_3' - 9b^4 a^2 e^{-2u} C_3' \\ &- 2ae^{-3u} b^5 C_3' - a^6 e^{2u} C_3' - 14b^2 a^4 C_3' - 16a^3 e^{-u} b^3 C_3') \cos(v) \\ &+ e^u (a^7 C_4' + 5a^5 bC_1') + e^{-u} (14a^3 C_1' b^3 + 10a^5 b^2 C_4') \\ &+ e^{-2u} (10b^3 a^4 C_4' + 11b^4 C_1' a^2) + e^{-3u} (5b^5 C_1' a + 5b^4 C_4' a^3) \\ &+ e^{-4u} (b^6 C_1' + b^5 C_4' a^2) + a^6 C_1' e^{2u} + 5a^6 bC_4' \\ &+ 11a^4 b^2 C_1']/(a^2 (ae^u + b)^3). \end{aligned} \qquad (13)$$

In this case the component of the KY tensor is given by

$$f_{12} = C_1'(ae^u + b)e^{-2u}. \qquad (14)$$

Here C_1', C_2', C_3' and C_4' are constants.

3. THE MOTION ON A SPHERE AND ITS INDUCED GEOMETRIES

It was proved in [12] that the motion on a sphere admits four constants of motion, the Hamiltonian and three components of the angular momentum. In the following it is started with a free Lagrangian theory and a surface term involving the components of the angular momentum is added. By a suitable choice of Lagrange multipliers it is obtained a quadratic Lagrangian, which may be, or may be not singular [23].

In this way four-dimensional manifolds will be generated. In our case the extended Lagrangian is given by [23, 9]

$$
L' = \frac{1}{2}(1+\frac{x^2}{u})\dot{x}^2 + \frac{1}{2}(1+\frac{y^2}{u})\dot{y}^2 + \frac{xy}{u}\dot{x}\dot{y} - \frac{xy}{\sqrt{u}}\lambda_1\dot{x} + (\frac{x^2}{\sqrt{u}}+\sqrt{u})\lambda_2\dot{x}
$$
$$
- (\frac{y^2}{\sqrt{u}}+\sqrt{u})\lambda_1\dot{y} + \frac{xy}{\sqrt{u}}\lambda_2\dot{y} + x\lambda_3\dot{y} - y\lambda_3\dot{x}, \tag{15}
$$

where $u = 1 - x^2 - y^2$. From (15) we identify the singular matrix a_{ij} as

$$
a_{ij} = \begin{pmatrix} 1+\frac{x^2}{u} & \frac{xy}{u} & -\frac{xy}{\sqrt{u}} & \frac{x^2}{\sqrt{u}}+\sqrt{u} & -y \\ \frac{xy}{u} & 1+\frac{y^2}{u} & -\frac{y^2}{\sqrt{u}}-\sqrt{u} & \frac{xy}{\sqrt{u}} & x \\ -\frac{xy}{\sqrt{u}} & -\frac{y^2}{\sqrt{u}}-\sqrt{u} & 0 & 0 & 0 \\ \frac{x^2}{\sqrt{u}}+\sqrt{u} & \frac{xy}{\sqrt{u}} & 0 & 0 & 0 \\ -y & x & 0 & 0 & 0 \end{pmatrix}. \tag{16}
$$

Because (16) is a singular matrix of rank 4, three symmetric minors of order four are identified. If these minors are treated as a metric it is observed that they are not conformaly flat but their scalar curvatures are zero.

The first metric is given by

$$
g^{(1)}_{\mu\nu} = \begin{pmatrix} 1+\frac{x^2}{u} & \frac{xy}{u} & \sqrt{u}+\frac{x^2}{\sqrt{u}} & -y \\ \frac{xy}{u} & 1+\frac{y^2}{u} & \frac{xy}{\sqrt{u}} & x \\ \sqrt{u}+\frac{x^2}{\sqrt{u}} & \frac{xy}{\sqrt{u}} & 0 & 0 \\ -y & x & 0 & 0 \end{pmatrix}. \tag{17}
$$

A Killing vector satisfies the following equation

$$
V_{\mu;\lambda} + V_{\lambda;\mu} = 0. \tag{18}
$$

Using (18) the Killing vector of (17) has the following components

$$V_1 = (y, -x, 0, 0),$$

$$V_2 = \left(\sqrt{1-x^2-y^2} + \frac{x^2}{1-x^2-y^2}, \frac{xy}{1-x^2-y^2}, 0, 0\right),$$

$$V_3 = \left(-\frac{xy}{1-x^2-y^2}, -\sqrt{1-x^2-y^2} - \frac{y^2}{1-x^2-y^2}, 0, 0\right). \qquad (19)$$

The next step is to investigate (KY) tensors corresponding to the metric (17). Solving (1) we obtain the following set of solutions:

a. One-by-one solution is $f_{21} = \frac{C_1}{\sqrt{1-x^2-y^2}}$, others zero.

b. Two-by-two solution has the form: $f_{31} = f_{42} = C$,

c. Three-by-three solution is $f_{21} = \frac{C_1}{\sqrt{-1+x^2+y^2}}$ and $f_{31} = f_{42} = C$, where C and C_1 are constants.

Using (15) another two metrics can be identified as

$$g^{(2)}_{\mu\nu} = \begin{pmatrix} 1 + \frac{x^2}{u} & \frac{xy}{u} & -\frac{xy}{\sqrt{u}} & -y \\ \frac{xy}{u} & 1 + \frac{y^2}{u} & -\sqrt{u} - \frac{y^2}{\sqrt{u}} & x \\ -\frac{xy}{\sqrt{u}} & -\sqrt{u} - \frac{y^2}{\sqrt{u}} & 0 & 0 \\ -y & x & 0 & 0 \end{pmatrix} \qquad (20)$$

and

$$g^{(3)}_{\mu\rho} = \begin{pmatrix} 1 + \frac{x^2}{u} & \frac{xy}{u} & -\frac{xy}{\sqrt{u}} & \frac{x^2}{\sqrt{u}} + \sqrt{u} \\ \frac{xy}{u} & 1 + \frac{y^2}{u} & -\frac{y^2}{\sqrt{u}} - \sqrt{u} & \frac{xy}{\sqrt{u}} \\ -\frac{xy}{\sqrt{u}} & -\frac{y^2}{\sqrt{u}} - \sqrt{u} & 0 & 0 \\ \frac{x^2}{\sqrt{u}} + \sqrt{u} & \frac{xy}{\sqrt{u}} & 0 & 0 \end{pmatrix} \qquad (21)$$

respectively. By direct calculations we obtained that (20) and (21) admit the same Killing vector as in (19).

Solving (1) corresponding to (20) and (21) we find one non-zero component of (KY) tensor as

$$f_{21} = \frac{C_1}{\sqrt{1-x^2-y^2}}. \qquad (22)$$

ACKNOWLEDGMENTS

One of the authors (D.B.) would like to thank M. Henneaux, B. Edgar and N. Montesinos for their helpful discussions.

REFERENCES

1. Arnold, V. I., Mathematical Methods of Classical Mechanics, Springer-Verlag, New York (1989).
2. Baleanu, D., New non-generic symmetries on extended Taub-NUT space-time. *Helv. Acta. Phys.* **71(3)** (1998), 343-351.
3. Baleanu, D., Geodesic motion on extended Taub-NUT spinning space. *Gen. Rel. Grav.* **30(2)** (1998), 195-207 .
4. Baleanu, D., and Başkal, S., Geometrization of the Lax pair tensors. *Mod. Phys. Lett. A* **15(24)** (2000), 1503-1510.
5. Baleanu, D., Symmetries of NUT-Kerr-Newman dual metrics. *Nuovo Cimento B* **116(2)** (2001), 205-213.
6. Baleanu, D., and Başkal, S., Dual metrics for a class of radiative space-times. *Mod. Phys. Lett. A* **16(3)** (2001), 135-142.
7. Baleanu, D., Symmetries of the dual metrics. *Int. Journ. Mod. Phys. D* **11(3)** (2002), 405-416.
8. Baleanu, D., and Başkal, S., Dual metrics and nongeneric supersymmetries for a class of Siklos space-times. *Int. Journ. Mod. Phys. A* **17** (2002), 3737-3747.
9. Baleanu, D., and Defterli, Ö., Killing-Yano tensors and angular momentum. *Czech. Journ. Phys.* **54** (2004), 157-165.
10. Bogoyavlensky, O. I., On perturbations of the periodic Toda lattice. *Commun. Math. Phys.* **51** (1976), 210-235.
11. Carter, B., and McLenaghan, R. G.,Generalized total angular momentum operator for the Dirac equation in curved space-time . *Phys. Rev. D* **19** (1979), 1093-1097.
12. Curtright, T. L., and Zachos, C. K., Deformation quantization of superintegrable systems in Nambu mechanics. *New J. Phys.* **4** (2002), 83-101.
13. Daskaloyannis, C., Quadratic Poisson algebras of two-dimensional classical superintegrable systems and quadratic associative algebras of quantum superintegrable systems. *J. Math. Phys.* **42** (2001), 1100-1127.
14. Dorizzi, B., Grammaticos, B., Padjen, R., and Papageorgiou, V., Integrals of motion for Toda systems with unequal masses. *J. Math. Phys.* **25** (1984), 2200-2211.
15. Evans, N. W., Superintegrability in classical mechanics. *Phys. Rev. A* **41** (1990), 5666-5676.
16. Fokas, A. S., and Lagerstrom, P. A., Quadratic and cubic invariants in classical mechanics. *J. Math. Anal. Appl.* **74** (1980), 325-342.
17. Fris, J., Mandrosov, V., Smorodinsky, Ya. A., Uhlir, M., and Winternitz, P., On higher symmetries in quantum mechanics. *Phys. Lett.* **16** (1965), 354-360.
18. Gibbons, G. W., Rietdijk, R. H., and van Holten, J. W., SUSY in the sky. *Nucl. Phys. B* **404** (1993), 42-64.
19. Goldstein, H., Classical Mechanics, Addison-Wesley, Massachusetts (1980).
20. Grammaticos, B., Dorizzi, B., and Ramani, A., Hamiltonians with higher order integrals and the "weak-Painleve" concept. *J. Math. Phys.* **25(12)** (1984), 3470-3473.
21. Grosche, C., Pogosyan, G. S., and Sissakian, A. N., Path-integral discussion for Smorodinsky-Winternitz potentials.2. The 2-dimensional and 3-dimensional sphere. *Fortschr. Phys.* **43(6)** (1995), 523-563.
22. Grosche, C., Pogosyan, G. S., and Sissakian, A. N., Path-integral discussion for Smorodinsky-Winternitz potentials.1. 2-dimensional and 3-dimensional Euclidean-space. *Fortschr. Phys.* **43(6)** (1995), 453-521.
23. Güler, Y., Baleanu, D., and Cenk, M., Surface terms, angular momentum and Hamilton - Jacobi formalism. *Nuovo Cimento B* **118** (2003), 293-306.
24. Hietarinta, J., Integrable families of Hénon-Heiles type Hamiltonians and a new duality. *Phys. Rev. A* **28** (1983), 3670-3672.
25. Hietarinta, J., Classical versus quantum integrability. *J. Math. Phys.* **25(6)** (1984), 1833-1840.
26. Hietarinta, J., Direct methods for the search of the second invariant. *Phys. Rep.* **147** (1987), 87-154.
27. Hinterleitner, F., Killing tensors as space-time metrics. *Ann. Phys.* **271** (1999), 23-30.
28. Holt, C. R., Construction of new integrable Hamiltonians in two degrees of freedom. *J. Math. Phys.* **23** (1982), 1037-1046.
29. Kalnins, E. G., Miller, W. Jr., and Pogosyan, G. S., Completeness of multiseparability on the complex 2-sphere. *J. Phys. A: Math. Gen.* **33** (2000), 6791-6806.

30. Kalnins, E. G., Kress, J. M., Pogosyan, G. S., and Miller, W. Jr., Completeness of multiseparability in two-dimensional constant curvature spaces. *J. Phys. A: Math. Gen.* **34** (2001), 4705-4720.
31. Kalnins, E. G., Kress, J. M., and Winternitz, P., Superintegrability in a two-dimensional space of non-constant curvature. *J. Math. Phys.* **43** (2002), 970-996.
32. Koenigs, G., Sur les géodésiques a intégrales quadratiques. A note appearing in *Leçons sur la théorie générale des surfaces* G. Darboux, (Chelsea Publishing, vol.4 1972), 368-404.
33. Kramer, D., Stephani, H., Herlt, E., and MacCallum, M., *Exact Solutions of Einstein's Field Equations* (Cambridge Univ.Press, 1980).
34. Létourneau, P., and Vinet, L., Superintegrable systems: polynomial algebras and quasi-exactly solvable Hamiltonians. *Ann. Phys.* **243** (1995), 144-162.
35. Moser, J., Three integrable Hamiltonian systems connected with isospectral deformations. *Adv. Math.* **16** (1975), 197-210.
36. Rañada, M. F., and Santander, M., Superintegrable systems on two-dimensional sphere S-2 and the hyperbolic plane H-2. *J. Math. Phys.* **40(10)** (1999), 5026-5057.
37. Rañada, M. F., Superintegrable n=2 systems, quadratic constants of motion, and potentials of Dirach. *J. Math. Phys.* **38(8)** (1997), 4165-4178.
38. Rañada, M. F., On a second Lax structure for the Calogero-Moser system: time-dependent constants and superintegrability. *Phys. Lett. A* **277** (2000), 219-222.
39. Rietdijk, R. H., and van Holten, J. W., Killing tensors and a new geometric duality. *Nuc. Phys. B* **472** (1996), 427-466.
40. Sheftel, M. B., Tempesta, P., and Winternitz, P., Superintegrable systems in quantum mechanics and classical Lie theory. *J. Math. Phys.* **42(2)** (2001), 659-673.
41. Tempesta, P., Turbiner, A. V., and Winternitz, P., Exact solvability of superintegrable systems. *J. Math. Phys.* **42** (2001), 4248-4261.
42. Yoshida, H., Integrability of generalized Toda lattice systems and singularities in the complex t-plane. In *Nonlinear Integrable Systems - Classical Theory and Quantum Theory* edited by M. Jimbo, and T. Miwa, (World Science Publishing Co., Singapore, 1983a), 273-289.
43. Wojciechowski, S., Superintegrability of the Calogero - Moser system. *Phys. Lett. A* **95** (1983), 279-281.

Compatibility of non-generic supersymmetries and geometric duality for a subclass of generalized pp-wave metrics

D. Baleanu* and S. Başkal [†]

*Department of Mathematics and Computer Sciences, Çankaya University, 06530 Ankara, Turkey,
dumitru@cankaya.edu.tr
and
On leave of absence from Institute of Space Sciences, R 76900 Magurele-Bucharest, Romania,
baleanu@venus.nipne.ro
†Department of Physics, Middle East Technical University, 06531 Ankara, Turkey
baskal@newton.physics.metu.edu.tr

Abstract. Spinning point particle theories accommodate non-generic supercharges in connection with the existence of Killing-Yano tensors. Killing-Yano tensors of order two and three and their corresponding Killing tensors are found for a subclass of generalized pp-wave metrics. These metrics include the pp-wave itself, its possible generalizations and the Siklos metric which is conformal to that. The compatibility between geometric duality and non-generic symmetries is discussed within the context of the metric solutions. It is found that some of the metric solutions admit anti-de Sitter spacetimes while some are found to be purely radiative.
Key words. Killing-Yano symmetry, exact solutions.
PACS. 04.20.-q, 02.40.Ky.
MSC. 83C20.

1. INTRODUCTION

Killing-Yano (KY) tensors were first introduced by Yano [11] from a mathematical point of view, long before they became an object of interest for relativists [5]. Later on, it was found that the separability of the Dirac equation is connected with the existence of Killing-Yano tensors on curved backgrounds [2]. Likewise, some specific KY tensors produce symmetries for Dirac type operators [3]. Recently, it is discovered that they are closely related to non-generic supersymmetries in spinning point particle theories [4], [10].

Killing tensors are related to constants of motion and therefore are crucial in solving the geodesic equation. It is known that Killing tensors can be constructed from KY tensors, without actually solving the Killing equation [2]. Furthermore, and as interesting as that, dual spacetimes and consequently geometric dualities in the phase space, can be realized from non-degenerate Killing tensors [6]. Since there isn't any well-defined method with which one can conclude in advance, whether the mere existence of KY

CP729, *Global Analysis and Applied Mathematics: International Workshop on Global Analysis*,
edited by K. Taş, D. Krupka, O. Krupková, and D. Baleanu
© 2004 American Institute of Physics 0-7354-0209-4/04/$22.00

tensors is sufficient to produce geometric dualities, analyzing specific examples can be considered as a step forward to a better understanding of this compatibility problem.

In this context, it is worthwhile to investigate metrics which are also interesting in their own right. We will consider the pp-wave metric [7] and the Siklos metric [9] which is conformal to it, in addition to the generalized pp-wave metric and its conformal form. Solutions to the KY tensor equations will restrict the form of the metrics [1]. We will also investigate whether those restrictions can be connected to some interesting features of those spacetimes, such as the radiation property.

2. GENERALITIES

In this section we will sketch the path which starts from the existence of supercharges in connection with KY tensors, going through the formation of non-degenerate Killing tensors, and finally ending up with dual geometries.

2.1. Killing-Yano tensors and non-generic symmetries

The spinning particle model was constructed to be supersymmetric [4], therefore independent of the form of the metric there is always a conserved supercharge $Q_0 = \Pi_\mu \psi^\mu$. Here, Π_μ are the covariant momenta and ψ^μ are odd Grassmann variables.

The existence of Killing-Yano tensors $f_{v_1 v_2 \cdots v_r}$ of rank r is related to non-generic supersymmetries described by the following supercharge

$$Q_f = f_{v_1 v_2 \cdots v_r} \Pi^{v_1} \psi^{v_2} \cdots \psi^{v_r}, \tag{1}$$

where $f_{v_1 v_2 \cdots v_r}$ is the solution of the KY equation

$$f_{v_1 v_2 \cdots (v_r;\lambda)} = 0, \tag{2}$$

and Q_f is a superinvariant: $\{Q_0, Q_f\} = 0$.

The Jacobi identities and (2) guarantee that Q_f is also a constant of motion: $\{Q_f, H\} = 0$, with

$$H = \frac{1}{2} g^{\mu v} \Pi_\mu \Pi_v \tag{3}$$

and with the appropriate definitions of the brackets.

2.2. Killing tensors obtained from Killing-Yano tensors

Second rank Killing tensors can be constructed from any rank KY tensors by appropriate constructions. For instance for a second rank KY tensor the Killing tensor is expressed as

$$K_{\mu v} = g^{\alpha \beta} f_{\mu \alpha} f_{\beta v} \tag{4}$$

and for the third rank we have

$$K_{\mu\nu} = g^{\alpha\delta} g^{\beta\gamma} f_{\mu\alpha\beta} f_{\gamma\delta\nu}. \tag{5}$$

The symmetric second rank tensors of (4) and (5) satisfy the Killing equation:

$$K_{(\mu\nu;\alpha)} = 0. \tag{6}$$

2.3. Dual metrics obtained from Killing tensors

If the Killing tensor $K^{\mu\nu}$ is non-degenerate, then through the relation

$$K^{\mu\alpha} k_{\alpha\nu} = \delta^{\mu}{}_{\nu}, \tag{7}$$

the second rank tensor $k_{\mu\nu}$, which itself is obviously non-degenerate can be viewed as the metric on the "dual" space. It has been shown in [6] that $K^{\mu\nu}$ and $g^{\mu\nu}$ are reciprocally the contravariant components of the Killing tensors with respect to each other.

The phase space accommodates the notion of geometric duality in the following manner: The constant of motion $K = \frac{1}{2}K^{\mu\nu}\Pi_{\mu}\Pi_{\nu}$, generates symmetry transformations on the phase space linear in momentum: $\{x^{\mu}, K\} = K^{\mu\nu}\Pi_{\nu}$, and in view of (6) the Poisson brackets satisfy $\{H, K\} = 0$, where H is as in (3). Thus, in the phase space there is a reciprocal model with constant of motion H and the Hamiltonian K.

3. EXAMPLES

Given any spacetime the path described in the previous section can not always be completed. In this section we will work through some specific spacetimes. We will give solutions to the KY equations of rank two and three for the pp-wave metric, the Siklos metric, the generalized pp-wave metric and its conformal form. In general there are 24 independent equations, to be solved for the six independent components of the Killing-Yano tensor of rank two and 15 independent equations correspond to four independent components of the KY tensor of rank three, respectively.

Although we have done a through analysis to find all solutions to (2), we will not give them all. From all the solutions we will choose those admitting both the KY tensor and the dual metrics. If no such solution exist, then we will give the cases when the spacetime have some particular interesting properties, or else we give at least one solution.

3.1. The pp-wave metric

The pp-wave metric describes plane fronted waves with parallel rays, admits a non-expanding shear-free and twist-free null- geodesic congruence, and can be expressed in the form [7]:

$$ds^2 = 2dudv + dx^2 + dy^2 + h(x, y, u)du^2. \tag{8}$$

This metric admits a two-component KY tensor with no restrictions on the metric as:

$$f_{24} = c_1, \qquad f_{34} = c_2. \tag{9}$$

One-component Killing tensor is calculated from (4)

$$K_{44} = c_1^2 + c_2^2. \tag{10}$$

A KY tensor of order three has three non-vanishing components:

$$\begin{aligned} f_{123} &= 0, & f_{124} &= c_1, \\ f_{134} &= c_2, & f_{234} &= q(u), \end{aligned} \tag{11}$$

and the metric function is found to be

$$h(x,y,u) = F(c_1 x + c_2 y) + q(u)_{,u} \left(\frac{y}{c_1} - \frac{x}{c_2} \right). \tag{12}$$

Here $q(u)$ and F are arbitrary functions of their arguments. From (5) and from above the Killing tensor is calculated as

$$\begin{aligned} K_{14} &= -2(c_2^2 + c_2^2), & K_{22} &= -2c_1^2, \\ K_{23} &= -2c_1 c_2, & K_{24} &= -2c_2 q(u), \\ K_{33} &- 2c_2^2, & K_{34} &= 2c_1 q(u), \\ K_{44} &= 2(q(u)^2 - (c_1^2 + c_2^2)h(x,y,u)). \end{aligned} \tag{13}$$

3.2. The Siklos metric

The Siklos metric is expressed as

$$ds^2 = \frac{\beta^2}{x^2} \left[2\,du\,dv + dx^2 + dy^2 + h(x,y,u)\,du^2 \right], \tag{14}$$

where $\beta = \sqrt{-3/\Lambda}$ and Λ is the negative cosmological constant. The presence of a negative cosmological constant implies that the spacetime is not asymptotically flat. It is demonstrated in [9], that the Siklos metric represents the only non-trivial Einstein space conformal to non-flat pp-waves. As in the case of the pp-wave metric the Siklos metric also has vanishing optical parameters.

The KY tensor is found as

$$\begin{aligned} f_{12} &= 0, & f_{13} &= 0, & f_{14} &= (2u - y)/2x^3, \\ f_{23} &= -1/2x^2, & f_{24} &= 1/x^2, & f_{34} &= (v + y + s)/x^3, \end{aligned} \tag{15}$$

and for the metric we have an anti-de Sitter solution

$$h(x,y,u) = (y - 2u)^2 + x^2. \tag{16}$$

As it is well-known, an anti-de Sitter spacetime is conformally flat and obeys Einstein's equations with a cosmological constant.

From (15) and (4) the Killing tensors are obtained as

$$K_{14} = (y - 2u)^2/(4\beta^2 x^4), \qquad K_{22} = k_{33} = -1/(4\beta^2 x^4),$$
$$K_{24} = -(v + y + s + x(y - 2u))/(2\beta^2 x^4),$$
$$K_{34} = (-(v + y + s)(y - 2u) + x)/(2\beta^2 x^4),$$
$$K_{44} = ((x^2 + y^2 + 4u^2 - 4uy)(y - 2u)^2 - 4x^2 - 4(v + y + s)^2)/(4\beta^2 x^4). \tag{17}$$

The dual metric for this Killing tensor is calculated and found to be:

$$k_{14} = 4\beta^6/(y - 2u)^2, \qquad k_{22} = k_{33} = -4\beta^6,$$
$$k_{24} = -8\beta^6 (v + y + s + x(y - 2u))/(y - 2u)^2,$$
$$k_{34} = 8\beta^6 (-(v + y + s)(y - 2u) + x)/(y - 2u)^2,$$
$$k_{44} = -4\beta^6[4((v + s)^2 + y(2v + u + 2s) - u^2) + 3(x^2 + y^2))]/(y - 2u)^2. \tag{18}$$

Although the initial anti-de Sitter spacetime specified by (16) is conformally flat, the above dual metric is more general and is not conformally flat.

This example is crucial in the sense that it exhibits compatibility between the non-generic symmetries and the dual metrics.

Third rank KY tensors are found to be

$$f_{123} = 0, \qquad\qquad f_{124} = 1/x^3,$$
$$f_{134} = (r(u) + y)/x^4, \qquad f_{234} = r(u)_{,u}/x^3. \tag{19}$$

We have found all the metric solutions, however here we give only the solutions admitting an anti-de Sitter spacetime

$$h(x, y, u) = h3(u)(x^2 + y^2), \tag{20}$$

with

$$h3(u) = r(u)_{,uu}/r(u). \tag{21}$$

3.3. The generalized pp-wave metric

The pp-wave metric can be generalized as

$$ds^2 = 2\, dv\, du + A(x, y, u)\, (dx^2 + dy^2) + H(v, x, y, u)\, du^2 \tag{22}$$

where $A(x, y, u)$ and $H(v, x, y, u)$ are functions of their arguments, and fall into the subclass of metrics admitting a parallel field of null 1-plane. Such a spacetime consists in a recurrent field of null vectors. If l_μ is a basis for the plane we have:

$$\nabla_v l_\mu = \kappa_v l_\mu, \qquad l_\mu l^\mu = 0, \quad l_\mu \neq 0 \tag{23}$$

where l_μ is a basis and κ_μ is the recurrence vector of the plane.

From (23) and the Ricci identity we have

$$l^\nu R_{\mu\nu\alpha\beta} = l_\mu f_{\alpha\beta}, \tag{24}$$

where $f_{\alpha\beta} = \partial_\alpha \kappa_\beta - \partial_\beta \kappa_\alpha$. Also from (23) one can observe that the principal null vector $l_\mu = \delta_\mu{}^4 = \partial_\mu u$, is hypersurface-orthogonal and that the recurrence vector for the PN1P is $\kappa_\mu = -\Gamma^4{}_{\mu 4}$ [8]. The KY tensor is found as

$$f_{14} = c_3, \qquad f_{23} = c_4 A(x,y), \tag{25}$$

where c_3 and c_4 are constants. The metric function $A(x,y)$ is independent of u, and $H(v,u) = v h1(u) + h2(u)$ where $h1, h2$ are independent of x, y. The components of the corresponding Killing tensor are

$$K_{14} = c_3^2, \qquad K_{22} = K_{33} = -c_4^2 A(x,y), \qquad K_{44} = c_3^2 H(v,u). \tag{26}$$

and one can construct the dual metric as

$$k_{14} = 1/c_3^2, \qquad k_{22} = k_{33} = -A(x,y)/c_4^2 \qquad k_{44} = H(v,u)/c_3^2. \tag{27}$$

This is the second example we have found which admits non-generic symmetries and dual metrics simultaneously.

We have found several two component third rank KY tensors. Here we introduce two distinct solutions only when the spacetime is pure radiative, i.e.,

$$G_{\mu\nu} = \rho(x^\sigma) l_\mu l_\nu, \tag{28}$$

where $G_{\mu\nu}$ is the Einstein tensor. The scalar curvature becomes zero, restricting the form of the metric function A to

$$A(x,y,u) = a(u)\exp\{\frac{c_1}{2}[(y-c_2)^2 - (x-c_2)^2]\}. \tag{29}$$

We have the third rank KY as

$$f_{124} = A(u)^{\frac{1}{2}}, \qquad f_{234} = \frac{1}{3} y A(u)^{\frac{3}{2}}{}_{,u}. \tag{30}$$

The metric functions $A(u)$ and $H = v h_1(u) + h_2(x,u)$ are subject to the solutions of

$$2A(u)A(u)_{,uu} - A(u)_{,u}^2 + A(u)A(u)_{,u} h1(u) = 0. \tag{31}$$

The components of the corresponding Killing tensor are

$$K_{14} = 2, \qquad K_{22} = 2A(u), \qquad K_{34} = -yA(u)_{,u},$$
$$K_{44} = 2H(v,x,u) - y^2 A(u)_{,u}^2/2A(u). \tag{32}$$

The other solution is

$$f_{134} = A(u)^{\frac{1}{2}}, \qquad f_{234} = -\frac{1}{2} x A(u)^{\frac{1}{2}} A(u)_{,u} \tag{33}$$

111

where the metric functions $A(u)$ and $h1(u)$ are related as in (31). The components of the corresponding Killing tensor are

$$K_{14} = 2, \qquad K_{24} = -xA(u)_{,u}, \qquad K_{33} = 2A(u),$$
$$K_{44} = 2H(v,y,u) - x^2 A(u)^2_{,u}/2A(u).$$

(34)

3.4. Conformal g-pp metric

The generalized pp-wave metric assumes a conform factor as in the Siklos metric and therefore takes the form

$$ds^2 = \frac{\beta^2}{x^2}(2\,dv\,du + A(x,y,u)\,(dx^2 + dy^2) + H(x,y,u)\,du^2).$$

(35)

It is known that not all metrics admit KY tensors. This metric admits at least a one-component KY tensor

$$f_{34} = A(u)^{\frac{1}{2}}x^{-3}.$$

(36)

where A is a function of u only, and $H(x,y,u)$ is a function of its arguments. Further analysis should be carried out to investigate whether it admits solutions more than one component.

The corresponding one-component Killing tensor is:

$$K_{44} = \frac{-1}{\beta^2 x^4}.$$

(37)

4. CONCLUSION

We have found KY tensors of order two and three for several metrics included in a subclass of a more general type from the pp-wave metric and their conformal forms of a particular type. We have found the spacetimes simultaneously admitting the KY tensors and the dual metrics. Although we have investigated all third rank KY tensors, none of those yielded dual metrics. We believe that this aspect of the third rank KY tensors should be investigated from a more general point of view. The metric functions have been restricted to obtain KY tensors. Consequently, it is observed that some metric solutions admitted pure radiative spacetimes, and some others are found to admit anti-de Sitter solutions.

ACKNOWLEDGMENTS

One of the authors (SB) would like to thank the organizers of IWGA for financial support.

REFERENCES

1. D. Baleanu and S. Başkal, *Dual Metrics for a Class of Radiative Space-times*, Mod. Phys. Lett., **A 16**, (2001), 135-142;
   D. Baleanu and S. Başkal, *Dual Metrics and Nongeneric Supersymmetries for a Class of Siklos Space-times*, Int. J. Mod. Phys., **A 17**, (2002), 3737-3747;
   D. Baleanu and S. Başkal, *Killing-Yano Symmetry for a Class of Space-times Admitting Parallel Null 1-Planes*, Il Nuovo Cimento, **B 117**, (2002), 501-510.
2. B. Carter, R. G. McLenaghan, *Generalized Total Angular Momentum Operator for the Dirac Equation in Curved Space-time*, Phys. Rev., **D19**, (1979), 1093-1097.
3. I. I. Cotaescu, M. Visinescu, *Symmetries of the Dirac Operators Associated with Covariantly Constant Killing-Yano Tensors*, Class. Quantum Grav., **21**, (2004), 11-28.
4. G. W. Gibbons, R. H. Rietdijk and J. W. van Holten, *SUSY in the Sky*, Nucl. Phys., **B 404**, (1993), 42-64.
5. G. S. Hall, *Killing-Yano Tensors in General Relativity*, Int. J. Theor. Phys., **26**, (1987), 71-81.
6. J. W. van Holten, *Supersymmetry and the Geometry of Taub-NUT*, Phys. Lett., **B 342**, (1995), 47-52;
   R. H. Rietdijk and J. W. van Holten, *Killing Tensors and a New Geometric Duality*, Nucl. Phys., **B 472**, (1996), 427-466.
7. W. Kundt, *The Plane-fronted Gravitational Waves*, Z. Phys., **163**, (1961), 77-86.
8. F. Öktem, *Parallel Planes in Space-Time and Possible Origin of Particles*, Nuovo Cimento, **A 15**, (1973), 189-208;
   F. Öktem, *On Parallel Null 1-Planes in Space-Time*, Nuovo Cimento, **B 34**, (1976), 169-181.
9. S. T. C. Siklos, *Lobatchevski Plane Gravitational Waves*, Galaxies, Axisymmetric Systems and Relativity, ed. M. A. H. MacCallum, pp:247-274 (Cambridge Univ. Press, 1985).
10. D. Vaman and M. Visinescu, *Genaralized Killing Equations and Taub-NUT spinning Space*, Phys. Rev., **D54**, (1996), 1398-1402;
    M. Visinescu, *Generalized Taub-NUT Metrics and Killing-Yano Tensors*, J. Math. Phys. A, **33**, (2001), 4383-4392.
11. K. Yano, *On Harmonic and Killing Vector Fields*, Ann. of Math., **55**, (1952), 38-45.

Collineations of (2+1)-Dimensional Friedmann-Robertson-Walker Spacetimes

Ugur Camci

Department of Physics, Faculty of Arts and Sciences, Çanakkale Onsekiz Mart University,
17100, Çanakkale, Turkey
E-mail: ucamci@comu.edu.tr

Abstract. Conformal Killing and Ricci collineation equations for (2+1)-dimensional Friedmann-Robertson-Walker (FRW) spacetimes are solved. These spacetimes are classified according to their Ricci conformal collineations (RCCs) and Ricci collineations (RCs). In the non-degenerate and degenerate cases of the Ricci tensor (the cases $det(R_{ab}) \neq 0$ and $det(R_{ab}) = 0$, respectively), the general forms of the vector fields generating RCCs and RCS are obtained. When the Ricci tensor is degenerate, the special cases are classified and it is shown that there are many cases of RCCs and RCs with infinite degrees of freedom. Furthermore, it is found that when the Ricci tensor is non-degenerate, the groups of RCCs and RCs are finite-dimensional, and we have always 10-parameter group of RCCs and 6-parameter group of RCs which are the maximal possible dimension for three-dimensional spacetime manifold. The results obtained are compared with conformal Killing vectors and Killing vectors.

Key words. Ricci collineation, Killing vector, conformal Killing vector, homothetic vector, (2+1)-dimensional Friedmann-Robertson-Walker spacetime.

PACS. 04.20.J.
MSC. 83C15, 83C20.

1. INTRODUCTION

Let M be a three-dimensional smooth (C^∞) manifold with smooth Lorentz metric g of signature $(-,+,+)$. The associated Riemann and Ricci tensors are denoted by their respective components R_{abcd} and $R_{ab}(\equiv R^c_{acb})$ and a covariant derivative with respect to the Levi-Civita connection associated with g is denoted by a semicolon and a partial derivative by a comma.

During the recent years the three-dimensional gravity has been received much attention. It is believed that (2+1)-dimensional gravity will provide new insights towards a better understanding of the physically relevant (3+1)-dimensional Einstein gravity. The addition to the Einstein-Hilbert action of a gravitational Chern-Simons term leads to topologically massive gravity (TMG) [2]. The field equations of TMG including matter are

$$G_{ab} + \frac{1}{\mu}C_{ab} = \kappa T_{ab}, \tag{1}$$

where $G_{ab} \equiv R_{ab} - \frac{1}{2}R g_{ab} + \Lambda g_{ab}$ is the Einstein tensor, $C^{ab} = \varepsilon^{acd}\left(R^b_c - \frac{1}{4}R\delta^b_c\right)_{;d}$ is the Cotton tensor which is symmetric, traceless and identically conserved, ε^{abc} is completely antisymmetric three-dimensional Levi-Civita tensor, μ is the topological mass constant

CP729, *Global Analysis and Applied Mathematics: International Workshop on Global Analysis,*
edited by K. Taş, D. Krupka, O. Krupková, and D. Baleanu
© 2004 American Institute of Physics 0-7354-0209-4/04/$22.00

at $\Lambda = 0$, and T_{ab} is the matter (or energy-momentum) tensor. It is well known that the Einstein gravity in (2+1)-dimensions is dynamically trivial, without propagating degrees of freedom. In three dimensions the Riemann tensor does not have any other components as those given by the Ricci tensor. Furthermore, for the three-dimensional spacetimes the Weyl tensor is always zero and the vanishing of the Cotton tensor depends on the kind of relation of the Ricci tensor to matter. The Cotton tensor is also identically zero when we use the Einstein-Hilbert action for gravity. In most situations of physical interest, we have spacetime symmetries which reduce further the number of unknown functions [5]. Therefore, understanding the symmetry structure of spacetimes in any dimensions provide an important contribution to finding exact solution of the field equations.

The line element of (2+1)-FRW spacetimes, in coordinates $x^a = \{t, x, y\}$, is given by

$$ds^2 = -dt^2 + \frac{S^2}{\left(1 + \frac{K}{4}r^2\right)^2}(dx^2 + dy^2), \tag{2}$$

where $S(t)$ is arbitrary (non-zero) scale factor, $K = 0, \pm 1$ and $r^2 = x^2 + y^2$. The spacetimes with constant curvature are a spacial case of the FRW spacetimes. Also, it is well-known that the FRW metrics are conformally flat and hence these metrics admit the maximum number of independent symmetry vectors. The Ricci tensor R_{ab} and the scalar curvature $R(= R_a^a)$ for (2+1)-FRW metric (2) are given by

$$R_{00} \equiv R_0 = -2\frac{\ddot{S}}{S}, \quad R_{11} = R_{22} = \frac{R_1(t)}{\left(1 + \frac{K}{4}r^2\right)^2}, \tag{3}$$

$$R = \frac{2}{S^2}\left(2S\ddot{S} + \dot{S}^2 + K\right), \tag{4}$$

where $R_1(t) = S\ddot{S} + \dot{S}^2 + K$ and the dot denotes derivative with respect to t.

In the next section, we shall present some general results on conformal symmetries and collineations.

2. GENERAL RESULTS ON COLLINEATIONS

Symmetries of the geometrical/physical relevant quantities of the gravity theory are known as *collineations*. In general, these can be represented as $\pounds_\xi A = B$, where A and B are the geometric/physical objects, ξ is the vector field generating the symmetry, and $\pounds_\xi$ signifies the Lie derivative operator along the vector field ξ.

A one-parameter group of conformal motions generated by a *conformal Killing vector* (CKV) ξ is defined by[4]

$$\pounds_\xi g_{ab} = 2\psi(x)g_{ab} \Leftrightarrow g_{ab,c}\xi^c + g_{ac}\xi^c_{,b} + g_{cb}\xi^c_{,a} = 2\psi(x)g_{ab}, \tag{5}$$

where the indices $a, b, c, \ldots$ run from 0 to 2, and ", a" represents differentiation with respect to x^a. If ψ is not a constant on M then ξ is called a *proper* CKV. If $\psi_{;ab} = 0$, then ξ is called a *special conformal Killing Vector* (SCKV). If ψ is constant on M

then ξ is called a *homothetic vector* (HV) and if $\psi = 0$ then ξ is said to be a *Killing vector* (KV) which is the well known definition of a spacetime symmetry which is based on the concept of *isometry*. An SCKV is called a proper SCKV if $\psi_{;a} \neq 0$ and an HV is called a proper HV if $\psi \neq 0$. The set of all CKV (respectively, HV and KV) form a finite-dimensional Lie algebra denoted by C (respectively, H and K). The dimensions of the Lie algebras C, H and K on M satisfy the well known results that $dim(C) \leq 10$, $dim(H) \leq 7$ and $dim(K) \leq 6$ which are recently described by Hall and Capocci [3]. The maximum dimension of the algebra of CKV on M is achieved if M is conformally flat. If the spacetime is not conformally flat, then the conformal algebra C satisfies $dim(C) \leq 4$. If $dim(K) \geq 5$, then M has constant curvature.

A vector field $\mathbf{X} = X^a \frac{\partial}{\partial x^a}$ is called a *Ricci conformal collineation* (RCC) if it satisfies the equation

$$\pounds_{\mathbf{X}} R_{ab} = 2\phi(x) R_{ab} \Leftrightarrow R_{ab,c} X^c + R_{ac} X^c_{,b} + R_{cb} X^c_{,a} = 2\phi(x) R_{ab}. \tag{6}$$

If $\phi = 0$, then $\pounds_{\mathbf{X}} R_{ab} = 0$ and $\mathbf{X}$ is said to generate a *Ricci collineation* (RC) [4]. We notice that Eq. (6) has the same form as Eq.(5) but with the metric tensor is replaced by the Ricci tensor. Thus, if the Ricci tensor is *non-degenerate*, that is, $det(R_{ab}) \neq 0$, we may apply standard results on CKVs (KVs) to RCCs (RCs) in three-dimensions. Then, it follows that the family of RCCs (and RCs) is Lie algebra of smooth vector fields on M of *finite dimensions* ≤ 10 (and ≤ 6) if the R_{ab} is non-degenerate. The set of all RCCs and RCs on M is a vector space, but it may be infinite dimensional and may not be a Lie algebra. If the R_{ab} is non-degenerate, the Lie algebra of RCCs and RCs is finite-dimensional. If the R_{ab} is *degenerate*, that is, $det(R_{ab}) = 0$, we cannot guarantee the finite dimensionality of the RCCs and RCs.

If any spacetime admit a foliation by non-null three-dimensional conformally flat hypersurfaces, then it can be obtained all the KVs and CKVs of the full spacetimes (if any exist) by "lifting" the CKVs in three-dimensional hypersurfaces. This technique has been applied recently by the author and Barnes to obtain all RCCs and RCs of the (3+1)-FRW metrics [1].

In the following section, we will obtain the solutions of Killing, homothetic and conformal Killing equations using (5).

3. KILLING, HOMOTHETIC AND CONFORMAL KILLING VECTORS

For line element (2) of (2+1)-FRW spacetimes, the conformal Killing equations (5) in expanded form are

$$\tilde{g}_{\alpha\alpha,\gamma} \xi^\gamma + \dot{\tilde{g}}_{\alpha\alpha} \xi^0 + 2\tilde{g}_{\alpha\alpha} \xi^\gamma_{,\alpha} = 2\psi \tilde{g}_{\alpha\alpha}, \tag{7}$$

$$\xi^0_{,t} = \psi, \quad \tilde{g}_{\alpha\gamma} \xi^\gamma_{,\beta} + \tilde{g}_{\gamma\beta} \xi^\gamma_{,\alpha} = 0, (\alpha \neq \beta) \tag{8}$$

where the dot denotes derivative with respect to time t; α, β take values 1,2, and $\tilde{g}_{\alpha\beta}$ is the two-dimensional metric

$$d\tilde{s}^2 = \left(1 + \frac{K}{4}r^2\right)^{-2}(dx^2 + dy^2). \tag{9}$$

Then, we will obtain the solutions of the above equations (7) and (8) in the cases of Killing, homothetic and conformal Killing vector fields.

3.1. Killing Vectors

In the case of Killing equation, i.e. $\psi = 0$, we obtained that when $K = \pm 1$, the KVs are

$$\xi_{(1)} = x\partial_y - y\partial_x, \; \xi_{(2)} = \mathbf{Y}_{(1^+)}, \; \xi_{(3)} = \mathbf{Y}_{(2^+)}, \; \xi_{(4)} = \partial_t - \alpha_2(x\partial_x + y\partial_y), \tag{10}$$

where $S = \alpha_1 e^{\alpha_2 t}$, ($\alpha_1, \alpha_2$ are constants) and we have defined $\mathbf{Y}_{(1^\pm)}$ and $\mathbf{Y}_{(2^\pm)}$ as

$$\mathbf{Y}_{(1^\pm)} = \left[\mp\frac{2}{K} + \frac{1}{2}(-x^2 + y^2)\right]\partial_x - xy\partial_y, \tag{11}$$

$$\mathbf{Y}_{(2^\pm)} = \left[\mp\frac{2}{K} + \frac{1}{2}(x^2 - y^2)\right]\partial_y - xy\partial_x. \tag{12}$$

If the following constraint equation for $S(t)$ is satisfied,

$$(\dot{S}/S)^{\cdot} = K/S^2, \tag{13}$$

then in addition to the KVs given in (10), we have

$$\xi_{(4)} = \frac{x}{1 + \frac{K}{4}r^2}\partial_t + \frac{\dot{S}}{2S}\mathbf{Y}_{(1^-)}, \quad \xi_{(5)} = \frac{y}{1 + \frac{K}{4}r^2}\partial_t + \frac{\dot{S}}{2S}\mathbf{Y}_{(2^-)}. \tag{14}$$

Now, for $K = 0$, it is obtained from the Killing equations that

$$\xi_{(1)} = x\partial_y - y\partial_x, \quad \xi_{(2)} = \partial_x, \quad \xi_{(3)} = \partial_y, \quad \xi_{(4)} = \partial_t - \alpha_2(x\partial_x + y\partial_y),$$

$$\xi_{(5)} = x\partial_t + \left[\frac{\alpha_2}{2}(-x^2 + y^2) - \frac{e^{-2\alpha_2 t}}{2\alpha_1^2\alpha_2}\right]\partial_x - \alpha_2 xy\partial_y, \tag{15}$$

$$\xi_{(6)} = y\partial_t + \left[\frac{\alpha_2}{2}(x^2 - y^2) - \frac{e^{-2\alpha_2 t}}{2\alpha_1^2\alpha_2}\right]\partial_y - \alpha_2 xy\partial_x,$$

where $S = \alpha_1 e^{\alpha_2 t}$.

3.2. Homothetic Vectors

If ψ is a constant in Eq.(5), then solving the homothetic Killing equation we have found the HVs that when $K \neq 0$ in which the possibility is only $K = -1$,

$$\xi_{(1)} = x\partial_y - y\partial_x + \psi t\partial_t, \quad \xi_{(2)} = Y_{(1+)} + \psi t\partial_t, \quad \xi_{(3)} = Y_{(2+)} + \psi t\partial_t,$$

$$\xi_{(4)} = \left(\frac{x}{1-r^2/4} + \psi t\right)\partial_t + \frac{1}{2t}Y_{(1-)}, \; \xi_{(5)} = \left(\frac{y}{1-r^2/4} + \psi t\right)\partial_t + \frac{1}{2t}Y_{(2-}(16)$$

$$\xi_{(6)} = \left(\frac{2}{1-r^2/4} + \psi t\right)\partial_t - \frac{\psi}{\psi t + 1}(x\partial_x + y\partial_y), \quad \xi_{(7)} = (\psi t + 1)\partial_t,$$

where $\psi = \pm 1$, and $S = \psi t$ for $\xi_{(1)}, ..., \xi_{(5)}$ and $S = \psi t + 1$ for $\xi_{(6)}$ and $\xi_{(7)}$. In case $K = 0$, the obtained HVs are

$$\xi_{(1)} = x\partial_y - y\partial_x + \psi t\partial_t, \quad \xi_{(2)} = \partial_x + \psi t\partial_t, \quad \xi_{(3)} = \partial_y + \psi t\partial_t,$$
$$\xi_{(4)} = \psi t\partial_t + x\partial_x + y\partial_y, \quad \xi_{(5)} = (-x^2 + y^2)\partial_x - 2xy\partial_y + \psi t\partial_t, \quad (17)$$
$$\xi_{(6)} = (x^2 - y^2)\partial_y - 2xy\partial_x + \psi t\partial_t, \quad \xi_{(7)} = (\psi t + 1)\partial_t,$$

where $S = \psi t + 1$ for $\xi_{(7)}$ only and $S = \psi t$ for the remaining HVs.

3.3. Conformal Killing Vectors

When $K = \pm 1$, the proper CKVs and the related conformal factors are obtained as follows

$$\xi_{(4)} = \partial_\tau, \qquad \qquad \psi_{(4)} = \frac{S_{,\tau}}{S},$$

$$\xi_{(5)} = \frac{-Kxf(\tau)_{,\tau}}{1+\frac{K}{4}r^2}\partial_\tau - \frac{f(\tau)}{2}Y_{(1-)}, \qquad \psi_{(5)} = \frac{x}{1+\frac{K}{4}r^2}\left[f(\tau) - Kf(\tau)_{,\tau}\frac{S_{,\tau}}{S}\right],$$

$$\xi_{(6)} = \frac{xf(\tau)}{1+\frac{K}{4}r^2}\partial_\tau - \frac{f(\tau)_{,\tau}}{2}Y_{(1-)}, \qquad \psi_{(6)} = \frac{x}{1+\frac{K}{4}r^2}\left[f(\tau)_{,\tau} - Kf(\tau)\frac{S_{,\tau}}{S}\right],$$

$$\xi_{(7)} = \frac{-Kyf(\tau)_{,\tau}}{1+\frac{K}{4}r^2}\partial_\tau - \frac{f(\tau)}{2}Y_{(2-)}, \qquad \psi_{(7)} = \frac{y}{1+\frac{K}{4}r^2}\left[f(\tau) - Kf(\tau)_{,\tau}\frac{S_{,\tau}}{S}\right], \; (18)$$

$$\xi_{(8)} = \frac{yf(\tau)}{1+\frac{K}{4}r^2}\partial_\tau - \frac{f(\tau)_{,\tau}}{2}Y_{(2-)}, \qquad \psi_{(8)} = \frac{y}{1+\frac{K}{4}r^2}\left[f(\tau)_{,\tau} - Kf(\tau)\frac{S_{,\tau}}{S}\right],$$

$$\xi_{(9)} = \frac{-Kf(\tau)_{,\tau}\left(1-\frac{K}{4}r^2\right)}{\left(1+\frac{K}{4}r^2\right)}\partial_\tau + f(\tau)(x\partial_x + y\partial_y), \quad \psi_{(9)} = \frac{1-\frac{K}{4}r^2}{1+\frac{K}{4}r^2}\left[f(\tau) - Kf(\tau)_{,\tau}\frac{S_{,\tau}}{S}\right],$$

$$\xi_{(10)} = \frac{f(\tau)\left(1-\frac{K}{4}r^2\right)}{\left(1+\frac{K}{4}r^2\right)}\partial_\tau - Kf(\tau)_{,\tau}(x\partial_x + y\partial_y), \quad \psi_{(10)} = -\frac{\left(1-\frac{K}{4}r^2\right)}{\left(1+\frac{K}{4}r^2\right)}\left[f(\tau)_{,\tau} - Kf(\tau)\frac{S_{,\tau}}{S}\right],$$

where we have defined τ as $\tau = \int dt/S$ and the function $f(\tau) = \cos\tau, \cosh\tau$ for $K = +1, -1$, respectively. Now, when $K = 0$, the proper CKVs and scaling factors are given by

$$\xi_{(4)} = \partial_\tau, \ \psi_{(4)} = \frac{S_{,\tau}}{S}; \quad \xi_{(5)} = x\partial_\tau + \tau\partial_x, \qquad \psi_{(5)} = x\frac{S_{,\tau}}{S},$$

$$\xi_{(6)} = y\partial_\tau + \tau\partial_y, \ \psi_{(6)} = y\frac{S_{,\tau}}{S}; \ \xi_{(7)} = \tau\partial_\tau + x\partial_x + y\partial_y, \ \psi_{(7)} = 1 + \tau\frac{S_{,\tau}}{S},$$

$$\xi_{(8)} = \left[-x^2 + y^2 - \tau^2\right]\partial_x - 2x(\tau\partial_\tau + y\partial_y), \qquad \psi_{(8)} = -2x\left[1 + \tau\frac{S_{,\tau}}{S}\right], \quad (19)$$

$$\xi_{(9)} = \left[x^2 - y^2 - \tau^2\right]\partial_y - 2y(\tau\partial_\tau + x\partial_x), \qquad \psi_{(9)} = -2y\left[1 + \tau\frac{S_{,\tau}}{S}\right],$$

$$\xi_{(10)} = \tfrac{1}{2}\left(\tau^2 + r^2\right)\partial_\tau + \tau(x\partial_x + y\partial_y), \qquad \psi_{(10)} = \tau + \frac{1}{2}\left(\tau^2 + r^2\right)\frac{S_{,\tau}}{S}.$$

In the following section, we shall give RCC equations derived from (2+1)-FRW spacetime, and obtain a general classification of degenerate and non-degenerate RCCs for this spacetime.

4. COLLINEATIONS IN DEGENERATE AND NON-DEGENERATE CASES

For (2+1)-FRW spacetimes given by (2), using the non-zero Ricci tensor components (3), we can write the RCC equations (6), generated by an arbitrary vector field $X^a(t,x,y)$, in terms of $R_\alpha(t)$ as follows:

$$2R_0 X^0_{,t} + \dot{R}_0 X^0 = 2\phi R_0, \tag{20}$$

$$R_1 \pounds_X g_{\alpha\beta} = (2\phi R_1 - \dot{R}_1 X^0) g_{\alpha\beta}, \tag{21}$$

$$R_0 X^0_{,\alpha} + R_1 g_{\alpha\beta} X^\beta_{,t} = 0. \tag{22}$$

where $g_{\alpha\beta}$ is given by (9). Then we find that $det(R_{ab}) = R_0 R_1 \left(1 + \frac{K}{4}r^2\right)^{-2}$. Therefore, we will study the RCCs according to whether $det(R_{ab}) = 0$ (degenerate case) or $det(R_{ab}) \neq 0$ (non-degenerate case). Throughout the following part of this paper, if we use $\bar{t}$ instead of t, then we will consider the following definition of $\bar{t}$:

$$\bar{t} = \int \sqrt{|R_0/R_1|}\,dt. \tag{23}$$

Further, in this paper we will take the *proper* RC to denote an RC which is not a KV or HV.

4.1. Degenerate Ricci Tensor Case

In this section, due to degeneracy of the Ricci tensor, i.e. $det(R_{ab}) = 0$, we have the following possibilities:

119

Case (a): $R_0 = R_1 = 0$. In this case, we have that either $S(t) = S_0$ and $K = 0$ *or* $S(t) = t - t_0$ and $K = -1$. Hence, this case gives the flat spacetime, and so it corresponds to the vacuum case. Therefore, every vector is an RCC or RC.

Case (b): When only one of $R_i \neq 0$, $(i = 0, 1)$. This implies the following possibilities:

(b.i): $R_0 \neq 0$ and $R_1 = 0$. In this case, the solution of RCC equations yield

$$X^0 = \frac{1}{\sqrt{R_0}} \left[\int \phi \sqrt{R_0} \, dt + a_0 \right], \quad X^\alpha = X^\alpha(t, x, y), \quad \phi = X^0_{,t} + \frac{\dot{R}_0}{2R_0} X^0 \quad (24)$$

where a_0 is a constant and $\alpha = 1, 2$. For RCs, i.e., when $\phi = 0$, we find that $X^0 = a_0/\sqrt{R_0}$ and $X^\alpha = X^\alpha(t, x, y)$. Therefore, the groups of RCCs and RCs are both infinite dimensional. A first integral of equation $R_1 \equiv S\ddot{S} + \dot{S}^2 + K = 0$ is $\dot{S}^2 = S_0 S^{-2} - K$ which give $t - t_0 = \int S \, dS/(S_0 - KS^2)^{1/2}$ and hence $\ddot{S} = -S_0 S^{-3}$. For $K = 0$, we get $S(t) = c(t - t_0)^{1/2}$, $c = constant$. The solution when $K \neq 0$ is $S(t) = \left[K \left(S_0 - (t - t_0)^2 \right) \right]^{1/2}$.

(b.ii): $R_0 = 0$ and $R_1 \neq 0$. For this case, we get $S(t) = at + b$ and $R_1 = a^2 + K$, where $a \neq 1$ if $K = -1$ and $a \neq 0$ if $K = 0$, (a and b are integration constants). In this case, the spatial components X^α have the form

$$X^1 = a_1 - a_3 y + a_4 x + a_5(r^2 - 2x^2) - 2a_6 xy, \quad (25)$$
$$X^2 = a_2 + a_3 x + a_4 y - 2a_5 xy + a_6(r^2 - 2y^2), \quad (26)$$

where the a_j's ($j = 1, ..., 6$) are constants. The time component X^0 is completely arbitrary, that is, $X^0 = X^0(t, x, y)$ and the associated conformal factor is

$$\phi = \left(1 + \frac{K}{4} r^2\right)^{-1} \left[-\frac{K}{2}(a_x + a_2 y) + a_4 \left(1 - \frac{K}{4} r^2\right) - 2(a_5 x + a_6 y) \right]. \quad (27)$$

For RCs, the X^α's are given by

$$X^1 = a_1 \left[1 - \frac{K}{4}(-x^2 + y^2) \right] + a_2 Kxy/2 - a_3 y, \quad (28)$$

$$X^2 = a_1 Kxy/2 + a_2 \left[1 - \frac{K}{4}(x^2 - y^2) \right] + a_3 x, \quad (29)$$

and the time component X^0 is again completely arbitrary. Thus, both of the groups of RCCs and RCs are infinite dimensional.

4.2. Non-degenerate Ricci Tensor Case

In this section, all components of the Ricci tensor are different from zero, i.e. $R_i \neq 0$, $(i = 0, 1)$ since $det(R_{ab}) \neq 0$. Thus, it follows from the RCC equations (20)-(22) that

$$X^0 = -\frac{\sqrt{R_1/R_0}}{(1 + Kr^2/4)} \left[x f_{1,\bar{i}} + y f_{2,\bar{i}} - \frac{2}{K} f_{4,\bar{i}} \right] + h,$$
$$X^1 = f_1 - y f_3 + x f_4 + (-x^2 + y^2) f_5 - 2xy f_6, \quad (30)$$
$$X^2 = f_2 + x f_3 + y f_4 - 2xy f_5 + (x^2 - y^2) f_6,$$

and

$$\phi = \left(1+\frac{K}{4}r^2\right)^{-1}\left[-\frac{K}{2}(xf_1+yf_2)+\left(1-\frac{K}{4}r^2\right)f_4-2(xf_5+yf_6)\right]+\frac{\dot{B}}{2B}X^0 \quad (31)$$

where $f_j(\bar{t})$'s $(j=1,...,6)$ and $h(\bar{t})$ for $K=+1$ are given by

$$f_1 = a_1\cosh\bar{t}+b_1\sinh\bar{t}-2c_1, \quad f_2 = a_2\cosh\bar{t}+b_2\sinh\bar{t}-2c_2, \quad (32)$$

$$f_3 = c_3, \quad f_4 = a_3\cosh\bar{t}+b_3\sinh\bar{t}, \quad f_5 = \frac{1}{4}[a_1\cosh\bar{t}+b_1\sinh\bar{t}]+c_1/2, \quad (33)$$

$$f_6 = \frac{1}{4}[a_2\cosh\bar{t}+b_2\sinh\bar{t}]+c_2/2, \quad h = \sqrt{\frac{R_1}{R_0}}[-a_3\sinh\bar{t}-b_3\cosh\bar{t}+\ell], \quad (34)$$

and for $K=-1$,

$$f_1 = a_1\cos\bar{t}+b_1\sin\bar{t}+2c_1, \quad f_2 = a_2\cos\bar{t}+b_2\sin\bar{t}+2c_2, \quad (35)$$

$$f_3 = c_3, \quad f_4 = a_3\cos\bar{t}+b_3\sin\bar{t}, \quad f_5 = -\frac{1}{4}[a_1\cos\bar{t}+b_1\sin\bar{t}]+c_1/2, \quad (36)$$

$$f_6 = -\frac{1}{4}[a_2\cos\bar{t}+b_2\sin\bar{t}]+c_2/2, \quad h = \sqrt{\frac{R_1}{R_0}}[-a_3\sin\bar{t}+b_3\cos\bar{t}+\ell]. \quad (37)$$

When $K=0$, the obtained proper RCC vector fields and scaling factors are as follows

$$\mathbf{X}_{(4)} = \partial_{\bar{t}}, \ \phi_{(4)} = \frac{B_{,\bar{t}}}{2B}; \quad \mathbf{X}_{(5)} = \bar{t}\partial_{\bar{t}}+x\partial_x+y\partial_y, \quad \phi_{(5)} = 1+\bar{t}\frac{B_{,\bar{t}}}{2B},$$

$$\mathbf{X}_{(6)} = -y\partial_{\bar{t}}+\bar{t}\partial_y, \ \phi_{(6)} = -y\frac{B_{,\bar{t}}}{2B}; \quad \mathbf{X}_{(7)} = -x\partial_{\bar{t}}+\bar{t}\partial_x, \quad \phi_{(7)} = -x\frac{B_{,\bar{t}}}{2B},$$

$$\mathbf{X}_{(8)} = (-x^2+y^2+\bar{t}^2)\partial_x - 2x(\bar{t}\partial_{\bar{t}}+y\partial_y), \quad \phi_{(8)} = -2x\left(1+\bar{t}\frac{B_{,\bar{t}}}{2B}\right)(38)$$

$$\mathbf{X}_{(9)} = (x^2-y^2+\bar{t}^2)\partial_y - 2y(\bar{t}\partial_{\bar{t}}+x\partial_x), \quad \phi_{(9)} = -2y\left(1+\bar{t}\frac{B_{,\bar{t}}}{2B}\right),$$

$$\mathbf{X}_{(10)} = \tfrac{1}{2}(\bar{t}^2-r^2)\partial_{\bar{t}}+\bar{t}(x\partial_x+y\partial_y), \quad \phi_{(10)} = \bar{t}+(\bar{t}^2-r^2)\frac{B_{,\bar{t}}}{2B}.$$

For proper RCs in non-degenerate case, we have the following possibilities:
Case(i): $R_0 \neq 0 \neq R_1$. In this case, we have that $B = K\bar{t}^2$ and proper RCs for $K \neq 0$ are obtained as

$$\mathbf{X}_{(4)} = -x\left(1+\frac{K}{4}r^2\right)^{-1}\partial_{\bar{t}} - \frac{1}{2\bar{t}}\mathbf{Y}_{(1-)}, \quad \mathbf{X}_{(5)} = -y\left(1+\frac{K}{4}r^2\right)^{-1}\partial_{\bar{t}} - \frac{1}{2\bar{t}}\mathbf{Y}_{(2-)},$$

$$\mathbf{X}_{(6)} = (1/K)\left(1-\frac{K}{4}r^2\right)\left(1+\frac{K}{4}r^2\right)^{-1}\partial_{\bar{t}} - \frac{1}{K\bar{t}}(x\partial_x+y\partial_y).$$

Case(ii): $R_0 \neq 0$ and $\dot{R}_1 = 0$, i.e. $B =$ constant. For this case, the proper Rc is only
$\mathbf{X}_{(4)} = (1/\sqrt{R_0})\partial_t.$

121

Case (iii): $R_0 \neq 0$ and $R_{11,x} = 0$ (or $R_{11,y} = 0$) which yields $K = 0$. In this case, the proper RCs are given by

$$X_{(4)} = -2\alpha_2 \partial_{\bar{t}} + x\partial_x + y\partial_y,$$

$$X_{(5)} = 4\alpha_2 x \partial_{\bar{t}} - 2xy\partial_y + \left[-x^2 + y^2 + (4\alpha_2^2/\alpha_1)e^{-\bar{t}/\alpha_2} \right] \partial_x,$$

$$X_{(6)} = 4\alpha_2 y \partial_{\bar{t}} - 2xy\partial_x + \left[x^2 - y^2 + (4\alpha_2^2/\alpha_1)e^{-\bar{t}/\alpha_2} \right] \partial_y,$$

where $B = \alpha_1 e^{\bar{t}/\alpha_2}$.

Case (iv): $\dot{R}_0 = 0$ and $R_1 \neq 0$. In this case, we have $R_0 = A_0 = $ constant which gives $\ddot{S} + A_0 S/2 = 0$ and $R_1 = K - A_0 S/2 + \dot{S}^2$. If $A_0 > 0$, then we get that $S = k\cos\left(\sqrt{(A_0/2)}\,t\right) + \ell\sin\left(\sqrt{(A_0/2)}\,t\right)$ and $B = 2K - A_0 S^2$ which yield the property $k^2 + \ell^2 = 2K/A_0$. If $A_0 < 0$, then we have that $S = k\cosh\left(\sqrt{(|A_0|/2)}\,t\right) + \ell\sinh\left(\sqrt{(|A_0|/2)}\,t\right)$ and $B = 2K + |A_0| S^2$ which gives the property $\ell^2 - k^2 = 2K/|A_0|$. For this case, the obtained RCs are

$$X_{(4)} = -x\left(1 + \frac{K}{4}r^2\right)^{-1} \partial_t - \frac{\dot{B}}{4B}Y_{(1-)}, \quad X_{(5)} = -y\left(1 + \frac{K}{4}r^2\right)^{-1} \partial_t - \frac{\dot{B}}{4B}Y_{(2-)},$$

$$X_{(6)} = -\left(1 - \frac{K}{4}r^2\right)\left(1 + \frac{K}{4}r^2\right)^{-1} \partial_t + \frac{\dot{B}}{2B}(x\partial_x + y\partial_y).$$

5. CONCLUDING REMARKS

In this study, we have provided that the general solutions of Killing, homothetic and conformal Killing equations for (2+1)-FRW metrics. The number or obtained KVs are 4, 5 and 6. In the case of homothety, there is 7 parameter group of HVs. As we expected, the number of CKVs is 10. Later, the most general form of a vector generating a RCC or RC has been obtained for both degenerate and non-degenerate cases. When $det(R_{ab}) = 0$ (degenerate case), the groups of RCs and RCCs of (2+1)-FRW metric are both infinite dimensional. The FRW scale factors leading to these infinite dimensional groups have been obtained in closed forms. When $det(R_{ab}) \neq 0$ (non-degenerate case), RCs and RCCs are respectively KVs and CKVs of the Ricci tensor metric. For (2+1)-FRW metrics, the Ricci tensor metric also assumes FRW form and so it follows immediately that the group of RCCs is of dimension 10 and for a generic (2+1)-FRW metric, the group of RCs is of dimension 6.

ACKNOWLEDGMENTS

The author thanks the workshop organization for their nice organizing.

122

REFERENCES

1. U. Camcı and A. Barnes, *Ricci collineations in Friedmann-Robertson-Walker spacetimes*, *Class. Quant. Grav.*, **19**, (2002), 393-404.
2. S. Deser, R. Jackiw and S. Templaton, *3-Dimensional Massive Gauge-Theories*, *Phys. Rev. Lett.***48**, 975-978 (1982); S. Deser, R. Jackiw and S. Templaton, *Topologically Massive Gauge-Theories*, *Ann. Phys.*, NY, **140**, (1982), 372-411.
3. G. S. Hall and M. S. Capocci, *Classification and conformal symmetry in three-dimensional space-times*, *J. Math Phys.*, **40**, (1999), 1466-1478.
4. G. H. Katzin, J. Levine, and W. R. Davis, *Curvature collineations a fundamental symmetry property of the space-times of general relativity defined by the vanishing Lie derivative of the Riemannian curvature tensor*, *J. Math. Phys.*, **10**, (1969), 617-629.
5. D. Kramer, H. Stephani, M. A. H. MacCallum, and E. Herlt, *Exact Solutions of Einstein Field Equations*, (VEB Deutscher Verlag der Wissenschaften, Berlin/Cambridge U.P., Cambridge, 1980).

Symmetries and supersymmetries of Dirac-type operators on Euclidean Taub-NUT space

Ion I. Cotăescu* and Mihai Visinescu[†]

*West University of Timişoara, V. Pârvan Ave. 4, RO-1900 Timişoara, Romania
[†]Department of Theoretical Physics, National Institute for Physics and Nuclear Engineering,
P.O.Box M.G.-6, Magurele, Bucharest, Romania

Abstract. The role of the Killing-Yano tensors in the construction of the Dirac-type operators is pointed out. The general results are applied to the case of the four-dimensional Euclidean Taub-NUT space. Three new Dirac-type operators, equivalent to the standard Dirac one, are constructed from the covariantly constant Killing-Yano tensors of this space. In this space there is also a non covariantly constant Killing-Yano tensor connected with hidden symmetries. The Runge-Lenz operator for the Dirac equation in this background is written down pointing out its algebraic properties.
Key words. Killing vectors, Killing-Yano tensors, Dirac-type operators, isometries, symmetries, supersymmetries.
PACS. 04.62.+v.
MSC. 83C47.

1. INTRODUCTION

Involved in many modern studies in physics [14], the metric of the Euclidean Taub-Newman-Unti-Tamburino (Taub-NUT) space is a gravitational instanton solution [15] of the Euclidean Einstein equations without cosmological constant. The Taub-NUT space is of interest since beside isometries there are hidden symmetries giving rise to conserved quantities associated to Stäckel-Killing (S-K) tensors. There is a conserved vector, analogous to the Runge-Lenz vector of the Kepler type problem, whose existence is rather surprising in view of the complexity of the equations of motion [13]. These hidden symmetries are related to the existence of four Killing-Yano (K-Y) tensors generating the S-K ones [13, 16]. The K-Y tensor here is a 2-form $f_{\mu\nu} = -f_{\nu\mu}$ which satisfies the equation $f_{\mu\nu;\lambda} + f_{\mu\lambda;\nu} = 0$.

The quantum theory in the Euclidean Taub-NUT background has also interesting specific features in the case of the scalar fields [7] as well as for fields with spin $\frac{1}{2}$ [9]-[4]. In both cases there exist large algebras of conserved observables [5] including the components of the angular momentum and three components of the Runge-Lenz operator that lead to six-dimensional dynamical algebras [4, 5].

CP729, *Global Analysis and Applied Mathematics: International Workshop on Global Analysis*,
edited by K. Taş, D. Krupka, O. Krupková, and D. Baleanu

Carter and McLenaghan [2] showed that in the theory of Dirac fermions for any isometry with Killing vector k_μ there is an appropriate operator

$$X_k = -i(k^\mu \hat{\nabla}_\mu - \frac{1}{4}\gamma^\mu \gamma^\nu k_{\mu;\nu}), \tag{1}$$

which commutes with the *standard* Dirac operator. Moreover, each Killing-Yano tensor $f_{\mu\nu}$ produces a *non-standard* Dirac operator of the form

$$D_f = -i\gamma^\mu (f_\mu{}^\nu \hat{\nabla}_\nu - \frac{1}{6}\gamma^\nu \gamma^\rho f_{\mu\nu;\rho}) \tag{2}$$

which also anticommutes with the standard Dirac operator.

2. THE EUCLIDEAN TAUB-NUT SPACE

Let us consider the Euclidean Taub-NUT space and the chart with Cartesian coordinates x^μ $(\mu, \nu, \ldots = 1,2,3,4)$ where x^i $(i, j, \ldots = 1,2,3)$ are the *physical* space coordinates while x^4 is the Cartesian extra-coordinate. We use the three-dimensional vector notations, $\vec{x} = (x^1, x^2, x^3)$, $r = |\vec{x}|$ and $dl^2 = d\vec{x} \cdot d\vec{x}$, for writing the line element

$$ds^2 = \frac{1}{V(r)}dl^2 + V(r)[dx^4 + A_i^{em}(\vec{x})dx^i]^2, \tag{3}$$

defined by the specific functions

$$\frac{1}{V} = 1 + \frac{\mu}{r}, \quad A_1^{em} = -\frac{\mu}{r}\frac{x^2}{r+x^3}, \quad A_2^{em} = \frac{\mu}{r}\frac{x^1}{r+x^3}, \quad A_3^{em} = 0. \tag{4}$$

The real number μ is a parameter of the theory. If one interprets $\vec{A}^{em}$ as the vector potential (or gauge field) it results the field of a magnetic monopole $\vec{B}^{em} = \mu \frac{\vec{x}}{r^3}$.

In the Cartesian charts one can choose a diagonal gauge with the gauge fields $\hat{e}^{\hat{\alpha}}$ and $e_{\hat{\alpha}}$ having the non-vanishing components [1]

$$\hat{e}^i_j = \frac{1}{\sqrt{V}}\delta_{ij}, \quad \hat{e}^4_i = \sqrt{V}A_i^{em}, \quad \hat{e}^4_4 = \sqrt{V},$$

$$e^i_j = \sqrt{V}\delta_{ij}, \quad e^4_i = -\sqrt{V}A_i^{em}, \quad e^4_4 = \frac{1}{\sqrt{V}}. \tag{5}$$

In this gauge the Killing vectors $k_{(i)}$ of the Taub-NUT space have the components

$$k^i_{(j)} = \frac{1}{\sqrt{V}}\varepsilon_{ijk}x^k, \quad k^4_{(i)} = -\mu\frac{x^i}{r}\sqrt{V}, \tag{6}$$

which covariantly transform under linear $SO(3)$ rotations. In this context one can correctly define the three-dimensional physical momentum $\vec{P}$ whose components in the above defined local frames are

$$P_i = -i\frac{1}{\sqrt{V}}e_i^\mu\partial_\mu = -i(\partial_i - A_i^{em}\partial_4)\,, \tag{7}$$

and the fourth component of the momentum operator is $P_4 = -i\partial_4$. In addition, the angular momentum can be written in covariant form as

$$\vec{L} = \vec{x}\times\vec{P} - \mu\frac{\vec{x}}{r}P_4\,. \tag{8}$$

They obey $[P_i, P_j] = i\varepsilon_{ijk}B_k^{em}P_4$, $[P_i, P_4] = 0$ and $[L_i, P_j] = i\varepsilon_{ijk}P_k$ which indicate that $\vec{P}$ behaves as a vector under rotations.

The scalar quantum mechanics in the Taub-NUT geometry is based on the static operator

$$\Delta = -\nabla_\mu g^{\mu\nu}\nabla_\nu = V\vec{P}^2 + \frac{1}{V}P_4^2\,, \tag{9}$$

which is either proportional with the Hamiltonian operator of the Schrödinger theory or represents the static part of the Klein-Gordon operator [5]. In both cases we are interested to find operators commuting with Δ since these give rise to the conserved quantities with physical significance.

The Euclidean Taub-NUT space is a hyper-Kähler manifold possessing a triplet of covariantly constant Killing-Yano tensors, $\mathbf{f} = \{f^{(1)}, f^{(2)}, f^{(3)}\}$, defined as

$$f^{(i)} = f_{\hat{\alpha}\hat{\beta}}^{(i)}\hat{e}^\alpha\wedge\hat{e}^\beta = 2\hat{e}^i\wedge\hat{e}^4 - \varepsilon_{ijk}\hat{e}^j\wedge\hat{e}^k\,. \tag{10}$$

In addition, there exists a fourth K-Y tensor,

$$f^Y = f_{\hat{\alpha}\hat{\beta}}^Y\hat{e}^\alpha\wedge\hat{e}^\beta = \frac{x^i}{r}f^{(i)} + \frac{2x^i}{\mu V}\varepsilon_{ijk}\hat{e}^j\wedge\hat{e}^k\,, \tag{11}$$

which is not covariantly constant. The presence of f^Y is due to the existence of the hidden symmetries of the Euclidean Taub-NUT geometry which are encapsulated in three non-trivial S-K tensors and interpreted as the components of the so-called Runge-Lenz vector of the Euclidean Taub-NUT problem. These S-K tensors can be expressed as symmetrized products of K-Y tensors [16],

$$k_{(i)\mu\nu} = \frac{\mu}{4}(f_{\mu\lambda}^Y f_{\ \nu}^{(i)\lambda} + f_{\nu\lambda}^Y f_{\ \mu}^{(i)\lambda}) + \frac{1}{2\mu}(k_{(4)\mu}k_{(i)\nu} + k_{(4)\nu}k_{(i)\mu})\,, \tag{12}$$

and with their help one defines the vector operator

$$\vec{K} = -\frac{1}{2}\nabla_\mu \vec{k}^{\mu\nu}\nabla_\nu = \frac{1}{2}\left(\vec{P}\times\vec{L} - \vec{L}\times\vec{P}\right) - \mu\frac{\vec{x}}{r}\left(\frac{1}{2}\Delta - P_4^2\right)\,, \tag{13}$$

which play the same role as the Runge-Lenz vector operator in the usual quantum mechanical Kepler problem. This transforms as a vector under $SO(3)$ rotations and one can write the following complete system of commutation relations

$$[L_i, L_j] = i\varepsilon_{ijk}L_k\,, \quad [L_i, K_j] = i\varepsilon_{ijk}K_k\,, \quad [K_i, K_j] = i\varepsilon_{ijk}L_k B^2\,, \tag{14}$$

126

where $B^2 = P_4{}^2 - \Delta$. The operators L_i and K_i commute with B since they commute with Δ and P_4. In addition, we observe that the new operators

$$C_1 = \vec{L}^2 B^2 + \vec{K}^2 = \frac{\mu^2}{4}\left(B^2 + P_4{}^2\right)^2 - B^2, \quad C_2 = \vec{L}\cdot\vec{K} = 0, \tag{15}$$

play the role of Casimir operators for the open algebra (14).

3. DIRAC-TYPE OPERATORS

For building the Dirac theory we consider the Cartesian chart, the usual four-dimensional space of the Dirac spinors, Ψ, and the Dirac matrices $\gamma^{\hat{\alpha}}$, that satisfy $\{\gamma^{\hat{\alpha}}, \gamma^{\hat{\beta}}\} = 2\delta_{\hat{\alpha}\hat{\beta}}$, in the following representation

$$\gamma^i = -i\begin{pmatrix} 0 & \sigma_i \\ -\sigma_i & 0 \end{pmatrix}, \quad \gamma^4 = \begin{pmatrix} 0 & \mathbf{1}_2 \\ \mathbf{1}_2 & 0 \end{pmatrix}, \tag{16}$$

where σ_i are the Pauli matrices and denote by $\gamma^5 = \gamma^1\gamma^2\gamma^3\gamma^4 = \mathrm{diag}(\mathbf{1}_2, -\mathbf{1}_2)$.

Let us start with the *standard* Dirac operator without explicit mass term defined now as $D_s = \gamma^\alpha \nabla_\alpha$. This is related to the Hamiltonian operator $H = \gamma^5 D_s$ [9, 4] which can be expressed in terms of $\pi = \sigma_P - iV^{-1}P_4$ and $\pi^* = \sigma_P + iV^{-1}P_4$ depending on $\sigma_P = \vec{\sigma}\cdot\vec{P}$:

$$H = \gamma^5 D_s = \begin{pmatrix} 0 & V\pi^*\frac{1}{\sqrt{V}} \\ \sqrt{V}\pi & 0 \end{pmatrix} = \begin{pmatrix} 0 & \alpha^* \\ \alpha & 0 \end{pmatrix}. \tag{17}$$

The Klein-Gordon operator (9) can be written as:

$$\Delta = \alpha^*\alpha = V\pi^*\pi. \tag{18}$$

We mention that here the star superscript is a mere notation that does not coincide with the Hermitian conjugation at the level of the Pauli operators which enter in the structure of the basic Dirac operators. The Hamiltonian operator that is the central piece of the Dirac theory has the remarkable property that the spectrum of the Hamiltonian operator coincides with the energy spectrum of the operator Δ. The main consequence is that the operator H is *invertible* since there are no zero modes.

The operators we intend to study here are operators of the Dirac theory which *commute* with the Hamiltonian operator (17) and we say that the Dirac operators which commute with H are conserved.

For the construction of conserved Dirac operators, it is useful to define the diagonal operators

$$D(\hat{X}) = \begin{pmatrix} \hat{X} & 0 \\ 0 & \alpha\hat{X}\Delta^{-1}\alpha^* \end{pmatrix}, \tag{19}$$

where $\hat{X}$ is a Pauli operator commuting with Δ. Particularly, for $\hat{X} = \mathbf{1}_2$ we obtain the projection operator

$$I = D(\mathbf{1}_2) = \begin{pmatrix} \mathbf{1}_2 & 0 \\ 0 & \alpha\Delta^{-1}\alpha^* \end{pmatrix}, \tag{20}$$

In [9] we introduced also the Q-operators defined as

$$Q(\hat{X}) = \left\{ H, \begin{pmatrix} \hat{X} & 0 \\ 0 & 0 \end{pmatrix} \right\} = \begin{pmatrix} 0 & \hat{X}\alpha^* \\ \alpha\hat{X} & 0 \end{pmatrix}, \tag{21}$$

where $\hat{X}$ may be any Pauli operator.

According to Eq. (1), the $U_4(1)$ generator is P_4 while other three Killing vectors give the $SO(3)$ generators connected with the components of the total angular momentum

$$\mathscr{J}_i = L_i + S_i, \tag{22}$$

where $S_i = \frac{1}{2}\varepsilon_{ijk}S^{jk} = \frac{1}{2}\mathrm{diag}(\sigma_i, \sigma_i)$ with $S^{jk} = -\frac{i}{4}[\gamma^j, \gamma^k]$.

The triplet $\mathbf{f}$ defined by Eq. (10) gives rise to the spin-like operators

$$\Sigma^{(i)} = \frac{i}{4}\hat{f}^{(i)}_{\hat{\alpha}\hat{\beta}}\gamma^{\hat{\alpha}}\gamma^{\hat{\beta}} = \begin{pmatrix} \sigma_i & 0 \\ 0 & 0 \end{pmatrix}, \tag{23}$$

and, according to Eq. (2), produces the Dirac-type operators [9]

$$D^{(i)} = -f^{(i)}_{\mu,\nu}\gamma^\nu\nabla^\mu = i[D_s, \Sigma^{(i)}] = -i\begin{pmatrix} 0 & \sigma_i\alpha^* \\ \alpha\sigma_i & 0 \end{pmatrix} = -iQ(\sigma_i), \tag{24}$$

which anticommute with D_s and γ^5. The operators D_s and $D^{(i)}$, $i = 1, 2, 3$, form the basis of the $\mathscr{N} = 4$ superalgebra with the same anticommutation relations

$$\left\{ D^{(i)}, D^{(j)} \right\} = \delta_{ij}D_s^2, \quad \left\{ D_s, D^{(j)} \right\} = 0. \tag{25}$$

When we are not interested to exploit the $\mathscr{N} = 4$ superalgebra, it is indicated to use the simpler operators

$$Q_i = iH^{-1}D^{(i)} = H^{-1}Q(\sigma_i) = D(\sigma_i), \tag{26}$$

instead of $D^{(i)}$. These operators form a representation of the quaternion algebra (or of the algebra of Pauli matrices)

$$Q_iQ_j = \delta_{ij}I + i\varepsilon_{ijk}Q_k, \tag{27}$$

producing an evident $\mathscr{N} = 3$ superalgebra.

The corresponding Dirac-type operator of the last K-Y tensor, f^Y, calculated according to the general rule (2) with a suitable phase factor (i), was obtained in [6]. This has the form

$$D^Y = -Q(\sigma_r) + \frac{2i}{\mu\sqrt{V}}\begin{pmatrix} 0 & \lambda \\ -\lambda & 0 \end{pmatrix}, \tag{28}$$

where $\sigma_r = \vec{\sigma} \cdot \vec{x}/r$ and $\lambda = \vec{\sigma} \cdot (\vec{x} \times \vec{P}) + 1_2 = \sigma_L + 1_2 + \mu\sigma_r P_4$.

One can verify that D^Y commutes with H and P_4 and anticommutes with D_s and γ^5.

As in the case of the Klein-Gordon theory, we can define the components of the conserved Runge-Lenz operator of the Dirac theory [6, 4]

$$\mathcal{K}_i I = \frac{i\mu}{4}\left\{ HD^Y, H^{-1}D^{(i)} \right\} + \frac{i}{2}(\mathcal{B} - P_4)H^{-1}D^{(i)} - \mathcal{J}_i I P_4 \,, \qquad (29)$$

where $\mathcal{B}^2 = P_4{}^2 - H^2$. Since $\mathcal{B}^2 I = \mathcal{D}(B^2)$, we can express $\mathcal{K}_i I = \mathcal{D}(\hat{K}_i)$, $\hat{K}_i = K_i + \frac{\sigma_i}{2}B$.

Furthermore, we obtain the following commutation relations

$$\left[\mathcal{J}_i, \mathcal{J}_j\right] = i\varepsilon_{ijk}\mathcal{J}_k, \quad \left[\mathcal{J}_i, \mathcal{K}_j\right] = i\varepsilon_{ijk}\mathcal{K}_k, \quad \left[\mathcal{K}_i, \mathcal{K}_j\right] = i\varepsilon_{ijk}\mathcal{J}_k\mathcal{B}^2 \qquad (30)$$

and the commutators with the operators Q_i [5],

$$\left[\mathcal{J}_i, Q_j\right] = i\varepsilon_{ijk}Q_k, \qquad \left[\mathcal{K}_i, Q_j\right] = i\varepsilon_{ijk}Q_k\mathcal{B}. \qquad (31)$$

Now we may ask how could be organized this very rich set of conserved Dirac operators. There are many commutation and anticommutation relations that can not be ignored such that it seems that the suitable structure may be a superalgebra. The main pieces here are the operators Q_i generating an usual $\mathcal{N} = 3$ superalgebra and the operators $\mathcal{J}_i$ and $\mathcal{K}_i$ giving dynamical algebras under suitable re scalings. This suggests us that the appropriate algebraic structure involving all the above ingredients may be an *infinite loop* superalgebra constructed as in [11].

Here we propose a version of a such superalgebra as an argument for further investigations of the new types of infinite algebraic structures [12]. Let us start with the definitions of the *bosonic* operators

$$I_n = \mathcal{B}^n I\,, \quad J_n^i = \mathcal{J}_i\mathcal{B}^n I\,, \quad K_n^i = \mathcal{K}_i\mathcal{B}^n I\,, \qquad (32)$$

and the supercharges of the *fermionic* sector

$$Q_n = Q\mathcal{B}^n I\,, \qquad Q_n^i = Q_i\mathcal{B}^n I\,, \qquad (33)$$

for all $n = 0, 1, 2, \ldots$ and $Q = 2\vec{\mathcal{J}} \cdot \vec{\mathcal{K}} - \frac{1}{2}\mathcal{B}$. Then, according to Eqs. (30) and (32), we obtain the following commutators of the bosonic sector

$$[I_n, I_m] = 0\,, \qquad [J_n^i, J_m^j] = i\varepsilon_{ijk}J_{n+m}^k\,, \qquad (34)$$

$$[I_n, J_m^i] = 0\,, \qquad [J_n^i, K_m^j] = i\varepsilon_{ijk}K_{n+m}^k\,, \qquad (35)$$

$$[I_n, K_m^i] = 0\,, \qquad [K_n^i, K_m^j] = i\varepsilon_{ijk}J_{n+m+2}^k\,, \qquad (36)$$

while from Eqs. (25) and (27) and we deduce the anticommutators of the fermionic sector,

$$\{Q_n^i, Q_m^j\} = 2\delta_{ij}I_{n+m}\,, \qquad (37)$$

$$\{Q_n, Q_m^i\} = 2(K_{n+m}^i + J_{n+m+1}^i)\,, \qquad (38)$$

$$\{Q_n, Q_m\} = c_0 I_{n+m} + c_1 I_{n+m+2} + c_2 I_{n+m+4}\,, \qquad (39)$$

where $c_0 = \frac{\mu^2 \hat{q}^4}{2}$, $\quad c_1 = \mu^2 \hat{q}^2$, $\quad c_2 = \frac{\mu^2}{2}$. In these structure constants the eigenvalue $\hat{q}$ of P_4 is fixed. Moreover the commutations relations between the bosonic and fermionic operators are

$$[Q_n, I_m] = 0, \qquad [Q_n^i, I_m] = 0, \tag{40}$$

$$[Q_n, J_m^j] = 0, \qquad [Q_n^i, J_m^j] = i\varepsilon_{ijk}Q_{n+m}^k, \tag{41}$$

$$[Q_n, K_m^j] = 0, \qquad [Q_n^i, K_m^j] = i\varepsilon_{ijk}Q_{n+m+1}^k. \tag{42}$$

From this infinite loop superalgebra one can extract the algebras or subalgebras one needs, including the dynamical algebras governing the quantum modes corresponding to the different spectral domains of H.

ACKNOWLEDGMENTS

M.V. thanks the organizers of the *IWGA 2004* for the invitation to present this work and for financial support.

REFERENCES

1. H. Boutaleb - Joutei and A. Chakrabarti, *Gauge field configurations in curved spacetimes. IV*, Phys. Rev., **D21**, (1980), 2280-2285.
2. B. Carter and R. G. McLenaghan, *Generalized total angular momentum operator for the Dirac equation in curved space-time*, Phys. Rev., **D19**, (1979), 1093-1097.
3. I. I. Cotăescu and M. Visinescu, *Dirac operators on Taub-NUT space: relationship and discrete transformations*, Gen. Relat. Grav., **35**, (2003), 389-400.
4. I. I. Cotăescu and M. Visinescu, *Dynamical algebra and Dirac quantum modes in the Taub-NUT background*, Class. Quantum Grav., **18**, (2001), 3383-3393.
5. I. I. Cotăescu and M. Visinescu, *Hierarchy of Dirac, Pauli, and Klein-Gordon conserved operators in Taub-NUT background*, J. Math. Phys., **43**, (2002), 2978-2987.
6. I. I. Cotăescu and M. Visinescu, *Runge-Lenz operator for Dirac field in Taub-NUT background*, Phys. Lett., **B502**, (2001), 229-234.
7. I. I. Cotăescu and M. Visinescu, *Schrödinger quantum modes in the Taub-NUT background*, Mod. Phys. Lett. A, **15**, (2000), 145-157.
8. I. I. Cotăescu and M. Visinescu, *Symmetries of the Dirac operators associated with covariantly constant Killing-Yano tensors*, Class. Quantum. Grav., **21**, (2004), 11-28.
9. I. I. Cotăescu and M. Visinescu, *The Dirac field in the Taub-NUT background*, Int. J. Mod. Phys. A, **16**, (2001), 1743-1758.
10. I. I. Cotăescu and M. Visinescu, *The induced representation of the isometry group of the Euclidean Taub-NUT space and new spherical harmonics*, Mod. Phys. Lett. A, **19**, (2004), 1397-1409.
11. J. Daboul, P. Slodowy and C. Daboul , *The hydrogen algebra as centerless twisted Kac-Moody algebra*, Phys. Lett., **B317**, (1993), 321-328.
12. L. Gy. Feher, private communication.
13. G. W. Gibbons and P. J. Ruback, *The hidden symmetries of multi-centre metrics*, Commun. Math. Phys., **115**, (1988), 267-300.
14. D. J. Gross and M. J. Perry, *Magnetic monopoles in Kaluza-Klein theories*, Nucl. Phys., **B226**, (1983), 29-48; R. D. Sorkin, *Kaluza-Klein monopoles*, Phys. Rev. Lett., **51**, (1983), 87-90.
15. S. W. Hawking, *Gravitational instantons*, Phys. Lett., **60A**, (1977), 81-85.
16. D. Vaman and M. Visinescu, *Supersymmetries and constants of motion in Taub-NUT spinning space*, Fortschr. Phys., **47**, (1999), 493-514.

Variational non-holonomic systems in physics

Lenka Czudková* and Jana Musilová*

*Faculty of Science, Masaryk University, Kotlářská 2, 611 37 Brno, Czech Republic

Abstract. In this paper conditions of variationality for mechanical systems subjected to non-holonomic constraints (constraint Helmholtz conditions), recently obtained by Krupková and Musilová, are examined in some concrete physical situations. Influence of the choice of constraints on the variationality is studied and corresponding constraint forces are obtained.

Key words. Fibred manifold, dynamical form, first-order mechanical system, constraint manifold, non-holonomic constraints, variational system.

PACS. 02.30.Xx, 45.20.Jj.

MSC. 49S05, 37J60, 70F25.

1. INTRODUCTION

Within the variational physical theories constrained systems are frequently studied. However, only the theory of holonomic constraints is satisfactorily elaborated. Recently, there were published a lot of papers concerning the geometry of non-holonomic systems with different approaches to this topic (see e.g. [1-3], [5], [7-18] and references therein). In this paper we adopt the geometrical theory of first-order mechanical systems with non-holonomic constraints in jet manifolds, developed by Krupková in 1990's ([5], [7]). A few years later, a new concept of variationality of non-holonomic systems was introduced and conditions of variationality (called the constraint Helmholtz conditions) were obtained (Krupková and Musilová, [10]). We recall the above mentioned results (Sections 2-4), and apply them to physical examples (Section 5).

Throughout this paper, standard notation is used. Let $\pi : Y \to X$ be a fibred manifold, $\dim Y = m+1$, $\dim X = 1$, $\pi_1 : J^1 Y \to X$, $\pi_2 : J^2 Y \to X$ its first and second jet prolongation and $\pi_{r,s} : J^r Y \to J^s Y$, $0 \leq r,s \leq 2$, $s < r$, $J^0 Y = Y$ canonical projections. Denote by (V, ψ), $\psi = (t, q^\sigma)$, $1 \leq \sigma \leq m$, a fibred chart on Y. There arise the associated fibred charts (V_1, ψ_1), $V_1 = \pi_{1,0}^{-1}(V)$, $\psi_1 = (t, q^\sigma, \dot{q}^\sigma)$ on $J^1 Y$ and (V_2, ψ_2), $V_2 = \pi_{2,0}^{-1}(V)$, $\psi_2 = (t, q^\sigma, \dot{q}^\sigma, \ddot{q}^\sigma)$ on $J^2 Y$.

Let γ be a section of the fibred manifold π and $J^r \gamma$, $1 \leq r \leq 2$, its r-th jet prolongation. A section $\delta : X \to J^r Y$ of π_r is called *holonomic* if there exists a section γ of π such that $\delta = J^r \gamma$.

A vector field ξ on $J^s Y$, $0 \leq s \leq 2$, is called π_s-*projectable* if there exists a vector field ξ_0 on X such that $T \pi_s \xi = \xi_0 \circ \pi_s$, and is called π_s-*vertical* if $T \pi_s \xi = 0$. By analogy, a $\pi_{r,s}$-*projectable* (respectively, $\pi_{r,s}$-*vertical*) vector field is defined. A form η on $J^s Y$ is called

CP729, *Global Analysis and Applied Mathematics: International Workshop on Global Analysis*,
edited by K. Taş, D. Krupka, O. Krupková, and D. Baleanu

π_s-*horizontal* if $i_\xi \eta = 0$ for every π_s-vertical vector field ξ on $J^s Y$, and is called *contact* if $J^s \gamma^* \eta = 0$, $s > 0$, for every section γ of π. Similarly a $\pi_{r,s}$-*horizontal* form is defined. In what follows, we use a basis $(\mathrm{d}t, \omega^\sigma, \mathrm{d}\dot{q}^\sigma)$ of 1-forms on $J^1 Y$ (respectively, a basis $(\mathrm{d}t, \omega^\sigma, \dot{\omega}^\sigma, \mathrm{d}\ddot{q}^\sigma)$ of 1-forms on $J^2 Y$) where $\omega^\sigma = \mathrm{d}q^\sigma - \dot{q}^\sigma \mathrm{d}t$ (respectively, $\dot{\omega}^\sigma = \mathrm{d}\dot{q}^\sigma - \ddot{q}^\sigma \mathrm{d}t$) are contact 1-forms. By Krupka's Decomposition Theorem ([4]), every k-form η on $J^s Y$ can be uniquely decomposed as follows: $\pi_{s+1,s}^* \eta = p_{k-1} \eta + p_k \eta$ where $p_{k-1} \eta$ (respectively, $p_k \eta$) is called $(k-1)$-*contact* (respectively, k-*contact*) *component* of η. The form η is called k-*contact* (respectively, $(k-1)$-*contact*) if $p_{k-1}\eta = 0$ (respectively, $p_k \eta = 0$). If $k = 1$ we denote $p_{k-1}\eta = p_0 \eta = h\eta$ the *horizontal component* of η.

A distribution $\mathscr{D}$ on $J^s Y$ is a mapping $J^s Y \ni x \to \mathscr{D}(x)$ assigning to every point $x \in J^s Y$ a vector subspace $\mathscr{D}(x)$ of the tangent space $T_x J^s Y$. A distribution is generated either by a system of local vector fields ξ_ι, $\iota \in \mathscr{I}$, on $J^s Y$ or by its *annihilators* η_κ, $\kappa \in \mathscr{K}$, such that $i_{\xi_\iota} \eta_\kappa = 0$ for every $\iota \in \mathscr{I}$, $\kappa \in \mathscr{K}$. We write: $\mathscr{D} = \mathrm{span} \{\xi_\iota \mid \iota \in \mathscr{I}\}$, $\mathscr{D}^0 = \mathrm{span} \{\eta_\kappa \mid \kappa \in \mathscr{K}\}$. A section δ of π_s is called an *integral section* of a distribution $\mathscr{D}$ if $\delta^* \eta = 0$ for every $\eta \in \mathscr{D}^0$.

2. UNCONSTRAINED MECHANICAL SYSTEMS

In this section we briefly recall without proofs fundamental concepts and ideas of a geometrical theory of (not necessarily Lagrangian) first-order mechanical systems on fibred manifolds (for more details see [5], [6]).

First, under a *dynamical form* we understand a 1-contact $\pi_{2,0}$-horizontal 2-form on $J^2 Y$, i.e. in every fibred chart it holds

$$E = E_\sigma (t, q^\rho, \dot{q}^\rho, \ddot{q}^\rho) \, \mathrm{d}q^\sigma \wedge \mathrm{d}t, \qquad 1 \le \sigma, \rho \le m.$$

A section γ of π is called a *path* of dynamical form E if

$$E \circ J^2 \gamma = 0.$$

Having on mind physical applications, only dynamical forms affine in the second derivatives will be considered here, i.e.

$$E_\sigma (t, q^\rho, \dot{q}^\rho, \ddot{q}^\rho) = A_\sigma (t, q^\rho, \dot{q}^\rho) + B_{\sigma \nu} (t, q^\rho, \dot{q}^\rho) \ddot{q}^\nu.$$

Then, the equations of paths take the form

$$A_\sigma + B_{\sigma \nu} \ddot{q}^\nu = 0 \quad \text{along} \quad J^2 \gamma$$

and are called *(unconstrained) equations of motion*.

By a *Lepagean 2-form associated to E* we mean a 2-form α on $J^1 Y$ such that $p_1 \alpha = E$. A direct calculation gives

$$\alpha = A_\sigma \, \omega^\sigma \wedge \mathrm{d}t + B_{\sigma \nu} \, \omega^\sigma \wedge \mathrm{d}\dot{q}^\nu + F_{\sigma \nu} \, \omega^\sigma \wedge \omega^\nu.$$

where $F_{\sigma v} = -F_{v\sigma}$ are some free functions on $J^1 Y$. Forms $i_\xi \alpha$ where ξ runs over the set of all π_1-vertical vector fields on $J^1 Y$ generate a distribution Δ_α on $J^1 Y$, called the *dynamical distribution of α*.

Put $\alpha_1 \sim \alpha_2$ iff $\alpha_1 - \alpha_2$ is a 2-contact 2-form on $J^1 Y$. The corresponding equivalence class $[\alpha]$ is called a *Lepagean class of E*, or a *first-order mechanical system associated to E*.

Proposition 1.
(a) *If $\alpha_1, \alpha_2 \in [\alpha]$ then the sets of holonomic integral sections of the distributions Δ_{α_1}, Δ_{α_2} coincide.*
(b) *A section γ of π is a path of $E = p_1\alpha$ iff $J^1\gamma$ is an integral section of any dynamical distribution Δ_α.*
(c) *A section γ of π is a path of $E = p_1\alpha$ iff $J^1\gamma^* i_\xi \alpha = 0$ for every π_1-vertical vector field ξ on $J^1 Y$.*

According to Proposition 1, a path of a dynamical form $E = p_1\alpha$ is also called a *path of the mechanical system $[\alpha]$ associated to E*.

Now, we are ready to recall the concept of variationality of E. A dynamical form E is *locally variational* if in a neighbourhood of every point $x \in J^1 Y$ there exists a Lagrangian $\lambda = L(t, q^\sigma, \dot{q}^\sigma)\, dt$ such that

$$E_\sigma = \frac{\partial L}{\partial q^\sigma} - \frac{d}{dt}\frac{\partial L}{\partial \dot{q}^\sigma}.$$

Theorem 1. *A dynamical form E is locally variational iff the corresponding mechanical system $[\alpha]$ contains a closed representative. Such a representative (called the Lepagean equivalent of E, and denoted by α_E) is unique and global.*

Remark 1. In a neighbourhood of every point $x \in J^1 Y$ it holds $\alpha_E = d\theta$ where $\theta = L\, dt + \frac{\partial L}{\partial \dot{q}^\sigma}\, \omega^\sigma$ is the *Cartan form* (or *Lepagean equivalent*) of *Lagrangian* $\lambda = L\, dt$.

Theorem 2. *A dynamical form E is locally variational iff for every fibred chart on $J^1 Y$ the following conditions (called Helmholtz conditions) are satisfied:*

$$\text{HC--1:} \quad (B_{\sigma v})_{\mathrm{alt}(\sigma v)} = 0, \qquad \text{HC--3:} \quad \left(\frac{\partial A_\sigma}{\partial \dot{q}^v} - \frac{d' B_{\sigma v}}{dt}\right)_{\mathrm{sym}(\sigma v)} = 0,$$

$$\text{HC--2:} \quad \left(\frac{\partial B_{\sigma v}}{\partial \dot{q}^\rho}\right)_{\mathrm{alt}(v\rho)} = 0, \qquad \text{HC--4:} \quad \left[\frac{\partial A_\sigma}{\partial q^v} - \frac{1}{2}\frac{d'}{dt}\left(\frac{\partial A_\sigma}{\partial \dot{q}^v}\right)\right]_{\mathrm{alt}(\sigma v)} = 0$$

where $\frac{d'}{dt} = \frac{\partial}{\partial t} + \dot{q}^\sigma \frac{\partial}{\partial q^\sigma}$.

3. CONSTRAINED MECHANICAL SYSTEMS

Now, let us consider the theory of constrained systems as introduced in [5].

By a *constraint manifold Q* we mean a submanifold of $J^1 Y$ fibred over Y, $\dim Q = 2m + 1 - k$, $1 \leq k \leq m - 1$. Q is locally defined by a system of first-order ordinary differential equations, called *non-holonomic constraints*, as follows:

$$f^i(t, q^\sigma, \dot{q}^\sigma) = 0, \qquad \text{rank}\left(\frac{\partial f^i}{\partial \dot{q}^\sigma}\right) = k, \qquad 1 \le i \le k.$$

These equations can be (locally) solved and rewritten into a *normal form*

$$\dot{q}^{m-k+i} - g^i\left(t, q^\sigma, \dot{q}^l\right) = 0, \qquad 1 \le l \le m - k.$$

Consider a neighbourhood $\mathcal{N}$ of Q, i.e. a union of open sets U_ι (in J^1Y) covering Q, and, on U_ι put

$$\varphi^i = f^i dt + \frac{\partial f^i}{\partial \dot{q}^\sigma} \omega^\sigma, \qquad 1 \le i \le k.$$

The above forms are called *(local) constraint 1-forms on* U_ι. Together with the forms df^i they generate the *constraint distribution* $\mathcal{C}_{U_\iota}$ on U_ι, i.e. $\mathcal{C}_{U_\iota}^0 = \text{span}\{\varphi^i, df^i \mid 1 \le i \le k\}$.

From the physical point of view, constraint forms give rise to a force

$$\Phi_{U_\iota} = \mu^i \frac{\partial f^i}{\partial \dot{q}^\sigma} dq^\sigma \wedge dt,$$

called a *Chetaev force*. Functions μ^i on U_ι are called *Lagrange multipliers*.

Remark 2. Chetaev force satisfies the *principle of virtual work* $i_\xi \Phi_{U_\iota} = 0$ for every π_1-vertical vector field ξ belonging to $\mathcal{C}_{U_\iota}$.

By means of a Chetaev force we define a *deformation of dynamical form* E, and a *deformation of a mechanical system* $[\alpha]$ *by the force* Φ_{U_ι} as follows:

$$E_{\Phi_{U_\iota}} = E - \Phi_{U_\iota}, \qquad \alpha_{\Phi_{U_\iota}} = \alpha - \Phi_{U_\iota}.$$

In keeping with the previous section one obtains the concept of a path of a deformed dynamical form:

Definition 1. A section γ of π is called a *path of a deformed dynamical form* $E_{\Phi_{U_\iota}}$ (i.e. a *path of a deformed mechanical system* $[\alpha_{\Phi_{U_\iota}}]$) if $f^i \circ J^1\gamma = 0$ together with one of the equivalent conditions:

$$
\begin{aligned}
E_{\Phi_{U_\iota}} \circ J^2\gamma &= 0, \\
J^1\gamma^* i_\xi \alpha_{\Phi_{U_\iota}} &= 0 \quad \text{for every } \pi_1 - \text{vertical vector field } \xi \text{ on } U_\iota \subset J^1Y, \\
A_\sigma + B_{\sigma v} \ddot{q}^v &= \mu^i \frac{\partial f^i}{\partial \dot{q}^\sigma} \quad \text{along } J^2\gamma.
\end{aligned}
$$

These equations are called *deformed equations of motion*. They represent a system of $(m + k)$ ordinary differential equations for $q^\sigma \circ \gamma(t)$ and $\mu^i \circ J^1\gamma(t)$.

There is an alternative (more geometrical) way of description of a constrained system which does not use Lagrange multipliers. For this purpose we will need the *canonical embedding* $\iota : Q \rightarrow J^1Y$, $\iota\left(t,q^\sigma,\dot{q}^l\right) = \left(t,q^\sigma,\dot{q}^l,g^i\left(t,q^v,\dot{q}^s\right)\right)$.

Proposition 2.
(a) Let $\mathscr{C}^0 = \text{span}\left\{\iota^*\varphi^i\right\}$ where φ^i runs over the set of all constraint 1-forms on a neighbourhood of Q. Then $\mathscr{C}$ is a distribution of corank k (with respect to Q) on Q.
(b) A section γ of π satisfies the condition $f^i \circ J^1\gamma = 0$ iff $J^1\gamma$ is a holonomic integral section of $\mathscr{C}$.

Distribution $\mathscr{C}$ (on Q) is called *canonical distribution*. By direct calculation we obtain

$$\iota^*\varphi^i = -\frac{\partial g^i}{\partial \dot{q}^l}\,\omega^l + dq^{m-k+i} - g^i dt .$$

This means that we have a basis $\left(dt, \omega^l, \iota^*\varphi^i, d\dot{q}^l\right)$ of 1-forms on Q.

Put $\iota^*\alpha_1 \sim \iota^*\alpha_2$ iff $\iota^*\alpha_1 - \iota^*\alpha_2 = \bar{F}_{ls}\,\omega^l \wedge \omega^s + \iota^*\varphi^i \wedge \chi_i$ where $\bar{F}_{ls} = -\bar{F}_{sl}$ are some functions on Q and χ_i is a 1-form on Q. The equivalence class $[\alpha_Q] = [\iota^*\alpha]$ is called a *constrained system on Q related to the mechanical system* $[\alpha]$. A direct calculation leads to the expression

$$\alpha_Q = \bar{A}_l\,\omega^l \wedge dt + \bar{B}_{ls}\,\omega^l \wedge d\dot{q}^s + \bar{F}_{ls}\,\omega^l \wedge \omega^s + \iota^*\varphi^i \wedge \chi_i$$

where

$$
\begin{aligned}
\bar{A}_l \;=\; & \left(A_l + \sum_{j=1}^{k} A_{m-k+j}\frac{\partial g^j}{\partial \dot{q}^l} + \sum_{i=1}^{k}\left(B_{l,m-k+i} + \sum_{j=1}^{k} B_{m-k+j,m-k+i}\frac{\partial g^j}{\partial \dot{q}^l}\right)\right. \\
& \left. \cdot\left(\frac{\partial g^i}{\partial t} + \frac{\partial g^i}{\partial q^\sigma}\dot{q}^\sigma\right)\right) \circ \iota ,
\end{aligned}
$$

$$
\bar{B}_{ls} \;=\; \left(B_{ls} + \sum_{i=1}^{k}\left(B_{l,m-k+i}\frac{\partial g^i}{\partial \dot{q}^s} + B_{m-k+i,s}\frac{\partial g^i}{\partial \dot{q}^l}\right) + B_{m-k+j,m-k+i}\frac{\partial g^j}{\partial \dot{q}^l}\frac{\partial g^i}{\partial \dot{q}^s}\right) \circ \iota .
$$

As above, the definition of a path of a constrained system is now straightforward:
Definition 2. A section γ is called a *path of a constrained system* $[\alpha_Q]$ if $f^i \circ J^1\gamma = 0$ together with one of the equivalent conditions:

$$J^1\gamma^*i_{\xi_\mathscr{C}}\alpha_Q \;=\; 0 \quad \text{for every } \pi_1 - \text{vertical vector field } \xi_\mathscr{C} \text{ belonging to } \mathscr{C},$$

$$\bar{A}_l + \bar{B}_{ls}\ddot{q}^s \;=\; 0 \quad \text{along } J^2\gamma.$$

These equations are called *reduced equations of motion*. They represent a system of m ordinary differential equations for $q^\sigma \circ \gamma(t)$.

Proposition 3. *A local section* γ *of* π *is a path of a constrained system iff it is a path of the corresponding deformed system.*

135

4. VARIATIONAL CONSTRAINED SYSTEMS

Now, using the properties of variational (unconstrained) system (see Theorem 1), the concept of a variational constrained system can be introduced, and the corresponding constraint Helmholtz conditions can be found (see [10]).

Definition 3. A constrained system $[\alpha_Q]$ is called *variational* if there exists a closed representative in the class $[\alpha_Q]$.

Remark 3. The constrained system arising from a variational one is obviously variational (in the sense of Definition 3), the converse is in general not true. Non-trivial situation of variational system with non-holonomic constraint concerns e.g. a particle in the special relativity theory, studied recently within the theory of non-holonomic systems on fibred manifolds (see [9]).

For the simplicity of calculations we use the following notations:

$$\frac{d_c' F}{dt} = \frac{\partial F}{\partial t} + \frac{\partial F}{\partial q^l}\dot{q}^l + \frac{\partial F}{\partial q^{m-k+j}}g^j, \qquad \frac{\partial_c F}{\partial q^l} = \frac{\partial F}{\partial q^l} + \frac{\partial F}{\partial q^{m-k+j}}\frac{\partial g^j}{\partial \dot{q}^l}$$

where F is a function on Q.

Theorem 3. *Let $[\alpha]$ be a mechanical system on J^1Y, and $[\alpha_Q]$ a corresponding constrained system. The constrained system $[\alpha_Q]$ is variational iff there exist functions c_{il}, b_{il}, b_i, γ_{ij} on Q such that*

CHC–1: $\quad (\bar{B}_{ls})_{\mathrm{alt}(ls)} = 0,$

CHC–2: $\quad \left(\dfrac{\partial \bar{B}_{ls}}{\partial \dot{q}^r} - \dfrac{\partial^2 g^i}{\partial \dot{q}^l \partial \dot{q}^r}c_{is} \right)_{\mathrm{alt}(sr)} = 0,$

CHC–3: $\quad \left(\left[\dfrac{\partial \bar{A}_l}{\partial \dot{q}^s} - \left(\dfrac{\partial_c g^i}{\partial q^l} - \dfrac{d_c'}{dt}\dfrac{\partial g^i}{\partial \dot{q}^l} \right)c_{is} \right] - \dfrac{d_c'\bar{B}_{ls}}{dt} - \dfrac{\partial^2 g^i}{\partial \dot{q}^l \partial \dot{q}^s}b_i \right)_{\mathrm{sym}(ls)} = 0,$

CHC–4: $\quad \left(\dfrac{1}{2}\dfrac{d_c'}{dt}\left[\dfrac{\partial \bar{A}_l}{\partial \dot{q}^s} - \left(\dfrac{\partial_c g^i}{\partial q^l} - \dfrac{d_c'}{dt}\dfrac{\partial g^i}{\partial \dot{q}^l} \right)c_{is} \right] - \dfrac{\partial_c \bar{A}_l}{\partial \dot{q}^s} + \left(\dfrac{\partial_c g^i}{\partial q^l} - \dfrac{d_c'}{dt}\dfrac{\partial g^i}{\partial \dot{q}^l} \right)b_{is} + \right.$

$\qquad\qquad \left. + b_i\dfrac{\partial_c}{\partial \dot{q}^s}\left(\dfrac{\partial g^i}{\partial \dot{q}^l} \right) \right)_{\mathrm{alt}(ls)} = 0,$

CHC–5: $\quad \dfrac{\partial \bar{A}_l}{\partial q^{m-k+i}} + 2\gamma_{ij}\left(\dfrac{\partial_c g^j}{\partial q^l} - \dfrac{d_c'}{dt}\dfrac{\partial g^j}{\partial \dot{q}^l} \right) - \dfrac{\partial_c b_i}{\partial q^l} - b_j\dfrac{\partial^2 g^j}{\partial \dot{q}^l \partial q^{m-k+i}} + \dfrac{d_c' b_{il}}{dt} + \dfrac{\partial g^j}{\partial q^{m-k+i}}b_{jl} = 0,$

CHC–6: $\quad \dfrac{\partial \bar{B}_{ls}}{\partial q^{m-k+i}} - 2\gamma_{ij}\dfrac{\partial^2 g^j}{\partial \dot{q}^l \partial \dot{q}^s} + \dfrac{\partial b_{il}}{\partial \dot{q}^s} - \dfrac{\partial_c c_{is}}{\partial q^l} - \dfrac{\partial^2 g^j}{\partial \dot{q}^l \partial q^{m-k+i}}c_{js} = 0$

where functions b_i, b_{il}, c_{il} and $\gamma_{ij} = -\gamma_{ji}$ on Q satisfy the conditions

CHC–7: $\quad b_{il} = \dfrac{\partial b_i}{\partial \dot{q}^l} - \dfrac{d_c' c_{il}}{dt} - \dfrac{\partial g^j}{\partial q^{m-k+i}}c_{jl},$

CHC–8: $\quad \left(-\dfrac{\partial b_i}{\partial q^{m-k+j}} + \dfrac{d_c'\gamma_{ij}}{dt} - 2\gamma_{ja}\dfrac{\partial g^a}{\partial q^{m-k+i}} \right)_{\mathrm{alt}(ij)} = 0,$

136

CHC–9: $c_{il} = \frac{\partial C_i}{\partial \dot{q}^l} + \bar{c}_{il}\left(t, q^{\sigma}\right)$,

CHC–10: $\gamma_{ij} = \left(\frac{\partial \Gamma}{\partial q^{m-k+j}}\right)_{\mathrm{alt}(ij)} + \bar{\gamma}_{ij}\left(t, q^l, \dot{q}^l\right)$, $\bar{\gamma}_{ij} = -\bar{\gamma}_{ji}$,

CHC–11: $\left(\frac{\partial(\Gamma_i - C_i)}{\partial q^{m-k+j}} + \bar{\gamma}_{ij}\right)_{\mathrm{alt}(ij)} = \left(\frac{\partial \bar{c}_{il}}{\partial q^{m-k+j}}\right)_{\mathrm{alt}(ij)} \dot{q}^l + \phi_{ij}\left(t, q^{\sigma}\right)$.

Above, the $C_i\left(t, q^{\sigma}, \dot{q}^l\right)$, $\Gamma\left(t, q^{\sigma}, \dot{q}^l\right)$, $\bar{c}_{il}\left(t, q^{\sigma}\right)$, $\phi_{ij}\left(t, q^{\sigma}\right)$ and $\gamma_{ij} = \left(t, q^l, \dot{q}^l\right)$ are some functions on Q.

Definition 4. The conditions CHC–1 to CHC–11 in Theorem 3 are called *constraint Helmholtz conditions*.

Proposition 4. *Let $[\alpha_Q]$ be a variational constrained system.*
(a) There exists a representative α_Q and a local 1-form

$$\bar{\theta} = \bar{L}\mathrm{d}t + \frac{\partial \bar{L}}{\partial \dot{q}^l}\omega^l + \bar{L}_{m-k+i}\iota^*\varphi^i$$

such that $\alpha_Q = \mathrm{d}\bar{\theta}$ where $\bar{L}$ and $\bar{L}_{m-k+i}$ are some functions on Q.
(b) Let ε_l be the operator defined as follows:

$$\varepsilon_l = \frac{\partial_c}{\partial q^l} - \frac{\mathrm{d}_c^l}{\mathrm{d}t}\frac{\partial}{\partial \dot{q}^l} - \ddot{q}^s\frac{\partial^2}{\partial \dot{q}^l \partial \dot{q}^s}.$$

Then the reduced equations of motion are of the form

$$f^i \circ J^1\gamma = 0, \qquad \varepsilon_l\left(\bar{L}\right) - \bar{L}_{m-k+i}\varepsilon_l\left(g^i\right) = 0.$$

Remark 4. Note that the form $\bar{\theta}$ is not unique. It is called a *constraint Cartan form* (cf [8]).

5. EXAMPLE

The aim of this paper is to study some physically interesting constrained systems and their variationality properties. In the following example the influence of the choice of a constraint on the constraint variationality, as well as on the form of the Chetaev force, is shown.

A point particle of mass m performs a two-dimensional motion under the Stokes frictional force $F_{\mathrm{Stokes}} = -k\left(\dot{x}, \dot{y}\right)$. Such a motion will be studied on the fibred manifold $\pi : \left(\mathbf{R} \times \mathbf{R}^2\right) \to \mathbf{R}$ and its prolongations. Unconstrained equations of motion

$$m\ddot{x} + k\dot{x} = 0, \qquad m\ddot{y} + k\dot{y} = 0$$

are non-variational (see Theorem 2).

(a) Consider the constraint

$$\frac{1}{2}m\dot{x}^2 + \frac{1}{2}m\dot{y}^2 = \bar{E}_0$$

representing the law of conservation of the kinetic energy ([10]).
Reduced equations of motion on $Q_+ = \{a \mid a \in Q, \dot{y}(a) > 0\}$ are

$$g^1 = \dot{y} = \sqrt{E_0 - \dot{x}^2}, \qquad \frac{mE_0\ddot{x}}{E_0 - \dot{x}^2} = 0 \quad \text{where} \quad E_0 = \frac{2\bar{E}_0}{m}.$$

The constrained system is variational (conditions CHC–1 to CHC–11 are satisfied) with
the functions

$$b_1 = 0, \qquad \frac{d'_c b_{11}}{dt} = 0, \qquad \gamma_{11} = 0, \qquad \frac{\partial b_{11}}{\partial \dot{x}} - \frac{\partial_c c_{11}}{\partial x} = 0, \qquad b_{11} = -\frac{d'_c c_{11}}{dt}$$

and with the constraint Cartan form (see Proposition 4)

$$\bar{\theta} = -\left(m\sqrt{E_0 - \dot{x}^2}\right) \imath^*\varphi^1 + dh$$

where $h(t, q^l)$ is a function on Q_+. Reduced equations have a solution

$$x(t) = x_0 + v_{0x}t, \qquad y(t) = y_0 + \sqrt{E_0 - v_{0x}^2}\, t$$

and Chetaev force takes the form

$$\Phi = (k\dot{x}, k\dot{y}) = -F_{\text{Stokes}}.$$

Interpretation of the results is evident: the constraint representing the law of conservation
of the kinetic energy provides a constrained variational system. Stokes force is eliminated by Chetaev force and the particle motion is uniform and straightforward. Note that
the form $\Phi^1 dx + \Phi^2 dy$ representing the "classical" work of Chetaev force is non-zero.

(b) For the same mechanical system consider the constraint

$$g^1 = \dot{y} = \alpha\dot{x}, \qquad \alpha = const.$$

Reduced equations of motion

$$g^1 = \dot{y} = \alpha\dot{x}, \qquad m(1 + \alpha^2)\ddot{x} + k(1 + \alpha^2)\dot{x} = 0$$

are non-variational (condition CHC–3 is not satisfied). The solution of reduced equations
is

$$x(t) = a + b\exp\left(-\frac{k}{m}t\right), \qquad y(t) = c + \alpha b\exp\left(-\frac{k}{m}t\right)$$

where a, b and c are constants given by initial conditions. Chetaev force becomes

$$\Phi = (0, 0).$$

(c) Consider the constraint

$$g^1 = \dot{y} = \beta x, \qquad \beta = const.$$

Reduced equations of motion take the form

$$g^1 = \dot{y} = \beta x, \qquad m\ddot{x} + k\dot{x} = 0.$$

Although the second reduced equation is the same (up to the constant) as in the previous situation (b), due to different constraint the system is variational, now with the functions

$$c_{11} = \frac{k}{\beta}, \qquad -\frac{\partial_c b_1}{\partial x} + \frac{d'_c b_{11}}{dt} = 0, \qquad \gamma_{11} = 0, \qquad \frac{\partial b_{11}}{\partial \dot{x}} = 0, \qquad b_{11} = \frac{\partial b_1}{\partial \dot{x}}$$

and with the constraint Cartan form

$$\bar{\theta} = -\frac{m\dot{x}^2}{2} dt - m\dot{x}\omega^1 - \frac{k\dot{x}}{\beta} \iota^* \varphi^1 + dh$$

where $h(t, q^I)$ is a function on Q_+. Reduced equations have a solution

$$x(t) = a + b\exp\left(-\frac{k}{m}t\right), \qquad y(t) = c + \beta a t - \frac{m}{k}\beta b \exp\left(-\frac{k}{m}t\right)$$

where a, b and c are constants given by the initial conditions. Chetaev force becomes

$$\Phi = (0, k\beta a).$$

Note that if $\beta = -\frac{k}{m}\alpha$, $a = 0$ and $c = 0$ then the solution of this (constrained) variational system is the same as in the previous non-variational case (b).

ACKNOWLEDGMENTS

Research is supported by grant 201/03/0512 of the Czech Grant Agency and by the grant MSM 143100006 of the Ministry of Education, Youth and Sports of the Czech Republic.

The authors are indebted to Professor Olga Krupková for her interest, numerous discussions and consultations.

REFERENCES

1. Cariñena, J. F., and Rañada, M. F., Lagrangian systems with constraints: a geometric approach to the method of Lagrange multipliers. *J. Phys. A: Math. Gen.* **26** (1993), 1335-1351.
2. Giachetta, G., Jet methods in nonholonomic mechanics. *J. Math. Phys.* **33** (1992), 1652-1665.
3. Koon, W. S., and Marsden, J. E., The Hamiltonian and Lagrangian approaches to the dynamics of nonholonomic system. *Reports on Mat. Phys.* **40** (1997), 21-62.

4. Krupka, D., Lepagean forms in higher order variational theory. In *Modern Developments in Analytical Mechanics I: Geometrical Dynamics*. Proc. IUTAM-ISIMM Symposium, Torino, Italy, 1982, edited by S. Benenti, M. Francaviglia, and A. Lichnerowicz (Accad. delle Science di Torino, Torino, 1983), 197-238.

5. Krupková, O., Mechanical systems with nonholonomic constraints. *J. Math. Phys.* **38** (1997), 5098-5126.

6. Krupková, O., *The Geometry of Ordinary Variational Equations*. Lecture Notes in Mathematics **1678**, Springer, Berlin 1997.

7. Krupková, O., On the geometry of non-holonomic mechanical systems. In *Differential Geometry and Applications*. Proc. conf., Brno, Czech Republic, 1998, edited by I. Kolář, O. Kowalski, D. Krupka, and J. Slovák (Masaryk University, Brno, 1999), 533-546.

8. Krupková, O., Recent results in the geometry of constrained systems. *Reports on Math. Phys.* **49** (2002), 269-278.

9. Krupková, O., and Musilová, J., The relativistic particle as a mechanical system with non-holonomic constraints. *J. Phys. A: Math. Gen.* **34** (2001), 3859-3875.

10. Krupková, O., and Musilová, J., *Constraint Helmholtz conditions*. Preprint 6/2002, Institute of Theoretical Physics and Astrophysics, Masaryk University Brno, Czech Republic, 2002.

11. de León, M., Marrero, J. C., and de Diego, D. M., Non-holonomic Lagrangian systems in jet manifolds. *J. Phys. A: Math. Gen.* **30** (1997), 1167-1190.

12. Massa, E., and Pagani, E., Classical mechanics of non-holonomic systems: a geometric approach. *Ann. Inst. Henri Poincaré* **55** (1991), 511-544.

13. Massa, E., and Pagani, E., A new look at classical mechanics of constrained systems. *Ann. Inst. Henri Poincaré* **66** (1997), 1-36.

14. Morando, P., and Vignolo, S., A geometric approach to constrained mechanical systems, symmetries and inverse problems. *J. Phys. A.: Math. Gen.* **31** (1998), 8233-8245.

15. Rañada, M. F., Time-dependent Lagrangian systems: A geometric approach to the theory of systems with constraints. *J. Math. Phys.* **35** (1994), 748-758.

16. Sarlet, W., A direct geometrical construction of the dynamics of non-holonomic Lagrangian systems. *Extracta Mathematicae* **11** (1996), 202-212.

17. Sarlet, W., Cantrijn, F., and Saunders, D. J., A geometrical framework for the study of non-holonomic Lagrangian systems. *J. Phys. A.: Math. Gen.* **28** (1995), 3253-3268.

18. Saunders, D. J., Sarlet, W., and Cantrijn, F., A geometrical framework for the study of non-holonomic Lagrangian systems II. *J. Phys. A.: Math. Gen.* **29** (1996), 4265-4274.

A new natural Hamiltonian system on T^*S^2 admitting an integral of degree 3 in momenta

Holger R. Dullin[*] and Vladimir S. Matveev[†]

[*]*Department of Mathematical Sciences, Loughborough University, LE11 3TU UK*
[†]*Mathematisches Institut, Universität Freiburg, 79104 Germany*

Abstract. We present a new integrable natural Hamiltonian system on S^2. The integral is cubic in momenta.
Key words. Integrable geodesic flows, polynomial in momenta integrals, locally-Hamiltonian vector fields, obstructions to integrability.
PACS. 02.30.Ik, 45.20.Jj, 45.40.Cc, 45.50.Dd.
MSC. 37J35, 58F07, 58F17, 70H06, 70E40.

1. RESULT

A Hamiltonian system is called natural if its Hamiltonian is a sum of a positive-definite kinetic energy and a potential. We present a new Liouville-integrable natural Hamiltonian system on the (cotangent bundle of the) sphere S^2. The second integral is cubic in the momenta.

Consider the sphere $S^2 \subset \mathbf{R}^3$ of radius 1 with the spherical coordinates

$$(x, y, z) = (-\sin\theta\cos\phi, -\sin\theta\sin\phi, \cos\theta)$$

and the following two functions H (the Hamiltonian) and F (the second integral) on the cotangent bundle of the sphere without poles $z = \pm 1$. Let $A, c, s \in \mathbf{R}$ be parameters with $s > 1$ and define

$$W(z) = z + s, \quad P(z) = 3z^2 + 4sz + 1, \quad Q(z) = 3z^2 + 2sz - 1$$

and

$$G(z) = \frac{P(z)}{(2W(z))^2}.$$

Then, the function H is given by $H := K + V$, where

$$
\begin{aligned}
K &:= \frac{1}{2}\left(\frac{1}{\sin^2\theta} + G(\cos\theta)\right)p_\phi^2 + \frac{1}{2}p_\theta^2 \\
V &:= A\frac{\sin\theta}{\sqrt{W(\cos\theta)}}\cos\phi + \frac{c}{W(\cos\theta)},
\end{aligned}
$$

CP729, *Global Analysis and Applied Mathematics: International Workshop on Global Analysis*,
edited by K. Taş, D. Krupka, O. Krupková, and D. Baleanu
© 2004 American Institute of Physics 0-7354-0209-4/04/$22.00

and F is defined as

$$F := 2H p_\phi - p_\phi^3 + A\cos(\phi)\frac{Q(\cos\theta)}{\sqrt{W(\cos\theta)}\sin\theta}p_\phi + 2A\sin\phi\sqrt{W(\cos\theta)}p_\theta.$$

Proposition: *The functions H and F can be analytically continued to the cotangent bundle of the whole sphere. The continuation is also polynomial in momenta (of degree 2 for H and of degree 3 for F).*

We will denote the continuations of H, K, F by the same letters H, K, F; in particular in the theorem below we mean the continued functions defined on the cotangent bundle of the whole sphere.

Theorem: *The following statements hold:*

1. *The functions H and F commute with respect to the standard Poisson bracket on T^*S^2 and are functionally independent*
2. *The kinetic energy K is positive definite*
3. *If $A \neq 0$, the Hamiltonian H does not admit a (smooth) nontrivial integral which is polynomial in velocities of degree less than three and which is linearly independent of H.*

In other words, we found a new Liouville integrable, natural Hamiltonian system on the sphere with an integral cubic in momenta.

The rigorous proof of Proposition and Theorem will appear elsewhere. Since H and F are explicitly given, the proof is not very complicated: indeed, in order to prove Proposition, one should rewrite the Hamiltonian and the integral in a local coordinate system near poles, and see that they have no singularity at the poles. The first half of the first statement of Theorem can be checked by straightforward calculations; one need to check that

$$\{H,F\} := \frac{\partial H}{\partial\phi}\frac{\partial F}{\partial p_\phi} + \frac{\partial H}{\partial\theta}\frac{\partial F}{\partial p_\theta} - \frac{\partial H}{\partial p_\phi}\frac{\partial F}{\partial\phi} - \frac{\partial H}{\partial p_\theta}\frac{\partial F}{\partial\theta}$$

is identically zero. The second half is trivial, since the integral is of degree 3 in momenta and is not the product of the Hamiltonian and of an integral of degree one in momenta. The second statement is also straightforward.

Our proof of the third statement is somehow tricky, but also not very complicated. It is based on the classification of metrics on closed surfaces whose geodesic flows admit integrals of degree 1 and 2 in momenta. The classification can be found in [2], [3], [10] or [12].

The nontrivial part of the result was to find the metric and the integral. We will explain how we found them in the last part of the paper.

2. MOTIVATION

Natural Hamiltonian systems on (cotangent bundles of) closed surfaces admitting integrals polynomial in momenta are interesting for a number of reasons:

1. They are classical: The formulation of the problem is due at least to Darboux 1896 [5].
2. If a natural Hamiltonian system admits a real-analytic integral, then it admits an integral polynomial in momenta (see, for example, Whittaker [14]).
3. In particular, all known natural integrable Hamiltonian systems on surfaces have integrals polynomial in momenta.
4. The existence of the integral polynomial in momenta of degree one or two has a very clear geometric background: the existence of an integral of degree one implies the existence of a one-parametric family of symmetries of the system. The existence of an integral of degree two implies the existence of so-called separating variables.
5. The problem of finding such systems is generally very hard: the last explicitly-given natural Hamiltonian system admitting an integral of degree three in momenta was found in 1916 [8].

Let us recall the known results about natural systems on closed surfaces admitting an integral polynomial in momenta.

First of all, in view of results of Kolokoltsov [12], a natural system on a surfaces of genus greater than two cannot admit a nontrivial integral polynomial in momenta. Then, an orientable surfaces admitting such systems must be the sphere or the torus.

We collect the main results about existence and classification in the following table:

	Sphere S^2	Torus T^2
Degree 1	All is known	All is known
Degree 2	All is known	All is known
Degree 3	Series of examples	Partial negative results
Degree 4	Series of examples	Partial negative results
Degree ≥ 5	Nothing is known	Nothing is known

The following notation is used in the table. "Degree" means the smallest degree of an integral polynomial in velocities. "All is known" means that there exists an effective description and classification.

The most valuable "partial negative results" are due to Byalyǐ [4] and Denisova and Kozlov [6]. They proved that if a natural Hamiltonian system on the torus whose kinetic energy is given by a flat metric admits an integral polynomial in momenta of degree three, it admits an integral linear in momenta.

There exists only one explicit "Series of examples" for degree three. It comes from the Goryachev-Chaplygin case of rigid body motion (by applying symplectic reduction), see [1] for the details. There exist two inexplicit series of examples, of Selivanova [13] and of Kiyohara [11]. It is possible to show that our systems are different from the Goryachev-Chaplygin and from Kiyohara examples, and are different from Selivanova examples at least for $c \neq 0$.

Thus, our system gives one more series of examples of integrable natural system on the sphere whose integral is polynomial in momenta of degree 3.

3. HOW WE FOUND IT

First of all, by the Maupertuis's principle, the existence of a natural Hamiltonian system admitting an integral of degree 3 in momenta implies the existence of a metric whose geodesic flow admits an integral which is homogeneous polynomial of degree 3 in momenta, see [1] for details. We were looking for metrics on the sphere whose geodesic flows admit integrals of degree three in momenta.

The first steps are standard: locally the existence of an integral of degree 3 for a geodesic flow is equivalent to a solution following system of partial differential equations:

$$\begin{cases} 2\lambda a_x + 3\lambda_x a + \lambda_y b & = 0 \\ \lambda b_x + \lambda_x b + \lambda a_y + \lambda_y c & = 0 \\ \lambda_x c + \lambda_y b + 2\lambda c_x + 2\lambda b_y + 3\lambda_x a + 3\lambda_y d & = 0 \\ \lambda c_y + \lambda_y c + \lambda_x b + \lambda d_x & = 0 \\ 2\lambda d_y + \lambda_x c + 3\lambda_y d & = 0 \end{cases} \quad (1)$$

In the system, $\lambda > 0$, a, b, c and d are unknown functions of variables x, y.

Let us explain where the system comes from. First of all, locally every metric is conformally equivalent to the flat one, i.e. there always exist coordinates where the metric is $\lambda(x,y)(dx^2 + dy^2)$. Suppose an integral F has the form $a(x,y)p_x^3 + b(x,y)p_x^2 p_y + c(x,y)p_x p_y^2 + d(x,y)p_y^3$. Then,

$$\{H,F\} = \left\{ \frac{p_x^2 + p_y^2}{\lambda(x,y)}, a(x,y)p_x^3 + b(x,y)p_x^2 p_y + c(x,y)p_x p_y^2 + d(x,y)p_y^3 \right\} = 0.$$

We see that $\{H,F\}$ is a homogeneous polynomial in momenta of degree 4. If it is identically 0, all coefficients must vanish. The left-hand side of the equations (1) are the coefficients of $\{H,F\}$.

We see that the system is nonlinear (in particular, it is almost no hope to explicitly obtain all its solutions), and that it satisfies Cauchy-Kowalewskaya conditions (in particular, locally it always has a lot of solutions).

The problem is to find an appropriate ansatz such that

- it is possible to explicitly solve the system, and
- the solution has chances to be defined globally, on the whole sphere.

Let us explain how we found the appropriate ansatz. The general idea is to assign geometrical objects to the system (1).

144

Lemma: *The following statements hold:*

1. *Consider the complex coordinate* $z \overset{\text{def}}{=} x + iy$. *Then,* $\left(\frac{1}{(a-c)+i(b-d)} \right) dz \otimes dz \otimes dz$ *is a meromorphic (3,0)-form without zeros.*

2. $T \overset{\text{def}}{=} ((3a+c)\lambda, (b+3d)\lambda)$ *is a Hamiltonian (with respect to the volume form of the metric) vector field.*

The first statement of Lemma is due to Kolokoltsov [12], the second was obtained in Dullin, Matveev and Topalov [7].

There exist dynamical arguments (which we will not explain here) that suggest, that the Kolokoltsov's (3,0)-form on the sphere should have 2 poles of order 3. Without loss of generality, we can put the poles in the north and south poles of the sphere. Then, the coefficient of the form is constant in the coordinates $\log(z)$, where the coordinate z is the standard coordinate on the Riemann sphere $\bar{\mathbf{C}} \cong S^2$. We constructed the Hamiltonian function for most known metrics admitting an integral of third degree in momenta (a list of such metrics can be taken, for example, in [9]), and found out that (for most systems) the Hamiltonian function of the vector field T has the form $\lambda^2 \cdot \kappa(y)$. These two conditions essentially give us an ansatz; after assuming them the system (1) is equivalent to the following ordinary differential equation for κ:

$$-4\kappa\kappa'' + 3\kappa'^2 - 10\kappa' + 3 = 16\omega\kappa^2.$$

We solved this equation and found our metric and our integral.

ACKNOWLEDGMENTS

The second author thanks DFG-programm 1154 (Global Differential Geometry) and Ministerium für Wissenschaft, Forschung und Kunst Baden-Württemberg (Eliteförder-programm Postdocs 2003) for partial support.

Note added in proof: After the paper was accepted, H. Yehia told us that, locally, our integrable system was obtained in his paper *On certain two-dimensional conservative mechanical systems with cubic second integral*, J. Phys. A: Math. Gen., **35**(2002) 9469-9487.

REFERENCES

1. A. V. Bolsinov, V. V. Kozlov, A. T. Fomenko, *The Maupertuis's principle and geodesic flows on S^2 arising from integrable cases in the dynamics of rigid body motion*, Russ. Math. Surv. **50**(1995) 473–501.
2. A. V. Bolsinov, V. S. Matveev, A. T. Fomenko, *Two-dimensional Riemannian metrics with an integrable geodesic flow. Local and global geometries*, Sb. Math. **189**(1998), no. 9-10, 1441–1466.
3. A.V. Bolsinov, A. T. Fomenko, *Integrable geodesic flows on two-dimensional surfaces*, Monographs in Contemporary Mathematics. Consultants Bureau, New York, 2000.
4. M. L. Byalyĭ, *First integrals that are polynomial in the momenta for a mechanical system on the two-dimensional torus*, Functional Anal. Appl. **21** (1987) no. 4, 310 – 312.

5. G. Darboux, *Lecons sur la théorie générale des surfaces,* Vol. 3, Chelsea Publishing 1894.
6. N. V. Denisova, V. V. Kozlov, *Polynomial integrals of reversible mechanical systems with a two-dimensional torus as configuration space,* Sb. Math. **191**(2000) no. 2, 43–63.
7. H. R. Dullin, V. S. Matveev, P. J. Topalov, *On the integrals of third degree in momenta,* Regul. Chaot. Dyn. **4**(1999) no. 3, 35–44.
8. D.N. Goryachev, *New cases of integrability of dynamical equations of Euler,* Warsaw. Univ. Izv., **3**(1916), 1–15.
9. J. Hijetarinta, *Direct methods for the search of the second invariant,* Phys. Rep. **147**(1987) no. 2, 87–154.
10. K. Kiyohara, *Compact Liouville surfaces,* J. Math. Soc. Japan **43**(1991), 555-591.
11. K. Kiyohara, *Two-dimensional geodesic flows having first integrals of higher degree,* Math. Ann. **320**(2001) no 3, 487–505.
12. V. N. Kolokol'tzov, *Geodesic flows on two-dimensional manifolds with an additional first integral that is polynomial with respect to velocities,* Math. USSR-Izv. **21**(1983), no. 2, 291–306.
13. E. N. Selivanova, *New examples of Integrable Conservative Systems on S^2 and the Case of Goryachev-Chaplygin,* Comm. Math. Phys. **207**(1999) 641–663.
14. E. T. Whittaker, *A Treatrise on the Analytical Dynamics of Particles and Rigid Bodies,* Cambridge University Press 1937.

Linear Operators on Generalized Bergman Spaces

Chiara de Fabritiis

Dipartimento di Scienze Matematiche, Universita' Politecnica delle Marche, Via Brecce Bianche, 60131, Ancona, ITALY

Abstract. In this paper we discuss several features of Hilbert spaces of holomorphic functions on domains of $\mathbf{C}^n$. We study composition and multiplication operators on generalized Bergman spaces and give results on the dynamical behaviour (*i.e.* cyclicity, hypercyclicity, compactness) of the first ones and on the algebraic properties of the space that the second one interprets. In particular we underline the analogies and differences between the case of bounded and unbounded domains in $\mathbf{C}$ and $\mathbf{C}^n$.

Key words. Hilbert spaces of holomorphic functions, Composition operators.
PACS. 02.30.Fn, 02.30.Sa.
MSC. 30H05, 32A35, 32A36.

1. INTRODUCTION

The aim of this paper is to illustrate the many different features of Hilbert spaces of holomorphic functions defined on domains in $\mathbf{C}$ and $\mathbf{C}^n$, spaces which are obtained as L^2-spaces of holomorphic functions with respect to suitables "domains of integration" or with respect to suitable measures which are obtained by multiplying the Lebesgue measure with a measurable, positive, L^1_{loc} function.

For instance, in the case of the Hardy space the "domain of integration" is in a certain generalized sense the boundary of the disc where the functions are defined, while in the case of the Bergman space the domain is the disc itself, while in the case of the generalized Bergman spaces either on $\mathbf{C}^n$ or on $\mathbf{C}^*$ it is a different measure which comes into play.

The interest on these topics goes back to the early twenties of the last century with the papers by Hardy (see [13] which is usually considered the starting point of the theory of these spaces) and Littlewood (see [19] where the so-called Littlewood Subordination principle was proved). For a wide account on Hardy spaces see, for example, the classical books by Duren ([12]) and Rudin ([22]).

The first part of this paper will be devoted to a general illustration of the features of Hilbert spaces of holomorphic functions in which we introduce composition and multiplication operators and we define their main dynamical properties, that is cyclicity and hypercyclicity.

CP729, *Global Analysis and Applied Mathematics: International Workshop on Global Analysis,*
edited by K. Taş, D. Krupka, O. Krupková, and D. Baleanu
© 2004 American Institute of Physics 0-7354-0209-4/04/$22.00

The second section is reduced to a short summary of the main properties of Hardy and Bergman spaces both on the unit disc of $\mathbf{C}$ and the unit ball of $\mathbf{C}^n$.

This will give a comparison term to the contents of the third section which contains the detailed study of generalized Bergman spaces in the unbounded case, that is both in the cases of $\mathbf{C}^n$ and $\mathbf{C}^*$. After noting that the Hilbert topology is stronger than the topology of convergence on compact subsets, we give several examples of generalized Bergman spaces and we introduce the representing kernel which will help us to detect some features of these spaces.

Then we illustrate several results on composition operators on generalized Bergman spaces (in particular in the case of rotational invariant ons) analysing their dynamical properties and we give an accurate study of the behaviour of multiplication operators and the algebraic structure of these spaces.

2. HILBERT SPACES OF HOLOMORPHIC FUNCTIONS

Let D be a domain in $\mathbf{C}^n$. We consider the linear space $\mathrm{Hol}(D)$ of holomorphic functions on D; it is well known that if we endow $\mathrm{Hol}(D)$ with the topology of uniform convergence on compacta of D, then this space is a Fréchet (non-Banachable) space and it is a closed subspace of the space of continuous functions with values in $\mathbf{C}$ because of Weierstrass theorem.

One of our aims (very generally up to now) is to investigate the behaviour of linear operators on linear subspaces of $\mathrm{Hol}(D)$, subspaces which in general can be endowed with topologies that can be different from the subspace topology.

In general, our setting will be as follows: H will be a linear subspace of $\mathrm{Hol}(D)$ (usually a Hilbert or a Banach space) and T will be a linear operator which acts (not continuously, *a priori*) on H.

In the two more interesting cases (the case of composition operators and the case of multiplication operators) T will be a linear operator which acts on $\mathrm{Hol}(D)$ and such that $T(H) \subset H$.

We will prove that, up to reasonable topological hypothesis, these operators are automatically continuous on H.

Definition 2.1. Let Ψ be a holomorphic self-map of the domain D. We call *composition operator* of *symbol* Ψ the linear operator C_Ψ defined by

$$\mathrm{Hol}(D) \ni f \mapsto C_\Psi(f) = f \circ \Psi \in \mathrm{Hol}(D).$$

If H is a linear subspace of $\mathrm{Hol}(D)$ such that $C_\Psi(H) \subset H$ we say that C_Ψ (or Ψ for short) acts on H.

One of the main problems in the theory of composition operators is to classify which are the function spaces $H \subset \mathrm{Hol}(D)$ and the holomorphic self-maps Ψ of D which give rise to such actions. In particular, in the sequel of the paper we will examine what happens in the case of Hardy and Bergman space on the unit disc $\Delta \subset \mathbf{C}$ and in the unit ball of $\mathbf{C}^n$ for bounded domains and what happens in the case of generalized Bergman spaces on $\mathbf{C}$, C^n and $\mathbf{C}^* = \mathbf{C} \setminus \{0\}$ for unbounded domains.

As for the continuity behaviour of the composition operators, a simple closed-graph argument gives the following result (see [11] for a proof):

Proposition 2.2. *Let H be a linear subspace of* $\mathrm{Hol}(D)$ *which is a Banach space and suppose that the Banach topology is stronger than the topology of pointwise convergence on D. If Ψ is a holomorphic self-map of D such that $C_\Psi(H) \subset H$, then $C_\Psi : H \to H$ is a continuous linear operator.*

As a first trivial consequence of the definition, we have the following:

Remark 2.3. *Any composition operator whose symbol is non-constant is one-to-one.*

Indeed, if $\Psi : D \to D$ is non-constant, then Ψ is open. So if $C_\Psi(f) = 0$, then the zero set of f must have non-empty interior and therefore f is zero.

The second class of linear operators we introduce are the so-called multiplication operators.

Definition 2.4. Let h a holomorphic function defined on D. We call *multiplication operator* of *ratio h* the linear operator M_h defined by

$$\mathrm{Hol}(D) \ni f \mapsto M_h(f) = f \cdot h \in \mathrm{Hol}(D).$$

If H is a linear subspace of $\mathrm{Hol}(D)$ such that $M_h(H) \subset H$ we say that M_h (again, h for short) acts on H.

Notice that the action of multiplication operators is connected to the algebraic structure of the space H. Indeed, if H is an algebra (a quite rare event, as we will see in the following) and very mild topological requirements on H are satisfied, then for any $h \in H$ the multiplication operator of ratio h acts continuously on H.

Anyway, there are many cases in which H is not an algebra and there is a lot of multiplication operators acting on it (as an example, we can consider the case of the Hardy or of the Bergman space on the unit disc $\Delta \subset \mathbf{C}$ for which any bounded holomorphic function on Δ gives rise to a continuous multiplication operator). In particular, we will focus our attention to the algebraic structure of these spaces, in the case of the generalized Bergman spaces on $\mathbf{C}$ and $\mathbf{C}^*$.

As in the case of composition operators, a mild topological requirement, yields a continuity result for multiplication operators, too.

Proposition 2.5. *Let H be a linear subspace of* $\mathrm{Hol}(D)$ *which is a Banach space and suppose that the Banach topology is stronger than the topology of pointwise convergence on D. If h is a holomorphic function on D such that $M_h(H) \subset H$, then $M_h : H \to H$ is a continuous linear operator.*

Proof. By closed-graph theorem, in order to prove that M_h is continuous, it is enough to show that its graph Γ_{M_h} is closed.

Now suppose that the sequence $\{f_n : n \in \mathbf{N}\} \subset H$ converges in (the Banach topology of) H to f and that the sequence $\{M_h(f_n) : n \in \mathbf{N}\} \subset H$ converges in (the Banach topology) H to g. In particular, since the Banach topology on H is stronger than the topology of pointwise convergence, then for any $z \in D$ we have that $f_n(z)$ converges to $f(z)$ and $M_h(f_n)(z) = f_n(z)h(z)$ converges to $g(z)$.

This implies that $g(z) = f(z)h(z)$ for any $z \in D$ and therefore $g = M_h(f)$, which ensures the closure of Γ_{M_h} and concludes the proof of the Proposition. ∎

Now we focus our attention on the dynamical behaviour of bounded linear operators on H.

Definition 2.6. Let T be a bounded linear operator on the Banach space H. We say that T is *cyclic* if there exists $h \in H$ (called a *cyclic vector*) such that the span of the orbit $\mathrm{Orb}(T, h) = \{T^n(h) \; : \; n \in \mathbf{N}\}$ is dense in H, *i.e.*, that $\{p(T)(h) \; : \; p(t) \in \mathbf{C}[t]\}$ is dense in H. We say that T is *hypercyclic* if there exists $h \in H$ (called a *hypercyclic vector*) such that the orbit $\mathrm{Orb}(T, h) = \{T^n(h) \; : \; n \in \mathbf{N}\}$ is dense in H.

In dynamical systems this last property is also called *topological transitivity* and it is a well know result that it can hold only if H has infinite dimension.

A very simple remark shows that, if H is a Hilbert space and T is a cyclic bounded linear operator, then the orthogonal complement of the range of T has dimension at most one, since the orthogonal projection of any cyclic vector onto the complement spans the complement itself.

3. THE BOUNDED CASE

3.1. The One-dimensional Case

We denote by $\mathbf{B}_n$ the unit ball in $\mathbf{C}^n$ for the standard Hermitian norm, that is

$$\mathbf{B}_n = \{z \in \mathbf{C}^n \; : \; ||z|| < 1\}.$$

When $n = 1$ we denote $\mathbf{B}_1$ as Δ, too. Let $f \in \mathrm{Hol}(\Delta)$. We write the power series expansion of f at the origin as $f(z) = \sum_{n \in \mathbf{N}} \hat{f}(n) z^n$ and we say that f belongs to the Hardy space $H^2(\Delta)$ if the sequence of its power series coefficients is square-summable. We denote by $||f||^2$ the sum $\sum_{n \in \mathbf{N}} |\hat{f}(n)|^2$.

It is easily seen that for any $f \in H^2(\Delta)$ and $z \in \Delta$ the following inequality holds

$$|f(z)| \leq ||f||(1 - |z|)^{-1/2}.$$

As a consequence of this result, the (pre)-Hilbert topology of $H^2(\Delta)$ is stronger than the topology of uniform convergence on compacta of Δ and hence (because of Weierstrass theorem) $H^2(\Delta)$ is a closed subspace of $L^2(\Delta)$ and therefore a separable Hilbert space. In particular all polynomials in z belong to $H^2(\Delta)$ and monomials of different degree are mutually orthogonal.

A different way of introducing the Hardy space $H^2(\Delta)$ simplifies several questions about its structure: indeed, it is difficult to say anything about the behaviour of holomorphic bounded functions or the action of composition operators on $H^2(\Delta)$ by means of the above presentation.

A well known result shows that for any $f \in \mathrm{Hol}(\Delta)$ we have that $f \in H^2(\Delta)$ if and only if

$$\lim_{r \to 1^-} \frac{1}{2\pi} \int_{-\pi}^{\pi} |f(re^{i\theta})|^2 d\theta$$

is bounded and in this case the limit is equal to $||f||^2$.

150

As a first, useful consequence, we obtain that all bounded holomorphic functions on Δ belong to $H^2(\Delta)$ and that, for any $f \in H^2(\Delta)$ and $h \in H^\infty(\Delta)$ the product fh belongs to $H^2(\Delta)$ and the norm of the linear operator $M_h : H^2(\Delta) \to H^2(\Delta)$ given by the multiplication by h is bounded by $||h||_\infty = \sup_{z \in \Delta} |h(z)|$.

This way of presenting the Hardy space, also gives the possibility to prove one of the most interesting results on composition operators, in a certain sense it was the first result from which the theory of composition operators originated: Littlewood's Subordination Principle.

Theorem 3.1. *Suppose Ψ is a holomorphic self-map of Δ with $\Psi(0) = 0$. Then for each $f \in H^2(\Delta)$ we have $C_\Psi(f) \in H^2(\Delta)$ and*

$$||C_\Psi(f)|| \leq ||f||.$$

A trivial corollary gives that for any holomorphic self-map Ψ of Δ the composition operator C_Ψ acts on $H^2(\Delta)$ and an estimate of the norm of C_Ψ can be given in terms of the image of the origin under Ψ, to be more precise we have

$$||C_\Psi|| \leq \left(\frac{1 + |\Psi(0)|}{1 - |\Psi(0)|} \right)^{1/2}$$

for any holomorphic self-map Ψ of Δ.

An accurate study of the main features (cyclicity, hypercyclicity, compactness) of composition operators on Hardy spaces is contained in [24] and [3].

Analogously to the introduction of the Hardy space, another Hilbert space can be defined on Δ: the so-called Bergman space.

For any $f \in \text{Hol}(\Delta)$ we say that f belongs to the Bergman space $A^2(\Delta)$ if and only if $\sum_{n \in \mathbb{N}} \frac{|\hat{f}(n)|^2}{n+1}$ is finite, where $f(z) = \sum_{n \in \mathbb{N}} \hat{f}(n) z^n$. In this case we write $||f||^2_{A^2} = \sum_{n \in \mathbb{N}} \frac{|\hat{f}(n)|^2}{n+1}$. As an immediate consequence of the definition we obtain that $H^2(\Delta) \subset A^2(\Delta)$ and the inclusion is continuous (a contraction, to be more precise).

If we denote by $d\omega(z)$ the measure $\frac{1}{i\pi} dz \wedge d\bar{z} = \frac{1}{\pi} dx \wedge dy$ and we take any $f \in \text{Hol}(\Delta)$ a simple computation shows that f belongs to $A^2(\Delta)$ if and only if

$$\int_\Delta |f(z)|^2 d\omega(z) < +\infty,$$

moreover $||f||^2_{A^2} = \int_\Delta |f(z)|^2 d\omega(z)$.

As in the Hardy space case, it is not difficult to prove that the Hilbert topology on $A^2(\Delta)$ is stronger than the topology of uniform convergence on compacta of Δ, then we obtain that the Bergman space is a separable Hilbert space. Moreover, by means of the same methods used in the proof of the Littlewood's Subordination Principle one can also prove that composition operators act continuously on $A^2(\Delta)$.

3.2. The Several Dimensional Case

Now we give a very short account of the generalizations of these topics to the several complex variables setting.

First of all we have to give a definition of the Hardy space in $\mathbf{C}^n$ for $n \geq 1$. The Hardy space in the unit ball of $\mathbf{C}^n$ is defined as follows

$$H^2(\mathbf{B}_n) = \{f \in \mathrm{Hol}(\mathbf{B}_n) \ : \ \lim_{r \to 1^-} \int_S |f(rz)|^2 d\sigma(z)\}$$

where S is the boundary of $\mathbf{B}_n$ and $d\sigma$ the rotational invariant positive Borel measure on S of total mass 1. A well known result shows that

$$H^2(\mathbf{B}_n) = \{f \in \mathrm{Hol}(\mathbf{B}_n) \ : \ |f|^2 \text{ has a harmonic majorant}\};$$

moreover one can prove the existence (a.e.) of limit on Korányi regions (in a certain sense a stronger radial limit) and this gives the possibility to prove that $H^2(\mathbf{B}_n)$ is the $L^2(\sigma)$ closure of the analytic polynomials.

Again, we have an inequality between the value of a function $f \in H^2(\mathbf{B}_n)$ at some point $z \in \mathbf{B}_n$ and the norm of f which is analogous to the one obtained in the one complex variable case, to be more precise

$$|f(z)| \leq 2^{n/2} ||f|| (1 - ||z||)^{-n/2}$$

for any $f \in H^2(\mathbf{B}_n)$ and $z \in \mathbf{B}_n$.

This entails that for any $n \geq 1$ the Hardy space is a Hilbert space whose Hilbert topology is stronger than the topology of uniform convergence on compact subsets of $\mathbf{B}_n$. For a more detailed study of the Hardy space in the several complex variable case see Rudin [22].

Anyway, if $n > 1$, there is no result like Littlewood's Subordination Principle. Indeed, following results by Hörmander and MacCluer (see [15] and [20]), Wogen described a wide class of holomorphic self-maps Ψ of $\mathbf{B}_n$ such that the composition operator C_Ψ is unbounded on $H^2(\mathbf{B}_n)$ (see [28]).

Some positive results on the boundedness of composition operators on Hardy spaces in several complex variables were obtained by Cowen and MacCluer in [9] where they prove that linear fractional maps of $\mathbf{B}_n$ into itself induce bounded composition operators on $H^2(\mathbf{B}_n)$.

Further results were obtained by Bisi and Bracci in [2] where they give a classification of the cyclic and hypercyclic behaviour of composition operators induced by linear fractional maps according to their fixed point sets in the closure of the unit ball. In particular, they prove that if Ψ is a linear fractional map which has more than two fixed points in $\bar{\mathbf{B}}_n$ then C_Ψ is not cyclic and that a linear fractional map Ψ with exactly two boundary fixed points induced a hypercyclic composition operator if and only if df is invertible at one–and hence any–point of $\bar{\mathbf{B}}_n$.

4. THE UNBOUNDED CASE

4.1. Definitions and First Properties

We denote by $d\omega(z)$ the (n,n)-form given by

$$d\omega(z) = \frac{i^n}{2^n} dz_1 \wedge \cdots \wedge dz_n \wedge d\bar{z}_1 \wedge \cdots \wedge d\bar{z}_n = dx_1 \wedge \cdots \wedge dx_n \wedge dy_1 \wedge \cdots \wedge dy_n$$

where we set $z_j = x_j + iy_j$ for any $j = 1, \ldots, n$, so that $d\omega(z)$ is the (n,n)-form associated to the Lebesgue measure on $\mathbf{C}^n$.

We denote by D either $\mathbf{C}^n$ (for $n \geq 1$) or $\mathbf{C}^*$.

Definition 4.1. Let φ be a Lebesgue-measurable positive function on D such that $\varphi \in L^1_{\mathrm{loc}}(D)$. The *weighted (or generalized) Bergman space associated to φ* is the complex vector space of holomorphic functions on D which are square-integrable with respect to the weight φ, that is

$$\mathscr{F}_\varphi = \left\{ f \in \mathrm{Hol}(D, \mathbf{C}) \; : \; \int_D \varphi(z) |f(z)|^2 d\omega(z) < \infty \right\}.$$

As usual, we denote by $\| \; \|_\varphi$ the norm on $\mathscr{F}_\varphi$ and by $(\cdot, \cdot)_\varphi$ the scalar product on $\mathscr{F}$; if no confusion can arise on the weight φ then we drop the subscript φ to simplify notations. For the same reason, we often drop the adjective "weighted".

The following estimate drops light on the relation between the (pre)-Hilbert topology and the topology of uniform convergence on compacta of D.

Proposition 4.2. *Let L be a compact subset of D. Then there exists a positive constant c_L such that $\max\{|f(z)|, \; z \in L\} \leq c_L \|f\|_\varphi$ for any $f \in \mathscr{F}_\varphi$.*

As a corollary we immediately obtain that the topology of $\mathscr{F}_\varphi$ as a (pre)-Hilbert space is stronger than the topology of the uniform convergence on compact subsets of the domain D.

Now, consider a Cauchy sequence $\{f_n\} \subset \mathscr{F}_\varphi$: in particular this is a Cauchy sequence in $L^2_\varphi(D)$, and therefore it converges to a function $g \in L^2_\varphi(D)$. Moreover, by the above corollary and the Weierstrass theorem, this sequence converges to a function f which is holomorphic on D. Then g coincides a.e. with the holomorphic function f and hence $\mathscr{F}_\varphi$ is a closed subspace of $L^2_\varphi(D)$; this proves that $\mathscr{F}_\varphi$ is a separable Hilbert space.

As a consequence of Proposition 2.2. we immediately obtain that any composition operator and any multiplication operator acting on a generalized Bergman space is continuous.

The two following examples show that the Bergman space associated to a given weight can be "very small".

Example 4.3. *If $\varphi(z) = 1$ for all $z \in \mathbf{C}$, then $\mathscr{F}_\varphi$ contains only the function which is identically zero.*

Example 4.4. *For $n \in \mathbf{N}$, let us consider the Bergman space on $\mathbf{C}$ associated to the weight $\varphi_n(z) = (|z|^2 + 1)^{-(n+2)}$ for any $z \in \mathbf{C}$. Then $\mathscr{F}_{\varphi_n}$ is the space of polynomials of degree less than or equal to n.*

The following example describes a "classical" example of generalized Bergman spaces, the so-called Fock space.

Example 4.5. The Fock space is the generalized Bergman space on $\mathbf{C}^n$ associated to the weight $\varphi_0(z) = e^{-||z||^2}$, where $||z||$ denotes the standard Euclidean norm on $\mathbf{C}^n$.

It is not difficult to prove that, for any multi-index $m \in \mathbf{N}^n$ the square norm of z^m is equal to $\pi^n m!$ and that $\{h_m = z^m \cdot (\pi^n m!)^{-1/2} : m \in \mathbf{N}^n\}$ is a complete orthogonal system for the Fock space.

Moreover, this space has a special interest, since it is isometric with the space $L^2(\mathbf{R}^n)$ of functions which are square integrable with respect to the Lebesgue measure (we outline the proof for $n = 1$ in order to simplify notations).

Indeed, consider the probability measure $d\sigma(t) = (2\pi)^{1/2} e^{-t^2/2} dt$ on $\mathbf{R}$; it is immediate to check that $L^2(\mathbf{R})$ is isometric to $L^2(\mathbf{R}, d\sigma)$. An complete orthonormal system in $L^2(\mathbf{R}, d\sigma)$ is given by the Hermite polynomials

$$He_m(t) = (-1)^m (m!)^{-1/2} e^{t^2/2} \frac{d^m}{dt^m} e^{-t^2/2}.$$

For any $z \in \mathbf{C}$ and any $\psi \in L^2(\mathbf{R}, d\sigma)$ we set

$$\Theta\psi(z) = \frac{1}{\sqrt{\pi}} \int_{\mathbf{R}} e^{-z^2/2 + zt} \psi(t) d\sigma(t).$$

Because of Bessel equality we obtain that $\Theta\psi$ is an entire function and that Θ is a linear isometry of $L^2(\mathbf{R}, d\sigma)$ into the Fock space. As $\Theta He_m = h_m$ for any $m \in \mathbf{N}$ we obtain that Θ is onto and therefore $L^2(\mathbf{R}, d\sigma)$ and the Fock space are isometric.

Recent results by Carswell, MacCluer and Schuster classified all composition operators acting on it and among them identifies the ones which are compact (see [6], where the weight used is $\varphi_1(z) = e^{-||z||^2/2}$ which of course gives rise to a space which is trivially isometric to the one associated to φ_0).

In particular, they proved that a composition operator C_Ψ acts on the Fock space only if $\Psi(z) = Az + B$, where $A \in M_{n,n}(\mathbf{C})$ is such that $||A|| \leq 1$ and $B \in \mathbf{C}^n$. *Vice versa* if $\Psi(z) = Az + B$, where $A \in M_{n,n}(\mathbf{C})$ is such that $||A|| \leq 1$ and $B \in \mathbf{C}^n$ is such that $\langle A\zeta, B \rangle = 0$ for any $\zeta \in \mathbf{C}^n$ with $||A\zeta|| = ||\zeta||$, then C_Ψ acts on the Fock space. Moreover the bounded linear operator C_Ψ is compact iff $||A|| < 1$.

Definition 4.6. Let $z_0 \in D$; we denote by $\varepsilon_{z_0} : \mathscr{F}_\varphi \ni f \mapsto f(z_0) \in \mathbf{C}$ the evaluation at the point z_0. The comparison between the Hilbert topology and the topology of uniform convergence on compacta of D yields the following

Corollary 4.7. *For any $z_0 \in D$ the evaluation map ε_{z_0} is a continuous linear form on $\mathscr{F}_\varphi$ endowed with the Hilbert topology.*

The above Corollary gives us the possibility to introduce a useful tool, which will be used in several places.

Definition 4.8. A *reproducing kernel* on $\mathscr{F}$ is a map $K : D \times D \to \mathbf{C}$ such that

- (i) for any $w \in D$ the function $K_w = K(\cdot, w)$ belongs to $\mathscr{F}$;
- (ii) for any $f \in \mathscr{F}$ and any $w \in D$ we have $f(w) = (f, K_w)_\varphi$.

The following result is a very interesting consequence of the continuity of evaluation maps at any point and it was first noticed in the general case of Hilbert spaces of holomorphic functions by Aronszajn, see [1], generalizing Bergman original idea.

Corollary 4.9. *There exists a unique reproducing kernel K on $\mathscr{F}$, in particular we have $K(z,w) = \overline{K(w,z)}$ for any $z,w \in D$. Moreover the set $\{K_z : z \in D\}$ spans a dense subspace of $\mathscr{F}$.*

The fact that $\mathscr{F}$ is a separable Hilbert spaces gives us the possibility to write K in a fairly explicit way. Let $\{e_n : n \in \mathbf{N}\}$ be a complete orthonormal system in $\mathscr{F}$. Then it is not difficult to prove that for any $z,w \in D$ the following equality holds

$$K(z,w) = \sum_{n=0}^{+\infty} \overline{e_n(w)}e_n(z); \tag{1}$$

in particular $\|K_z\|^2 = K(z,z)$ for any $z \in D$. As a trivial consequence we obtain that for any $z,w \in D$ we have

$$|K(z,w)| \leq \frac{1}{2}\big(K(z,z) + K(w,w)\big)$$

and that

$$\sup\{|f(z)| : f \in \mathscr{F}, \ \|f\| \leq 1\} = \|K_z\| = K(z,z)^{1/2}$$

for any $z \in D$ which can also be written as

$$|f(z)| \leq \|K_z\|\|f\| \tag{2}$$

for any $z \in D$ and any $f \in \mathscr{F}$.

A first, simple remark concerning composition operators on generalized Bergman spaces is the following one, which gives us the possibility to get rid of the trivial case in which the symbol of a composition operator is constant.

Remark 4.10. *If $\Psi : D \to D$ is a constant map, then C_Ψ acts on $\mathscr{F} \neq \{0\}$ iff $\mathscr{F}$ contains the constant functions, that is, iff $\varphi \in \mathrm{L}^1(D)$.*

As we will see in the next section, there are many cases in which constant functions do not belong to generalized Bergman spaces (even if the space is finite dimensional) and hence the above result shows that there are cases in which composition operators associated to constant functions do not act on generalized Bergman spaces. Anyway, notice that, if they act, then these composition operators are compact, since their range has dimension 1.

4.2. Composition Operators on Generalized Bergman Spaces

We now give several results on composition operators on Bergman spaces on $\mathbf{C}$ and $\mathbf{C}^*$. In particular we will focus our attention to the cases of finite dimensional Bergman spaces and rotational invariant Bergman spaces (for a more detailed list of results and complete proofs see [11]). Here $D = \mathbf{C}$ or $\mathbf{C}^*$.

Proposition 4.11. *Let $\mathscr{F}$ be a Bergman space on D of dimension $d+1$. There exists $\beta \in \mathrm{Hol}(D,D)$ such that $\mathscr{F}_\varphi = \mathrm{Span}\{\beta(z)z^j : 0 \le j \le d\}$.*

By means of the above proposition one can classify the composition operators acting on finite dimensional Bergman spaces. here, as an example, we give the classification result in the case $D = \mathbf{C}^*$. **Theorem 4.12.** *Let $\mathscr{F}$ be a Bergman space of dimension 1 and suppose that Ψ acts on $\mathscr{F}$. If $\mathscr{F}$ is not the space of constant functions, then $\Psi \in \mathrm{Aut}\mathbf{C}^*$. Moreover if Ψ is linear, say $\Psi(z) = cz$, then either there exists $m \in \mathbf{Z}$ such that $\mathscr{F} = \mathbf{C}z^m$ or c is a root of 1 of finite order q and there exists $\tau \in \mathrm{Hol}(\mathbf{C}^*, \mathbf{C})$ and $m \in \mathbf{Z}$ such that $\mathscr{F} = \mathbf{C}z^m e^{\tau(z^q)}$. If Ψ is non-linear, say $\Psi(z) = cz^{-1}$, then we can find $\tau \in \mathrm{Hol}(\mathbf{C}, \mathbf{C})$ such that $\mathscr{F} = \mathbf{C}e^{\tau(z+cz^{-1})}$.*

The above results and a few more computations allow us to classify cyclic composition operators (till we are concerned with finite dimensional vector spaces, hypercyclicity is always excluded).

Proposition 4.13. *Let $\mathscr{F}$ be a finite dimensional Bergman space on $\mathbf{C}^*$ with $\dim\mathscr{F} = d+1 > 1$. If $\Psi(z) = cz$ for any $z \in \mathbf{C}^*$ and c is not a root of 1, then C_Ψ is always cyclic; if c is a root of 1 of order q then C_Ψ is cyclic if and only if $q > d$. If $\Psi(z) = c/z$ for any $z \in \mathbf{C}^*$, then C_Ψ is never cyclic.*

Now we turn our attention to the rotational invariant Bergman spaces both in the case of $\mathbf{C}^n$ and $\mathbf{C}^*$.

Definition 4.14. A Bergman space $\mathscr{F}_\varphi$ is said to be *rotationally invariant (r.i.)* if the composition operators whose symbols are the rotations map the space $\mathscr{F}_\varphi$ into itself. A weight φ is said to be *rotationally invariant (r.i.)* if $\varphi(z) = \varphi(e^{i\theta}z)$ for all $\theta \in \mathbf{R}$ and $z \in \mathbf{C}^*$.

Trivially, if a weight is r.i. then the Bergman space associated to this weight is r.i., too. *Vice versa*, we will prove that if the Bergman space $\mathscr{F}$ is r.i. then we can find a r.i. weight ϕ such that $\mathscr{F} = \mathscr{F}_\phi$. Of course, this change of weight could affect the Hilbert structure of the space, but does not affect the induced Hilbert topology. In he one dimensional case, this change of weight gives us the possibility to classify r.i Bergman spaces.

Definition 4.15. Let $\mathscr{F}$ be a r.i. Bergman space either on $\mathbf{C}$ or on $\mathbf{C}^*$ and set

$$n_1(\mathscr{F}) = \inf\{n \in \mathbf{Z} : z^n \in \mathscr{F}\} \qquad \text{and} \qquad n_2(\mathscr{F}) = \sup\{n \in \mathbf{Z} : z^n \in \mathscr{F}\}.$$

We denote by $\mathscr{N}(\mathscr{F})$ the set of integer numbers which lie between $n_1(\mathscr{F})$ and $n_2(\mathscr{F})$. When no confusion can arise about the Bergman space we are considering, we drop the $\mathscr{F}$ in order to simplify notation. In a certain sense, to be clarified by the following result, the set $\mathscr{N}(\mathscr{F})$ is the one we need to describe the structure of $\mathscr{F}$.

Proposition 4.16. *Let $\mathscr{F}$ be a r.i. Bergman space either on $\mathbf{C}$ or on $\mathbf{C}^*$. Then the monomial z^m belongs to $\mathscr{F}$ iff $m \in \mathscr{N}$. In particular, $\dim\mathscr{F} < +\infty$ if and only if n_2 and n_1 are both finite, in that case we have $\mathscr{F} = \mathrm{Span}\{z^m : m \in \mathbf{Z} \text{ and } n_1 \le m \le n_2\}$.*

The following results summarize which composition operators act on r.i. Bergman spaces on $\mathbf{C}^*$ and classifies their dynamical behaviour according to the structure of the Bergman space itself.

Theorem 4.17. *Suppose that $\mathscr{F}_\varphi$ is an infinite dimensional r.i. Bergman space and that the composition operator C_Ψ acts continuously on it. Then either Ψ is constant or there exists $c \in \mathbf{C}^*$ such that $\Psi(z) = cz$ or $\Psi(z) = c/z$. Moreover if $\Psi(z) = c/z$ then $n_1 = -\infty$ and $n_2 = +\infty$.*

A trivial remark ensures that composition operators associated with symbols of the form $\Psi(z) = c/z$ are never cyclic, since Ψ is an involution and therefore C_Ψ is an involution, too.

The following result classifies all cyclic composition operators on infinite dimensional r.i. Bergman spaces on $\mathbf{C}^*$ whose symbol is of the form $\Psi(z) = cz$ with $|c| = 1$.

Proposition 4.18. *Suppose $\mathscr{F}_\varphi$ is a r.i. weighted Bergman space of infinite dimension. Given $c \in \mathbf{C}$ with $|c| = 1$, the composition operator associated to the symbol $\Psi(z) = cz$ for any $z \in \mathbf{C}^*$ is cyclic if and only if c is not a root of 1. In particular, if c is not a root of 1 and the weight φ is r.i. the function K_z is a cyclic vectors for C_Ψ for any $z \in \mathbf{C}^*$.*

Corollary 4.19. *Let $\mathscr{F}$ be a r.i. Bergman space with $n_1 = -\infty$ and $n_2 = +\infty$. The composition operator C_Ψ acts cyclically on $\mathscr{F}$ if and only if $\Psi(z) = cz$ for some $c \in \mathbf{C}^*$ with $|c| = 1$ and c is not a root of 1.*

Since for any $c \in \mathbf{C}^*$ with modulus equal to 1, the composition operator associated to the symbol $\Psi(z) = cz$ is an isometry, we immediately obtain that none of these operators is compact.

Remark 4.20. *Let $\mathscr{F}$ be a r.i. Bergman space and $c \in \mathbf{C}^*$ with $|c| = 1$. Then the composition operator C_Ψ associated to $\Psi(z) = cz$ is not a compact operator.*

Now we turn to the study of infinite dimensional r.i. Bergman spaces on which composition operators associated to symbols of the form $\Psi(z) = cz$ with $0 < |c| < 1$ act (in this case we must have $n_2 = +\infty$ and $n_1 \in \mathbf{Z}$); the following result shows that they are all cyclic.

Proposition 4.21. *Suppose $\mathscr{F}$ is a r.i. Bergman space such that $n_2 = +\infty$ and $n_1 \in \mathbf{Z}$. Then for any $c \in \mathbf{C}$ with $0 < |c| < 1$ the composition operator associated to the symbol $\Psi(z) = cz$ is cyclic. In particular the function $h \in \mathscr{F}$ is a cyclic vector for C_Ψ if and only if the coefficient of z^m in its Laurent expansion are different from 0 for any $m \geq n_1$.*

In this case, since the action on the monomials contained in $\mathscr{F}$ is diagonal and the action of $\Psi(z) = cz$ on z^n is a multiplication by c^n we have that all composition operators associated to the symbol $\Psi(z) = cz$ with $0 < |c| < 1$ are compact (in order to approximate C_Ψ with a finite range operator, it is enough to "truncate" it to zero starting from a suitable degree onward).

Remark 4.22. *Let $\mathscr{F}$ be a r.i. Bergman space $n_2 = +\infty$ and $n_1 \in \mathbf{Z}$ and $c \in \mathbf{C}^*$ with $0 < |c| < 1$. Then the composition operator C_Ψ associated to $\Psi(z) = cz$ is a compact operator.*

Corollary 4.23. *Let $\mathscr{F}$ be a r.i. Bergman space with $n_1 \in \mathbf{Z}$ and $n_2 = +\infty$. The composition operator C_Ψ acts cyclically on $\mathscr{F}$ if and only if $\Psi(z) = cz$ for some $c \in \mathbf{C}^*$ with $|c| = 1$ and c is not a root of 1 or $0 < |c| < 1$. Let $\mathscr{F}$ be a r.i. Bergman space with $n_1 = -\infty$ and $n_2 \in \mathbf{Z}$. The composition operator C_Ψ acts cyclically on $\mathscr{F}$ if and only if $\Psi(z) = cz$ for some $c \in \mathbf{C}^*$ with $|c| = 1$ and c is not a root of 1 or $|c| > 1$.*

Proposition 4.24. *Let $\mathscr{F}$ be a r.i. Bergman space and let Ψ be a holomorphic self-map of $\mathbf{C}^*$ acting on $\mathscr{F}$. Then the composition operator C_Ψ is never hypercyclic.*

As a last topic of this paragraph we develop in full details an example of a generalized Bergman space which is not rotationally invariant, that is, on which not all composition operators whose symbol is a rotation act.

Example 4.25. Set $\varphi(z) = \begin{cases} e^{-\sqrt{2}\mathrm{Re}z} & \text{if } -\frac{\pi}{4} \leq \mathrm{arg}z \leq \frac{\pi}{4}, \\ e^{-|z|} & \text{otherwise.} \end{cases}$

A trivial computation shows that $\varphi(z) \leq e^{-|z|}$ for any $z \in \mathbf{C}$ and therefore any polynomial belongs to $\mathscr{F}_\varphi$. Moreover it is easily seen that $e^{z/c}$ belongs to $\mathscr{F}_\varphi$ for any $c \in \mathbf{C}$ such that $|c| > 2$.

Now we prove that the entire function $e^{z/2}$ belongs to $\mathscr{F}_\varphi$ while $e^{-iz/2}$ does not, so that the composition operator whose symbol is the rotation of an angle equal to $-\pi/2$ does not act on $\mathscr{F}_\varphi$.

Indeed, we have the following chain of inequalities which ensures that the function $e^{z/2}$ belongs to $\mathscr{F}_\varphi$:

$$\int_{\mathbf{C}} |e^{z/2}|^2 \varphi(z) d\omega(z) = \int_{\mathbf{C}} e^{\mathrm{Re}z} \varphi(z) d\omega(z)$$

$$= \int_0^{+\infty} \int_{-\pi/4}^{\pi/4} e^{(1-\sqrt{2})\mathrm{Re}z} d\omega(z) + \int_0^{+\infty} \int_{\pi/4}^{7\pi/4} e^{\mathrm{Re}z-|z|} d\omega(z).$$

Since $|\mathrm{Re}z| \geq |\mathrm{Im}z|$ when $-\frac{\pi}{4} \leq \mathrm{arg}z \leq \frac{\pi}{4}$, we obtain that the first addendum is bounded above by

$$\int_0^{+\infty} \int_{-\pi/4}^{\pi/4} e^{(1-\sqrt{2})|z|/\sqrt{2}} d\omega(z).$$

As for the second addendum we have that it is bounded above by

$$\int_0^{+\infty} \int_{\pi/4}^{7\pi/4} e^{(1/\sqrt{2}-1)|z|} d\omega(z).$$

Since

$$\int_0^{+\infty} \int_0^{2\pi} e^{(1/\sqrt{2}-1)|z|} d\omega(z) < +\infty,$$

we obtain that $e^{z/2}$ belongs to $\mathscr{F}_\varphi$.

Now we show that $e^{-iz/2}$ does not belong to the generalized Bergman space associated to φ. Indeed, suppose that

$$\int_{\mathbf{C}} |e^{-iz/2}|^2 \varphi(z) d\omega(z) = \int_{\mathbf{C}} e^{\mathrm{Im}z} \varphi(z) d\omega(z)$$

is finite. In particular we obtain that

$$\int_0^{+\infty} \int_{\pi/4}^{\pi/2} e^{\mathrm{Im}z} e^{-|z|} d\omega(z)$$

has to be finite. Then we should have

$$\lim_{\varepsilon \to 0^+} \int_0^{+\infty} \int_{\pi/4}^{\pi/2-\varepsilon} e^{\rho(\sin\theta - 1)} \rho \, d\rho \, d\theta < +\infty$$

and a simple estimate shows that this does not hold.

4.3. Multiplication Operators on Generalized Bergman Spaces

In this paragraph we give some results on multiplication operators on generalized Bergman spaces and on the algebraic structure of these spaces.

A first result on multiplication operators underlines the great difference between the case of the Hardy and Bergman spaces on Δ and the case of the generalized Bergman spaces on $\mathbf{C}^n$ or $\mathbf{C}^*$.

As we will see, there are no multiplication operators on generalized Bergman spaces unless the space is trivial, that is, is contained in the space of constant functions. This is in sharp contrast with the case of multiplication operators on Hardy and Bergman spaces on Δ, where there is a large amount of multiplication operators (in particular we have seen that all the bounded holomorphic functions on Δ give rise to multiplication operators both on Hardy and Bergman spaces on Δ).

In the proof of the following Proposition it we become clear that this is due in a certain sense to the fact that the only bounded holomorphic functions on $\mathbf{C}^n$ or $\mathbf{C}^*$ are the constant ones.

Proposition 4.26. *Let $\mathscr{F}$ be a generalized Bergman space either on $\mathbf{C}^n$ or on $\mathbf{C}^*$. If the multiplication operator M_h acts on $\mathscr{F}$, then h is constant.*

Proof. By Proposition 2.5. the multiplication operator M_h is continuous, then there exists $C > 0$ such that $||M_h(f)|| \leq C||f||$ for any $f \in \mathscr{F}$.

Therefore for any $z \in \mathbf{C}^n$ ($\mathbf{C}^*$), the definition of the representing kernel yields

$$|f(z)h(z)| = |M_h(f)(z)| = |(M_h(f), K_z)| \leq ||M_h(f)|| \cdot ||K_z|| \leq C||f|| \cdot ||K_z||.$$

In particular, if we apply this inequality to $f = K_z$, we have

$$|K_z(z)h(z)| \leq C||K_z||^2$$

for any $z \in \mathbf{C}^n$ ($\mathbf{C}^*$). As $K_z(z) = (K_z, K_z) = ||K_z||^2$ we therefore obtain $|h(z)| \leq C$ and hence the holomorphic function h defined on $\mathbf{C}^n$ ($\mathbf{C}^*$) is constant. ∎

This result also answers another interesting question on the algebraic structure of generalized Bergman spaces which asks to classify the cases in which $\mathscr{F}$ is an algebra. The following Corollary shows that the only possibility is the trivial case in which $\mathscr{F}$ coincides with the space of constant functions.

Corollary 4.27. *If $\mathscr{F}$ is a generalized Bergman space on $\mathbf{C}^n$ or $\mathbf{C}^*$ which is an algebra, then $\mathscr{F}$ is contained in the space of constant functions.*

Indeed, if $\mathscr{F}$ is an algebra, then for any $h \in \mathscr{F}$ the multiplication operator M_h acts on $\mathscr{F}$ and therefore h has to be constant.

An interesting question which is still partially open is whether there exists a never-vanishing holomorphic function in a given generalized Bergman space $\mathscr{F}$.

Due to the classification of finite dimensional Bergman spaces on $\mathbf{C}$ and $\mathbf{C}^*$, the answer is positive if $\dim \mathscr{F} < +\infty$ and the same happens if $\mathscr{F}$ is a rotationally invariant Bergman space either on $\mathbf{C}$ or on $\mathbf{C}^*$, but the question is still open both in the case when $\mathscr{F}$ is not r.i. and when $D = \mathbf{C}^n$. Notice that, even in the case in which $D = \mathbf{C}$, approximation methods as the ones used in the proofs of Runge or Weierstrass theorems do not work, because of Rouché theorem and the fact that the Hilbert convergence implies the convergence on compacta of $\mathbf{C}$.

More generally, almost no results (except the ones which are obtained in the easy case of finite dimensional spaces) are known in the study of the zero sets of functions which belong to generalized Bergman spaces on $\mathbf{C}$ and $\mathbf{C}^*$, in sharp contrast with the ones obtained in the cases of Hardy and Bergman spaces on Δ (see [12] for a detailed account of these results).

ACKNOWLEDGMENTS

Many warm thanks go to Prof. Sergio Venturini (Università di Bologna) for a fruitful discussion which simplified the structure of the paper.

It is a pleasure to thank the whole staff of Cankaya University, in particular Prof. Dr. Kenan Tas and Prof. Dr. Dumitru Baleanu for making my staying in Ankara for the Workshop both fruitful and pleasant.

REFERENCES

1. Aronszajn, N. *Theory of reproducing kernels*, Trans. Amer. Math. Soc., 68 (1950), 337–404.
2. Bisi C., Bracci F., *Linear fractional maps of the unit ball: a geometric study*, Adv. Math. 167, 2, (2002), 265–287.
3. Bourdon P., Shapiro J., *Cyclic Phenomena for Composition Operators*, Memoirs A.M.S., Providence, 1997.
4. Bourdon P., Shapiro J., *Hardy Spaces that Support no Compact Composition Operators*, preprint.
5. Bourdon P., Levi D., Narayan S., Shapiro J., *Which Linear-Fractional Composition Operators are Essentially Normal?*, preprint.
6. Carswell B., MacCluer B., Schuster A., *Composition Operators on the Fock Space*, preprint.
7. Chan K., Shapiro J., *The Cyclic Behaviour of Translation Operators On Hilbert Spaces of Entire Functions*, Indiana Math. Jour., 40 (1991), 1421–1449.
8. Cowen C., MacCluer B., *Composition Operators on Spaces of Analytic Functions*, CRC Press, Studies in Advanced Mathematics Serie, 1995.
9. Cowen C., MacCluer B., *Linear Fractional Maps of the Ball and Their Composition Operators*, Acta Sci. Math. (szeged) 66 (2000), 351–376.
10. de Fabritiis C., *Actions of Linear Groups on Fock Spaces*, Seminari di Geometria della Università di Bologna, XI (1995), 81–116.
11. de Fabritiis C., *Composition Operators on Hilbert Spaces of Holomorphic Functions*, Preprint Università Politecnica delle Marche, April 2003.
12. Duren P.L., *Theory of H^p spaces*, Academic Press, New York, 1970.
13. Hardy G.H., *The mean value of the modulus of an analytic function*, Proc. London Math. Soc. 14 (1915), 269–277.
14. Herrero D., *Limits of Hypercyclic and Supercyclic Operators*, Jour. Funct. Anal. 99 (1991), 79–190.

15. Hörmander L., *L^p-estimates for (pluri)-subharmonic functions*, Math. Scand. 20 (1967), 65–78.
16. Horowitz C., *Zeroes of Functions in the Bergman Spaces*, Duke Math. Jour. 41 (1974), 693–710.
17. Kriete T., MacCluer B., *Composition Operators in large weighted Bergman spaces*, Indiana Univ. Math., 41 (1992), 755–788.
18. Kriete T., MacCluer B., *Composition operators between Hardy and weighted Bergman spaces on convex domains in* C^n, Proc. Amer. Math. Soc., 123 (1995), 2093–2102.
19. Littlewood J.E., *On inequalities in the theory of functions*, Proc. London Math. Soc., 23 (1925), 481–519.
20. MacCluer B., *Compact Composition Operators on* $H^p(B_n)$, Michigan Math. J., 32 (1985), 237–248.
21. MacCluer B., Zeng X., Zorboska N., *Composition Operators on small weighted Hardy spaces*, Ill. J. Math., 40 (1996), 662–677.
22. Rudin W., *Function Theory in the Unit BAll of* C^n, Springer, New York, 1980.
23. Shapiro J., *The Essential Norm of a Composition Operator*, Annals of Math. 125 (1987), 375–404.
24. Shapiro J., *Composition operators and classical function theory*, Springer, New York, 1993.
25. Shapiro J., *Notes on Dynamics of Linear Operators*, notes of a course held at the Universities of Florence, Rome and Padua, 2001.
26. Shapiro J., Smith W., *Hardy Spaces that Support no Compact Composition Operators*, preprint.
27. Shapiro J., Taylor P., *Compact, Nuclear and Hilbert -Schmidt Operators*, Indiana Univ. Math. J. 23 (1976), 471–496.
28. Wogen W., *Composition Operators Acting on Spaces of Holomorphic Functions on Domains of* C^n, Proc. Symp. pure Math., 51 (1990), 361–366.

On the multi-component NLS type systems and their gauge equivalent: Examples and reductions

V. S. Gerdjikov* and G. G. Grahovski*

*Institute for Nuclear Research and Nuclear Energy, Bulgarian Academy of Science, 72
Tzarigradsko chaussèe, 1784 Sofia, Bulgaria

Abstract. Some recent results concerning the multi-component Nonlinear Schrödinger (MNLS) type systems and their gauge equivalent are presented. On the example of MNLS system related to the $so(5)$-algebra the derivation of the corresponding generating (recursion) operator is given using a gauge covariant approach. A nontrivial reduction of the $so(5)$ MNLS model with a compatible dynamics is also reported.
Key words. Lax representation, Lie algebras, Reductions of soliton equations.
PACS. 03.65.Ge, 02.20.Sv, 02.30.Zz.
MSC. 37Q51, 37K05, 37K40.

1. INTRODUCTION

The Lax representation $[L(\lambda), M(\lambda)] = 0$ is invariant with respect to the gauge group action [2]. The first nontrivial example is the gauge equivalence between the nonlinear Schrödinger (NLS) equation and the Heisenberg feromagnet (HF) equation:

$$iu_t + u_{xx} + 2|u|^2 u = 0, \qquad u = u(x,t), \qquad (1)$$

$$iS_t + [S, S_{xx}] = 0, \qquad S^2 = \mathbb{1}. \qquad (2)$$

Both models are exactly solvable by applying the inverse scattering method to the Zakharov-Shabat system (see e.g. [2, 14]). The multi-component generalization of the NLS model (1) is associated to the generalized Zakharov-Shabat system

$$L(\lambda)\psi \equiv \left(i\frac{d}{dx} + q(x,t) - \lambda J \right) \psi(x,t,\lambda) = 0, \qquad (3)$$

related to a simple Lie algebra $\mathfrak{g}$ of rank $r > 1$. Here J is **not** a regular element of $\mathfrak{h}$ and the corresponding subalgebra $\mathfrak{g}_J$ (the kernel of the operator ad_J) is a non-commutative one. This makes more difficult the derivations of: i) the fundamental analytic solutions (FAS) of (3); ii) the construction of the related recursion operators and iii) the application of the gauge transformation, i.e. the transition to the pole gauge in which:

$$\tilde{L}\tilde{\psi}(x,t,\lambda) \equiv \left(i\frac{d}{dx} - \lambda S(x,t) \right) \tilde{\psi}(x,t,\lambda) = 0, \qquad (4)$$

CP729, *Global Analysis and Applied Mathematics: International Workshop on Global Analysis*,
edited by K. Taş, D. Krupka, O. Krupková, and D. Baleanu

where

$$\tilde{\psi}(x,t,\lambda) = g^{-1}(x,t)\psi(x,t,\lambda), \qquad S(x,t) \equiv \mathrm{Ad}_g \cdot J = g^{-1}(x,t)Jg(x,t).$$

$$g(x,t) = \psi(x,t,\lambda = 0), \qquad \text{i.e.,} \qquad \left(i\frac{d}{dx} + q(x,t) \right) g(x,t) = 0, \qquad (5)$$

The interpretation of the inverse scattering method (ISM) as a generalized Fourier transform and the expansions over the so called "squared solutions" (see e.g. [9] for regular J and [4] for non-regular J) allows one to study all the fundamental properties of the relevant NLEE's. These include:

(i) the description of the whole class of NLEE related to the Lax operator $L(\lambda)$ (3) solvable by the ISM;

(ii) derivation of the infinite family of integrals of motion and

(iii) the Hamiltonian formulation of the NLEE's.

Both classes of NLEE possess hierarchies of Hamiltonian structures. The phase spaces $\mathcal{M}_J$ and $\mathcal{M}_S^*$ corresponding to the standard and the pole gauge are

$$\mathcal{M}_J \equiv \{q(x,t), \quad \pi_0 q(x,t) = 0\}, \qquad \widetilde{\mathcal{M}}_S \equiv \{S(x,t), \quad S(x,t) = g^{-1}Jg(x,t)\}, \qquad (6)$$

where π_0 is the projector on the subalgebra $\mathfrak{g}_0 \subset \mathfrak{g}$ of the elements $X \in \mathfrak{g}$ commuting with J, i.e. $[J, \pi_0 X] = 0$ and in addition we assume that $q(x,t)$ and $S(x,t) - J$ are smooth functions tending to zero fast enough for $|x| \to \infty$. The hierarchies of symplectic structures defined on $\mathcal{M}_J$ and $\widetilde{\mathcal{M}}_S$ are generated by the corresponding recursion operators and are given by the following families of compatible two-forms:

$$\Omega_q^{(k)} = i\int_{-\infty}^{\infty} dx \left\langle \delta q \wedge \Lambda^k \mathrm{ad}_J^{-1}\delta q(x,t) \right\rangle, \qquad \tilde{\Omega}_S^{(k)} = i\int_{-\infty}^{\infty} dx \left\langle \delta \mathcal{S} \wedge \tilde{\Lambda}^k \mathrm{ad}_{\mathcal{S}}^{-1}\delta \mathcal{S}(x,t) \right\rangle.$$

Note also that the gauge transformation relates nontrivially the symplectic structures, i.e. $\Omega_q^{(k)} \simeq \tilde{\Omega}_S^{(k+2)}$ (for NLS–HFE equivalence, see [12, 10]). These two hierarchies are dynamically equivalent.

Here we outline the explicit gauge covariant approach for the construction of the recursion operators $\Lambda_\pm$ related to L (3) for non-regular elements J. In order to avoid a number of technicalities we consider as a basic example only the Lax operators (3) related to the algebra $so(5)$. Our ideas, however, can be generalized to any simple Lie algebra thus allowing one to treat the corresponding class of NLEE. This allows one to reformulate everything also for the pole gauge extending the results in [4, 10].

In Section 2 we derive the general form of the equations and concrete example of such models related to $so(5)$-algebra. In Section 3 we resolve the technical problems in constructing the recursion operators $\Lambda_\pm$ for the $so(5)$ MNLS type equations by using the spectral decomposition of the operator ad_J. In Section 4 we briefly outline the reductions of $so(5)$ MNLS model and present an example of reductions with automorphism belonging to the corresponding Weyl group of $\mathfrak{g}$.

2. MNLS AND MHF TYPE EQUATIONS

The Lax operator for the MNLS equations has the form (3) and J is a non-regular element of $\mathfrak{h}$. The dispersion law of the MNLS eq. is quadratic in λ: $f_{MNLS} = 2\lambda^2 J$. The MNLS eq., its gauge equivalent MHF eq. and their M-operators have the form:

$$i\frac{\partial q}{\partial t} + 2\mathrm{ad}_J^{-1}\frac{\partial^2 q}{\partial x^2} + [q, \pi_0[q, \mathrm{ad}_J^{-1}q]] - 2i(\mathbb{1} - \pi_0)[q, \mathrm{ad}_J^{-1}q_x] = 0, \qquad (7)$$

$$i\frac{\partial S}{\partial t} + 2\frac{\partial}{\partial x}\left(\mathrm{ad}_S^{-1}\frac{\partial S}{\partial x}\right) = 0, \qquad (8)$$

$$M(\lambda)\psi \equiv \left(i\frac{d}{dt} - V_0^d + 2i\mathrm{ad}_J^{-1}q_x(x,t) + 2\lambda q(x,t) - 2\lambda^2 J\right)\psi(x,t,\lambda) = 0, \qquad (9)$$

$$\tilde{M}\tilde{\psi}(x,t,\lambda) \equiv \left(i\frac{d}{dt} - 2i\lambda\,\mathrm{ad}_S^{-1}S_x - 2\lambda^2 S\right)\tilde{\psi}(x,t,\lambda) = 0, \qquad (10)$$

where $V_0^d = \pi_0\left([q, \mathrm{ad}_J^{-1}q_x]\right)$ and π_0 is the projector onto $\mathfrak{g}_J$.

Basic example: For the sake of brevity from now on we will consider the "maximal degenerate" case of $so(5)$ MNLS model.

This algebra has 4 positive roots: $\alpha_1 = e_1 - e_2$, $\alpha_2 = e_2$, $\alpha_3 = \alpha_1 + \alpha_2$ and $\alpha_4 = \alpha_1 + 2\alpha_2$. Let we choose also J to be a degenerate Cartan element ($\alpha_1(J) = 0$) so the set of roots $\Delta_1^+ = \{\alpha_2, \alpha_3, \alpha_4\}$ of $so(5)$, for which $\alpha(J) \neq 0$ labels the coefficients of the potential $q(x,t)$:

$$q(x,t) \equiv \sum_{\alpha\in\Delta_1^+}(q_\alpha E_\alpha + p_\alpha E_{-\alpha}) = \begin{pmatrix} 0 & 0 & q_{11} & q_{12} & 0 \\ 0 & 0 & q_1 & 0 & q_{12} \\ p_{11} & p_1 & 0 & q_1 & -q_{11} \\ p_{12} & 0 & p_1 & 0 & 0 \\ 0 & p_{12} & -p_{11} & 0 & 0 \end{pmatrix}; \qquad (11)$$

$$J = \mathrm{diag}\,(a,a,0,-a,-a).$$

Here q_1 and p_1 are related to the root α_2; the labels mn in $q_{mn}(x,t)$ and $p_{mn}(x,t)$ refer to the roots $(mn) \leftrightarrow m\alpha_1 + n\alpha_2$. Then the corresponding MNLS type system is of the form:

$$i\frac{\partial q_{12}}{\partial t} + \frac{1}{2a}\frac{\partial^2 q_{12}}{\partial x^2} + \frac{1}{a}q_{12}(q_1p_1 + q_{11}p_{11} + q_{12}p_{12}) + \frac{i}{a}q_1q_{11,x} - \frac{i}{a}q_{11}q_{1,x} = 0,$$

$$i\frac{\partial q_{11}}{\partial t} + \frac{1}{a}\frac{\partial^2 q_{11}}{\partial x^2} + \frac{1}{a}q_{11}(q_1p_1 + q_{11}p_{11} + \frac{1}{2}q_{12}p_{12}) + \frac{i}{a}q_{12}p_{1,x} + \frac{i}{2a}q_{12,x}p_1 = 0,$$

$$i\frac{\partial q_1}{\partial t} + \frac{1}{a}\frac{\partial^2 q_1}{\partial x^2} + \frac{1}{a}q_1(q_1p_1 + q_{11}p_{11} + \frac{1}{2}q_{12}p_{12}) - \frac{i}{a}q_{12}p_{11,x} - \frac{i}{2a}q_{12,x}p_{11} = 0, \quad (12)$$

$$i\frac{\partial p_1}{\partial t} - \frac{1}{a}\frac{\partial^2 p_1}{\partial x^2} - \frac{1}{a}p_1(q_1p_1 + q_{11}p_{11} + \frac{1}{2}q_{12}p_{12}) - \frac{i}{a}p_{12}q_{11,x} - \frac{i}{2a}p_{12,x}q_{11} = 0,$$

$$i\frac{\partial p_{11}}{\partial t} - \frac{1}{a}\frac{\partial^2 p_{11}}{\partial x^2} - \frac{1}{a}p_{11}(q_1p_1 + q_{11}p_{11} + \frac{1}{2}q_{12}p_{12}) + \frac{i}{a}p_{12}q_{1,x} + \frac{i}{2a}p_{12,x}q_1 = 0,$$

164

$$i\frac{\partial p_{12}}{\partial t} - \frac{1}{2a}\frac{\partial^2 p_{12}}{\partial x^2} - \frac{1}{a}p_{12}(q_1p_1 + q_{11}p_{11} + q_{12}p_{12}) + \frac{i}{a}p_1p_{11,x} - \frac{i}{a}p_{11}p_{1,x} = 0.$$

3. GENERATING (RECURSION) OPERATORS FOR THE MNLS TYPE SYSTEMS

The (generating) recursion operator appeared first in the AKNS-approach [1] as a tool to generate the class of all M-operators as well as the NLEE related to the given Lax operator. Next I. M. Gel'fand and L. A. Dickey [3] discovered that the class of these M-operators is contained in the diagonal of the resolvent of L. The kernel of the resolvent of L can be explicitly defined in terms of the fundamental analytic solutions $\chi^\pm(x,\lambda)$ of (3), see [9, 4].

The "squared solutions" that appeared first in [1, 11] were later generalized in [9, 4] to Lax operators of the type (3); they can also be viewed as natural generalizations of the usual exponentials and are introduced by:

$$e_\alpha^\pm(x,t,\lambda) = (1 - \pi_0)\left(\chi^\pm(x,t,\lambda)E_\alpha\hat{\chi}^\pm(x,t,\lambda)\right) \tag{13}$$

In fact their completeness relations [9, 4] provide us the spectral decompositions of the recursion operators:

$$\Lambda_+ e_{\pm\alpha}^\pm(x,t,\lambda) = \lambda e_{\pm\alpha}^\pm(x,t,\lambda), \quad \Lambda_- e_{\mp\alpha}^\pm(x,t,\lambda) = \lambda e_{\pm\alpha}^\pm(x,t,\lambda), \quad \alpha \in \Delta_+. \tag{14}$$

Our choice of J (11) means that $K = \mathrm{ad}_J$ has five different eigenvalues: $-2a, -a, 0, a$ and $2a$. Then the minimal characteristic polynomial for K is:

$$K(K^2 - a^2)(K^2 - 4a^2) = 0. \tag{15}$$

Let us also introduce the projectors onto the eigensubspaces of K as follows:

$$\pi_{\pm2} = \frac{K(K^2 - a^2)(K \pm 2a)}{24a^4}, \quad \pi_{\pm1} = -\frac{K(K^2 - 4a^2)(K \pm a)}{6a^4},$$
$$\pi_0 = \frac{(K^2 - a^2)(K^2 - 4a^2)}{4a^4}.$$

Using the characteristic equation (15) it is easy to check that π_j are orthogonal projectors, i.e. they satisfy: $\pi_j\pi_k = \delta_{jk}\pi_j$ for all $j,k = \pm2, \pm1, 0$ and:

$$K\pi_{\pm2} = \pm2a\pi_{\pm2}, \quad K\pi_{\pm1} = \pm a\pi_{\pm1}, \quad K\pi_0 = 0, \tag{16}$$

Thus any function $f(K)$ can be expressed in terms of these projectors:

$$f(K) = f(2a)\pi_2 + f(a)\pi_1 + f(0)\pi_0 + f(-a)\pi_{-1}f(-2a)\pi_{-2} \tag{17}$$

provided $f(\lambda)$ is regular for $\lambda = \pm2a, \pm a$ and 0.

The phase space $\mathcal{M}_J$ of the MNLS equations is the co-adjoint orbit of the $so(5)$ determined by J; its elements are matrices $q(x,t)$ satisfying $\pi_0q(x,t) = 0$. Note also that $\mathrm{ad}_J = K$ introduces a grading on $\mathfrak{g} = \overset{2}{\underset{j=-2}{\oplus}} \mathfrak{g}_j$ and the projectors π_j project precisely onto $\mathfrak{g}_j$. Obviously $\mathfrak{g}_j = \pi_j \mathfrak{g}$, $\mathfrak{g}_0 \equiv \mathfrak{g}_J$ and $\mathcal{M}_J \simeq \mathfrak{g} \backslash \mathfrak{g}_J$.

The "squared solutions" again have the form (13), but now the corresponding FAS and the projectors π_0 are different due to the choice of J:

$$\chi^{\pm}(x,t,\lambda) = \phi(x,t,\lambda) S_J^{\pm}(t,\lambda) = \psi(x,t,\lambda) T_J^{\mp}(t,\lambda) D_J^{\pm}(\lambda). \tag{18}$$

Here $S_J^{\pm}$, $T_J^{\pm}$ and $D_J^{\pm}$ are the factors in the generalized Gauss decompositions of $T(t,\lambda)$ [4]:

$$T(t,\lambda) = T_J^-(t,\lambda) D_J^+(\lambda) \hat{S}_J^+(t,\lambda) = T_J^+(t,\lambda) D_J^-(\lambda) \hat{S}_J^-(t,\lambda), \tag{19}$$

$$S_J^{\pm}(t,\lambda) = \exp\left(\sum_{\alpha \in \Delta_1^+} s_{J,\alpha}^{\pm}(t,\lambda) E_{\pm\alpha}\right), \quad T_J^{\pm}(t,\lambda) = \exp\left(\sum_{\alpha \in \Delta_1^+} t_{J,\alpha}^{\pm}(t,\lambda) E_{\pm\alpha}\right) \tag{20}$$

$$D_J^{\pm}(\lambda) = \exp\left(\pm d_1^{\pm}(\lambda) H_1 \pm 2 d_2^{\pm}(\lambda) H_2 + d_{\alpha_1}^{\pm}(\lambda) E_{\alpha_1} + d_{-\alpha_1}^{\pm}(\lambda) E_{-\alpha_1}\right). \tag{21}$$

If $q(x,t)$ evolves according to the MNLS (12) then

$$i\frac{dS_J^{\pm}}{dt} - 2\lambda^2[J, S_J^{\pm}(t,\lambda)] = 0, \quad i\frac{dT_J^{\pm}}{dt} - 2\lambda^2[J, T_J^{\pm}(t,\lambda)] = 0, \quad \frac{dD_J^{\pm}}{dt} = 0. \tag{22}$$

This means that the MNLS eq. (12) has four series of integrals of motion as compare to the two series of integrals for the $so(5)$ 4-wave system [6]. This is due to the special (degenerate) choice of the dispersion law $f_{\mathrm{MNLS}} = 2\lambda^2 J$. We have to remember, however, that only two of these four series are in involution, which in turn is related to the non-commutativity of the subalgebra $\mathfrak{g}_J$.

As a result we get the following expression for the recursion operator:

$$\Lambda_{\pm} Z = K^{-1}(\mathbb{1} - \pi_0)\left(i\frac{dZ}{dx} + [q(x), Z(x)] + i\left[q(x), \int_{\pm\infty}^x dy\, \pi_0[q(y), Z(y)]\right]\right), \tag{23}$$

where we assume that $Z \in \mathcal{M}_J$, i.e. $\pi_0 Z(x) = 0$.

In order to evaluate the gauge equivalent recursion operator we again make use of the gauge covariant approach. First it allows one easily to recalculate the projectors on the eigensubspaces of $\mathrm{ad}_{S(x)} \equiv \widetilde{K}(x) = g_0^{-1} K g_0(x,t)$:

$$\widetilde{\pi}_{\pm 2} = \frac{\widetilde{K}(\widetilde{K}^2 - a^2)(\widetilde{K} \pm 2a)}{24 a^4}, \quad \widetilde{\pi}_{\pm 1} = -\frac{\widetilde{K}(\widetilde{K}^2 - 4a^2)(\widetilde{K} \pm a)}{6 a^4},$$

$$\widetilde{\pi}_0 = \frac{(\widetilde{K}^2 - a^2)(\widetilde{K}^2 - 4a^2)}{4 a^4}.$$

In particular these formulae allow us to cast the MHF system (8) in the form:

$$iS_t - \frac{5}{4 a^2}[S, S_{xx}] + \frac{1}{4 a^4}\left((\mathrm{ad}_S)^3 S_x\right)_x = 0, \tag{24}$$

where S is constrained by $S(S^2 - a^2)^2 = 0$. In addition the operator $\widetilde{K}(x,t)$ satisfy the equation (15).

One can derive the explicit form of the recursion operators $\widetilde{\Lambda}_\pm$ by applying the gauge transformation to $\Lambda_\pm$ (23). For degenerate choice of J this is more difficult because $\mathfrak{g}_J$ is non-abelian. Here we provide the explicit form of the operators inverse to $\widetilde{\Lambda}_\pm$

$$\widetilde{\Lambda}_\pm^{-1}\widetilde{Z} = \frac{1}{i}(\mathbb{1} - \widetilde{\pi}_0)\int_{\pm\infty}^{x} dy\,[S(y),\widetilde{Z}(y)] \tag{25}$$

where we restrict $\widetilde{Z}$ by $\widetilde{\pi}_0\widetilde{Z}(x) = 0$.

4. ON THE REDUCTIONS OF MNLS TYPE SYSTEMS

The numerous Z_2-reductions have been recently classified for the N-wave equations [6, 8] using the reduction group introduced by A. V. Mikhailov [13]. They can be classified for the MNLS models and for their gauge equivalent systems. Here we briefly outline the main steps in this direction on the following

Example: Let us impose on the potential $U(x,t,\lambda)$ of (3) a reduction

$$w_{e_1+e_2}(U^*(x,t,\lambda^*)) = -U(x,t,\lambda), \quad U(x,t,\lambda) = q(x,t) - \lambda J, \quad w_{e_1+e_2}(K) = -K,$$

where $w_{e_1+e_2}$ is the Weyl reflection with respect to the maximal root $e_1 + e_2$ of the $so(5)$-algebra. For the matrix elements of $q(x,t)$ one gets

$$p_1(x,t) = -q_{11}^*(x,t), \qquad q_1(x,t) = -p_{11}^*(x,t), \qquad p_{12}(x,t) = -q_{12}^*(x,t),$$

so this leads to the following new 3-component NLS type model with compatible dynamics:

$$i\frac{\partial q_{12}}{\partial t} + \frac{1}{2a}\frac{\partial^2 q_{12}}{\partial x^2} - \frac{1}{a}q_{12}(q_1 q_{11}^* + q_1^* q_{11} + |q_{12}|^2) + \frac{i}{a}q_1 q_{11,x} - \frac{i}{a}q_{11}q_{1,x} = 0$$

$$i\frac{\partial q_{11}}{\partial t} + \frac{1}{a}\frac{\partial^2 q_{11}}{\partial x^2} - \frac{1}{a}(q_1 q_{11}^* + q_1^* q_{11} + \frac{1}{2}|q_{12}|^2) - \frac{i}{a}q_{12}q_{11,x}^* - \frac{i}{2a}q_{12,x}q_{11}^* = 0$$

$$i\frac{\partial q_1}{\partial t} + \frac{1}{a}\frac{\partial^2 q_1}{\partial x^2} - \frac{1}{a}q_1(q_1 q_{11}^* + q_1^* q_{11} + \frac{1}{2}|q_{12}|^2)) + \frac{i}{a}q_{12}q_{1,x}^* + \frac{i}{2a}q_{12,x}q_1^* = 0$$

For the scattering matrix $T_J(t,\lambda)$ and for its Gauss factors $T_J^\pm(t,\lambda)$, $S_J^\pm(t,\lambda)$ and $D_J^\pm(\lambda)$ this reduction leads to:

$$w_{e_1+e_2}(T_J(t,\lambda^*))^* = T_J(t,\lambda), \qquad w_{e_1+e_2}(T_J^+(t,\lambda^*))^* = T_J^-(t,\lambda),$$
$$w_{e_1+e_2}(S_J^+(t,\lambda^*))^* = S_J^-(t,\lambda), \qquad (D_J^+(\lambda^*))^* = D_J^-(\lambda), \tag{26}$$

Finally for the gauge equivalent MHF systems (24) one gets: $w_{e_1+e_2}(S(x,t)) = -S^*(x,t)$.

5. CONCLUSIONS

Here we derived new results about the construction of the recursion operators for MNLS models and their gauge equivalent MHF systems by using the gauge covariant approach.

The scattering data for the gauge equivalent MHF equations $\tilde{S}_J^{\pm}(t,\lambda)$, $\tilde{T}_J^{\pm}(t,\lambda)$ and $\tilde{D}_J^{\pm}(\lambda)$ are related to the scattering data of the "canonical" systems as follows:

$$\tilde{S}_J^{\pm}(t,\lambda) = T_J(0)S_J^{\pm}(t,\lambda)\hat{T}_J(0), \qquad \tilde{T}_J^{\pm}(t,\lambda) = T_J^{\pm}(t,\lambda), \qquad \tilde{D}_J^{\pm}(\lambda) = D_J^{\pm}(\lambda)\hat{T}_J(0),$$

and $\tilde{T}_J(0)$ is an element of the subgroup $\mathscr{G}_J$ of $SO(5)$. On the real λ axis again $\tilde{G}_{J,0}(t,\lambda) = \hat{\tilde{S}}_J^{-}(t,\lambda)\tilde{S}_J^{+}(t,\lambda)$ can be considered as a *minimal set of scattering data* for the gauge equivalent systems.

In order to obtain the soliton solutions for the gauge equivalent MHF systems one needs to apply the Zakharov–Shabat dressing method to a regular FAS (see e.g. [7, 5]).

These results can be naturally extended in several directions. First, one can apply the reduction group method [13] in order to investigate the Z_2 and other finite order reduction of the MNLS systems thus extending the results of [6, 8] for the N-wave systems and for their gauge equivalent [7, 5]. Second, it is natural to extend these reductions also to the gauge equivalent systems (8). A third open problem is to study reductions of the gauge equivalent systems and the spectral decompositions for the relevant recursion operators.

ACKNOWLEDGMENTS

One of us (GGG) thanks the organizers of the International Workshop on Global Analysis (Çankaya University) for the financial support and for the warm hospitality in Ankara.

REFERENCES

1. M. Ablowitz, D. Kaup, A. Newell, H. Seegur, *The inverse scattering transform – Fourier analysis for nonlinear problems,* Stud. Appl. Math., **53** (1974), 249–315.
2. L. D. Faddeev, L. A. Takhtadjan, *Hamiltonian approach in the theory of solitons*, Springer, Berlin, (1987).
3. I. M. Gelfand, L. A. Dickey, *Asymptotic behaviour of the resolvent of SturmLiouville equations and the algebra of the Korteweg-de Vries equations,* Usp. Mat. Nauk **30** (1977), 67–100; *A Lie algebra structure in a formal variational calculation,* Funk. Anal. Pril. **10** (1976), 13–29.
4. V. S. Gerdjikov, *The Generalized Zakharov–Shabat System and the Soliton Perturbations,* Theor. Math. Phys. **99**, No. 2, 292–299 (1994).
5. V. S. Gerdjikov, G. G. Grahovski, *On N-wave and NLS type systems: generating operators and the gauge group action: the so(5) case,* Proc. of IM of NAS of Ukraine **50**, 388–395 (2004).
6. V. S. Gerdjikov, G. G. Grahovski, R. I. Ivanov, N. A. Kostov, *N-wave interactions related to simple Lie algebras. $\mathbb{Z}_2$- reductions and soliton solutions,* Inverse Problems **17**, 999–1015 (2001).

7. V. S. Gerdjikov, G. G. Grahovski, N. A. Kostov, *On N-wave Type Systems and Their Gauge Equivalent,* Eur. Phys. J. B **29**, 243–248 (2002).
8. V. S. Gerdjikov, G. G. Grahovski, N. A. Kostov, *Reductions of N-wave interactions related to low-rank simple Lie algebras. I: $\mathbb{Z}_2$–reductions,* J. Phys. A: Math and Gen. **34**, 9425–9461 (2001).
9. V. S. Gerdjikov, P. P. Kulish, *The generating operator for the $n \times n$ linear system,* Physica **D3**, 549–564, (1981); V. S. Gerdjikov, *Generalized Fourier transforms for the soliton equations. Gauge covariant formulation,* Inverse Problems **2**, 51, (1986).
10. V. S. Gerdjikov, A B. Yanovski, *Gauge covariant formulation of the generating operator. II. Systems on homogeneous spaces,* Phys. Lett. A, **110**, 53–58 (1985); *Gauge covariant formulation of the generating operator. I.* Commun. Math. Phys., **103**, 549–568 (1986).
11. D. J. Kaup, *Closure of the squared Zakharov–Shabat eigenstates,* J. Math. Annal. Appl. **54**, n. 3, 849–864, 1976.
12. P. P. Kulish, A. G. Reyman, *Hierarchy of Symplectic forms for the Schrödinger and the Dirac equations on a line,* Zapiski nauchnich seminarov LOMI **77**, 134–147 (1978).
13. A. V. Mikhailov, *The reduction problem and the inverse scattering problem,* Physica D, **3**, 73–117 (1981).
14. V. E. Zakharov, S. V. Manakov, S. P. Novikov, L. I. Pitaevskii, *Theory of solitons. The inverse scattering method,* Plenum, N.Y. (1984).

Discrete Calculus of Variations

Gusein Sh. Guseinov

Department of Mathematics, Atılım University, 06836 Incek, Ankara, Turkey
E-mail: guseinov@atilim.edu.tr

Abstract. The continuous calculus of variations is concerned mainly with the determination of minima or maxima of certain definite integrals involving unknown functions. In this paper, a discrete calculus of variations for sums is treated, including the discrete Euler-Lagrange equation.
Key words. Discrete variational problem, admissible functions.
PACS. 02.60.Lj.
MSC. 65N22.

1. MOTIVATING EXAMPLES

Let $\mathbf{T} = \{t_0, t_1, \ldots, t_N\}$ be a finite set of real numbers (points)

$$t_0 < t_1 < \ldots < t_N.$$

Let, further, α and β be given real numbers. Each function $y : \mathbf{T} \to \mathbb{R}$ with $y(t_0) = \alpha$ and $y(t_N) = \beta$ defines a uniquely determined polygonal line Γ joining the points (t_0, α) and (t_N, β), with vertices at the points $(t_n, y(t_n))$, $n = 0, 1, \ldots, N$.

Let F be the set of *admissible functions* defined by

$$F = \{y : \mathbf{T} \to \mathbb{R} \,|\, y(t_0) = \alpha \text{ and } y(t_N) = \beta\}.$$

In the case $\alpha \geq 0$ and $\beta \geq 0$ we define also the set of admissible functions F_+ by

$$F_+ = \{y \in F \,|\, y(t) \geq 0 \text{ for } t \in \mathbf{T}\}.$$

Example 1 *(The problem of minimum length). Find the shortest polygonal line joining the points (t_0, α) and (t_N, β).*

Using the distance formula for the length of the line segment joining the points $(t_{n-1}, y(t_{n-1}))$ and $(t_n, y(t_n))$ we find that the length of the polygonal line Γ is equal to

$$J_1[y] = \sum_{n=1}^{N} \sqrt{(t_n - t_{n-1})^2 + [y(t_n) - y(t_{n-1})]^2}. \tag{1}$$

CP729, *Global Analysis and Applied Mathematics: International Workshop on Global Analysis,*
edited by K. Taş, D. Krupka, O. Krupková, and D. Baleanu

Thus we have to minimize the quantity (functional) $J_1[y]$ defined by (1), over all functions $y \in F$.

Example 2 *(The problem of minimum area). Let $\alpha \geq 0$ and $\beta \geq 0$. Determine a polygonal line Γ corresponding to a function $y \in F_+$, such that the area of the figure in the (t, y) plane bounded by the polygonal line Γ, by the t−axis, and by the straight lines $t = t_0$ and $t = t_N$, is of minimum area.*

Using the formula for the area of a trapezoid we find that the area of the figure under the polygonal line Γ is

$$J_2[y] = \frac{1}{2} \sum_{n=1}^{N} (t_n - t_{n-1}) [y(t_{n-1}) + y(t_n)]. \tag{2}$$

Thus we have to minimize the quantity $J_2[y]$ defined by (2), over all functions $y \in F_+$.

Example 3 *(The problem of minimum surface of revolution). Let $\alpha \geq 0$ and $\beta \geq 0$. Determine a polygonal line Γ corresponding to a function $y \in F_+$, such that by revolving this line around the t−axis we get a surface of minimum area.*

Using the formula for the lateral area of a frustrum of a cone we find that the area of the surface of revolution is

$$J_3[y] = \pi \sum_{n=1}^{N} [y(t_{n-1}) + y(t_n)] \sqrt{(t_n - t_{n-1})^2 + [y(t_n) - y(t_{n-1})]^2}. \tag{3}$$

Thus we have to minimize the quantity $J_3[y]$ defined by (3), over all functions $y \in F_+$.

Example 4 *(The problem of minimum solid of revolution). Let $\alpha \geq 0$ and $\beta \geq 0$. Determine a polygonal line Γ corresponding to a function $y \in F_+$, such that the volume of the solid of revolution bounded by the planes $t = t_0$ and $t = t_N$ and by the surface of revolution generated by the rotation of the polygonal line Γ about the t−axis, is of minimum volume.*

Using the formula for the volume of a frustrum of a cone we find that the volume of the solid of revolution is

$$J_4[y] = \frac{1}{3}\pi \sum_{n=1}^{N} (t_n - t_{n-1}) [y^2(t_n) + y(t_n)y(t_{n-1}) + y^2(t_{n-1})]. \tag{4}$$

Thus we have to minimize the quantity $J_4[y]$ defined by (4), over all functions $y \in F_+$.

In order to formulate a discrete variational problem that corresponds reasonably well with continuous variational problems, we set

$$\Delta y(t_{n-1}) = y(t_n) - y(t_{n-1}), \qquad \Delta t_{n-1} = t_n - t_{n-1}$$

so that Δ is the forward difference operator.

Then the above functionals J_1, J_2, J_3, J_4 can be written as

$$J_1[y] = \sum_{n=1}^{N} \sqrt{1 + \left[\frac{\Delta y(t_{n-1})}{\Delta t_{n-1}}\right]^2} \, \Delta t_{n-1}, \tag{5}$$

$$J_2[y] = \frac{1}{2} \sum_{n=1}^{N} [y(t_{n-1}) + y(t_n)] \, \Delta t_{n-1}, \tag{6}$$

$$J_3[y] = \pi \sum_{n=1}^{N} [y(t_{n-1}) + y(t_n)] \sqrt{1 + \left[\frac{\Delta y(t_{n-1})}{\Delta t_{n-1}}\right]^2} \, \Delta t_{n-1}, \tag{7}$$

$$J_4[y] = \frac{1}{3}\pi \sum_{n=1}^{N} [y^2(t_n) + y(t_n)y(t_{n-1}) + y^2(t_{n-1})] \, \Delta t_{n-1}, \tag{8}$$

respectively.

Note that the continuous counterparts of the functionals J_1, J_2, J_3, J_4 given in (5)-(8) are

$$I_1[y] = \int_a^b \sqrt{1 + [y'(x)]^2} \, dx,$$

$$I_2[y] = \int_a^b y(x) \, dx,$$

$$I_3[y] = 2\pi \int_a^b y(x)\sqrt{1 + [y'(x)]^2} \, dx,$$

$$I_4[y] = \pi \int_a^b y^2(x) \, dx,$$

respectively.

2. THE DISCRETE VARIATIONAL PROBLEM

In order to unify and extend the above considered examples, suppose that $f(s,t,u,v,w)$ is a function defined on $\mathbf{T} \times \mathbf{T} \times \mathbb{R} \times \mathbb{R} \times \mathbb{R}$ which has continuous first order partial derivatives in the variables $u, v,$ and w. Let F be the set of functions $y : \mathbf{T} \to \mathbb{R}$ defined in Section 1. The "simplest discrete variational problem" is to extremize (minimize or maximize)

$$J[y] = \sum_{n=1}^{N} f\left(t_{n-1}, t_n, y(t_{n-1}), y(t_n), \frac{\Delta y(t_{n-1})}{\Delta t_{n-1}}\right) \Delta t_{n-1} \tag{9}$$

subject to y belonging to the set F.

Note that the above problem is a discrete model of the continuous variational problem

$$I[y] = \int_a^b f\left(x, y(x), y'(x)\right) dx$$

on a class of functions $y(x)$ which satisfy fixed end conditions $y(a) = \alpha$ and $y(b) = \beta$.

We say that $\hat{y}$ in F minimizes the simplest discrete variational problem if

$$J[y] \geq J[\hat{y}] \tag{10}$$

for all y in F. We say that J has a local minimum at $\hat{y}$ provided there is a $\delta > 0$ such that (10) is satisfied for all y in F with

$$|y(t) - \hat{y}(t)| < \delta, \qquad t \in \{t_0, t_1, \ldots, t_N\}.$$

Now we establish a necessary condition for the simplest discrete variational problem to have a local extremum. Let us use the subscript notation $y_n = y(t_n)$.

Theorem 5 *(Discrete Euler- Lagrange equation). If the simplest discrete variational problem has a local extremum at $\hat{y}$, then $\hat{y}$ satisfies the Euler-Lagrange equation*

$$f_u\left(t_n, t_{n+1}, y_n, y_{n+1}, \frac{\Delta y_n}{\Delta t_n}\right)\frac{\Delta t_n}{\Delta t_{n-1}} + f_v\left(t_{n-1}, t_n, y_{n-1}, y_n, \frac{\Delta y_{n-1}}{\Delta t_{n-1}}\right)$$

$$= \frac{\Delta f_w\left(t_{n-1}, t_n, y_{n-1}, y_n, \frac{\Delta y_{n-1}}{\Delta t_{n-1}}\right)}{\Delta t_{n-1}} \qquad \text{for} \qquad n \in \{1, 2, \ldots, N-1\}. \tag{11}$$

Since $\hat{y}$ belongs to F, we have $\hat{y}_0 = \alpha$ and $\hat{y}_N = \beta$.

Proof. Assume J has a local extremum on F at $\hat{y}$. Fix $\eta : \mathbf{T} \to \mathbb{R}$ as an admissible variation with zero end conditions $\eta_0 = \eta_N = 0$. Define a function $\varphi : \mathbb{R} \to \mathbb{R}$ by

$$\varphi(\varepsilon) = J[\hat{y} + \varepsilon\eta]$$

$$= \sum_{n=1}^N f\left(t_{n-1}, t_n, \hat{y}_{n-1} + \varepsilon\eta_{n-1}, \hat{y}_n + \varepsilon\eta_n, \frac{\Delta(\hat{y}_{n-1} + \varepsilon\eta_{n-1})}{\Delta t_{n-1}}\right)\Delta t_{n-1},$$

where $-\infty < \varepsilon < \infty$. Note that $\hat{y} + \varepsilon\eta$ belongs to F for all real numbers ε. Since φ has a local extremum at $\varepsilon = 0$ we have that $\varphi'(0) = 0$. Calculation of the derivative of $\varphi(\varepsilon)$ gives

$$\varphi'(\varepsilon) = \sum_{n=1}^N \left(f_u\eta_{n-1} + f_v\eta_n + f_w\frac{\Delta\eta_{n-1}}{\Delta t_{n-1}}\right)\Delta t_{n-1}.$$

Therefore

$$\sum_{n=1}^N f_u\left(t_{n-1}, t_n, \hat{y}_{n-1}, \hat{y}_n, \frac{\Delta\hat{y}_{n-1}}{\Delta t_{n-1}}\right)\eta_{n-1}\Delta t_{n-1}$$

173

$$+ \sum_{n=1}^{N} f_v \left(t_{n-1}, t_n, \hat{y}_{n-1}, \hat{y}_n, \frac{\Delta \hat{y}_{n-1}}{\Delta t_{n-1}} \right) \eta_n \Delta t_{n-1}$$

$$+ \sum_{n=1}^{N} f_w \left(t_{n-1}, t_n, \hat{y}_{n-1}, \hat{y}_n, \frac{\Delta \hat{y}_{n-1}}{\Delta t_{n-1}} \right) \Delta \eta_{n-1} = 0.$$

Now we change in the first sum n to $n+1$, and in the third sum apply the summation by parts formula

$$\sum_{n=1}^{N} a_{n-1} \Delta b_{n-1} = a_n b_n \big|_0^N - \sum_{n=1}^{N} (\Delta a_{n-1}) b_n.$$

Taking into account that $\eta_0 = \eta_N = 0$, we obtain

$$\sum_{n=1}^{N-1} \left\{ f_u \left(t_n, t_{n+1}, \hat{y}_n, \hat{y}_{n+1}, \frac{\Delta \hat{y}_n}{\Delta t_n} \right) \Delta t_n + f_v \left(t_{n-1}, t_n, \hat{y}_{n-1}, \hat{y}_n, \frac{\Delta \hat{y}_{n-1}}{\Delta t_{n-1}} \right) \Delta t_{n-1} \right.$$

$$\left. - \Delta f_w \left(t_{n-1}, t_n, \hat{y}_{n-1}, \hat{y}_n, \frac{\Delta \hat{y}_{n-1}}{\Delta t_{n-1}} \right) \right\} \eta_n = 0.$$

Since the values of η_n are arbitrary for $n = 1, 2, \ldots, N-1$, the Euler-Lagrange equation (11) holds along $\hat{y}$. ∎

Remark 6 *The Euler-Lagrange equation (11) can be derived directly noting that the functional $J[y]$ in (9) can be considered as a function of variables $y_n = y(t_n)$, $n = 1, 2, \ldots, N-1$, and therefore we determine the values of these variables which extremize $J[y]$, by the equations*

$$\frac{\partial J[y]}{\partial y_n} = 0, \qquad n = 1, 2, \ldots, N-1.$$

The latter equations yield (11).

Now let us examine the problem formulated above in Example 1. Obviously, the solution to this problem must be a straight line. However, we use it to illustrate the Euler-Lagrange equation (11). Writing (5) in the form of (9), we have

$$f(s, t, u, v, w) = \sqrt{1 + w^2}.$$

Next,

$$f_u(s, t, u, v, w) = f_v(s, t, u, v, w) = 0, \qquad f_w(s, t, u, v, w) = \frac{w}{\sqrt{1 + w^2}}.$$

Therefore the Euler-Lagrange equation (11) becomes

$$\frac{\Delta f_w \left(t_{n-1}, t_n, y_{n-1}, y_n, \frac{\Delta y_{n-1}}{\Delta t_{n-1}} \right)}{\Delta t_{n-1}} = 0, \qquad n \in \{1, 2, \ldots, N-1\}.$$

Consequently, there exists a constant $c \in \mathbb{R}$ such that

$$f_w\left(t_{n-1}, t_n, y_{n-1}, y_n, \frac{\Delta y_{n-1}}{\Delta t_{n-1}}\right) \equiv c, \qquad n \in \{1, 2, \ldots, N\},$$

i.e.,

$$\frac{\Delta y_{n-1}}{\Delta t_{n-1}} = c\sqrt{1 + \left(\frac{\Delta y_{n-1}}{\Delta t_{n-1}}\right)^2}, \qquad n \in \{1, 2, \ldots, N\} \tag{12}$$

holds. Therefore

$$\left(\frac{\Delta y_{n-1}}{\Delta t_{n-1}}\right)^2 = c^2\left[1 + \left(\frac{\Delta y_{n-1}}{\Delta t_{n-1}}\right)^2\right]$$

and

$$(1 - c^2)\left(\frac{\Delta y_{n-1}}{\Delta t_{n-1}}\right)^2 = c^2 \qquad \text{for all} \qquad n \in \{1, 2, \ldots, N\}. \tag{13}$$

From (13) it follows that $|c| \neq 1$ and $\frac{c^2}{1-c^2} \geq 0$. Hence

$$\left(\frac{\Delta y_{n-1}}{\Delta t_{n-1}}\right)^2 = \frac{c^2}{1 - c^2} \qquad \text{for all} \qquad n \in \{1, 2, \ldots, N\}$$

so that

$$\left|\frac{\Delta y_{n-1}}{\Delta t_{n-1}}\right| = \sqrt{\frac{c^2}{1 - c^2}} \qquad \text{for all} \qquad n \in \{1, 2, \ldots, N\}.$$

But, by (12), $\frac{\Delta y_{n-1}}{\Delta t_{n-1}}$ is always positive (if $c > 0$), always zero (if $c = 0$), or always negative (if $c < 0$). Therefore we finally have

$$\frac{\Delta y_{n-1}}{\Delta t_{n-1}} = (\operatorname{sgn} c)\sqrt{\frac{c^2}{1 - c^2}} \qquad \text{for all} \qquad n \in \{1, 2, \ldots, N\}, \tag{14}$$

where $\operatorname{sgn} c = 1$ if $c > 0$, $\operatorname{sgn} c = 0$ if $c = 0$, and $\operatorname{sgn} c = -1$ if $c < 0$. Now (14) shows that the slopes of all the line segments constituting the polygonal line Γ corresponding to $y : \mathbf{T} \to \mathbb{R}$ are the same. This means that the polygonal line Γ is a straight line.

For the further results concerning discrete variational problems we refer to [1-5]. Note that our Examples 2-4 presented in Section 1 are new and our functional $J[y]$ given in (9) is more general than that considered in the mentioned literature. Note also that the minimum value of functionals $J_2[y]$ and $J_4[y]$ corresponding to Examples 2 and 4 is attained on the function $\hat{y} \in F_+$ defined by $\hat{y}_0 = \alpha$, $\hat{y}_N = \beta$, $\hat{y}_n = 0$ $(1 \leq n \leq N-1)$. Solution of Example 3 is more involved.

REFERENCES

1. R. P. Agarwal, C. D. Ahlbrandt, M. Bohner, and A. Peterson, Discrete linear Hamiltonian Systems: A survey, *Dynam. Systems Appl.* **8**, (1999), 307-333.
2. C. D. Ahlbrandt, Equivalence of discrete Euler equations and discrete Hamiltonian systems, *J. Math. Anal. Appl.* **180**, (1993), 498-517.
3. C. D. Ahlbrandt and A. Peterson, *Discrete Hamiltonian Systems: Difference Equations, Continned Fractions, and Riccati Equations*, Kluwer Academic, Boston, MA, 1996.
4. R. Hilscher and V. Zeidan, Nonnegativity of a discrete quadratic functional in terms of the (strengthened) Legendre and Jacobi conditions, *Comput. Math. Appl.* **45**, (2003),1369-1383.
5. W. G. Kelley and A. C. Peterson, *Difference Equations: An Introduction with Applications,* Academic Press, San Diego, CA, 1991.

176

Symmetries and Orbit Theory in 4-dimensional Lorentz Manifolds

G. S. Hall

Department of Mathematical Sciences,
University of Aberdeen,
Aberdeen AB24 3UE, Scotland, U.K.

Abstract. A brief survey is presented of some mathematical details of the study of symmetry in 4-dimensional Lorentz manifolds with applications to general relativity. The main symmetry studied is local isometry but the other symmetries encountered in Einstein's theory are also briefly discussed.
Key words. Symmetry, Killing vector, orbits, isotropy.
PACS. 04.20.-q, 02.40.-k.
MSC. 83C20.

1. WHAT IS A SPACE-TIME SYMMETRY?

Let M be a space-time so that M is a smooth connected Hausdorff 4-dimensional manifold admitting a smooth Lorentz metric g. Let Γ denote the Levi-Civita connection associated with g and R the corresponding curvature tensor. Throughout this paper, a comma, a semi-colon and the symbol $\mathscr{L}$ will denote a partial, a covariant (with respect to Γ) and a Lie derivative, respectively. The Lie bracket of vector fields is denoted by $[\ ]$.

Now suppose, for simplicity, that T is some tensor on M and that one asks what is meant by a "symmetry" of T. First let $f : M \to M$ be a diffeomorphism of M, so that f is a smooth bijective map with smooth inverse. Then call f a "symmetry" of T if $f^*T = T$ (where f^* is the pullback of f). To see what this means let U be some coordinate domain of M with coordinates x^a and let $f(U)$ be the (open) image of U under f. Then put a coordinate system y^a on $f(U)$ by attaching the same coordinates to $f(p) \in f(U)$ that were originally given to the point p in the x^a system, that is, $y^a = x^a \circ f^{-1}$ (see Figure 1). The condition $f^*T = T$ is then equivalent to the condition that the tensor T has identical components, numerically, at p (with coordinates x^a) and at $f(p)$ (with coordinates y^a) for each $p \in U$.

This definition of a symmetry f requires f to be a diffeomorphism of M. This is a very restrictive definition from the point of view of physics. Physical observation is essentially a local phenomenon and one is inclined to suggest that a symmetry in physics should also be local. In mathematics also, there is a well-defined and rather

CP729, *Global Analysis and Applied Mathematics: International Workshop on Global Analysis,*
edited by K. Taş, D. Krupka, O. Krupková, and D. Baleanu

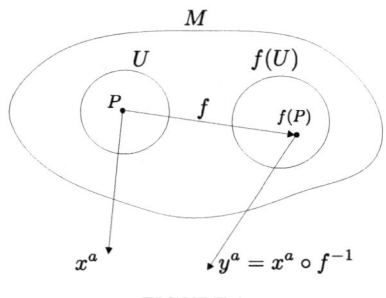

FIGURE 1.

useful concept of local symmetry and one is tempted to say that it is sometimes avoided in favour of the type of "global" symmetry described above because of the difficulties involved. So suppose that the above concept of symmetry is modified so that f could be a *local diffeomorphism* of M (that is, f is smooth, bijective, and with smooth inverse, and defined on some open subset U of M, $f : U \to f(U)$) and still retaining the condition $f^*T = T$. The problem now is that, whilst the set of all (global) diffeomorphisms which are symmetries of T have a group structure, the set of local diffeomorphisms which are symmetries of T has a more complicated structure. Whilst much progress in this latter direction has been achieved by Stefan [20, 21] (see also the important paper by Sussmann [23]) the usual approach to symmetries in general relativity is by specifying a certain family of "symmetry vector fields". These vector fields give rise in a natural way to a well structured family of local diffeomorphisms of M which are then postulated to have the symmetry property described above.

2. VECTOR FIELDS AND LOCAL FLOWS ON M

Let X be a globally defined smooth vector field on M. Associated with X is a family of local diffeomorphisms of M which arise in the following way [16, 3]. Let $p \in M$. Then the theory of differential equations shows that there exists an open neighbourhood U of p and a real number $\varepsilon > 0$ such that for any $q \in U$ the integral curve c of X which begins at q (i.e. $c(0) = q$) is defined on the interval $(-\varepsilon, \varepsilon)$ (i.e. $c(t)$ is defined for $|t| < \varepsilon$). Then each $q \in U$ can be "moved" a "parameter distance" t along c, provided $|t| < \varepsilon$, and this defines, in an obvious way, a map, denoted by φ_t, with domain U. The map φ_t is, in fact, a local diffeomorphism of M [16, 3] for each $t \in (-\varepsilon, \varepsilon)$ and the above construction can be applied at any $p \in M$ since X is globally defined. In this way X creates a family of local diffeomorphisms and each point of M is in the domain of such a map. The maps φ_t are sometimes referred to as the *local flows* of X.

The idea now is to define the symmetries of some geometrical object on M through vector fields whose corresponding local flows satisfy, for example, the conditions given earlier for being a symmetry of the tensor T. It should be pointed out that this example (of a symmetry of the tensor T) is not the only type of symmetry encountered in general relativity. One may, for example, consider symmetries which preserve the *conformal* *structure* of the space-time M (i.e. vector fields whose local flows φ_t satisfy $\varphi_t^* g = \alpha g$ for some appropriate local function α). In this paper, symmetries of the metric tensor will mainly be considered although a brief discussion of other types of symmetry will be included for completeness.

The act of transferring the problem of a symmetry defined initially through a local diffeomorphism to one defined by vector fields through their local flows has the advantage that the difficult problem of manipulating local diffeomorphisms with the appropriate symmetry property is replaced by one of handling vector fields whose components satisfy certain differential equations (depending on the symmetry) and which often turn out to be easier.

3. LOCAL ISOMETRIES AND KILLING VECTOR FIELDS

Perhaps the most important symmetry in general relativity concerns the space-time metric g. A smooth global vector field X on M is a *symmetry of g* if for each local flow φ_t of X the condition $\varphi_t^* g = g$ holds. Thus each such φ_t is a *local isometry* and X is called a *Killing vector field*. This condition on each φ_t resulting from X being Killing, together with the definition of the Lie derivative, can then be shown equivalent to the condition [16]

$$\mathscr{L}_X g = 0 . \tag{1}$$

This condition is easily evaluated in terms of the components of X in some (any) coordinate system, as mentioned at the end of the last section, and yields

$$X_{a;b} + X_{b;a} = 0 , \tag{2}$$

which are *Killing's equations*. They are equivalent to X being a Killing vector field. They can trivially, but usefully, be rewritten as

$$X_{a;b} = F_{ab} \qquad (F_{ab} = -F_{ba}) , \tag{3}$$

where F is a skew-symmetric tensor (bivector) called the *Killing bivector*.

Now let $K(M)$ denote the set of all Killing vector fields on M. Then the relations

$$a\mathscr{L}_X + b\mathscr{L}_Y = \mathscr{L}_{aX+bY}, \quad \mathscr{L}_{[X,Y]} = \mathscr{L}_X\mathscr{L}_Y - \mathscr{L}_Y\mathscr{L}_X \tag{4}$$

for vector fields X, Y on M and $a, b \in \mathbb{R}$, show that $K(M)$ is a *real vector space* which is also a *real Lie algebra* under the Lie bracket operation (i.e. $X, Y \in K(M) \Rightarrow aX + bY$ and $[X, Y] \in K(M)$). It also follows from (3) and the Ricci identity that

$$F_{ab;c} = R_{abcd}X^d . \tag{5}$$

Equations (3) and (5) together constitute a first order differential system for the ten independent quantities contained in the pair X_a and F_{ab}. From this it follows that a global Killing vector field on M is uniquely determined by its components together with those of its first covariant derivatives (i.e. those of its Killing bivector) at some (any) point of M. To see this, briefly, let $X, Y \in K(M)$, let $p \in M$ and suppose $X_a(p) = Y_a(p)$, $X_{a;b}(p) = Y_{a;b}(p)$. Let q be any other point of M. Since M is connected, it is path connected, and so there is a smooth path c from p to q with parameter s. Equations (3) and (5) give first order differential equations along c with independent variable s for the values of X_a and F_{ab}. It follows from Picard's theorem that if the values of X and Y and their first derivatives agree at any point p' on the curve (i.e. on the image of c) they will do so in some open interval on the curve containing p' and that if they disagree at p', they will, by continuity, disagree in some open interval on this curve containing p'. Since the image of c is connected and since agreement occurs at p, agreement occurs at each point of the curve and hence at q. Since q is arbitrary $X \equiv Y$ on M. From this result it immediately follows that dim $K(M) \leq 10$ (and, incidentally, if this maximum is achieved, M has constant curvature). The converse is false (but is true locally in a well defined way). It also follows that if $X \in K(M)$ and X vanishes on some non-empty open subset of M, then $X \equiv 0$ on M.

Recalling some remarks in the final paragraph of the first section, one now asks whether the local (flow) isometries φ_t arising from the members of $K(M)$ could be global and whether $K(M)$ leads to some (finite-dimensional Lie) group action on M. In spite of many remarks in the literature suggesting this is always the case, it is not. In fact, such a global (Lie) group action arises if and only if $K(M)$ is finite-dimensional (and this has already been established) and consists entirely of complete (Killing) vector fields [18] (see also [3]).

4. ORBIT STRUCTURE AND ZEROS OF KILLING VECTOR FIELDS

An important concept relating to the Lie algebra $K(M)$ is that of an orbit. These orbits are not only interesting geometrically but are also useful for constructing natural coordinate systems in space-times admitting the symmetry described by $K(M)$.

Let $X_1, \ldots, X_k \in K(M)$ and let $\varphi_t^1, \ldots, \varphi_t^k$ be the respective local flows (local isometries of M). Then one can construct from them a "combined" local isometry of the form

$$p \to \varphi_{t_1}^1 \left(\varphi_{t_2}^2 \left(\cdots \varphi_{t_k}^k(p) \cdots \right) \right) \tag{6}$$

for appropriate $t_1, \ldots, t_k$ and (recalling that the local flows are only local maps) provided (6) makes sense. The set S of all such local isometries of M is sometimes called the "local group of local isometries of M". This set leads to an equivalence relation $\sim$ on M given for $p_1, p_2 \in M$ by $p_1 \sim p_2 \Leftrightarrow \exists f \in S$ such that $f(p_1) = p_2$. The relation $\sim$ partitions M into equivalence classes and each such equivalence class is called an orbit of $K(M)$. Thus the orbit of $K(M)$ containing $p \in M$ contains all the points which p can be mapped into using members of S.

Also associated with $K(M)$ is the following "generalised" type of distribution on M, denoted by Δ, and which maps $p \in M$ to the subspace $\Delta(p)$ of the tangent space T_pM to M at p defined by

$$\Delta(p) = \{X(p) : X \in K(M)\} . \tag{7}$$

If $\dim \Delta(p)$ is the same for each $p \in M$ the classical Fröbenius theorem applies and reveals that (see e.g. [3])

(a) each orbit of $K(M)$ is a submanifold of M,
(b) if O is an orbit of $K(M)$ and $p \in O$ then the tangent space to O at p equals $\Delta(p)$.

Condition (a) ensures that (b) makes sense and (b) says that the orbits are *integral manifolds* of Δ. In general, however, the dimension of $\Delta(p)$ may not be constant over M and so the Fröbenius theorem is not applicable. However, a theorem due to Hermann [15] (and which has been extended to more general situations by Stefan [20, 21] and Sussmann [23]) can be used to show that, *because $K(M)$ is finite-dimensional, the conditions (a) and (b) above still hold*. Thus the orbit structure for the Lie algebra of Killing vector fields on M is, in this sense, well-behaved. Another potential technical problem was also solved in [20, 21]. If $f : M_1 \to M_2$ is a smooth map between (smooth) manifolds M_1 and M_2 and if the range of f lies inside a (smooth) submanifold N of M_2 then the obvious associated map $f : M_1 \to N$ is not necessarily smooth (or even continuous; if it is continuous it is smooth). If the (sub)manifold topology of N coincides with its subspace topology from M_2 (so that N is a regular submanifold of M_2) then, of course, $f : M_1 \to N$ is smooth [3]. It turns out that if N is an orbit arising from a Lie algebra of vector fields as described here then, even though N may not be a regular submanifold, the above associated map $f : M_1 \to N$ is smooth [20, 21].

Although the above definition of an orbit of $K(M)$ is natural and satisfactory, certain Lorentz manifolds may admit orbits with undesirable properties. Certainly for any orbit O of dimension 1, 2 or 3 the *nature* of O (i.e. spacelike, timelike or null) is the same at each of its points and if O is not null it admits an induced metric $h = i^* g$ where $i : O \to M$ is the natural inclusion map. Then any $X \in K(M)$ induces a vector field $\tilde{X}$ on O such that $i_* \tilde{X} = X$ and it is easily checked that if O is not null $\mathscr{L}_{\tilde{X}} h = 0$, so that $\tilde{X}$ is a Killing vector field in the induced geometry on O (that is a member of $K(O)$) and that the map $f : K(M) \to K(O)$ given by $f(X) = \tilde{X}$ is a Lie algebra homomorphism. Whilst examples can easily be constructed with f failing to be onto, it is less obvious whether f can fail to be one-to-one, that is, fail to map independent members of $K(M)$ into independent members of $K(O)$. To see this latter failure consider the space-time M with metric

$$ds^2 = -dt^2 + 2dt\,dz + t\,dz^2 + e^{x^2+y^2}(dx^2 + dy^2) \qquad (8)$$

for $t > -1$, $-\infty < x, y, z < \infty$. Then $K(M)$ is 2-dimensional, being spanned by $x\frac{\partial}{\partial y} - y\frac{\partial}{\partial x}$ and $\frac{\partial}{\partial z}$. Also every orbit of $K(M)$ is 2-dimensional except those containing points with $x = y = 0$ which are 1-dimensional. Clearly the map f for these latter orbits is not one-to-one. Such a behaviour can be sometimes undesirable from both a mathematical and physical viewpoint. So if O is an orbit of $K(M)$ of dimension 1, 2 or 3, call O *stable* (respectively, *dimensionally stable*) if for each $p \in O$ there exists a neighbourhood U of p in the manifold topology on M such that for each $p' \in U$ the $K(M)$ orbit through p' is of the same nature and dimension (respectively, the same dimension) as O [9, 10]. It turns out that if O is 3-dimensional and not null then it is necessarily stable and that for any dimensionally stable orbit the associated map f described earlier is one-to-one. Also, the existence of a dimensionally stable orbit O of $K(M)$ gives a tighter control of the dimension of $K(M)$ (depending on the dimension of O) [9, 10]. In fact, if $K(M)$ is not trivial, M can be disjointly decomposed as $M = M_1 \cup M_2$ where M_1 is open and dense in M and is the union of all the stable orbits of $K(M)$ whilst M_2 is closed with empty interior and consists of the unstable orbits of $K(M)$.

Now let X be a non-trivial member of $K(M)$ and let $p \in M$ such that $X(p) = 0$. Then p is a *zero* of X and, as can be easily checked, p is a *fixed point* of any local flow φ_t of p, $\varphi_t(p) = p$. Zeros of Killing vector fields will appear at any point on an orbit O if $\dim O < \dim K(M)$. To see this by means of an example let $\dim O = 3$ and $\dim K(M) = 4$ (such as occurs in the Schwarzchild and Reissner-Nordstrøm metrics). Let $X_1, \ldots, X_4$ be a basis of global vector fields in $K(M)$ and let $p \in O$. Consider the vectors $X_1(p)$, $\ldots$, $X_4(p)$ in the 3-dimensional tangent space T_pO to O at p. Clearly there exists real numbers $a_1, \ldots, a_4$ not all zero such that $\Sigma_{i=1}^4 a_i X_i(p) = 0$. Now define the vector field X on M by $X = \Sigma_{i=1}^4 a_i X_i$. Clearly, $X \in K(M)$ with X non-trivial (since $X_1, \ldots, X_4$ form a basis for $K(M)$) and $X(p) = 0$. One can obtain useful information about a space-time at a zero of a Killing vector field. In fact, suppose p is a zero of $X \in K(M)$. Then

(i) either p is isolated or lies in a coordinate domain U such that the zeros of X in U constitute a 2-dimensional submanifold of U [8],
(ii) the Petrov type of the Weyl tensor at p is either N, D or O [6],
(iii) the Ricci (and energy-momentum) tensor has an eigenvalue degeneracy at p [4, 22].

In each of these three results further information is obtained if the algebraic type of the Killing bivector at p is known [6]. The above results can be rather useful. For example, they can be used to show rather quickly that in a cosmological model admitting the standard F.R.W. symmetry the Weyl tensor is necessarily zero and the energy-momentum tensor must be of the perfect fluid type. Similarly any metric admitting "plane-wave" symmetry must be of Petrov type N or O and at those points where the energy-momentum tensor is not zero it must be of the null fluid type. Further details can be found in [10].

5. OTHER TYPES OF SPACE-TIME SYMMETRY

Conformal symmetry.

A global smooth vector field X on M is called *conformal* if each of its local flows φ_t is a *local conformal map* (i.e. if $\varphi_t^* g = \alpha g$ for some appropriate local function α). This is equivalent to $\mathcal{L}_X g = \Psi g$ for some function $\Psi : M \to \mathbb{R}$). If X is conformal and Ψ is constant on M, X is called *homothetic* (and clearly X is Killing if $\Psi \equiv 0$ on M). The study of conformal symmetry is related in a natural way to the study of isometries [10],[1]–[14].

Projective symmetry.

A global smooth vector field X on M is called *projective* if the associated local flows φ_t

map geodesics to geodesics (not necessarily preserving affine parameters). This can be shown to lead to the (equivalent) condition that by decomposing $X_{a;b}$ into its symmetric and skew-symmetric parts as

$$X_{a;b} = \frac{1}{2}h_{ab} + G_{ab} \qquad h_{ab} = h_{ba}(= \mathscr{L}_X g_{ab}), \quad G_{ab} = -G_{ba} \qquad (9)$$

then

$$h_{ab;c} = 2g_{ab}\chi_c + g_{ac}\chi_b + g_{bc}\chi_a \qquad (10)$$

for some 1-form χ_a which is closed (i.e. locally a gradient). In the case that $\chi_a \equiv 0$ on M (equivalently $h_{ab;c} = 0$ on M) X is called *affine* (and then the local flows preserve not only the geodesics of M but also their affine parameters). If $h_{ab} = cg_{ab}$ ($c \in \mathbb{R}$) then X is homothetic (and Killing if, in addition, $c = 0$).

Let $C(M)$, $H(M)$, $A(M)$ and $P(M)$ denote the sets of all conformal, homothetic, affine and projective vector fields, respectively. Then each can be shown to be a *finite-dimensional Lie algebra of vector fields on M* under the Lie bracket operator. In fact, $\dim C(M) \leq 15$, $\dim H(M) \leq 11$, $\dim A(M) \leq 20$ and $\dim P(M) \leq 24$. It follows that the nice orbit results (a) and (b) of the previous section hold for each of these Lie algebras just as for $K(M)$.

Curvature symmetry.

A global *smooth* vector field X on M is called a *curvature collineation* if its local flows are symmetries of the curvature tensor of M, that is, if

$$\mathscr{L}_X R^a{}_{bcd} = 0 . \qquad (11)$$

The set of all such vector fields is denoted by $CC(M)$ and (4) shows that $CC(M)$ is a Lie algebra. The Lie algebras $K(M)$, $H(M)$ and $A(M)$ are subalgebras of $CC(M)$. However, $CC(M)$ is not necessarily finite-dimensional [11] and this fact makes their study more difficult. Fortunately one can say that, in a well-defined way, for "almost all" space-times $CC(M) = H(M)$ [11]–[19].

For the study of the Lie algebras $H(M)$ and $K(M)$ one can also introduce concepts of stability and dimensional stability, the results being essentially as for $K(M)$ [13]. It is also remarked that a study of the Lie algebras $A(M)$ and $P(M)$ is facilitated by use of holonomy group theory in space-times [10, 12].

It is briefly remarked that members of $K(M)$, $H(M)$, $A(M)$ and $P(M)$ were defined to be smooth and one might ask if such vector fields need be smooth. In fact, one can show that if X is in $A(M)$ and is C^2, or if X is in $C(M)$ or $P(M)$ and is C^3, then it is necessarily smooth. For $CC(M)$, however, one can construct a vector field X satisfying (11) of differentiability C^k (but not C^{k+1}) for any k, $1 \leq k < \infty$. If one does not insist on smoothness for $CC(M)$ it will remain a vector space but may fail to be a Lie algebra [11].

183

ACKNOWLEDGEMENTS

The author wishes to thank Profs Kenan Tas and Dumitru Baleanu in Ankara and Prof. Tekin Dereli in Istanbul for their hospitality and Dr. Ugur Camci for several valuable discussions.

REFERENCES

1. D.V. Alexeevski., Selfsimilar Lorentzian manifolds, *Ann. Global. Anal. Geom.* **3**, (1985), 59-84.
2. R. Bilyalov., Conformal transformation groups in gravitational fields, *Sov. Phys.* **8**, (1964), 878-880.
3. F. Brickell and R.S. Clark. "Differentiable Manifolds". Von Nostrand, London, (1970).
4. R.F. Crade and G.S. Hall., Energy-momentum tensors in locally isotropic space-times, *Phys. Lett.* **75A**, (1979), 17-18.
5. L. Defrise-Carter., Conformal groups and conformally equivalent isometry groups, *Comm. Math. Phys.* **40**, (1975), 273-281.
6. J. Ehlers and W. Kundt, Exact solutions of the gravitational field equations in "Gravitation; an introduction to current research" ed. L. Witten. Wiley, New York, (1962), 49-101.
7. G.S. Hall., Curvature collineations and the determination of the metric from the curvature in general relativity, *Gen. Rel. Grav.* **15**, (1983), 581-589.
8. G.S. Hall., Homothetic transformations with fixed points in space-time, *Gen. Rel. Grav.* **20**, (1988), 671-681.
9. G.S. Hall., On the theory of Killing orbits in space-time, *Class. Quant. Grav.* **20**, (2003), 4067-4084.
10. G.S. Hall. "Symmetries and Curvature Structure in General Relativity". World Scientific Pub. Co. (2004).
11. G.S. Hall and J. da Costa., Curvature collineations in general relativity I and II, *J. Math. Phys.* **32**, (1991), 2848-2853, and ibid **32**, (1991), 2854-2862.
12. G.S. Hall and D.P. Lonie., Holonomy and projective symmetry in space-times. To be submitted for publication (2004).
13. G.S. Hall and M.T. Patel., On the theory of homothetic and affine orbits in space-time. To be submitted for publication (2004).
14. G.S. Hall and J.D. Steele., Conformal vector fields in general relativity, *J. Math. Phys.* **32**, (1991), 1847-1853.
15. R. Hermann., On the accessibility problem in control theory, *Int. Symposium on non-linear differential equations and non-linear mechanics.* Academic Press, New York, (1963), 325-332.
16. S. Kobayashi and K. Numizu. "Foundations of Differential Geometry". Interscience, New York, Vol 1, (1963).
17. C.B.G. McIntosh and W.D. Halford., The Riemann tensor, the metric tensor and curvature collineations in general relativity, *J. Math. Phys.* **23**, (1982), 436-443.
18. R.S. Palais., A global formulation of the Lie theory of transformation groups, *Mem. Am. Math. Soc.* **22**, (1957).
19. A.D. Rendall., Curvature of generic space-times in general relativity, *J. Math. Phys.* **29**, (1988), 1569-1574.
20. P. Stefan., Accessible sets, orbits and foliations with singularities, *Proc. London. Math. Soc.* **29**, (1974), 699-713.
21. P. Stefan., Integrability of systems of vectorfields, *J. London. Math. Soc.* **21**, (1980), 544-556.
22. H. Stephani, D. Kramer, M.A.H. MacCallum, C. Hoenselaers and E. Herlt. "Exact Solutions to Einstein's Field Equations, Second Edition" C.U.P. (2003).
23. H.J. Sussmann., Orbits of families of vector fields and integrability of distributions, *Trans. Am. Math. Soc.* **180**, (1973), 171-188.

Finite Group Invariance and Solution of Jaynes-Cummings Hamiltonian

Derya Haydargil* and Ramazan Koç*

*Department of Physics, Faculty of Engineering University of Gaziantep, 27310 Gaziantep Turkey

Abstract. The finite group invariance of the $E \otimes \beta$ and Jaynes-Cummings models are studied. A method is presented to obtain finite group invariance of the $E \otimes \beta$ system. A suitable transformation of a Jaynes-Cummings Hamiltonian leads to equivalence of $E \otimes \beta$ system. Then a general method is applied to obtain the solution of Jaynes-Cummings Hamiltonian with Kerr nonlinearity. Number operator for this structure and the generators of $su(2)$ algebra are used to find the eigenvalues of the Jaynes-Cummings Hamiltonian for different states. By using the invariance of number operator the solution of modified Jaynes-Cummings Hamiltonian is also discussed.
Key words. Jaynes-Cummings model, Group theory, Exact solution.
PACS. 42.50.Ct, 03.65.Fd, 02.20.Bb.
MSC. 81R05.

1. INTRODUCTION

The Jaynes-Cummings(JC) model, which describes the interaction of two-level atoms with field [9], has been playing a very significant role in our understanding between radiation and matter that lies at the heart of many problems in laser physics and quantum optics. The model is an old one, dating back over thirty years, but it appears, however, its analysis not have been fully exploited. Despite the relatively large amount of work found in the literature [5, 8, 6, 2, 11, 7, 1, 12, 10], the eigenvalues and the eigenfunctions of the JC Hamiltonian are not known analytically in general. It has been solved by applying various approximation schemes [4].

In the first part of this article we investigate the symmetry properties of the JC Hamiltonian. Our investigation leads to an interesting result: Mathematical structure of JC Hamiltonian and $E \otimes \beta$ Jahn-Teller Hamiltonian are the same. We also show that, there exists an interrelation between the modified JC Hamiltonian and $E \otimes \beta$ Jahn-Teller Hamiltonian. This connection of the Hamiltonians plays a crucial role to the solution and analysis of the JC Hamiltonians, because in literature one can find some exact results for the eigenvalues and eigenstates of the Jahn-Teller Hamiltonians.

We also presented here exact solution of the approximated JC and modified JC Hamiltonians by algebraic method. With the help of deformed $su(2)$ algebraic structure it is easy to obtain the eigenvalues and eigenstates of the system.

CP729, *Global Analysis and Applied Mathematics: International Workshop on Global Analysis,*
edited by K. Taş, D. Krupka, O. Krupková, and D. Baleanu
© 2004 American Institute of Physics 0-7354-0209-4/04/$22.00

2. FINITE SUBGROUPS OF $SU(2)$ AND JC MODEL

In this section we investigate the symmetry properties of the JC model. It is well known that the finite subgroups of $su(2)$ have been focus of interest from the both physical and mathematical point of view, and it can be generated by discrete rotations around the properly chosen two axes. For example the two dimensional matrix generators of dihedral groups can be generated by using two basis generator matrices

$$A = \begin{pmatrix} \cos\phi & \sin\phi \\ -\sin\phi & \cos\phi \end{pmatrix} \qquad B = \begin{pmatrix} -\cos\phi & \sin\phi \\ \sin\phi & \cos\phi \end{pmatrix} \qquad (1)$$

where, $\phi = \frac{2\pi}{n}$ and the generation relations of the A and B are given by

$$A^n = B^2 = (AB)^2 = 1. \qquad (2)$$

Dihedral groups include one and two-dimensional irreducible presentations. The matrix realizations of the two-dimensional irreducible representations can be obtained from (1). It is obvious that the characters of the one-dimensional irreducible representations are ∓ 1. When a physical system possesses the symmetry of a finite group, the coupling between energy levels can be determined by the direct product of the corresponding representation. For dihedral groups the decomposition of the symmetric direct product of two-dimensional representation E can be expressed as

$$E \times E = e \qquad E \times E = a_0 + b_1 \qquad E \times E = b_1 + b_2 \qquad (3)$$

where e is two-dimensional representation and b_1 and b_2 are the one-dimensional representations, with the characters ± 1. Our task is now to construct the physical Hamiltonian which is invariant under the dihedral groups. Let us consider the following polynomial function

$$P(x_i, q) = \sum_{i=1}^{2}\sum_{j=1}^{2} \alpha_{ij} x_i x_j q_0 + \sum_{i=1}^{2}\sum_{j=1}^{2} \beta_{ij} x_i x_j q_1 + H.C. \qquad (4)$$

where x_i represents the coordinates of the field and q_i is the coordinate of the atom. When atom interacts with field then the coupling type may be classified according to (3) and the coordinates x_i transform with the two-dimensional representation of the group while q transform with the one-dimensional representations. In general we can write

$$P(x_i, q) = P(Ax_i, g_1 q_j) = P(Bx_i, g_2 q_j) \qquad (5)$$

where g_1 and g_2 are the characters of the b_i. The invariant matrix potential can be obtained from minimality condition:

$$V = \frac{\partial P}{\partial x_i x_j} \quad . \qquad (6)$$

It is interesting that the procedure presented above leads to the construction of the JC

TABLE 1. Coefficients of the polynomial $P(x_i, q)$ for dihedral groups D_2 and D_4.

n	g_1	g_2	α_{11}	α_{22}	α_{12}
2	1	1	α_{11}	α_{22}	0
	1	-1	0	0	α_{12}
4	1	1	α_{11}	α_{11}	0
	1	-1	0	0	α_{12}
	-1	-1	α_{11}	$-\alpha_{11}$	0

Hamiltonian. From the relations (1), (4) and (5) one can obtain the following expressions

$$P(Ax_i, g_1q) = \{\frac{1}{2}[\alpha_{11} + \alpha_{22} + (\alpha_{11} - \alpha_{22})\cos\frac{4\pi}{n} - \alpha_{12}\sin\frac{4\pi}{n}]x_1^2 +$$
$$\frac{1}{2}[\alpha_{11} + \alpha_{22} - (\alpha_{11} - \alpha_{22})\cos\frac{4\pi}{n} - \alpha_{12}\sin\frac{4\pi}{n}]x_2^2 +$$
$$(\alpha_{11} - \alpha_{22})\sin\frac{4\pi}{n} + \alpha_{12}\cos\frac{4\pi}{n}]x_1 x_2\}(g_1 q) \qquad (7)$$

$$P(Bx_i, g_2q) = \{\frac{1}{2}[\alpha_{11} + \alpha_{22} + (\alpha_{11} - \alpha_{22})\cos\frac{4\pi}{n} - \alpha_{12}\sin\frac{4\pi}{n}]x_1^2 +$$
$$\frac{1}{2}[\alpha_{11} + \alpha_{22} - (\alpha_{11} - \alpha_{22})\cos\frac{4\pi}{n} - \alpha_{12}\sin\frac{4\pi}{n}]x_2^2 -$$
$$(\alpha_{11} - \alpha_{22})\sin\frac{4\pi}{n} - \alpha_{12}\cos\frac{4\pi}{n}]x_1 x_2\}(g_2 q). \qquad (8)$$

The coefficients α_{ij} can be obtained by comparing (4), (7) and (8). The results are summarized in *Table*1. For the values of n except that $n = 2$ and 4, the generators g_1 and g_2 take the value 1 and $\alpha_{11} = \alpha_{22}, \alpha_{12} = 0$.

Let us turn our attention to the construction of the interaction Hamiltonian. Consider the $E \times E$ interaction is in the form

$$E \times E = a_0 + b_1 \qquad (9)$$

The Hamiltonian can be written as

$$H = H_0 + V(b_1) \qquad (10)$$

where H_0 is the purely harmonic part

$$H_0 = -\frac{\hbar^2}{2m}\frac{\partial^2}{\partial q^2} + \frac{1}{2}m\omega^2 q^2 \qquad (11)$$

and the interaction Hamiltonian $V(b_1)$ can be obtained by using the minimality condition (6) and *Table*1,

$$V(b_1) = \alpha_{12}q \begin{pmatrix} 0 & 1 \\ 1 & 0 \end{pmatrix} + \alpha_{11} \begin{pmatrix} 1 & 0 \\ 0 & -1 \end{pmatrix}. \qquad (12)$$

187

In order to construct JC Hamiltonian, we can bosonise the Hamiltonian H. Introducing usual realizations of the ladder operators,

$$a = \frac{1}{\sqrt{2}}(\frac{d}{dq}+q) \qquad\qquad a^+ = \frac{1}{\sqrt{2}}(-\frac{d}{dq}+q) \qquad (13)$$

The Hamiltonian (10) can be expressed as

$$H = a^+ a + \alpha_{11}\sigma_0 + \alpha_{12}(a^+ + a)(\sigma_+ + \sigma_-) \qquad (14)$$

Therefore we have shown that the JC Hamiltonian is invariant under the dihedral group D_2. The Hamiltonian (14) is also associated to the $E \otimes \beta$ JT Hamiltonian. There exist two type Hamiltonian invariant under D_4 group. These are

$$H = a^+ a + \alpha_{12}(\sigma_+ + \sigma_-) \qquad or \qquad H = a^+ a + \alpha_{11}(a^+ + a)\sigma_0. \qquad (15)$$

In this case the energy splitting term $\frac{1}{2}\mu\sigma_0$ is missing. One can also investigate the symmetry properties of the modified JC Hamiltonian. In this case we consider the interaction $E \otimes \varepsilon$. The polynomial function can be written as

$$P(x_i,q) = \sum_{i=1}^{2}\sum_{j=1}^{2}\alpha_{ij}x_i x_j + \sum_{i=1}^{2}\sum_{j=1}^{2}\sum_{k=1}^{2}\beta_{ijk}x_i x_j q_k \qquad (16)$$

The interaction exist in the dihedral group D_4. The matrix generators of D_4 are given by

$$A_1 = \begin{pmatrix} -1 & 0 \\ 0 & -1 \end{pmatrix} \qquad\qquad B_1 = \begin{pmatrix} 1 & 0 \\ 0 & -1 \end{pmatrix} \qquad (17)$$

which transform q_i, q_1 and x_i, x_1. In this case the invariant polynomial takes the form;

$$P(x,q) = \alpha_{112}x_1^2 + \alpha_{222}x_2^2 + \beta_{121}x_1 x_2 q_1 \qquad (18)$$

when the generators of D_8 are used

$$A_2 = \begin{pmatrix} 0 & 1 \\ -1 & 0 \end{pmatrix} \qquad\qquad B_2 = \begin{pmatrix} 0 & 1 \\ 1 & 0 \end{pmatrix} \qquad (19)$$

that transform the x_1 and x_2 we obtain

$$P(x,q) = \alpha_{112}x_1^2 + \alpha_{222}x_2^2 + \beta_{222}(x_1^2 q_2 - x_2^2 q_2) \qquad (20)$$

One can obtain other possible forms of the polynomial function which are invariant under the dihedral groups. In this section we have presented a procedure to investigate the symmetry of the JC model. In the following section we will discuss the solution of the Hamiltonians.

3. EIGENVALUES OF THE JAYNES-CUMMINGS HAMILTONIAN WITH KERR NONLINEARITY

In this section we discuss the eigenvalues of effective Hamiltonian which represent the Jaynes-Cummings model surrounded by a Kerr medium. This Hamiltonian has the form:

$$H = \omega a^+ a + \frac{1}{2}\omega_0 \sigma_0 + \kappa(a^+ \sigma_- + a\sigma_+) + \lambda a^+ a a^+ a \tag{21}$$

where is κ atom-field coupling constant, λ is nonlinearity of Kerr medium, a and a^+ are creation and annihilation operators, $\sigma_\pm$ are Pauli matrices. Number operator of this structure is,

$$N = a_1^+ a_1 + \frac{1}{2}\sigma_0 \tag{22}$$

The invariance algebra if the Hamiltonian(20) is generated by introducing the generators

$$J_+ = a_1 \sigma_+ , \qquad J_- = a_1^+ \sigma_- , \qquad J_0 = a_1^+ a_1 + \sigma_0 \tag{23}$$

yields the commutation relation,

$$[J_0, J_\pm] = \pm J_\pm \qquad\qquad [J_+, J_-] = (1 + 2J_0)(J_0 - N) - \frac{1}{2} \tag{24}$$

The commutation relation $[J_+, J_-]$ is a polynomial in J_0. Since the deformation is quadratic in J_0, we have a quadratic algebra. The algebras of type have been considered as deformed $su(2)$ algebra. We can easily express the Hamiltonian (21) in terms of the generators of the deformed $su(2)$ algebra,

$$H = \omega(2N - J_0) + \omega_0(J_0 - N) + \kappa(J_+ + J_-) + \lambda(2N - J_0)^2 \tag{25}$$

Deformed $su(2)$ algebra generators have the forms

$$\begin{aligned}
J_+ \mid j,m\rangle &= \sqrt{(j-m)(j+m+1)}\mid j,m+1\rangle \\
J_- \mid j,m\rangle &= \sqrt{(j+m)(j-m+1)}\mid j,m-1\rangle \\
J_0 \mid j,m\rangle &= m \mid j,m\rangle \\
N \mid j,m\rangle &= 2j \mid j,m\rangle
\end{aligned} \tag{26}$$

If equation (26) is inserted in the Hamiltonian (25) and thinking about its eigenvalue $H\Psi = E\Psi$, we get

$$\left(\omega(4j - m) + \omega_0(m - 2J) + \lambda(4j - m)^2 - E\right)\mid j,m\rangle +$$
$$\kappa\sqrt{(j-m)(j+m+1)}\mid j,m+1\rangle + \kappa\sqrt{(j+m)(j-m+1)}\mid j,m-1\rangle \;=\; 0 \tag{27}$$

Different eigenvalues can be found for the equation (27) when different j and m values are used. Starting from the condition $j = 0, m = 0$ it is seen that $E = 0$. If we write

$j = 1$, m will take the values of $-1, 0$ and 1. The matrix form of this condition can be constructed as

$$\begin{pmatrix} P(5) & \sqrt{2}\kappa & 0 \\ \sqrt{2}\kappa & P(4) & \sqrt{2}\kappa \\ 0 & \sqrt{2}\kappa & P(3) \end{pmatrix} \tag{28}$$

where $P(n) = n\omega + (n)^2\lambda - (n - 2j)\omega_0$. For $j = 2, m$ will take the values $-2, -1, 0, 1, 2$ this situation can be represented by the matrix,

$$\begin{pmatrix} P(10) & 2\kappa & 0 & 0 & 0 \\ 0 & P(9) & \sqrt{6}\kappa & 0 & 0 \\ 0 & \sqrt{6}\kappa & P(8) & \sqrt{6}\kappa & 0 \\ 0 & 0 & \sqrt{6}\kappa & P(7) & 0 \\ 0 & 0 & 0 & 2\kappa & P(6) \end{pmatrix} \tag{29}$$

For $j = N, m$ will take the values of $-N, ..., N$ and the general matrix has the $2N + 1$ dimension and it can be represented as

$$\begin{pmatrix} P(5N) & \sqrt{2N}\kappa & ... & ... \\ \sqrt{2N}\kappa & P(5N - 1) & \sqrt{4N - 2}\kappa & ... \\ ... & \sqrt{4N - 2}\kappa & P(5N - 2) & \sqrt{5N - 2}\kappa \\ ... & ... & \sqrt{5N - 2}\kappa & ... \end{pmatrix} \tag{30}$$

by equalling the determinant of this matrix to zero the results for the eigenvalues of the Jaynes-Cummings Hamiltonian with Kerr nonlinearity can be obtained.

4. THE MODIFIED JAYNES-CUMMINGS HAMILTONIAN

The modified Jaynes-Cummings Hamiltonian have been constructed to investigate single two level atom placed in the common domain of two cavities interacting with two quantizied modes. It is given by

$$H = \omega(a_1^+ a_1 + a_2^+ a_2) + \frac{\omega_0}{2}\sigma_0 + \lambda_1(a_1\sigma_+ + a_1^+\sigma_-) + \lambda_2(a_2\sigma_+ + a_2^+\sigma_-) \tag{31}$$

The number operator for such systems can be expressed in the form

$$N = sa_1^+ a_1 + pa_2^+ a_2 + r\sigma_0 \tag{32}$$

For Hamiltonian (31), taking $s = p = 2r$ and $r = 1$, the number operator is,

$$N = 2a_1^+ a_1 + 2a_2^+ a_2 + \sigma_0 \tag{33}$$

Before going further in the following we use the Bargmann-Fock representation, where creation and annihilation operators are replaced by multiplication and differentiation operators:

$$a_i^+ = z_i \qquad a_i = \frac{d}{dz_i} \tag{34}$$

190

with respect to complex variable z_i. The eigenfunction of the Hamiltonian in the form

$$n_1, n_2 \rangle = (z_1)^j \phi(x) |\uparrow\rangle + (z_1)^{j+1} \phi(x) |\downarrow\rangle \tag{35}$$

where $x = (z_1)^{-1} z_2$. The solution of this system describes a quantum mechanical state of H provided that $\phi(x)$ belong to the Bargmann-Fock space. The scalar product should be complete and normalizable. The eigenvalue equation of the modified Jaynes-Cummings Hamiltonian can be written as

$$H \mid n_1, n_2 \rangle = E \mid n_1, n_2 \rangle \tag{36}$$

Insertion of (34) and (35) into (36) the following two set of differential equation

$$[j\omega + \frac{\omega_0}{2} - E]\phi_1(x) + (j+1)\lambda_2\phi_2(x) + (\lambda_2 - x\lambda_1)\frac{d\phi_2(x)}{dx} = 0$$
$$[(j+1)\omega - \frac{\omega_0}{2} - E]\phi_2(x) + (\lambda_1 + x\lambda_2)\phi_1(x) = 0 \tag{37}$$

Bargmann-Fock space solution of the (37) can easily be obtained and they are given by

$$\phi_1(x) = C_0(\lambda_2 - x\lambda_1)^{j-n+1}(\lambda_1 + x\lambda_2)^{n-1}$$
$$\phi_2(x) = C_1(\lambda_2 - x\lambda_1)^{j-n+1}(\lambda_1 + x\lambda_2)^{n} \tag{38}$$

where n is an integer and eigenvalues of the Hamiltonian is given by

$$E = \frac{1}{2}((2j+1)\omega \pm \sqrt{4n(\lambda_1^2 + \lambda_2^2) + (\omega_0 - \omega)^2}) \tag{39}$$

Exact results for the modified Jaynes-Cummings Hamiltonian are obtained. The procedure given here can be applied to obtain eigenfunctions and eigenvalues of the various physical Hamiltonians.

5. CONCLUSIONS

We have considered the Jaynes-Cummings Hamiltonian under the effects of dihedral groups. It has permitted us to build the matrix generators of these groups by using (1). As a result of the interactions between the symmetric states and the E states Jaynes - Cummings Hamiltonian is associated to the $E \otimes \beta$ Jahn-Teller Hamiltonian.

In section 3 we present a method to solve the problem of the Jaynes-Cummings Hamiltonian with Kerr nonlinearity. A general matrix depending on j, m values is given. A procedure is recommended by using number operator and generators of deformed $su(2)$ algebra to solve the eigenvalues of JC Hamiltonian with Kerr nonlinearity. Modified Jaynes-Cummings-Hamiltonian is solved by using its differential realization in Bargmann-Fock state in section 4.

REFERENCES

1. Aniello, P., Porzio, A., Solimeno, S., *Evolution of the N ion Jaynes-Cummings model beyond the standard rotating wave approximation*, J.Opt. B: Quantum Semiclass. Opt., vol. **5**, (2003), 233-240.
2. Berlin, G. and Aliaga, J., *Validity of the rotating wave approximation in the deriven Jaynes-Cummings model*, J.Opt.B: Quantum Semiclass. Opt., vol. **6**, (2004), 231-237.
3. Du, Si-de, Gong, Shang-qing, Xu, Zhi-zhan, and Gong, Chang-de, *Unusual Rabi oscillations in the squeezed Jaynes - Cummings model: effects of the detuning and the Kerr nonlinearity*, Quantum Semiclass. Opt., vol. **9**, (1997), 941-952.
4. Gantsog, Ts., Joshi, A., and Tanas, R., *Phase properties of one- and two-photon Jaynes - Cummings models with a Kerr medium*, Quantum Semiclass. Opt., vol. **8**, (1996), 445-456.
5. Jie Quan-lin, Wang Shun-Jin, and Wei Lian-Fu, *Algebraic structure and analitical solutions of generalized three level Jaynes Cummings models*, J.Phys.A: Math.gen, vol. **30**, (1997) 6147-6154.
6. Ling, Z., He-Shan, S. and Li, Y., *The two-photon degenerate Jaynes-Cummings model with and without rotating-wave approximation*, Chinese Phys., vol. **10**, (2001), 413-417.
7. Marchiolli, M. A., *Nonclassical statistical properties of finite-coherent states in the framework of the Jaynes-Cummings model*, Physica A, vol. **319**, (2003), 355-361.
8. Ng, K. M, Lo, C.F., Liu, K.L., *Exact eigenstates of the intensity-dependent Jaynes-Cummings model with the counter-rotating term*, Physica A, vol. **275**, (2000), 463-474.
9. Tur, E. A., *Jaynes–Cummings Model: Solutions without Rotating-Wave Approximation*, Opt. Spectrosc. vol. **89** (4), (2000), 574-588.
10. Tütüncüler, H., and Koç, R., *Differential realizations of the two-mode bosonic and fermionic Hamiltonians: A unified approach*, Pramana J. Phys., vol. **53**, (2004), 993-1006.
11. Xie, Rui-hua, Wu, Xiao-hua, Liu, Dun-huan, Xu, Gong-ou, *Symmetric structure of the field and atomic squeezing in a quantum optical model*, Z. Phys. B, vol. **99**, (1996), 253-260.
12. Zheng, Shi-Biao, and Guo, Guang-Can, *Generation of superpositions of displaced Fock states via the driven Jaynes - Cummings model*, Quantum Semiclass. Opt., vol. **8**, (1996), 951-957.

Decomposition of ODEs with an sl$_2$ algebra of symmetries

C. V. Jensen

Department of Mathematics, University of Tromsø,
`cath@math.uit.no`

Abstract. Viewing linear ODEs as D-modules provides an algebraic apparatus which can be used to embed aspects of the theory of representations of Lie algebras into the quest of decomposing and solving equations. In particular we shall see examples of how equations of high order may be solved in terms of a small symmetry algebra. Our main example will be sl$_2$-equations.
Key words. Linear Ordinary Differential Equations, Symmetry Lie Algebras, Representations, Decomposition.
PACS. 02.30.Hq.
MSC. 34A30, 17B10.

1. EQUATIONS AS D-MODULES

1.1. Geometric image of equations in Jet space

Consider a vector bundle $B \xrightarrow{\beta} \mathbf{R}$ of rank m with sections $C^\infty(\beta)$. The bundle $J^k(\beta) \xrightarrow{\pi_k} \mathbf{R}$ of k-jets of sections of β is of rank $m(k+1)$ over R, and is equipped with the *Cartan distribution*. A system of linear kth order ordinary differential equations is a linear subbundle

$$\mathscr{E} \xrightarrow{\alpha} \mathbf{R} \subset J^k(\beta) \xrightarrow{\pi_k} \mathbf{R} \tag{1}$$

of codimension m such that the *Cartan distribution* on $J^k(\beta)$ when restricted to $\mathscr{E}$, denoted $\mathscr{C}_\mathscr{E}$, is 1-dimensional, and projects to $\mathbf{R}$ without singularities. The $C^\infty(\mathbf{R})$-module of sections $C^\infty(\alpha)$ is free and of rank m. We have a *linear connection* in the bundle α,

$$\nabla : D(\mathbf{R}) \longrightarrow Der(C^\infty(\alpha)) \tag{2}$$

where $Der(C^\infty(\alpha))$ denotes derivations of $C^\infty(\alpha)$ over $\frac{d}{dx}$, i. e. $\mathbf{R}$-linear maps

$$\Delta : C^\infty(\alpha) \to C^\infty(\alpha) \quad \text{with} \quad \Delta(fs) = f' \cdot s + f \cdot \Delta(s), \tag{3}$$

for any $f \in A, s \in C^\infty(\alpha)$. ∇ is defined by the requirement that it lifts $\frac{d}{dx}$ on the base $\mathbf{R}$ to a generator $X \in D(\mathscr{E})$ of $\mathscr{C}_\mathscr{E}$ on $\mathscr{E}$. Constant sections of ∇, i.e. sections s such that $\nabla_Y(s) = 0$ for all $Y \in D(\mathbf{R})$ are precisely the *integral curves* of $\mathscr{C}_E$.

CP729, *Global Analysis and Applied Mathematics: International Workshop on Global Analysis*,
edited by K. Taş, D. Krupka, O. Krupková, and D. Baleanu
© 2004 American Institute of Physics 0-7354-0209-4/04/$22.00

Thus, to any linear $n \times n$ system

$$\underline{h}' + B(x)\underline{h} = 0 \tag{4}$$

there is the associated pair $(C^\infty(\alpha), \delta = \nabla_{\frac{d}{dx}})$ such that

$$\ker \delta \cong \text{Solutions}(4) \tag{5}$$

Definition 1 *Let A be a commutative ring over a field k and δ_A a derivation of A. (E, δ) is a D-module over (A, δ_A) if E is a rank n free A-module, and δ is a derivation over δ_A. The latter means that δ is k-linear and satisfies a Leibniz property*

$$\delta(a \cdot e) = \delta_A(a) \cdot e + a \cdot \delta(e) \tag{6}$$

for $a \in A$, $e \in E$.

Hence, associated to any system (4) is a D-module (E, δ) over $(A = C^\infty(\mathbf{R}), \delta_A = \frac{d}{dx})$, with $\ker \delta \cong \text{Solutions}(4)$. We may treat these modules abstractly, without the geometric realization as sections in a jet sub bundle.

1.2. Algebraic properties

D-modules over $(A = C^\infty(\mathbf{R}), \frac{d}{dx})$ form a monoidal category with the product being tensor product of modules over A, and morphisms being homomorphisms that commute with the respective δ-s. Given D-modules (E_1, δ_1) (E_2, δ_2) their product is $(E_1 \otimes_A E_2, \delta)$ with

$$\delta(e_1 \otimes e_2) = \delta_1(e_1) \otimes e_2 + e_1 \otimes \delta_2(e_2) \tag{7}$$

on decomposable elements. This extends to symmetric and wedge products, hence we get induced D-modules $(S^n(E), \delta)$ and $(\bigwedge^l(E), \delta)$. Homomorphisms form a module $(Hom_A(E_1, E_2), \delta)$ with

$$\delta(F) \stackrel{def}{=} \delta \circ F - F \circ \delta \tag{8}$$

for $F \in Hom_A(E_1, E_2)$. We may define the dual module $E^* = Hom_A(E, A)$, and it yields exactly the adjoint equation of E. Denote the kernel $\ker \delta = E^\# \subset E$.

Lemma 1 *Let (E, δ) be a D-module over $(C^\infty(\mathbf{R}), \frac{d}{dx})$. Then any basis of $E^\#$ over $\mathbf{R}$ is a basis of E over A.*

Proof Let $\{e_1, ..., e_k\}$ be a basis of E with $\delta\underline{e} = B(x)^t\underline{e}$. Then the associated system is (4). Let $\underline{h}_1, ..., \underline{h}_k$ be linear independent solutions of the system. Take $H(x) = [\underline{h}_1, ..., \underline{h}_k]^t$. Then

$$\underline{\gamma} = H(x)\underline{e} \tag{9}$$

is a new basis of E over A, since $H(x)$ is the Wronskian of the system, and non-degenerate. Moreover, any basis of $E^\#$ over $\mathbf{R}$ is on the form in (9).

Note that taking the kernel commutes with the above algebraic operations, i.e. $(Hom_A(E_1, E_2))^\# = Hom_\mathbf{R}(E_1^\#, E_2^\#)$, $(E_1 \otimes_A E_2)^\# = E_2 \otimes_\mathbf{R} E_2$ etc.

2. SYMMETRIES AND REPRESENTATIONS

In the D-module setting a symmetry of an equation (E, δ) is an element $F \in (End_A(E))^{\#}$, and we denote the equation corresponding to the module $(End_A(E), \delta)$ the symmetry equation of E. Its solution generates linear symmetries of E. A Lie-algebra $\mathbf{g}$ is a *symmetry algebra* of E if there is a representation

$$\rho : \mathbf{g} \longrightarrow End_A(E) \tag{10}$$

such that

$$\rho(g) \circ \delta = \delta \circ \rho(g) \quad \forall g \in \mathbf{g} \tag{11}$$

i.e. ρ maps $\mathbf{g}$ into δ-invariant endomorphisms of E, $End_A(E)^{\#} \subset End_A(E)$. Note that this yields a representation into the $\mathbf{R}$-vector space $E^{\#}$, and combining this with Lemma 1 enables us to make use of results from the theory of representations of Lie algebras into vector spaces, and extend this to our modules.

Theorem 1 *Any equation E with a representation of symmetries*

$$\mathbf{sl}_2(\mathbf{R}) \to End_A(E)$$

is decomposable into a direct sum of D-modules

$$E = \bigoplus_{i=1}^{m} E_i \tag{12}$$

where each $E_i \cong S^{n_i}(M_i)$ for a rank 2 D-module M_i , and each E_i is an irreducible subrepresentation $\subset E$.

Proof The representation of $\mathbf{sl}_2$ into $V = E^{\#}$ decomposes into

$$V = \bigoplus_{i=1}^{m} V_i$$

where the V_i are irreducible subrepresentations of V, hence they must be isomorphic to symmetric powers $S^{n_i}(Y_i)$, where dim $Y_i = 2$, see [1]. The E_i-s in the theorem are precisely the sub-D-modules in E spanned by V_i over A, and $E_i \cong S^{n_i}(M_i)$, M_i spanned by V_i over A. They are really sub-D-modules of E, and $\oplus E_i = E$ due to Lemma 1, which states that any basis of V over $\mathbf{R}$ is a basis of E over A.

Corollary 1 *(1)   If $mult(n_i) = 1$ for $i = 1...m$ equation E is solved directly by algebraic operations and integration of the base second order equations.*

(2)   For each i such that $mult(n_i) = j$ the remaining problem is reduced to solving a new $j \times j$ system of first order equations.

Note that Theorem 1 deals purely with existence and theory, we shall turn to see how second order Schrödinger equations can be viewed as model equations of the Lie algebra $\mathbf{sl}_2$, in the sense that the rank 2 modules of Theorem 1 correspond to such equations.

2.1. Symmetries as operators

An alternative formulation of a linear symmetry of a kth order equation $L(y) = 0$ is the following. $\Delta = b_1(x) + b_2(x)\partial + ... + b_k(x)\partial^{k-1}$ is a symmetry if there is an operator ∇ such that

$$L \circ \Delta = \nabla \circ L \qquad (13)$$

Δ in this case obviously maps $kerL$ into itself. Note that Δ is a symmetry iff the associated function $\phi = b_1 p_0 + ... + b_k p_{k-1}$ in $J^k(\mathbf{R})$ is a generating function of a symmetry of the Cartan distribution on $\mathcal{E} \subset J^k(\mathbf{R})$.

2.2. Model equations for sl$_2$

Consider an equation of Schrödinger type, with potential $W(x)$.

$$y'' + W(x)y = 0 \qquad (14)$$

We recall that all irreducible representations of $\mathbf{sl}_2$ are symmetric powers of the standard two dimensional representation.
A general second order equation

$$y'' + fy' + gy = 0 \qquad (15)$$

has associated D-module (E, δ) with primitive element basis $\{e_1 = e, e_2 = \delta e\}$ with $\delta^2 e = (g - f')e + (f)\delta e$. The induced equation $(\wedge^2(E^*), \delta)$ is first order, and $\delta(e_1^* \wedge e_2^*) = -f e_1^* \wedge e_2^*$. Thus, Schrödinger equations are precisely the second order equations with a generic δ-invariant volume form. Conditions for

$$\Delta = a_1 + a_2\partial \qquad (16)$$

to be a symmetry of (14) in the sense of (13) are by direct calculation found to be that a_2 solves

$$z''' + 4Wz' + 2W'z = 0 \qquad (17)$$

and that $a_1 = \frac{-a_2'}{2}$. Thus symmetries are given by generating functions that solve (17), i.e.

$$\Delta_a = -\frac{a'}{2} + a\partial, \quad a \in \text{Sol}(17)$$

The symmetry equation actually coincides with the symmetric 2-power $S^2(E)$ of our basic equation (14), and we will see that it plays a special role in the whole hierarchy of $S^k(E)$ equations. Its solution space is isomorphic to $\mathbf{sl}_2$, where the bracket operation is defined by means of the corresponding operator commutators

$$[\Delta_a, \Delta_b] = \Delta_{<a,b>} \quad \text{i.e.} \quad < a,b >= ab' - a'b \qquad (18)$$

196

From our basic equation we derive a hierarchy of new equations $S^k(E^*)$. Choosing to work with the dual modules $S^k(E^*)$ merely simplifies some calculations, and generates exactly the same equations as the module E, as follows:

$$
\begin{aligned}
E^* \quad &: \quad y'' + Wy = 0 \\
S^2(E^*) \quad &: \quad y''' + 4Wy' + 2W'y = 0 \\
S^3(E^*) \quad &: \quad y^{(4)} + 10Wy'' + 10W'y' + (9W^2 + 3W'')y = 0 \\
S^4(E^*) \quad &: \quad y^{(5)} + 20Wy''' + 30W'y'' + [64W^2 + 18W'']y' + [64WW' + 4W''']y = 0 \\
&\quad \cdots
\end{aligned}
$$

For each k the kernel $S^k(E^*)^{\#}$ consists of elements

$$
\theta_y = y\alpha_1 + y'\alpha_2 + \sum_{l=3}^{k+1} g_l(y)\alpha_l \ , \quad y \in \mathrm{Sol}(k) \tag{19}
$$

where $g_l = \frac{1}{l-1}[(k-l+3)W \cdot g_{l-2} + g'_{l-1}]$ for $l = 3, .., k+1$ and

$$
\alpha_l = (e_1^*)^{k-l+1} \cdot (e_2^*)^{l-1}
$$

is the standard symmetric product basis of $S^k(E^*)$ induced by $\{e_1^*, e_2^*\}$ in E^*, dual to the primitive element basis.

Proposition 1 *There is a unique skew-symmetric bracket*

$$
[\cdot, \cdot] : S^m(E^*) \times S^n(E^*) \to S^{m+n-2}(E^*)
$$

for $m, n \geq 1$ which is

(i) A-linear.
(ii) $[f \cdot g, h] = f \cdot [g, h] + g \cdot [f, h] \quad \forall f, g, h \in S^{\cdot}(E^*)$
(iii) $[\cdot, \cdot] = < \cdot, \cdot > = e_1 \wedge e_2 = \Omega$ for $m = n = 1$.

Proof The δ-invariant skew-symmetric form $\Omega = e_1 \wedge e_2 \in \wedge^2(E)^{\#}$ defines the base bracket $< \cdot, \cdot > = \Omega : E^* \times E^* \to A$, and the properties $(i) - (ii)$ determine its extension to $[\cdot, \cdot]$.

We immediately observe that Ω being δ-invariant implies that so is the extended bracket $[\cdot, \cdot]$, and by restriction we get

Proposition 2 *The bracket in* **Proposition 1** *restricts to an* **R***-linear bracket*

$$
[\cdot, \cdot] : S^m(V) \times S^n(V) \longrightarrow S^{m+n-2}(V)
$$

where $V = (E^*)^{\#}$. This is equivalent to a bracket

$$
[\cdot, \cdot] : \mathrm{Sol}(m) \times \mathrm{Sol}(n) \to \mathrm{Sol}(m+n-2)
$$

where

$$
[\theta_y, \theta_z] = \theta_{[y,z]}
$$

We are now ready to observe that solutions of the $S^2(E^*)$ equation (17) produce symmetries of *all* equations $S^k(E^*)$, and not only E^*.

Theorem 2 *Any solution* $a \in Sol(S^2(E^*))$ *produces a symmetry*

$$\mathcal{O}^m_{\theta_a} \stackrel{\text{def}}{=} [\theta_a, \cdot] : S^m(E^*) \longrightarrow S^m(E^*)$$

The corresponding symmetry *operator* is

$$\mathcal{O} = \mathcal{O}^m_a : Sol(m) \longrightarrow Sol(m) \tag{20}$$

with the correspondence

$$\mathcal{O}^m_{\theta_a}(\beta_y) = \beta_{\mathcal{O}^m_a(y)}$$

The precise expression is

$$\mathcal{O}^m_a = \frac{1}{2}(-ma' + 2a\partial) \tag{21}$$

for any $m \geq 1, a \in Sol(S^2(E^*))$.

Now, let u, v span the solution space of the base equation E^*.

Theorem 3 *For any* $k \leq 1$ *the symmetries of* $S^k(E^*)$

$$X_+ = -\frac{1}{2c}\mathcal{O}^k_{v^2}, \quad X_- = \frac{1}{2c}\mathcal{O}^k_{u^2} \quad \text{and} \quad H = \frac{1}{c}\mathcal{O}^k_{uv} \tag{22}$$

where $c = <u, v> \in \mathbf{R}$ *constitute a basis of the* $\mathbf{sl_2}$*-algebra of symmetries* $\cong Sol(S^2(E^*))$ *with commutators*

$$[X_+, X_-] = H, \quad [H, X_+] = 2X_+, \quad [H, X_-] = -2X_- \tag{23}$$

$S^k(E^*)$ *decomposes into rank 1 sub-D-modules corresponding to different eigenvalues of* H

$$S^k(E^*) = <\theta_{u^k}>_A \oplus <\theta_{u^{k-1}v}>_A \oplus ... \oplus <\theta_{v^k}>_A$$

$-k, -k+2, ..., k-2, k$ *respectively, and the action is precisely* $S^k(V_0)$, *with* V_0 *being the standard representation of* $\mathbf{sl_2}$.

Thus, the action of $(S^2(E^*)^\#, [\cdot, \cdot])$ into $V \subset E^*$ is precisely the standard representation of $\mathbf{sl_2}$, and the Schrödinger equations serve as models for the second order base modules in Theorem 1.

Given an equation with an $\mathbf{sl_2}$ action, the first step should be to identify the standard basis of the action, as in Theorem 3, and decompose the module according to the eigenvalues of the diagonal element H. If, as in part (1) of Corollary 1, the multiplicities of the n_i-s are one, this is enough to identify the irreducible subrepresentations $E_i \cong S^{n_i}(M_i)$. Then to identify the corresponding base module M_i it is sufficient to take $w \in E_i$ with $w \in \ker X_+ \cap Eig_{n_i}(H)$ and recover the potential $W_i(x)$ of the corresponding

198

Schrödinger equation in terms of derivatives of fractions of coefficients of w and $X_- w$. Any $w \in (\ker X_+ \cap Eig_{n_i}(H)) \cap A \cdot \ker \delta$ will generate a sub-D-module $\cong S^{n_i}(M_i)$. For $mult(n_i) > 1$ we should thus add the requirement $\delta w = \lambda(x)w$ to find a generating element. The base second order equations can be integrated, they have a three dimensional symmetry algebra, and solutions of M_i produce all solutions of $S^{n_i}(M_i)$, which are again solutions of E.

ACKNOWLEDGMENTS

I would like to thank Prof. V. Lychagin for essential contributions through our discussions

REFERENCES

1. Fulton, W., and Harris, J., *Representation Theory. A First Course*, Springer GTM 129, New York, 1991, ISBN 0-387-97495-4.
2. Serre, J.-P., *Lie Algebras and Lie Groups. 1964 Lectures at Harvard University*, Springer, Berlin, 1992, ISBN 0-387-55008-9.

Dixmier-Douady Sheaves of Groupoids and Brownian Loops

Rémi Léandre

Département de Mathématiques. Faculté des Sciences Université de Bourgogne. 21000. Dijon. France

Rémi Léandre

Département de Mathématiques. Faculté des Sciences Université de Bourgogne. 21000. Dijon. France

Abstract. We build a line bundle on the Brownian bridge by using gerbes theory.
Key words. Brownian motion, Gerbes.
PACS. 02.50.Fz, 02.40.Ky.
MSC. 58J65, 55P47.

1. INTRODUCTION

Let us consider the smooth free loop space of a manifold. Unitary complex line bundles over it (or equivalently circle bundles over it) are very useful.

For instance, this allows to define a central extension of a loop group ([11], [24], [25]). It allows to Felder-Gawedzki-Kupiainen ([8]) to define formally the Hilbert space of a conformal field theory on a group, by using Deligne cohomology. Connected to Grothendieck's theory of gerbes, construction of line bundles on the loop space is done by Brylinski ([3]). By using bundle gerbes, Gawedzki-Reis ([9]) have constructed a line bundle on the loop space, by avoiding the theory of categories.

Conformal field theory predicts the existence of an Hilbert space associated to the loop space. Following Felder-Gawedzki-Kupiainen, it should be the Hilbert space of sections of a line bundle on the loop space. Moreover, the measure of physicists are purely formal.

There are several approaches to define a line bundle of the loop space, by using continuous loops. Namely the good understanding of it requires the introduction of stochastic integrals. Malliavin-Malliavin ([23]) have constructed the analogue of a Kac-Moody group for continuous loops. Léandre ([16]) has constructed the L^2 of a Kac-Moody group, by using Brownian bridge measure on a loop group. We refer to [13], [14], [15], [20] for various constructions of stochastic string structures on the loop space.

There are others measures on the loop space than the Brownian bridge measure: they are called following the terminology of Airault-Malliavin [1] heat kernel measures. See [4] for an axiomatic set-up. If we consider heat kernel measures, Brzezniak-Léandre ([5]) and Léandre ([17], [18], [19], [21]) have constructed various stochastic line bundles on the loop space, eventually continuous, with fiber almost surely defined. In particular, [21] used the construction of Gawedzki-Reis ([9]) of a line bundle over the loop space,

CP729, *Global Analysis and Applied Mathematics: International Workshop on Global Analysis*,
edited by K. Taş, D. Krupka, O. Krupková, and D. Baleanu
© 2004 American Institute of Physics 0-7354-0209-4/04/$22.00

by using bundle gerbes theory.

The constructions of Felder-Gawedzki-Kupiainen ([8]) and Gawedzki-Reis ([9]) of a line bundle on the loop space are purely combinatorial. Brylinski ([3]) purposed another approach, by using the theory of categories and the theory of gerbes of Grothendieck of a construction of a line bundle on the loop space.

The purpose of this work is to do a stochastic interpretation of the works of Brylinski, by using Brownian bridge measure.

2. A BRIEF OVERVIEW ON GERBES

Let X be a smooth compact manifold. Let $f : Y \to X$ be a local diffeomorphism of manifold. We don't suppose that f is onto.

We recall (See [3] p 192):

Definition II.1: A presheaf of categories C consists of the following datas:

-i)For any local diffeomorphism $f : Y \to X$ a category $C(f : Y \to X)$.

-ii)For any diagram $g : Z \to Y$ and $f : Y \to X$ of local diffeomorphisms, a functor $g^{-1} : C(f : Y \to X)$ into $C(fg : Z \to X)$.

-iii)For any diagram $h : W \to Z$, $g : Z \to Y$ and $f : Y \to X$ of local diffeomorphisms, a natural invertible transformation $\theta_{g,h} : h^{-1}g^{-1} \to (gh)^{-1}$.

We say that a presheaf of categories C is a sheaf of categories (or stack in the terminology of Grothendieck) if the descent properties (D_1) of [3] p 187 and (D_2) of [3] p 188 are satisfied.

Let us recall that a sheaf of groups is a collection of groups $A(f : Y \to X)$ associated to each local diffeomorphisms satisfying some glueing properties. The descent properties (D_1) and (D_2) of [3] p 187-188 say we have some glueing properties. The traditional example of local diffeomorphism is when Y is equal to the disjoint union of U_i where U_i are open subset of X. Eventually, the system of U_i constitute a cover of X. (D_1) says that if we consider two objects P_1 and P_2 of the category $C(f : Y \to X)$ with some morphism h_i between the restriction of P_1 to U_i to the restriction of P_2 to U_i such that $(h_i)|(U_{i,j}) = (h_j)|(U_{i,j})$, then the morphism h_i glue together in a global morphism from P_1 into P_2 ($U_{i,j} = U_i \cap U_j$ and $U_{i,j,k} = U_i \cap U_j \cap U_k$). (D_2) says that given objects Q_i of $C(U_i \to X)$ and isomorphisms $u_{i,j}$ between the restriction of Q_j to $U_{i,j}$ to the restriction of Q_i to $U_{i,j}$ which satisfy to $u_{i,k} = u_{i,j}u_{j,k}$ over $U_{i,j,k}$, then there is a unique way to glue the local objects Q_i into a global object of $C(X)$.

Let f be a smooth map $Y \to X$. Let us consider a sheaf of categories C on X. We can define the pullback sheaf of categories $f^{-1}C$ on Y. Let us consider $h : W \to X$ a local diffeomorphism and $g : Z \to Y$ a local diffeomorphism. We suppose that i is a smooth map from Z into W such that:

$$h \circ i = f \circ g \qquad (2.1)$$

such that $Z \to W \times_X Y$ is an inclusion of open subset ($X \times_X Y$ is the set of $(w,y) \in W \times Y$ such that $h(w) = f(y)$). Then $B(g : Z \to Y)$ is the direct limit of the categories $C(h : W \to X)$ under these assumptions (We don't write the way we assimilate two such diagrams: see [3] p 193). In other words, a generic object of $B(g : Z \to Y)$ is an object of

$C(h : W \rightarrow x)$ for the diagram associated to (2.1). These objects are submitted to some identifications.

The relation with the traditional theory of sheaves appears when we consider Y an open subset of X and when f is the inclusion from Y into X.

Let us recall the definition of a torsor (See [3] p 190):

Definition II.2: Let H be a sheaf of groups on X. An H-torsor on X is a sheaf F, together with an action $H \times F \rightarrow F$ of the sheaf of groups H on F such that every point x on X has an open neighborhood U with the property that $F(V)$ is a principal homogeneous space under $H(V)$ if V is an open subset of U.

We are ready to give the definition of a unitary Dixmier-Douady sheaf of groupoids:

Definition II.3: A unitary Dixmier-Douady sheaf of groupoids is a sheaf of categories such that:

-i)$C(f : Y \rightarrow X)$ consists of objects Q such that the sheaf $Aut(Q)$ is locally isomorphic to the sheaf S_Y^1 of smooth S^1 valued functions on Y.

-ii)Given two objects Q_1 and Q_2 of $C(f : Y \rightarrow X)$, there exists a surjective local diffeomorphism $g : Z \rightarrow Y$ such that $g^{-1}Q_1$ and $g^{-1}Q_2$ are isomorphic (This means that Q_1 and Q_2 are locally isomorphic).

-iii)There exists a surjective local diffeomorphism $f : Y \rightarrow X$ such that $C(f : Y \rightarrow X)$ is not empty.

Remark: We can suppose that in iii) Y is constituted by the disjoint union of U_i where the U_i are convex open subset of the Riemannian manifold X which constitute a cover of X. We say that the U_i constitute a good cover of X (See [3] Theorem 5.2.8).

We recall the following result: let f be a smooth application from Y into X and let C be a unitary sheaf of groupoids on X. $f^{-1}(C)$ is a unitary sheaf of groupoids on Y.

In the previous definition, we can replace S_Y^1 by the sheaf A_X of imaginary 1-forms on X.

These statements constituted a brief overview of Chapter 5.1 and Chapter 5. of [3]. The next statements constitute a brief overview of Chapter 5.3 of [3], where a differential geometry of gerbes is done, related to connective structures on gerbes.

Definition II.4: Let C be a unitary Dixmier-Douady sheaf of groupoids over the smooth manifold X. A connective structure for C is the assignment to a A_Y-torsor over $f : Y \rightarrow X$ to any objects of $C(Y) = C(f : Y \rightarrow X)$ submitted to the following requirements:

-i) For any diagram $g : Z \rightarrow Y$ and $f : Y \rightarrow X$ of local diffeomorphism and for any object P of $C(Y)$, an isomorphism $\alpha_g : g^*Co(P) \rightarrow Co(g^{-1}P)$ of A_Z torsors which have to satisfy (C_1) of [3] p 206.

-ii)Any isomorphism ϕ between two objects P_1 and P_2 of $C(f : Y \rightarrow X)$ induces an isomorphism ϕ_* between $Co(P_1)$ and $Co(P_2)$ of A_Y torsors. This isomorphism must satisfy the properties listed in (C_2) of [Br] p 206.

The example of unitary Dixmier-Douady sheaf of groupoids is when we consider as $C(f : Y \rightarrow X)$ the category of unitary line bundles on Y. We consider for $Co(P)$ the set of unitary connections of P over Y. They satisfy some natural conditions which have to be still satisfied in the general case. We say that we are in presence in this case of the trivial gerbe.

We get the definition of a curving:

Definition II.5: Let C be a unitary sheaf of groupoids on the smooth Riemannian manifold X, equipped with a connective structure $P \to Co(P)$. A curving of the connective structure is a rule which to any object $P \in C(f : Y \to X)$ and to any section ∇ (called a connection) of the A_Y torsor $Co(P)$ associates a purely imaginary 2-form $K(\nabla)$ on Y (called the curvature of ∇), which satisfies the following requirements:

-i)Let $P \in C(f : Y \to X)$ and ∇ be a section of $Co(P)$. Let us consider a local diffeomorphism $g : Z \to Y$, the curving of $K(\alpha_g(g^*\nabla))$ of the section $\alpha_g(g^*\nabla)$ of $Co(g^{-1}(P))$ is equal to $g^*K(\nabla)$.

-ii)Let (P, ∇) and let $\phi : P \to P'$ an isomorphism between two object of $C(f : Y \to X)$. Let $\phi_*(\nabla)$ be the corresponding element of $Co(P')$. Then we have $K(\nabla) = K(\phi_*(\nabla))$.

-iii)Let α be a complex-valued 1-form on Y. then we have:

$$K(\nabla + \alpha) = K(\nabla) + d\alpha \qquad (2.2)$$

We recall (See [3] p 213 and p 207): given a unitary Dixmier-Douady shead of groupoids C, there exists a connective structure Co associated to C as well as a curving associated to Co. If $\nabla \to K(\nabla)$ is such a curving, any other curving is written as $K(\nabla) + f^*\beta$ for some imaginary smooth 2-form on X (We consider the category $C(f : Y \to X)$).

We recall (See [3] p 214): there exists a unique 3-form Ω on X such that

$$dK(\nabla) = f^*\Omega \qquad (2.3)$$

for a curving K associated to the connective structure Co of C. ∇ is a connection on P which is an object of $C(f : Y \to X)$.

Moreover, if $f : Y \to X$ is a map of smooth manifold, if C is a unitary Dixmier-Douady sheaf of groupoids on X, we get another unitary Dixmier-Douady sheaf of groupoids f^*C on Y. An object of f^*C is given by (2.1). We have $Co(P) = i^*Co(Q)$ (See [3] p 211). The pullback curving is given by $(f^*K(i^*\nabla) = i^*K(\nabla)$. Moreover, the pullback operation is consistent with operation of composition: if $f_1 : Y_1 \to X$ and $f_2 : Y_2 \to Y_1$, then

$$(f_1 \circ f_2)^*C = f_2^*(f_1^*(C)) \qquad (2.4)$$

for the gerbes C on X and the connective structure and the associated curving.

Later we will take W associated to a good cover U_i of X such that $C(h : W \to X)$ is not empty. We consider in (2.1) $Y = S^1 \times X$, and f is the second projection map. We consider a cover of S^1 by small intervals $]s_j, s'_j[$. We take as Z the disjoint union of $]s_j, s'_j[\times U_i$. This constitute a good cover of Y. This satisfies to the requirements of (2.1). $Co(P) = i^*Co(Q)$ if Q is an object of $C(h : W \to X)$. The curving associated is $i^*K(\nabla)$ if ∇ is a connection of Q for $Co(Q)$. It depends in particular only on the second variables in X.

3. STOCHASTIC LINE BUNDLES

Let $s \to \gamma(s)$ be a smooth based loop in X. Let $L_{x,\infty}(X)$ be the set of smooth loops γ based in x. We can construct a complex line bundle over $L_{x,\infty}(X)$ following the lines of [3], chapter 6.

We consider a unitary Dixmier-Douady sheaf of groupoids C on X. Let Co be a connective structure on C. Let K be a curving on Co.

Let us recall the statements of [3] 6.2.1.. We can construct a line bundle over $L_{x,\infty}(X)$. The line bundle Λ consists of equivalence classes (γ, F, ∇, z) where γ is the generic element of $L_{x,\infty}(X)$, F is an object of $\gamma^*(C)$, ∇ is a section of $\gamma^*Co(F)$ and $z \in S^1$ (We identify an Hermitian line bundle with a circle bundle!). Moreover:

-i)We identify (γ, F, ∇, z) with (γ, F', ∇', z) if the pairs (F, ∇) and (F', ∇') are isomorphic.

-ii)We identify $(\gamma, F, \nabla + \alpha, z)$ with $(\gamma, F, \nabla, z \exp[-\int_{S^1} \alpha])$ if α is a imaginary 1-form on the circle.

Let $\gamma(Z)$ be a smooth loop and $\tau(Z)$ be the associated transport parallel for the Levi-Civita connection on the tangent bundle. Let us consider loop which are written of the shape

$$\gamma(s) = \exp_{\gamma(Z)(s)}[\tau(Z)(s)r_s] \tag{3.1}$$

where $s \to r_s$ is a smooth loop in a small open convex ball B_r and exp is the Riemannian exponential. We consider Z as the set of γ such that $\gamma(s)$ satisfies (3.1). Let $G(Z)$ be the map from $S^1 \times B_r$ into X

$$(s, r) \to \exp_{\gamma(Z)(s)}[\tau(Z)(s)(r)] \tag{3.2}$$

By theorem 5.2.8. of [3], $G(Z)^*C$ is the trivial gerbe.Let $F(Z)$ be an object associated to this gerbe. Let $\nabla(Z)$ .a section of the corresponding torsor for the pullback connective structure. We get a section of the bundle on (Z) by putting:

$$\phi_Z(\gamma) = (\gamma, \gamma^*(F(Z)), \gamma^*\nabla(Z), 1) \tag{3.3}$$

(See [3] p 237, formula (6-6)). Namely, we can assimilate $\gamma \in Z$ to a loop in $S^1 \times B_r$, where B_r is a small convex open ball in the tangent space of x. Moreover, by (2.4) this operation is consistent. So we say that the line bundle (or the circle bundle) is trivial on Z. But the system of Z constitute a basis of the topology of $L_{x,\infty}(X)$ for the uniform topology. We deduce that we have a smooth structure on Λ, because the intersection of Z and Z' coincide with the space of loops with values in a open connected subset of $S^1 \times B_r$.

Let K be a curving of Co for the gerbe C on X. Let us consider the pullback curving $G(Z)^*K$ on $G(Z)^*C$. We can consider a connection whose connection form on Z in the trivialization $\phi_Z(\gamma)$ is given by

$$-\int_\gamma i_X G(Z)^*K \tag{3.4}$$

where X is a tangent vector on γ.

Let $\gamma(Z)$ be the central element of Z. Let γ be a continuous element of Z. In the sequel, we endow the continuous loop space $L(X)$ with the Brownian bridge measure (See [2], [10]). We remark that Z is contractible. We can join γ to $\gamma(Z)$ by a curve $l_u(Z)$. If u in $[0,1]$,

$$l_u(Z)(s) = \exp_{\gamma(Z)(s)}[u(\gamma(s) - \gamma(Z)(s))] \tag{3.5}$$

where exp denotes the exponential chart on X for the product metric and $\gamma(s) - \gamma(Z)(s)$ is the vector of the unique geodesic joining $\gamma(Z)(s)$ to $\gamma(s)$. In the trivialization given by (3.3), there is a formal parallel transport, on Λ from the continuous curve $l_u(Z)$. It is given by

$$\exp[\int_{S^1 \times [0,1]} (G(Z)^* K)(l_u(z)(s))(d_u l_u(Z)(s), d_s l_u(Z)(s)] \qquad (3.6)$$

It is an ordinary integral in the direction u and a Stratonovitch stochastic integral in the direction s, which can be approximate in all the L^p by using the polygonal approximations of γ (see [10]).

We can join if we suppose that $L_x(M)$ is connected $\gamma(Z)$ to γ^0. We find a distinguished path joining γ to γ^0 through $\gamma(Z)$. We denote by $\alpha_Z(\gamma)$ the formal parallel transport along this curve for Λ. We remark if γ belongs to the intersection of Z_1 and Z_2, the quantity $\alpha_{Z_1,Z_2}(\gamma) = \alpha_{Z_1}(\gamma)\alpha_{Z_2}^{-1}(\gamma)$ is an element of the circle almost surely defined.

Moreover, over a double intersection $(Z_1 \cap Z_2)$, we have almost surely:

$$\alpha_{Z_1,Z_2}(\gamma)\alpha_{Z_2,Z_1}(\gamma) = 1 \qquad (3.7)$$

and on a triple intersection $Z_1 \cap Z_2 \cap Z_3$, we have almost surely:

$$\alpha_{Z_1,Z_2}(\gamma)\alpha_{Z_2,Z_3}(\gamma)\alpha_{Z_3,Z_1}(\gamma) = 1 \qquad (3.8)$$

Let Z_i be a countable family which constitutes cover of $L_x(X)$. We consider as we said the Brownian bridge measure on $L_x(X)$, the based loop space of X.

We consider the Hilbert space Ξ of the line bundle Λ on $L_x(X)$, with fiber almost surely defined, associated to C.

Definition III.1: an element ξ of Ξ is given .by a family of random variables ξ_i over Z_i such that over $Z_i \cap Z_j$:

$$\xi_j = \xi_i \alpha_{Z_i,Z_j} \qquad (3.9)$$

almost surely. Since α_{Z_i,Z_j} is almost surely of modulus 1, we can define the norm $|\xi|$ of the section and we put

$$\|\xi\|_{\Xi}^2 = E[|\xi|^2] \qquad (3.10)$$

REFERENCES

1. Airault H., Malliavin P., Integration on loop groups, Publication Univ. Paris VI., (1991).
2. Bismut J.M, *Large deviations and the Malliavin Calculus*, Progress in Maths., 45, Birkhauser, Basel, 1984.
3. Brylinski J.L., *Loop spaces, characteristic classes and geometric quantization*, Progress in Maths., 107. Birkhauser, Basel, 1992.
4. Brzezniak Z. Elworthy K.D., Stochastic differential equations on Banach manifolds, *Meth. Funct. Ana. Topo.*, **6**, (2000), 43-84.
5. Brzezniak Z. Léandre R., Horizontal lift of an infinite dimensional diffusion, *Potential Analysis.*, **12**, (2000), 249-280.
6. Carey A.L. Murray M.K., String structure and the path fibration of a group, *Com. Math. Phys.*, **141**, (1991), 441-452.
7. Coquereaux R. Pilch K., String structures and loop bundles, *Com. Math. Phys.*, **120**, (1989), 353-378.

8. Felder G. Gawedzki K. Kupiainen A., Spectra of Wess-Zumino-Witten model with arbitrary simple groups, *Com. Math.Phys.*, **117**, (1988), 127-159.
9. Gawedzki K. Reis N., WZW branes and gerbes, *Rev. Math. Phys.*, **14**, (2002), 1281-1334.
10. Ikeda N. Watanabe S., *Stochastic differential equations and diffusion processes.*, North-Holland, Amsterdam, 1981.
11. Kac V., *Infinite dimensional Lie algebras.* Cambridge University Press, Cambridge, 1985.
12. Killingback T., World-sheet anomalies and loop geometry, *Nucl. Phys. B.*, **288**, (1987), 578-588.
13. Léandre R., Hilbert space of spinor fields over the free loop space, *Rev. Math. Phys.*, **9**, (1997), 243-277.
14. Léandre R., Stochastic gauge transform of the string bundle, *Jour. Geo. Phys.*, **26**, (1998), 1-25.
15. Léandre R., String structure over the brownian bridge, *Jour. Math. Phys.*, **40**, (1999), 454-479.
16. Léandre R., A unitary representation of the basical central extension of a loop group, *Nagoya Math. Jour.*, **159**, (2000), 113-124.
17. Léandre R., Stochastic Wess-Zumino-Novikov-Witten model on the torus, *Jour. Math. Phys.*, **44**, (2003), 5530-5568.
18. Léandre R., Brownian surfaces with boundary and Deligne cohomology, *Rep. Math. Phys.*, **52**, (2003), 353-362.
19. Léandre R., Makov property and operads, In *Quantum limits and the second law of thermodynamics, Entropy.*, **6**, (2004), 180-215.
20. Léandre R., Stochastic processes on classifying spaces and string structures. To appear *Inf. Dim. Ana. Quant. Prob.*
21. Léandre R., Bundle gerbes and Brownian motion. To appear *Lie theory and its application to physics* (Varna. 2003).
22. Mac Laughlin D., Orientation and string structures on loop spaces, *Pac. Jour. Math.*, **155**, (1992), 143-156.
23. Malliavin M.P. Malliavin P., Integration on loop groups III, *Jour. Funct. Ana.*, **108**, (1992), 13-46.
24. Mickelsson J., *Current algebras and groups.* Plenum Press, N.Y., 1986.
25. Pressley A. Segal G., *Loop groups.* Oxford University Press, Oxford 1986.
26. Witten E., The index of the Dirac operator in loop space. *Elliptic curves and modular forms in algebraic topology.* Lect. Notes. Math. 1326, (1988), 161-181.

The geometry of the geodesic equation in the framework of jets of submanifolds

Gianni Manno

Dept. of Mathematics, King's College, London, `Gianni.Manno@kcl.ac.uk`
and
Dip. di Matematica "Ennio De Giorgi", Università di Lecce, `gianni@poincare.unile.it`

Abstract. The geometry of differential equations and their higher symmetries is studied by using spaces of jets of submanifolds. The geodesic equation $\mathscr{G}$ introduced in this framework is proved to be a third order polynomial with respect to first derivatives. Contact symmetries of this equation are discussed.
Key words. Jets of submanifolds, geodesic equation, contact symmetries.
PACS. 02.30.Hq, 02.40.Ky, 02.40.Vh.
MSC. 34C14, 34H26, 53C22, 58A20.

1. INTRODUCTION

Jet spaces are fundamental objects in differential geometry. Namely they are the basis for a geometrization of partial differential equations. The r-order jet of n-dimensional submanifolds $J^r(E,n)$ of a given manifold E is the space of equivalence classes of n-dimensional submanifolds of E having an r-th contact at a certain point. A differential equation of order r on n-dimensional submanifolds is a submanifold of a such jet space. Historically, less attention has been devoted to the theory of differential equations on n-dimensional submanifolds with respect to the theory of differential equations on a bundle [3, 9, 16]. Despite this fact, the geometry of differential equations in the framework of jet of submanifolds is important in problems where there is no distinction between independent and dependent variables, such as minimal submanifolds and general relativistic mechanics [6].

In this paper we shall discuss the geometry of the geodesic equation in the framework of jets of 1-dimensional submanifolds.

First, we give the basic notions about the geometry of differential equations from the viewpoint of jets of submanifolds. In particular, after introducing contact sequences associated with a formally integrable equation, we construct the operator of universal linearization. Then we give the generating sections formalism for symmetries of partial differential equations. This formalism has been largely developed in the case of equations on vector bundles [3, 9], and it was partially introduced in [18, 19, 15] in the case of jets of submanifolds.

CP729, *Global Analysis and Applied Mathematics: International Workshop on Global Analysis,*
edited by K. Taş, D. Krupka, O. Krupková, and D. Baleanu

Then we define intrinsically the geodesic equation $\mathscr{G}$ on 1-dimensional submanifolds of a given pseudo-Riemannian manifold E as the kernel of a suitable non-linear differential operator. A coordinate description of the equation $\mathscr{G}$ in the case of surfaces is present in [10] and generalized in [1]. Here, we also construct a global section $\ddot{\nabla} : J^1(E,1) \longrightarrow J^2(E,1)$ whose image is the geodesic equation $\mathscr{G}$ by introducing a connection on $J^1(E,1)$ associated with the Levi-Civita connection on TE. This result generalizes the result obtained in [6], where such a section is found in the case of manifolds which are orientable and time-like orientable.

At the end we briefly discuss symmetries of $\mathscr{G}$, and we show that all higher symmetries reduce to contact symmetries. Examples of computations in the case of the plane, of the sphere and of the Kepler equation are present in [14].

2. GEOMETRY OF DIFFERENTIAL EQUATIONS ON SUBMANIFOLDS

2.1. Preliminary Definition

Let E be an $(n+m)$-dimensional smooth manifold and L an n-dimensional embedded submanifold of E. Let (V, y^A) be a local chart on E. The coordinates (y^A) can be divided in two parts, as

$$(x^\lambda) = (y^1, y^2, \ldots, y^n) \text{ and } (u^i) = \left(u^1 = y^{n+1}, u^2 = y^{n+2}, \ldots, u^m = y^{n+m} \right)$$

such that the submanifold L is locally given by $u^i = f^i(x^1, x^2, \ldots, x^n)$, $i = 1, \ldots, m$.

The chart (x^λ, u^i) is said to be a *divided chart* which is *concordant* to L. Here, and in what follows, Greek indices run from 1 to n and Latin indices run from 1 to m. Also, all submanifolds are *embedded* submanifolds.

Let $\iota : L \hookrightarrow E$ and $\iota' : L' \hookrightarrow E$ be two submanifolds, and $p \in L \cap L'$. We say that L and L' have a *contact of order r* at p if ι and ι' have a contact of order r at p. Locally this means that the Taylor expansion of $(\iota - \iota')$ around p vanishes up to order r. This is an invariant property.

The above relation is an equivalence relation; an equivalence class is denoted by $[L]_p^r$. The set of such classes is said to be the *r-jet of n-dimensional submanifolds of E* and it is denoted by $J^r(E,n)$. If E has a bundle structure π on M, the *r-jet bundle $J^r\pi$* is defined to be the r-jet of submanifolds which are the image of local setions of π.

The set $J^r(E,n)$ has a natural manifold structure. Namely, let $\sigma = (\sigma_1, \sigma_2, \ldots, \sigma_r)$, with $1 \leq \sigma_i \leq n$ and $r \in \mathbb{N}$, be a multi-index, and $|\sigma| \overset{\text{def}}{=} r$. Any divided chart (V, x^λ, u^i) on E induces the local chart $\left(V_n^r, (x^\lambda, u_\sigma^i) \right)$ on $J^r(E,n)$ at $[L]_p^r$, where $V_n^r \overset{\text{def}}{=} \bigcup_{p \in V} J_p^r(E,n)$, and the functions u_σ^j are determined by $u_\sigma^i \circ j_r L = \partial^{|\sigma|} f^i / \partial x^\sigma$.

We have the following natural maps:

1. the embedding $j_r L : L \longrightarrow J^r(E,n)$, $p \longmapsto [L]_p^r$,
2. the projection $\pi_{k,h} : J^k(E,n) \longrightarrow J^h(E,n)$, $[L]_p^k \longmapsto [L]_p^h$ $\quad k \geq h$.

We denote by $L^{(r)}$ the image of j_rL. We call the tangent plane $T_{\theta_r}L^{(r)}$ an *R-plane*.

We denote by $\mathcal{F}_r$ the algebra $C^\infty(J^r(E,n))$. We also denote by $\chi(J^r(E,n))$ the module of vector fields.

In the rest of the paper we shall put $\theta_{r+1} = [L]_p^{r+1}$ and $\theta_r = \pi_{r+1,r}(\theta_{r+1})$.

The Cartan plane $\mathcal{C}_{\theta_r}^r$ on $J^r(E,n)$ at θ_r is defined as the span of the planes $T_{\theta_r}L^{(r)}$. The coorespondence $\theta_r \longmapsto \mathcal{C}_{\theta_r}^r$ is called the *Cartan distribution*. We denote by $\mathcal{C}^r(D)$ the set of vector fields lying in the Cartan distribution of $J^r(E,n)$. It is easy to realize that to each point θ_{r+1} there corresponds the *R*-plane $R_{\theta_{r+1}} = T_{\theta_r}L^{(r)}$, and that $\mathcal{C}_{\theta_r}^r = R_{\theta_{r+1}} \oplus \ker T_{\theta_r}\pi_{r,r-1}$. This direct sum is not canonical as we have many *R*-planes passing through a point θ_r. A vector field on $J^r(E,n))$ which preserves the Cartan distribution is called a *Lie field*. We denote by $D_{\mathcal{C}^r}$ the set of Lie fields of $J^r(E,n)$.

A *differential equation* $\mathcal{E}$ *of order* r *on* n-*dimensional submanifolds of a manifold* E is a submanifold of $J^r(E,n)$. The manifold $J^r(E,n)$ is called the *trivial equation*.

Let $F \xrightarrow{\eta} J^1(E,n)$ be a vector bundle and let $\eta_r = \pi_{r,1}^*(\eta)$. A differential equation $\mathcal{E} \subset J^r(E,n)$ can be described by $\phi = 0$ where $\phi \in \Gamma(\eta_r)$. We say that ϕ *represents* $\mathcal{E}$. We can associate with ϕ the following operation:

$$\Box_\phi : L \longmapsto (\eta^*(\pi_{r,1}) \circ \phi \circ j_rL) \in \Gamma(\eta|_L), \ L \subset E, \tag{1}$$

that we call a *non-linear differential operator* following the terminology of Vinogradov (see [20]). Conversely, any map $\Box : L \longmapsto \Box(L) \in \Gamma(\eta|_L)$ such that $\Box(L)(p) = \Box(\tilde{L})(p)$ when $[L]_p^r = [\tilde{L}]_p^r$, induces the following section:

$$\phi_\Box : J^r(E,n) \longrightarrow J^r(E,n) \times_{J^1(E,n)} F, \quad [L]_p^r \longmapsto \left([L]_p^r, \Box(L)(p)\right) \tag{2}$$

that we call the *jet morphism* associated with $\Box$. The correspondence $\phi \longmapsto \Box_\phi$ is bijective.

An *infinitesimal classical external symmetry* of the equation $\mathcal{E} \subset J^r(E,n)$ is a Lie field of $J^r(E,n)$ which is tangent to $\mathcal{E}$. An *infinitesimal classical internal symmetry* of the equation $\mathcal{E} \subset J^r(E,n)$ is a vector field on $\mathcal{E}$ which preserves the induced Cartan distribution $\mathcal{C}(\mathcal{E}) = \mathcal{C}^r \cap T\mathcal{E}$ on $\mathcal{E}$.

The *1-prolongation* $\mathcal{E}^1$ of an equation $\mathcal{E} \subset J^r(E,n)$ is the set

$$\mathcal{E}^1 = \{\theta_{r+1} \in J^{r+1}(E,n) \mid \theta_r \in \mathcal{E}, \ R_{\theta_{r+1}} \subset T_{\theta_r}\mathcal{E}\}.$$

By iteration we can define the *l-prolongation* $\mathcal{E}^l$. We denote by $\theta_r^{(l)}$ the *l*-prolongation of the point $\theta_r \in \mathcal{E}$. The equation $\mathcal{E}$ is said to be *formally integrable* if all the prolongations $\mathcal{E}^l$ are smooth manifolds and the maps $\pi_{r+l+1,r+l}|_{\mathcal{E}^{l+1}}$ are fibre bundles.

In the rest of the section we deal just with equations which are formally integrable.

A vector field which belongs to $\mathcal{C}^\infty(D)$ is called *trivial* as it is tangent to all the integral manifolds of $\mathcal{C}^\infty$. For this reason we call the quotient algebra $\mathrm{Sym} \overset{\mathrm{def}}{=} D_{\mathcal{C}^\infty}/\mathcal{C}^\infty(D)$ the algebra of *non-trivial symmetries* of the distribution $\mathcal{C}^\infty$.

An element X of $D_{\mathcal{C}^\infty}$ is an *infinitesimal higher external symmetry* of the equation $\mathcal{E}$ if X is tangent to $\mathcal{E}^\infty$, and the equivalence class $[X]$ of Sym is called a *non-trivial infinitesimal higher external symmetry* of $\mathcal{E}$. The set of such classes is denoted by $\mathrm{Sym}_{\mathrm{ext}}(\mathcal{E})$.

Let us restrict our attention to $\mathcal{E}^\infty$. An *infinitesimal higher internal symmetry* of the equation $\mathcal{E}$ is a symmetry of $\mathcal{C}(\mathcal{E}^\infty)$. The set of such symmetries is denoted by $D_{\mathcal{C}}(\mathcal{E}^\infty)$. We denote by $\mathcal{C}D(\mathcal{E}^\infty)$ the ideal of vector fields on $\mathcal{E}^\infty$ tangent to $\mathcal{C}(\mathcal{E}^\infty)$, and we call *trivial symmetries* such fields. We denote by $\mathrm{Sym}_{\mathrm{int}}(\mathcal{E})$ the algebra $D_{\mathcal{C}}(\mathcal{E}^\infty)/\mathcal{C}D(\mathcal{E}^\infty)$ of non-trivial infinitesimal higher internal symmetries of the equation $\mathcal{E}$.

Remark 1. The restriction map $\mathrm{Sym}_{\mathrm{ext}}(\mathcal{E}) \to \mathrm{Sym}_{\mathrm{int}}(\mathcal{E})$ is surjective. Despite this fact, the set of classical external symmetries does not project on the set of classical internal symmetries (see [3]).

2.2. Contact Sequences and Generating Sections of Symmetries

The method of generating sections of symmetries has been largely developed in the case of an equation defined as a submanifold of a suitable jet bundle (see [3, 9] for instance). In the case of jets of submanifolds, it was partially introduced in [18] and, more recently, in [15].

Here we construct the formalism of generating sections of symmetries of an equation $\mathcal{E} \subset J^r(E,n)$ by introducing new contact sequences defined on $\mathcal{E}$. Also, we give an intrinsic definition of the operator of universal linearization on $\mathcal{E}$, whose kernel coincides with internal higher symmetries.

For $r \geq 0$, let us consider the following bundles over $\mathcal{E}^1$: the pull-back bundle

$$T^{r+1,r}(\mathcal{E}) = \mathcal{E}^1 \times_{\mathcal{E}} T\mathcal{E},$$

the sub-bundle $H^{r+1,r}(\mathcal{E}) = \{(\theta_{r+1}, v) \mid \theta_{r+1} \in \mathcal{E}^1, v \in R_{\theta_{r+1}}\}$ and the quotient bundle $V^{r+1,r}(\mathcal{E}) = T^{r+1,r}(\mathcal{E})/H^{r+1,r}(\mathcal{E})$.

We shall denote by $T^{r+1,r}$, $H^{r+1,r}$ and $V^{r+1,r}$ respectively the above bundles in the case of the trivial equation $J^r(E,n)$.

The most important property of the bundles $T^{r+1,r}(\mathcal{E})$, $H^{r+1,r}(\mathcal{E})$ and $V^{r+1,r}(\mathcal{E})$ is the following *contact exact sequence*:

$$0 \longrightarrow H^{r+1,r}(\mathcal{E}) \xrightarrow{D^{r+1}} T^{r+1,r}(\mathcal{E}) \xrightarrow{\omega^{r+1}} V^{r+1,r}(\mathcal{E}) \longrightarrow 0, \qquad (3)$$

where D^{r+1} and ω^{r+1} are, respectively, the natural inclusion and quotient projection.

Let us introduce the exact sequence of the $\mathcal{F}_r$-modules:

$$0 \longrightarrow \chi^H_{r,r-1} \overset{\mathcal{I}^r}{\hookrightarrow} \chi_r \xrightarrow{\mathcal{P}^r} \varkappa_r \longrightarrow 0. \qquad (4)$$

where $\chi^H_{r,r-1}$ is the module of sections of $\pi^*_{r,1}(H^{1,0})$, χ_r the module of sections of $\pi^*_{r,1}(T^{1,0})$ and $\varkappa_r$ the quotient space.

Let $\mathcal{F} = \mathrm{inj}\lim \mathcal{F}_r$. There is a bijection between $D_{\mathcal{C}^\infty}$ and the $\mathcal{F}$-module $\chi = \mathrm{inj}\lim \chi_r$. This comes from the fact that any field belonging to $D_{\mathcal{C}^\infty}$ is completely characterized by its restriction to $\mathcal{F}_0$ (see [20] for further details). In view of our considerations, the direct limit of sequence (4) reads

$$0 \longrightarrow \mathscr{C}^{\infty}(D) \overset{\mathscr{I}^{\infty}}{\longrightarrow} D_{\mathscr{C}^{\infty}} \overset{\mathscr{P}^{\infty}}{\longrightarrow} \varkappa \longrightarrow 0.$$

The map $\mathscr{P}^{\infty}$ induces a map on the quotient $D_{\mathscr{C}^{\infty}}/\mathscr{C}^{\infty}(D)$, that we denote by $[\mathscr{P}^{\infty}]$.

Proposition 1. *The map* $[\mathscr{P}^{\infty}] \colon D_{\mathscr{C}^{\infty}}/\mathscr{C}^{\infty}(D) \longrightarrow \varkappa$ *is an* $\mathscr{F}$*-module isomorphism.*

The map $[\mathscr{P}^{\infty}]$ also provides $\varkappa$ with a Lie algebra structure. The element $[\mathscr{P}^{\infty}][X]$ with $X \in D_{\mathscr{C}^{\infty}}$ is called the *generating section of the non-trivial symmetry* $[X]$. Locally if $X = a^{\lambda}\frac{\partial}{\partial x^{\lambda}} + b^{i}\frac{\partial}{\partial u^{i}}$, $a^{\lambda}, b^{i} \in \mathscr{F}$, then $\mathscr{P}^{\infty}(X) = (b^{i} - a^{\lambda}u_{\lambda}^{i})\left[\frac{\partial}{\partial u^{i}}\right]$.

Conversely, to any section φ of the bundle $\varkappa$ is associated a non-trivial symmetry of the Cartan distribution $\mathscr{C}^{\infty}$ via the inverse map $[\mathscr{P}^{\infty}]^{-1}$. We denote such an inverse map by $\mathfrak{I}$. Locally, if $\varphi = \varphi^{j}[\partial/\partial u^{j}]$, we have $\mathfrak{I}_{\varphi} = D_{\sigma}(\varphi^{j})\left[\frac{\partial}{\partial u_{\sigma}^{j}}\right]$.

Below we introduce the operator of *universal linearization* $\ell_{\mathscr{E}}$. This operator has been defined in different ways in the case of an equation $\mathscr{E} \subset J^{r}(\pi)$ where π is a vector bundle (see [3, 9], for instance). Here we construct such an operator in the general case of a formally integrable equation $\mathscr{E} \subset J^{r}(E, n)$.

Let P be the set of local sections of some bundle. Let us denote by $J^{r}(P)$ the r-jet space of sections which belong to P. Let us now consider the module $\varkappa|_{\mathscr{E}}$ of sections of the bundle $J^{\infty}(E, n) \times_{J^{1}(E,n)} V^{1,0} \longrightarrow J^{\infty}(E, n)$ restricted to $\mathscr{E}^{\infty}$. In this module we can define the following equivalence relation:

$$\xi \sim_{r,\theta} \hat{\xi} \Leftrightarrow \bar{D}_{\sigma}(\xi^{j})(\theta) = \bar{D}_{\sigma}(\hat{\xi}^{j})(\theta), \; |\sigma| \leq r, \; \theta \in \mathscr{E}^{\infty}.$$

Let us denote by $\{\xi\}_{\theta}^{r}$ the equivalence class with respect to the previous relation. The set of such equivalence classes is called the *horizontal r-jet space of* $\varkappa|_{\mathscr{E}}$ and it is denoted by $\bar{J}^{r}(\varkappa|_{\mathscr{E}})$.

In a similar way as in the case of jet spaces, we can define the maps $\bar{\pi}_{r,r-1}$ and $\bar{j}_{r}\xi$ where $\xi \in \varkappa|_{\mathscr{E}}$.

The following fibred inclusion $i \colon \mathscr{E}^{\infty} \times_{\mathscr{E}^{1}} V^{r+1,r}(\mathscr{E}) \hookrightarrow \bar{J}^{r}(\varkappa|_{\mathscr{E}})$ on $\mathscr{E}^{\infty}$ induces the natural projection $\bar{J}^{r}(\varkappa|_{\mathscr{E}}) \overset{\text{pr}}{\longrightarrow} \bar{J}^{r}(\varkappa|_{\mathscr{E}})/\left(\mathscr{E}^{\infty} \times_{\mathscr{E}^{1}} V^{r+1,r}(\mathscr{E})\right)$.

We continue to denote by pr the induced morphism between the modules of their sections $\bar{\mathscr{J}}^{r}(\varkappa|_{\mathscr{E}}) \overset{\text{pr}}{\longrightarrow} P$.

Definition 1. The *universal linearization operator* $\ell_{\mathscr{E}}$ *of the equation* $\mathscr{E}$ is defined to be the following map:

$$\ell_{\mathscr{E}} \colon \varkappa|_{\mathscr{E}} \longrightarrow P, \quad \xi \longmapsto \text{pr}(\bar{j}_{r}\xi)$$

Let us evaluate the local expressions of the objects defined above. Let (ε) be an intrinsic system of coordinates on $\mathscr{E}$. The chart on $\bar{J}^{r}(\varkappa|_{\mathscr{E}})$ induced by the chart (ε, q^{j}) on $\mathscr{E}^{\infty} \times_{\mathscr{E}^{1}} V^{1,0}$ is

$$(\varepsilon, q_{\sigma}^{j}), \quad |\sigma| \leq r, \quad q_{\sigma}^{j}(\{\xi\}_{\theta}^{\infty}) = \bar{D}_{\sigma}\xi^{j}(\theta)$$

where $\bar{D}_{\sigma}$ denotes the restriction of the total derivative D_{σ} on $\mathscr{E}^{\infty}$.

Let the equation $\mathscr{E}$ be locally given by $F^1 = 0, \ldots, F^k = 0$. Then the tangent space $T\mathscr{E}$ is made up by the vectors $X = X^\lambda \frac{\partial}{\partial x^\lambda} + X_\tau^j \frac{\partial}{\partial u_\tau^j}$, $|\tau| \leq r$, which satisfy $\frac{\partial F^k}{\partial x^\lambda} X^\lambda + \frac{\partial F^k}{\partial u_\tau^j} X_\tau^j = 0$, $k = 1 \ldots l$. We can construct the fibre coordinates on $V^{r+1,r}(\mathscr{E})$ as follows. The condition that a vector belongs to some fibre of $V^{r+1,r}(\mathscr{E})$ amounts to putting $X^\lambda = 0$. More precisely a vector v which belongs to $V^{r+1,r}(\mathscr{E})$ can be expressed locally by:

$$v = \xi_\tau^j \left[\frac{\partial}{\partial u_\tau^j} \right] , \qquad \text{with} \qquad \frac{\partial F^k}{\partial u_\tau^j} \xi_\tau^j = 0.$$

So we can choose these ξ_τ^j as fibred coordinates on $\mathscr{E}^\infty \times_{\mathscr{E}^1} V^{r+1,r}(\mathscr{E})$. Then the fibred inclusion i is given locally by $i(\xi_\tau^j) = \bar{D}_\tau \xi^j$. So we have $\mathrm{pr}(\bar{D}_\tau \xi^j) = \frac{\partial F^k}{\partial u_\tau^j} \bar{D}_\tau \xi^j$, and the induced morphism $\ell_\mathscr{E}$ reads

$$\ell_\mathscr{E}(\xi) = \frac{\partial F^k}{\partial u_\tau^j} \bar{D}_\tau \xi^j. \tag{5}$$

Now our target is to describe the algebra of classical and higher symmetries by means of generating sections. We have that the evolutionary derivation $\Im_\varphi$ is an external higher symmetry of the equation $\mathscr{E}$ if $\Im_\varphi(I(\mathscr{E})) \subset I(\mathscr{E})$, where $I(\mathscr{E})$ is the ideal of functions on $J^\infty(E, n)$ which vanish on any prolongation of $\mathscr{E}$. If $\mathscr{E}$ is locally described by $F^1 = 0, \ldots, F^k = 0$, then the functions $F^1, \ldots, F^k$ are differential generators of $I(\mathscr{E})$ in view of the formal integrability. Then φ is a higher external symmetry if

$$\Im_\varphi(F^i)|_{\mathscr{E}^\infty} = \ell_\mathscr{E}(\bar{\varphi}) = 0, \tag{6}$$

where $\bar{\varphi}$ is the restriction of φ on the equation $\mathscr{E}^\infty$. In view of our discussion and of Remark 1, we have proved the following

Theorem 1. $\mathrm{Sym}_{\mathrm{int}}(\mathscr{E}) = \ker \ell_\mathscr{E}$.

If we want to find the classical symmetries of $\mathscr{E}$, we have to apply the Lie-Bäcklund theorem [3, 9, 14]. In the case $m > 1$, it will be sufficient to solve (6) where $\varphi = (\varphi^1, \varphi^2, \ldots, \varphi^m)$ with $\varphi^i = b^i - a^\lambda u_\lambda^i$ and $b^i, a^\lambda \in C^\infty(E)$. In the case $m = 1$ it will be sufficient to solve (6) for an arbitrary function $\varphi \in C^\infty(J^1(E, 1))$.

3. GEODESIC EQUATION

In this section we define the geodesic equation of a pseudo-Riemannian manifold E as a submanifold of $J^2(E, 1)$.

Let $\rho: F \longrightarrow E$ be a bundle. A connection $\square$ on ρ can be seen both as a tangent valued form and as a vertical valued form, respectively:

$$\mu_\square: F \longrightarrow T^*E \otimes_F TF \ , \quad \upsilon_\square: F \longrightarrow T^*F \otimes_F VF. \tag{7}$$

Let $(u^0, u^1, \ldots, u^m)$ be a chart on E and $(u^0, u^1, \ldots, u^m, \eta^1, \ldots, \eta^l)$ a chart on F. We have the local expressions:

$$\mu_\square = du^\varphi \otimes (\partial_{u^\varphi} + \square_\varphi^i \partial_{\eta^i}) \ , \quad \upsilon_\square = (d\eta^i - \square_\varphi^i du^\varphi) \otimes \partial_{\eta^i} \ , \quad \square_\varphi^i \in C^\infty(F).$$

212

The connection $\square$ is *linear* if is linear over E. [7, 8, 17]. Locally, this means that $\square^i_\varphi = \square^i_{\varphi\alpha}\eta^\alpha$, $\square^i_{\varphi\alpha} \in C^\infty(E)$.

For any map $f \colon Q \longrightarrow E$ we denote by $f^*(\square)$ the pull-back connection on the pull-back bundle $f^*(\rho)$.

From here Greek indices will run from 0 to m, and the Latin ones from 1 to m, where $m+1$ is the dimension of E. We denote by $(u^0, u^j, u^j_0, u^j_{00})$ a local chart on $J^2(E, 1)$.

Let ∇ be the Levi-Civita connection on the tangent bundle $\tau_E \colon TE \longrightarrow E$.

Let us consider a 1-dimensional submanifold $\iota \colon \gamma \hookrightarrow E$. We can restrict the domain of the pull-back connection $\iota^*(\nabla)$ to the module of vector fields $\chi(\gamma)$ as it is a submodule of $\Gamma(\iota^*(\tau_E))$. Let us denote by ∇^γ this restriction. Let us denote by $T^\perp\gamma$ the normal bundle of γ and by $\tau^\perp_\gamma$ the natural projection of $\iota^*(\tau_E)$ on $T^\perp\gamma$. Then we say that γ is a *geodesic* if $\tau^\perp_\gamma \circ \nabla^\gamma = 0$.

It is easy to realize that the operator $\tau^\perp_\gamma \circ \nabla^\gamma$ is $C^\infty(\gamma)$-linear, then $\tau^\perp_\gamma \circ \nabla^\gamma \in \Lambda^*\gamma \otimes \Lambda^*\gamma \otimes \Gamma(T^\perp\gamma)$, that implies that the operator $\Delta \colon \gamma \longmapsto \tau^\perp_\gamma \circ \nabla^\gamma$ defines a second order non-linear differential operator. It is worth noting that $T^*\gamma \otimes T^*\gamma \otimes T^\perp\gamma$ is the pull-back bundle, through the map $j_1 L$, of $(H^{1,0})^* \otimes (H^{1,0})^* \otimes V^{1,0}$, which is the vector bundle associated with the affine bundle $\pi_{2,1}$ ([13, 14]). In view of (1) and (2) we can associate to Δ the jet morphism ϕ_Δ. We have that $\phi_\Delta \in \mathscr{H}^1_{2,1} \otimes \mathscr{H}^1_{2,1} \otimes \varkappa_2$, where $\mathscr{H}^1_{2,1}$ is the module of sections of $\pi^*_{2,1}(H^{1,0})^*$.

Definition 2. The *geodesic equation* $\mathscr{G}$ on 1-dimensional submanifolds of E is the submanifold of $J^2(E, 1)$ represented by the operator ϕ_Δ.

Let us evaluate the geodesic equation in local coordinates. We have to calculate $\nabla^\mathscr{G}(D_{u^0}, D_{u^0}, \omega^h)$, where $D_{u^0} = \partial_{u^0} + u^j_0 \partial_{u^j} + u^j_{00}\partial_{u^j_0}$ and $\omega^h = du^h - u^h_0 du^0$. We obtain that the geodesic equation is locally described by:

$$\nabla^h_{00} + 2u^j_0\nabla^h_{0j} + u^i_0 u^j_0 \nabla^h_{ij} + u^h_{00} - u^h_0(\nabla^0_{00} + 2u^j_0\nabla^0_{0j} + u^i_0 u^j_0 \nabla^0_{ij}) = 0 \qquad (8)$$

This equation is the image of some section of $\pi_{2,1}$ as it can be solved with respect to the second order derivatives. In the next propositions we construct such a section with the help of contact sequences.

Theorem 2. *With any linear connection ∇ on TE it is associated a connection $\dot\nabla$ on* $\pi_{1,0}$.

Proof. Taking into consideration that $\ker T\pi_{1,0} \simeq (H^{1,0})^* \otimes_{J^1(E,1)} V^{1,0}$, the top row of the following commutative diagram:

$$
\begin{array}{ccc}
TJ^1(E,1) & \xrightarrow{\left(\mathrm{id}_{(H^{1,0})^*}\otimes\omega^1\right)\circ\left(\Xi\otimes(\pi^*_{1,0}(\nabla))\right)\circ TD^1} & (H^{1,0})^* \otimes_{J^1(E,1)} V^{1,0} \\
\Big\downarrow{\scriptstyle TD^1} & & \Big\downarrow{\scriptstyle \mathrm{id}_{(H^{1,0})^*}\otimes\omega^1} \\
T(H^{1,0*} \otimes_{J^1(E,1)} T^{1,0}) & \xrightarrow{\Xi\otimes(\pi^*_{1,0}(\nabla))} & (H^{1,0})^* \otimes_{J^1(E,1)} T^{1,0}
\end{array}
$$

213

characterizes the connection $\dot{V}$ as vertical valued form. Here Ξ is the metric flat connection on the real line bundle $(H^{1,0})^*$. This connection Ξ is constructed as follows. Taking into consideration that $T^{1,0} \simeq T^{1,0*}$ by virtue of the metric structure, we continue to denote by $\pi_{1,0}^*(V)$ the connection on $T^{1,0*}$. Then $\Xi = \left(\mathrm{id}_{T^*J^1(E,n)} \otimes TD^{1*}\right) \circ \mu_{\pi_{1,0}^*(V)}|_{H^{1,0*}}$.

Let $(u^0, u^j, u_0^j, \eta^\alpha)$ be the local chart on $H^{1,0*} \otimes T^{1,0}$. Locally we have:

$$\Xi \otimes \left(\pi_{1,0}^*(V)\right) = \left(d\eta^\delta + (\nabla_{\alpha 0}^\delta + \nabla_{\alpha i}^\delta u_0^i)du^\alpha\right) \otimes du^0 \otimes \partial_{u^\delta}.$$

and the local expression of $v_{\dot{V}}$ is

$$v_{\dot{V}} = \left(du_0^j + \dot{V}_\alpha^j du^\alpha\right) \otimes \partial_{u_0^j} \quad \text{where} \quad \dot{V}_\alpha^j = \nabla_{\alpha 0}^j + \nabla_{\alpha i}^j u_0^i - u_0^i(\nabla_{\alpha 0}^0 + \nabla_{\alpha h}^0 u_0^h).$$

$\square$

Theorem 3. *With any connection $\dot{V}$ on $\pi_{1,0}$ it is associated a section $\ddot{V}$ of $\pi_{2,1}$.*

Proof. In view of the following commutative diagram (see [13])

$$
\begin{array}{ccc}
J^2(E,n) & \xrightarrow{\ D^2\ } & (H^{1,0})^* \otimes_{J^1(E,n)} TJ^1(E,n) \\
\Big\downarrow{\scriptstyle \pi_{2,1}} & & \Big\downarrow{\scriptstyle \mathrm{id} \otimes T\pi_{1,0}} \\
J^1(E,n) & \xrightarrow{\ D^1\ } & (H^{1,0})^* \otimes_{J^1(E,n)} T^{1,0}
\end{array}
$$

we obtain a section of $\pi_{2,1}$ if we consider a section of $(H^{1,0})^* \otimes_{J^1(E,n)} TJ^1(E,n)$ and restrict it to the image of D^2. In order to obtain such a section it is sufficient to consider the map $\ddot{V} = \left(D^{1*} \otimes \mathrm{id}_{TJ^1(E,1)}\right) \circ \mu_{\dot{V}}$. The local expression of $\ddot{V}$ is given by:

$$\ddot{V} = du^0 \otimes \left(\partial_{u^0} + u_0^j \partial_{u^j} - (\dot{V}_0^j + u_0^i \dot{V}_i^j)\partial_{u_0^j}\right). \tag{9}$$

$\square$

Corollary 1. *The image of $\ddot{V}$ coincides with the geodesic equation*

Proof. It is sufficient to take into consideration the local expression of D^2 and (9). $\square$

The construction of the section $\ddot{V}$ is similar to that given in [6], but our result is more general. In fact, in that paper, such a section is found in the case of jets of 1-dimensional time-like oriented submanifolds. Here there is not such a restriction.

The section $\ddot{V}$ splits canonically the Cartan distribution $\mathscr{C}^1$ of $J^1(E,1)$. Namely, for each $\theta_1 \in J^1(E,1)$ we have that $\mathscr{C}_{\theta_1}^1 = R_{\ddot{V}(\theta_1)} \oplus \ker T_{\theta_1}\pi_{1,0}$. We denote by $R \circ \ddot{V}$ the distribution that to any point $\theta_1 \in J^1(E,n)$ associates the R-plane $R_{\ddot{V}(\theta_1)}$. Such a splitting induced the exact sequence

$$0 \longrightarrow \mathscr{H} \lhook\joinrel\longrightarrow \chi(J^1(E,1)) \xrightarrow{\ \mathscr{W}\ } \chi(J^1(E,1))/\mathscr{H} \longrightarrow 0, \tag{10}$$

214

where $\mathscr{H}$ is the module of sections of $H \overset{\text{def}}{=} \{(\theta_1, v) \in TJ^1(E,1) \mid v \in R_{\ddot{V}(\theta_1)}\}$. We note that H is nothing but the pull-back bundle of $H^{2,1}$ through $\ddot{V}$.

·4. SYMMETRIES OF THE GEODESIC EQUATION

In this section we discuss the symmetries the geodesic equation. We see that in this case the calculus of higher symmetries can be reduced to the calculus of classical symmetries.

Theorem 4. *All the l-prolongations $\mathscr{G}^l$ of the equation $\mathscr{G}$ are diffeomorphic to $J^1(E,1)$, and the induced Cartan distribution $\mathscr{C}(\mathscr{G}^l)$ is isomorphic to $R \circ \ddot{V}$.*

Proof. Let $\theta_1 \in J^1(E,1)$. For each point $\ddot{V}(\theta_1)$ there exists only one R-plane which contains such a point and which is contained in $T_{\ddot{V}(\theta_1)}\mathscr{G}$. In fact if we represent the point $\ddot{V}(\theta_1)$ as the pair $(\theta_1, R_{\ddot{V}(\theta_1)})$, then such an R-plane is given by $T_{\theta_1}\ddot{V}(R_{\ddot{V}(\theta_1)})$, and the existence follows. Also, if two R-planes R and $\tilde{R}$ contain the same point $\ddot{V}(\theta_1)$, then the difference between a vector $v \in R$ and a vector $\tilde{v} \in \tilde{R}$ belongs to the kernel of $T_{\ddot{V}(\theta_1)}\pi_{2,1}$. Of course we have that $T_{\ddot{V}(\theta_1)}\mathscr{G} \cap \ker T_{\ddot{V}(\theta_1)}\pi_{2,1} = 0$, and the uniqueness is proved.

From this discussion we have that $\mathscr{C}_{\ddot{V}(\theta_1)}(\mathscr{G}) = \mathscr{C}_{\ddot{V}(\theta_1)} \cap T_{\ddot{V}(\theta_1)}\mathscr{G} = T_{\theta_1}\ddot{V}(R_{\ddot{V}(\theta_1)}) \simeq R_{\ddot{V}(\theta_1)}$.

We notice that the pair $(\ddot{V}(\theta_1), T_{\theta_1}\ddot{V}(R_{\ddot{V}(\theta_1)}))$ is a point of $\mathscr{G}^1$, which is the image of the following section:

$$\ddot{V}^{(1)} : \theta_1 \in J^1(E,n) \longrightarrow (\ddot{V}(\theta_1))^{(1)} \equiv \left(\ddot{V}(\theta_1), T_{\theta_1}\ddot{V}(R_{\ddot{V}(\theta_1)})\right) \in \mathscr{G}^1.$$

We have that $\mathscr{C}_{(\ddot{V}(\theta_1))^{(1)}}(\mathscr{G}^1) = T_{\theta_1}\ddot{V}^{(1)}(R_{\ddot{V}(\theta_1)}) \simeq R_{\ddot{V}(\theta_1)}$.

By iterating the previous construction, we obtain that the l-prolongation of the equation $\mathscr{G}$ is diffeomorphic to $\mathscr{G}$, and so to $J^1(E,n)$, and that every Cartan plane $\mathscr{C}_{(\ddot{V}(\theta_1))^{(l)}}(\mathscr{G}^l)$ is isomorphic to $R_{\ddot{V}(\theta_1)}$.

$\square$

Corollary 2. *The algebra of trivial internal symmetries of $\mathscr{G}$ is isomorphic to $\mathscr{H}$. The algebra of higher internal symmetries of $\mathscr{G}$ is isomorphic to the algebra of classical internal symmetries of $\mathscr{G}$, and coincides with the algebra of vector fields on $J^1(E,1)$ preserving the 1-dimensional distribution $R \circ \ddot{V}$. The algebra of non-trivial higher internal symmetries of $\mathscr{G}$ is the projection through $\mathscr{W}$ of the algebra of internal symmetries.*

Proof. From Theorem 4 it follows that $(\mathscr{G}^\infty, \mathscr{C}(\mathscr{G}^\infty))$ is isomorphic to $(J^1(E,1), R \circ \ddot{V})$.

$\square$

Corollary 3. *The generating sections of the internal symmetries of $\mathscr{G}$ belong to $\varkappa_1$.*

Proof. It follows from Theorem 1 and from $\varkappa_{\mathscr{G}} \simeq \varkappa_1$.

$\square$

Examples of computations of symmetries of the geodesic equation in the case of the plane and of the sphere are contained in [14]. There it is applied our construction to the Kepler equation too, and we got symmetries whose coefficients of generating sections are polynomial with respect to first derivatives.

ACKNOWLEDGMENTS

I would like to thank Raffaele Vitolo who introduced me to the theory of jets of submanifolds, and for his constant support. I would like also to thank Alexander Verbovetsky, who clarified my doubts.

REFERENCES

1. A. V. Aminova, *Projective transformations and symmetries of differential equations*, Sbornik: Mathematics, **186**, (YEAR), 1711-1726.
2. D. V. Alekseevsky, A. M. Vinogradov and V. V. Lychagin, *Basic ideas and concepts of differential geometry*, Geometry I. Encycl. Math. Sci., Vol. 28, Springer-Verlag, 1991.
3. A. V. Bocharov, V. N. Chetverikov, S. V. Duzhin, N. G. Khor′kova, I. S. Krasil′shchik, A. V. Samokhin, Y. N. Torkhov, A. M. Verbovetsky, A. M. Vinogradov, *Symmetries and Conservation Laws for Differential Equations of Mathematical Physics*, Amer. Math. Soc., (YEAR), PAGES.
4. L. Fatibene, M. Ferraris, M. Francaviglia, R.G. McLenaghan, *Generalized symmetries in mechanics and field theories*, Journal of mathematical phisics, **43**, no.6, (2002), 3147-3161.
5. A. Jadczyk, M. Modugno, *A scheme for Galilei general relativistic quantum mechanics*, General relativity and gravitational physics (Bardonecchia, 1992), 319-337, World Sci. Publishing, River Edge, NJ, 1994.
6. J. Janyška, M. Modugno, *Classical particle phase space in general relativity*, Differential geometry and applications, (Brno, 1995), 573-602.
7. W. Klingenberg, *Riemannian geometry*, THE NAME OF PUBLISHER, (1982).
8. I. Kolář, P. Michor and J. Slovák, *Natural Operations in Differential Geometry*, Springer-Verlag, 1993.
9. I. S. Krasil′shchik and A. M. Verbovetsky, *Homological methods in equations of mathematical physics*, Open Education and Sciences, Opava (Czech Rep., 1998), math.DG/9808130.
10. E. Kreyszig, *Differential geometry*, University of Toronto Press, (1959)
11. D. Krupka, *Global variational functionals in fibered spaces*, Nonlinear analysis, **47**, (2001), 2633-2642.
12. G. Manno and R. Vitolo, *Variational sequences on finite order jets of submanifolds*, Diff. Geom. and Appl., (Opava, 2001).
13. G. Manno and R. Vitolo, *The geometry of finite order jets of submanifolds and the variational formmalism*, submitted to *Journal of London Math. Soc.*
14. G. Manno, *Jet methods for the finite order variational sequence and the geodesic equation*, Ph.D. thesis, September 2003.
15. M. Modugno, A. M. Vinogradov, *Some Variations on the Notion of Connection*, Ann. di Mat. Pura ed Appl. IV, Vol. CLXVII, (1994), 33-71.
16. P. J. Olver, *Application of Lie groups to differential equations*, Springer-Verlag, (YEAR).

17. D. J. Saunders, *The Geometry of Jet Bundles*, Cambridge Univ. Press, (1989).
18. A. M. Vinogradov, *Local symmetries and conservation laws*, Acta Applicandae Mathematicae **2**, (1981), 21-78.
19. A. M. Vinogradov, *An informal introduction to the geometry of jet spaces*, Rend. Seminari Fac. Sci. Univ. Cagliari, **58**, (1988), 301-333.
20. A. M. Vinogradov, *Cohomological Analysis of Partial Differential Equations and Secondary Calculus*, Amer. Math. Soc., (2001).

Doubly Warped Product Manifolds and Submanifolds

Koji Matsumoto

Department of Mathematics, Faculty of Education, Yamagata University, Yamagata, 990-8560, Japan

Abstract. In this report, we consider doubly warped product manifolds and we get fundamental properties of this manifold and consider these submanifolds.
Key words. Doubly warped product manifold, warping function, totally geodesic, Riemannian submanifold.
PACS. 02.40.Ky.
MSC. 53C40.

1. INTRODUCTION

Recently, B. Y. Chen introduced the notion of two kinds of warped product CR-submanifolds in Kaehler manifolds [4]. Then, we can see few papers about this notion [2, 5, 6]. In this paper, we mainly consider doubly warped product Riemannian manifolds and their submanifolds. Our target is to define a doubly warped product CR-submanifold in Hermitian manifolds which is a generalization of the notion of B. Y. Chen and to get differential geometrical properties of this submanifold. This report is the first step of our object.

2. DOUBLY WARPED PRODUCT RIEMANNIAN MANIFOLDS

Let (M_1, g_1) and (M_2, g_2) be two Riemannian manifolds and (M, g) be their product Riemannian manifold. Then by definition the Riemannian metric g of M is defined by

$$g(U,V) = g_1(\pi_{1*}U, \pi_{1*}V) + g_2(\pi_{2*}U, \pi_{2*}V) \tag{2.1}$$

for any $U,V \in TM$, where π_1 (resp. π_2) is the projection of M to M_1 (resp. M_2) and π_{1*} (resp. π_{2*}) is a differential map of π_1 (resp. π_2).

Next, let $f_1 > 0$ (resp. $f_2 > 0$) be a differentiable function on M_1 (resp. M_2). The *doubly warped product Riemannian manifold* $\bar{M} = M_1 \times_{(f_1, f_2)} M_2$ is the manifold M equipped with the Riemannian metric $\bar{g}$ such that

$$\bar{g}(U,V) = f_2^2 g_1(\pi_{1*}U, \pi_{1*}V) + f_1^2 g_2(\pi_{2*}U, \pi_{2*}V) \tag{2.2}$$

CP729, *Global Analysis and Applied Mathematics: International Workshop on Global Analysis*,
edited by K. Taş, D. Krupka, O. Krupková, and D. Baleanu
© 2004 American Institute of Physics 0-7354-0209-4/04/$22.00

for any $U, V \in TM$ [1]. Thus, we have $\bar{g} = f_2^2 g_1 + f_1^2 g_2$. The pair of the functions (f_1, f_2) is called the *the pair of the warping functions.*

Remark 1.1. In a doubly warped product Riemannian manifold, if one of the warping functions is positive constant, then the manifold is called a warped product Riemannian manifold [7].

We will write our two metrices g and $\bar{g}$ as follows;

$$
\begin{cases}
g = \begin{pmatrix} g_1 & 0 \\ 0 & g_2 \end{pmatrix}, \quad \bar{g} = \begin{pmatrix} \bar{g}_1 & 0 \\ 0 & \bar{g}_2 \end{pmatrix} = \begin{pmatrix} f_2^2 g_1 & 0 \\ 0 & f_1^2 g_2 \end{pmatrix}, \\[3mm]
\bar{g}^{-1} = \begin{pmatrix} \bar{g}_1^{-1} & 0 \\ 0 & \bar{g}_2^{-1} \end{pmatrix} = \begin{pmatrix} \frac{1}{f_2^2} g_1^{-1} & 0 \\ 0 & \frac{1}{f_1^2} g_2^{-1} \end{pmatrix}.
\end{cases}
\tag{2.3}
$$

Next, let $\tilde{M}$ be a Riemannian manifold with the Riemannian structure $\tilde{g}$.

Let us consider a submanifold M of $\tilde{M}$ such that M is a doubly warped product manifold of two submanifolds M_1 and M_2 of $\tilde{M}$ and the Riemannian metrices g_1 and g_2 are the induced metrices of $\tilde{g}$, that is,

$$
g_1(X, Y) = \tilde{g}(i_{1*}X, i_{1*}Y), \quad g_2(Z, W) = \tilde{g}(i_{2*}Z, i_{2*}W)
\tag{2.4}
$$

for any $X, Y \in TM_1$ and $Z, W \in TM_2$, where i_1 (resp. i_2) is the isometric immersion of M_1 (resp. M_2) in $\tilde{M}$. And we assume that two submanifolds are always orthogonal, that is,

$$
\tilde{g}(X, Z) = 0
\tag{2.5}
$$

for any $X \in TM_1$ and $Z \in TM_2$.

In the above submanifold, we will give a structure of a doubly warped product manifold with pair of warping functions (f_1, f_2) and we write it $\bar{M} = M_1 \times_{(f_1, f_2)} M_2$, too.

Remark 1.2. The submanifold (M, g) is a Riemannian submanifold of $(\tilde{M}, \tilde{g})$. But the submanifold $(\bar{M}, \bar{g})$ is not a Riemannian submanifold of $(\tilde{M}, \tilde{g})$.

Remark 1.3. As an another submanifold of a Riemannian manifold, we can consider the metric g is a induced metric of $\tilde{g}$ for certain Riemannian metrices g_1 and g_2 on M_1 and M_2, respectively. In this case, (M_1, g_1) and (M_2, g_2) are not Riemannian submanifolds of $\tilde{M}$. And warped product CR-submanifolds which are defined by B. Y. Chen are in this case [4].

Between a Riemannian manifold $(\tilde{M}, \tilde{g})$ and its Riemannian submanifold (M, g), the Weingarten formula is known as

$$
\tilde{R}(U, V, W, X) = R(U, V, W, Z) + \tilde{g}(\sigma(U, X), \sigma(V, W))
\tag{2.6}
$$

$$
- \tilde{g}(\sigma(U, W), \sigma(V, X))
$$

for any $U, V, W, Z \in TM$, where $\tilde{R}$ (resp R) is the curvature tensor with respect to $\tilde{g}$ (resp. g) and σ is the second fundamental form of M [4].

Hereafter, we assume that the manifolds M_1 and M_2 are orthogonal, that is, $\tilde{g}(X, Z) = 0$ for any $X \in TM_1$ and $Z \in TM_2$.

Now, we take a local coordinate system $\{x^1, x^2, ..., x^m\}$ in $\tilde{M}$ as follows:

(i) $\{x^1, x^2, ..., x^{n_1}\}$ is a local coordinate system of M_1,

(ii) $\{x^{n_1+1}, x^{n_1+2}, ..., n^{n_1+n_2}\}$ is a local coordinate system of M_2 and

(iii) $\{x^{n+1}, x^{n+2}, ..., x^m\}$ is a local coordinate system of the normal bundle $T^\perp M$ of M, where $n_1 = \dim M_1$, $n_2 = \dim M_2$, $n = n_1 + n_2 = \dim M$ and $m = \dim \tilde{M}$. We assume that the indices $\{i, j, .., k\}$ run over the range $\{1.2.,,.n_1\}$, $\{\lambda, \mu, ..., \nu\}$ run over the range $\{1, 2, ..., n\}$ and $\{a, b, ..., c\}$ run over the range $\{n_1+1, n_1+2, ..., n\}$. The induced Riemannian metric g_{ji} (resp. g_{ba}) are expressed by

$$g_{ji} = \tilde{g}_{BA} \frac{\partial x^B}{\partial x^j} \frac{\partial x^A}{\partial x^i}, \quad g_{ba} = \tilde{g}_{BA} \frac{\partial x^B}{\partial x^b} \frac{\partial x^A}{\partial x^a}, \tag{2.7}$$

where the suffices $\{A, B, ..., C\}$ run over the range $\{1, 2, ..., m\}$.

For this coordinate system,

$$\left\{ \frac{\partial}{\partial x^1}, ..., \frac{\partial}{\partial x^{n_1}} \right\}, \quad \left\{ \frac{\partial}{\partial x^{n_1+1}}, ..., \frac{\partial}{\partial x^{n_1+n_2}} \right\} \tag{2.8}$$

are the local bases of TM_1 and TM_2, respectively.

Next, in our submanifold $\tilde{M}$, the pair of warping functions are characterized as

$$\frac{\partial f_1}{\partial x^a} = 0, \quad \frac{\partial f_2}{\partial x^i} = 0 \tag{2.9}$$

for any $a \in \{n_1+1, ..., n\}$ and $i \in \{1, ..., n_1\}$.

Next, let us consider the relation between the Christoffel symbols $\overline{\{_v{}^\lambda{}_\mu\}}$ with respect to $\bar{g}$ and those $\{_v{}^\lambda{}_\mu\}$ with respect to g. In generally, the Christoffel symbols $\{_v{}^\lambda{}_\mu\}$ with respect to $g_{\mu\lambda}$ are locally written by

$$\{_v{}^\lambda{}_\mu\} = \frac{1}{2} g^{\lambda\varepsilon} \left(\frac{\partial g_{\varepsilon\mu}}{\partial x^v} + \frac{\partial g_{v\varepsilon}}{\partial x^\mu} - \frac{\partial g_{v\mu}}{\partial x^\varepsilon} \right), \tag{2.10}$$

where $(g^{\mu\lambda})$ is the inverse matrix of $(g_{\mu\lambda})$.

By virtue of (2.3), we get

$$\begin{cases} \overline{\{_j{}^h{}_i\}} = \{_j{}^h{}_i\}, \quad \overline{\{_j{}^h{}_a\}} = \{_j{}^h{}_a\} + \frac{\partial \log f_2}{\partial x^a} \delta_j{}^h, \\ \overline{\{_j{}^a{}_i\}} = -(\frac{f_2}{f_1})^2 (\{_j{}^a{}_i\} + \frac{\partial \log f_2}{\partial x^b} g^{ba} g_{ji}), \\ \overline{\{_b{}^h{}_a\}} = -(\frac{f_1}{f_2})^2 (\{_b{}^h{}_a\} + \frac{\partial \log f_1}{\partial x^j} g^{jh} g_{ba}), \\ \overline{\{_b{}^a{}_i\}} = \{_b{}^a{}_i\} + \frac{\partial \log f_1}{\partial x^i} \delta_b{}^a, \quad \overline{\{_c{}^a{}_b\}} = \{_c{}^a{}_b\} \end{cases} \tag{2.11}$$

Using (2.11), we can obtain the relation of the curvature tensor $\bar{R}_{\omega v\mu}{}^\lambda$ with respect to $\bar{g}_{\mu\lambda}$ and $R_{\omega v\mu}{}^\lambda$ with respect to $g_{\mu\lambda}$. The Riemannian curvature tensor $R_{\omega v\mu}{}^\lambda$ is defined by

$$R_{\omega v\mu}{}^\lambda = \frac{\partial \{_v{}^\lambda{}_\mu\}}{\partial x^\omega} - \frac{\partial \{_\omega{}^\lambda{}_\mu\}}{\partial x^v} + \{_v{}^\lambda{}_\varepsilon\} \{_\omega{}^\varepsilon{}_\mu\} - \{_v{}^\lambda{}_\varepsilon\} \{_\omega{}^\varepsilon{}_\mu\}.$$

By the straightforward calculation, we have

$$
\begin{cases}
\bar{R}_{kji}{}^h = R_{kji}{}^h - (\tfrac{f_2}{f_1})^2 \dfrac{\partial \log f_2}{\partial x^e}\dfrac{\partial \log f_2}{\partial x^b} g^{be}(\delta_k{}^h g_{ji} - \delta_j{}^h g_{ki}),\\[2mm]
\bar{R}_{kji}{}^a = (\tfrac{f_2}{f_1})^2 \dfrac{\partial \log f_2}{\partial x^c} g^{ba}(\dfrac{\partial \log f_1}{\partial x^k} g^{ji}\dfrac{\partial \log f_1}{\partial x^j} g_{ki}),\\[2mm]
\bar{R}_{kjb}{}^h = \dfrac{\partial \log f_2}{\partial x^b}\dfrac{\partial \log f_1}{\partial x^l}(\delta_j{}^l\delta_k{}^h - \delta_k{}^l\delta_j{}^h), \quad \bar{R}_{kjb}{}^a = 0, \quad \bar{R}_{kci}{}^h = 0,\\[2mm]
\bar{R}_{kci}{}^a = \{\nabla_k(\dfrac{\partial \log f_2}{\partial x^i}) + \dfrac{\partial \log f_1}{\partial x^k}\dfrac{\partial \log f_1}{\partial x^i}\}\delta_c{}^a\\[2mm]
\qquad\quad + (\tfrac{f_2}{f_1})^2\{\dfrac{\partial \log f_2}{\partial x^c}\dfrac{\partial \log f_2}{\partial x^b} + \nabla_c(\dfrac{\partial \log f_2}{\partial x^b})\}g^{ba}g_{ki},\\[2mm]
\bar{R}_{kcb}{}^a = \dfrac{\partial \log f_1}{\partial x^k}\dfrac{\partial \log f_2}{\partial x^d}(g^{da}g_{cb} - \delta_b{}^d\delta_c{}^a)\\[2mm]
\bar{R}_{kcb}{}^h = -(\tfrac{f_1}{f_2})^2\{\nabla_k\dfrac{\partial \log f_1}{\partial x^m} + \dfrac{\partial \log f_1}{\partial x^k}\dfrac{\partial f_1}{\partial x^m}\}g^{mh}g_{cb}\\[2mm]
\qquad\quad - (\nabla_c\dfrac{\partial \log f_2}{\partial x^b} + \dfrac{\partial \log f_2}{\partial x^c}\dfrac{\partial \log f_2}{\partial x^b})\delta_k{}^h,\\[2mm]
\bar{R}_{dci}{}^h = 0, \quad \bar{R}_{dci}{}^a = (\tfrac{f_1}{f_2})^2\dfrac{\partial \log f_1}{\partial x^i}\dfrac{\partial \log f_2}{\partial x^e}(\delta_c{}^e\delta_d{}^a - \delta_d{}^e\delta_c{}^a),\\[2mm]
\bar{R}_{dcb}{}^h = (\tfrac{f_1}{f_2})^2\dfrac{\partial \log f_1}{\partial x^k} g^{kh}(\dfrac{\partial \log f_2}{\partial x^d} g^{cb}\dfrac{\partial \log f_2}{\partial x^c}g_{db}),\\[2mm]
\bar{R}_{dcb}{}^a = R_{dcb}{}^a - (\tfrac{f_1}{f_2})^2\dfrac{\partial \log f_1}{\partial x^k}\dfrac{\partial \log f_1}{\partial_x^l} g^{kl}(\delta_d{}^a g_{cb} - \delta_c{}^a g_{db}),
\end{cases}
\tag{2.12}
$$

where $R_{kji}{}^h$ (resp. $R_{dcb}{}^a$) are the Riemannian curvature tensor with respect to g_{ji} (resp. g_{ba}).

Moreover, the Ricci tensors $\bar{R}_{\mu\lambda} = R_{\varepsilon\mu\lambda}{}^\varepsilon$ with respect to $\bar{g}_{\mu\lambda}$ are given by

$$
\begin{cases}
\bar{R}_{ji} = R_{ji} - \{n_1\dfrac{\partial \log f_2}{\partial x^c}\dfrac{\partial \log f_2}{\partial x^b} + (\tfrac{f_2}{f_1})^2\nabla_c(\dfrac{\partial \log f_2}{\partial x^b})\}g^{cb}g_{ji}\\[2mm]
\qquad - n_2\{\nabla_j(\dfrac{\partial \log f_1}{\partial x^i}) + \dfrac{\partial \log f_1}{\partial x^j}\dfrac{\partial \log f_1}{\partial x^i}\},\\[2mm]
\bar{R}_{ja} = (n_1 + n_2 - 2)\dfrac{\partial \log f_1}{\partial x^j}\dfrac{\partial \log f_2}{\partial x^a},\\[2mm]
\bar{R}_{cb} = R_{cb} - \{n_2\dfrac{\partial \log f_1}{\partial x^j}\dfrac{\partial \log f_1}{\partial x^k} + (\tfrac{f_1}{f_2})^2\nabla_k(\dfrac{\partial \log f_1}{\partial x^j})\}g^{jk}g_{cb}\\[2mm]
\qquad - n_1\{\nabla_j(\dfrac{\partial \log f_2}{\partial x^b}) + \dfrac{\partial \log f_2}{\partial x^c}\dfrac{\partial \log f_2}{\partial x^b}\},
\end{cases}
\tag{2.13}
$$

where R_{ji} (resp. R_{ba}) denotes the Ricci tensor with respect to g_{ji} (resp. g_{ba}).

Using (2.3) and (2.13), we calculate the scalar curvature $\bar{R}$ with respect to $\bar{g}$. Then it is given by

$$
\bar{R} = \frac{1}{f_2^2}R_1 - \frac{1}{f_2^2}\{n_1\frac{\partial \log f_2}{\partial x^c}\frac{\partial \log f_2}{\partial x^b} + (\tfrac{f_2}{f_1})^2\nabla_c(\frac{\partial \log f_2}{\partial x^b})\}g^{cb} + \tag{2.14}
$$

$$
\frac{1}{f_1^2}R_2 - \frac{1}{f_1^2}\{n_2\frac{\partial \log f_1}{\partial x^j}\frac{\partial \log f_1}{\partial x^k} + (\tfrac{f_1}{f_2})^2\nabla_k(\frac{\partial \log f_1}{\partial x^j})\}g^{jk},
$$

where R_1 (resp. R_2) denotes the scalar curvature with respect to g_1 (resp. g_2).

3. COVARIANT DIFFERENTIATIONS

In this section, we consider the relations of covariant differentiations between the metric tensors which defined in the last section.

For the covariant differentiations $\bar{\nabla}_{\partial_\nu}\partial_\lambda$ and those $\nabla_{\partial_\mu}\partial_\lambda$, we get

$$
\begin{cases}
\bar{\nabla}_{\partial_j}\partial_i = \nabla_{\partial_j}\partial_i - (\frac{f_1^2+f_2^2}{f_1^2})\{_j{}^a{}_i\}\partial_a - (\frac{f_2}{f_1})^2 \frac{\partial \log f_2}{\partial x^b} g^{ba} g_{ji}\partial_a, \\[2mm]
\bar{\nabla}_{\partial_j}\partial_b = \nabla_{\partial_j}\partial_b + \frac{\partial \log f_2}{\partial x^b}\partial_j + \frac{\partial \log f_1}{\partial x^j}\partial_b, \\[2mm]
\bar{\nabla}_{\partial_c}\partial_b = \nabla_{\partial_c}\partial_b - (\frac{f_1^2+f_2^2}{f_2^2})\{_c{}^h{}_b\}\partial_h - (\frac{f_1}{f_2})^2 \frac{\partial \log f_1}{\partial x^j} g^{jh} g_{cb}\partial_h,
\end{cases}
\tag{3.1}
$$

where $\partial_\lambda = \frac{\partial}{\partial x^\lambda}$.

Next, let for $X,Y \in TM_1$ and $Z,W \in TM_2$, we can write

$$ X = X^i\partial_i, \quad Y = Y^i\partial_i, \quad Z = Z^a\partial_a, \quad W = W^a\partial_a, $$

where X^i,Y^i (resp. Z^a,W^a) are differentiable functions of M_1 (resp. M_2). Then, by virtue of (3.1), we obtain

$$
\begin{cases}
\bar{\nabla}_X Y = \nabla_X Y - \frac{f_1^2+f_2^2}{f_1^2} X^j Y^i \{_j{}^a{}_i\}\partial_a - (\frac{f_2}{f_1})^2 \frac{\partial \log f_2}{\partial x^b} g^{ba} g_1(X,Y)\partial_a, \\[2mm]
\bar{\nabla}_X Z = \nabla_X Z + (Z\log f_2)X + (X\log f_1)Z, \\[2mm]
\bar{\nabla}_Z W = \nabla_Z W - \frac{f_1^2+f_2^2}{f_2^2} Z^c W^b \{_c{}^h{}_b\}\partial_h - (\frac{f_1}{f_2})^2 \frac{\partial \log f_1}{\partial x^j} g^{jh} g_2(Z,W)\partial_h.
\end{cases}
\tag{3.2}
$$

If we assume that the induced metric g_{ji} (resp. g_{ba}) is independent to the coordinate system $\{x^{n_1+1},...,x^{n_1+n_2}\}$ (resp. $\{x^1,...,x^{n_1}\}$). This means that

$$ \{_j{}^a{}_j\} = \{_c{}^h{}_b\} = 0 $$

for any $h,i,j \in \{1,...,n_1\}$ and $a,b,c \in \{n_1+1,...,n_1+n_2\}$. In this case, (3.2) becomes

$$
\begin{cases}
\bar{\nabla}_X Y = \nabla_X Y - (\frac{f_2}{f_1})^2 \frac{\partial \log f_2}{\partial x^b} g^{ba} g_1(X,Y)\partial_a, \\[2mm]
\bar{\nabla}_X Z = \nabla_X Z + (Z\log f_2)X + (X\log f_1)Z, \\[2mm]
\bar{\nabla}_Z W = \nabla_Z W - (\frac{f_1}{f_2})^2 \frac{\partial \log f_1}{\partial x^j} g^{jh} g_2(Z,W)\partial_h.
\end{cases}
\tag{3.2$'$}
$$

Now, we let $\nabla^1,\sigma^1,\bar{\nabla}^1,\bar{\sigma}^1,\nabla^2,\sigma^2,\bar{\nabla}^2,\bar{\sigma}^2$ be the Riemannian connections and put the second fundamental forms of $g_1,\bar{g}_1,g_2,\bar{g}_2$ of $M_1,\bar{M}_1,M_2,\bar{M}_2$ in $M,\bar{M}$, respectively. That is, they are expressed as

$$
\begin{cases}
\nabla_X Y = \nabla_X^1 Y + \sigma^1(X,Y), \quad \bar{\nabla}_X Y = \bar{\nabla}_X^1 Y + \bar{\sigma}^1(X,Y), \\[2mm]
\nabla_X Z = -A_Z^1 X + D_X^1 Z, \quad \bar{\nabla}_Z X = -\bar{A}_Z^1 X + \bar{D}_X^1 Z, \\[2mm]
\nabla_Z W = \nabla_Z^2 W + \sigma^2(X,Y), \quad \bar{\nabla}_Z W = \bar{\nabla}_Z^2 W + \bar{\sigma}^2(X,Y)
\end{cases}
\tag{3.3}
$$

for any $X,Y \in TM_1$ and $Z,W \in TM_2$. From (3.2) and (3.3), we easily obtain

$$
\sigma^1(X,Y) = X^i Y^j \{_i{}^a{}_j\} \partial_a,
\tag{3.4}
$$

and

$$
\bar{\sigma}^1(X,Y) = -(\frac{f_2}{f_1})^2 (X^i Y^j \{_i{}^a{}_j\} + \frac{\partial \log f_2}{\partial x^b} g^{ba} g_1(X,Y)) \partial_a .
\tag{3.5}
$$

Similarly, we have

$$
\sigma^2(Z,W) = Z^b W^c \{_b{}^h{}_c\} \partial_h,
\tag{3.6}
$$

and

$$
\bar{\sigma}^2(Z,W) = -(\frac{f_1}{f_2})^2 (Z^b W^c \{_b{}^h{}_c\} + \frac{\partial \log f_1}{\partial x^j} g^{jh} g_2(Z,W)) \partial_h .
\tag{3.7}
$$

Thus, we have from (3.4) and (3.6):

Proposition 2.1. *In a product submanifold $M = M_1 \times M_2$ of a Riemannian manifold $(\tilde{M}, \tilde{g})$, M_1 (resp. M_2) is totally geodesic if and only if $\{_i{}^a{}_j\} = 0$ (resp. $\{_b{}^h{}_c\} = 0$) for any $i,j,h \in \{1,...,n_1\}$ (resp. $\{a,b,c \in \{n_1+1,...,n_1+n_2\})$. Or equivalently g_{ij} (resp. g_{ab}) are functions of $\{x^1,...,x^{n_1}\}$ (resp. $\{x^{n_1+1},...,x^{n_1+n_2}\}$).*

Corollary 2.2. *In a doubly warped product submanifold $\bar{M} = \bar{M}_1 \times_{(f_1 \times f_2)} \bar{M}_2$ in a Riemannian manifold $(\tilde{M}, \tilde{g})$, if both of M_1 and M_2 are totally geodesic in M, then the second fundamental form $\bar{\sigma}^1$ (resp. $\bar{\sigma}^2$) satisfies*

$$
\bar{\sigma}^1(X,Y) = -\frac{\partial \log f_2}{\partial x^b} g^{ba} g_1(X,Y)) \partial_a
\tag{3.8}
$$

and

$$
\bar{\sigma}^2(Z,W) = -\frac{\partial \log f_1}{\partial x^j} g^{jh} g_2(Z,W)) \partial_h.
\tag{3.9}
$$

ACKNOWLEDGMENTS

The author wishes to express his hearty thanks to Professor L. Kozma at University of Debrecen (Hungary) for his very important information and valuable suggestions.

REFERENCES

1. Y. Agaoka, I. B. Kim, B.H. Kim and D. J. Yeon, *On doubly warped product manifolds,* Mem. Fac. Integrated Arts and Sci., Hiroshima Univ., Ser. IV, **24**, (1998), 1-10.
2. V. Bonanzinga and K. Matsumoto, *Warped product CR-submanifolds in locally conformal Kaehler manifolds,* Periodica Math. Hungarica **48** (1-2), (2004), 1-15.
3. B. Y. Chen, *Geometry of submanifolds,* New York: Marcel Dekker, (1973).
4. B. Y. Chen, *Geometry of warped product CR-submanifolds in Kaehler manifolds,* Monatsh. Math., **133**, (2001), 177-195.
5. K. Matsumoto and I. Mihai, *Warped product submanifolds in Sasakian space forms,* SUT J. Math., **38**, (2002), 135-144.
6. A. Mihai, *Warped product submanifolds in complex space forms,* Acta Sci. Math. Szeged, **20**, (2004), 311-319.
7. B. O'Neil, *Semi-Riemannian Geometry with Applications to Relativity*, New York: Academic Press, (1983).

On Einstein Lagrangian submanifolds of a complex projective space[1]

Yoshio Matsuyama

Department of Mathematics, Chuo University, 1-13-27 Kasuga, Bunkyo-ku,Tokyo 112-8551, Japan
e-mail address: matuyama@math.chuo-u.ac.jp

Abstract. In the present paper we study Einstein Lagrangian submanifolds. We show that if M is a complete Einstein Lagrangian minimal submanifold of a complex projective space $CP_n(c)$, then M is parallel and M is one of the following conditions holds: a) M is totally geodesic, b) $n = 2$ and M is a finite Riemannian covering of a torus minimally embedded in $CP^2(c)$ with parallel second fundamental form, c) $n > 2$ and M is an embedded submanifold congruent to the standard embedding of: $SU(3)/SO(3), n = 5; SU(3), n = 8; SU(6)/Sp(3), n = 14$ or $E_6/F_4, n = 26$. We also study a compact Einstein Kaehler submanifold of a complex projective space.
Key words. Complex projective space, Einstein Lagrangian submanifold, minimal submanifold, parallel submanifold.
PACS. 02.40.Ma.
MSC. Primary 53C40; Secondary 53B25.

1. INTRODUCTION

An $2n$-dimensional manifold $\overline{M}$ is called a symplectic manifold if it admits a nondegenerate closed 2-form Ω, i.e., $\Omega^n \neq 0$ everywhere. An n-dimensional submanifold M of a symplectic $2n$-manifold $(\overline{M}, \Omega)$ is called Lagrangian if the restriction of Ω on the tangent bundle of M vanishes identically. Thus one has $\Omega(X, Y) = 0$ for X, Y tangent to M. Symplectic manifolds and their Lagrangian submanifolds appear naturally in the context of classical mechanics and mathematical physics. For instance, the system of partial differential equations of Hamilton-Jacobi type leads to the study of Lagrangian submanifolds and foliations in the cotangent bundle. Furthermore, Lagrangian submanifolds are part of a growing list of mathematically rich special geometries that occur naturally in string theory. The study of Lagrangian submanifolds of Kaehler manifolds from Riemannian geometric point of view was initiated in about 1970 (See, for example, [1] and [12]). A submanifold M of a Kaehler manifold $\overline{M}$ is Lagrangian if the complex structure J of the ambient manifold $\overline{M}$ carries each tangent space of M onto the corresponding normal space of M, i.e., $J(T_xM) = T_x^{\perp}$ for any point $x \in M$. In detail, $g(JX, Y) = 0$ is satisfied, where g is the metric tensor of $\overline{M}$. Since every curve in a Kaehler curve is Lagrangian automatically, we only consider Lagrangian submanifolds of dimension greater than or equal to two.

[1] This was partially supported by Personal Research Grant of 2004 of Chuo Univ.

CP729, *Global Analysis and Applied Mathematics: International Workshop on Global Analysis,*
edited by K. Taş, D. Krupka, O. Krupková, and D. Baleanu
© 2004 American Institute of Physics 0-7354-0209-4/04/$22.00

Because the tangent bundle and the normal bundle of a Lagrangian submanifold are isomorphic via the complex structure J of the ambient manifold, the Lagrangian submanifold is a flat space if and only if the Lagrangian submanifold has flat normal connection.

Let $CP_{n+p}(c)$ (resp D_{n+p}) be an $(n+p)$-dimensional complex projective space with the Fubini-Study metric (resp. Bergman metric) of constant holomorphic sectional curvature $c > 0$ (resp. < 0). Let C_{n+p} be an $(n+p)$-dimensional complex Euclidean space with the metric of constant holomorphic sectional curvature 0.

Let M be an n-dimensional Lagrangian minimal submanifold isometrically immersed in $CP_n(c)$. Let h be the second fundamental form of M in $CP_n(c)$. Recently, Montiel, Ros and Urbano [8] proved the following: Let M be an n-dimensional compact Lagrangian minimal submanifold isometrically immersed in $CP_n(c)$. Then the Ricci curvature S of M satisfies

$$S \geq \frac{3(n-2)}{16}c$$

if and only if one of the following conditions holds: a) $S = \frac{n-1}{4}c$ and M is totally geodesic, b) $S = 0, n = 2$ and M is a finite Riemannian covering of a flat torus minimally embedded in $CP_2(c)$ with parallel second fundamental form, c) $S = \frac{3(n-2)}{16}c, n > 2$ and M is an embedded submanifold congruent to the standard embedding of: $SU(3)/SO(3), n = 5; SU(6)/Sp(3), n = 14; SU(3), n = 8;$ or $E_6/F_4, n = 26$.
The author generalized the above result (See [4], [5], [6], [7], [12], [16] and [17]): Let M be an n-dimensional compact Lagrangian minimal submanifold isometrically immersed in $CP_n(c)$. Under the assumption of the scalar curvature ρ satisfying

$$\rho \geq \frac{3n(n-2)}{16}c$$

it also remains true, i.e., the scalar curvature ρ of M satisfies the inequality if and only if one of the following conditions holds: a) $\rho = \frac{n(n-1)}{4}c$ and M is totally geodesic, b) $\rho = 0, n = 2$ and M is a finite Riemannian covering of a flat torus minimally embedded in $CP_2(c)$ with parallel second fundamental form, c) $\rho = \frac{3n(n-2)}{16}c, n > 2$ and M is an embedded submanifold congruent to the standard embedding of: $SU(3)/SO(3), n = 5; SU(6)/Sp(3), n = 14; SU(3), n = 8;$ or $E_6/F_4, n = 26$.

In the present paper we first would like to consider that M_n is Einstein.

Problem Is Einstein Lagrangian submanifold of a complex projective space parallel?

Next, let $M_{n+p}(c)$ be an $(n+p)$-dimensional complex space form with constant holomorphic sectional curvature c. We remark that if $c > 0$ (resp. $c < 0, c = 0$), then $M_{n+p}(c) = CP_{n+p}(c)$ (resp. D_{n+p}, C_{n+p}). Let M_n be an n-dimensional complex Kaehler submanifold of $M_{n+p}(c)$. There are a number of conjecture for Kaehler submanifolds in $CP_{n+p}(c)$ suggested by K. Ogiue ([11]); some have been resolved under a suitable topological restriction (e.g. M_n is complete) (cf. [3], [11], [14] and [15]). In this direction, one of the open problems so far is as follows.

Conjecture (K. Ogiue [11]). Let M be an n-dimensional Kaehler submanifold immersed in $M_{n+p}(c), c > 0$. If M is irreducible (or Einstein) and if the second fundamental form is parallel, is M one of the following: $M_n(c), M_n(\frac{c}{2})$ or locally the complex quadric $Q_n(c)$?

In the case that M_n is an Einstein Kaehler submanifold of the codimension two immersed in $CP_{n+2}(c)$, it was proved in [3] and [14] that such a submanifold M_n is totally geodesic in $CP_{n+2}(c)$ or the complex quadric $Q_n(c)$ in the totally geodesic hypersurface $CP_{n+1}(c)$ of $CP_{n+2}(c)$. Moreover, if M_n is an Einstein Kaehler submanifolds immersed in a complex linear or hyperbolic space, then M_n is totally geodesic ([15]). The main results are the following:

Theorem 1 *Let M^n be an n-dimensional complete Einstein Lagrangian minimal submanifold immersed in $CP_n(c)$. Then M^n is parallel, isotropic and M is one of the following conditions holds: a) M is totally geodesic, b) $n = 2$ and M is a finite Riemannian covering of a torus minimally embedded in $CP_2(c)$ with parallel second fundamental form, c) $n > 2$ and M^n is an embedded submanifold congruent to the standard embedding of: $SU(3)/SO(3), n = 5; SU(6)/Sp(3), n = 14; SU(3), n = 8$; or $E_6/F_4, n = 26$.*

Theorem 2 *Let M_n be an n-dimensional compact Einstein Kaehler submanifold immersed in $CP_{n+p}(c)$. If $p \leq \dfrac{n(n+1)}{2}$, then M_n is parallel and totally geodesic in $CP_{n+p}(c)$, $CP_n(\frac{c}{2})$, the complex quadric $Q_n(c)$ in the totally geodesic submanifold $CP_{n+1}(c)$ of $CP_{n+p}(c)$, $SU(\frac{n}{2} + 2)/SU(\frac{n}{2}) \times U(2), n > 6, p = \frac{n(n-6)}{8}$; $SU(10)/U(5), n = 10, p = 5$ or $E_6/Spin(10) \times S^1, n = 16, p = 10$.*

Remark In the case of which M is complex space form there is a the result as the following: ([2], [11]) Let M^n be a Kaehler submanifold with constant holomorphic sectional curvature in a complex space form $M_{n+p}(c)$. If $p \leq \dfrac{n(n+1)}{2}$, then M_n is parallel.

2. PRELIMINARIES

Let M^n be an n-dimensional Riemannian manifold. We denote by T_xM the tangent space at $x \in M$.

Now, we suppose that M^n is an n-dimensional minimal submanifold immersed in an $n + p$-dimensional Riemannian manifold $\overline{M}$. We denote by $<,>$ the metric of $\overline{M}$ as well as that induced on M. Let ∇ and h is the Riemannian connection and the second fundamental form of the immersion, respectively. A_ξ and $\nabla^\perp$ are the Weingarten endomorphism associated to a normal vector ξ and the normal connection. The first and the second covariant derivations of the normal valued h are given by

$$(\nabla h)(X,Y,Z) = \nabla_X^\perp(h(Y,Z)) - h(\nabla_X Y, Z) - h(Y, \nabla_X Z)$$

and

$$\begin{aligned}(\nabla^2 h)(X,Y,Z,W) &= \nabla_X^\perp((\nabla h)(Y,Z,W)) - (\nabla h)(\nabla_X Y, Z, W) \\ &\quad - (\nabla h)(Y, \nabla_X Z, W) - (\nabla h)(Y, Z, \nabla_X W),\end{aligned}$$

respectively, for any vector fields X, Y, Z and W tangent to M.

If S and ρ is the Ricci tensor of M and the scalar curvature of M, respectively, and M is minimally immersed in $\overline{M}$, then from the Gauss equation we have

$$S(v,w) = \sum_{i=1}^{n} \overline{R}(v,e_i,e_i,w) - \sum_{i=1}^{n} < A_{h(v,e_i)}e_i, w >, \tag{1}$$

$$\rho = \sum_{i,j=1}^{n} \overline{R}(e_j,e_i,e_i,e_j) - |h|^2, \tag{2}$$

where $\overline{R}$ is the curvature operator of $\overline{M}$.

Let M be an n-dimensional compact Riemannian manifold. We denote by UM the unit tangent bundle over M and by UM_x its fibre over $x \in M$. If dx, dv and dv_x denote the canonical measures on M, UM and UM_x, respectively, then for any continuous function $f : UM \to R$, we have:

$$\int_{UM} f dv = \int_M \{ \int_{UM_x} f dv_x \} dx.$$

If T is a k-covariant tensor on M and ∇T is its covariant derivative, then we have:

$$\int_{UM} \{ \sum_{i=1}^{n} (\nabla T)(e_i, e_i, v, \cdots, v) \} dv = 0,$$

where $e_1, \cdots, e_n$ is an orthonormal basis of $T_xM, x \in M$.

3. PROOF OF THEOREM 1

Since M is Einstein, we have

$$S = \frac{n-1}{4} cI - \sum_{i=1}^{n} A_{h(v,e_i)} e_i = \frac{\rho}{n} I, \tag{3}$$

where I denotes the identity transformation. From (1) and (2) the equation (13) yields

$$\sum_i A_{h(v,e_i)} e_i = \frac{|h|^2}{n} v. \tag{4}$$

Using the fact that $< h(u,v), Jw >=< h(w,v), Ju >$ for $u, v, w \in T_xM$, we see that

$$\sum_j A_{Je_j}^2 = \frac{|h|^2}{n} I. \tag{5}$$

This is equivalent to

$$\text{trace} A_{Je_i} A_{Je_j} = \frac{|h|^2}{n} \delta_{ij}. \tag{6}$$

Since M is Einstein, we see that $|h|^2 = \text{constant}$. From (4) and (5) we get

$$\sum < (\nabla h)(x, v, e_i), h(v, e_i) > = 0 \tag{7}$$

for any $x \in T_M$. Differentiating (6) by w, we obtain

$$\text{trace}(\nabla_v^* A_{Jv}) A_{Jw} + \text{trace} A_{Jv} \nabla_v^* A_{Jw} = 0, \tag{8}$$

where $\nabla_v^* A_{Jw} = \nabla_v A_{Jw} - \sum < \nabla^\perp Jw, Je_i > A_{Je_i}$. Combining (8) with (7), we have

$$\text{trace}(\nabla_v^* A_{Jv}) A_{Jw} = 0.$$

Noting that $(\nabla_v^* A_{Jv}) u = 0$ for $u \in T_x M$ which satisfies $A_{Jw} u = 0$ for any $w \in T_x M$, we get

$$(\nabla_v^* A_{Jv}) u = 0$$

for any $u \in T_x M$. Hence we see that M is parallel. By the results of Naitoh [9] we obtain the conclusion.

4. PROOF OF THEOREM 2

Let M_n is an n-dimensional compact Kaehler submanifold of complex dimension n, immersed in the complex projective space $CP_{n+p}(c)$. We denote by J and $<,>$ the complex structure and the Fubini-Study metric. Let ∇ and h be the Riemannian connection and the second fundamental form of the immersion, respectively.

Then we have the relations

$$h(JX, Y) = Jh(X, Y) \text{ and } A_{J\xi} = JA_\xi = -A_\xi J,$$

where ξ is a normal vector to M_n.

Now, let $v \in UM_x, x \in M$. If $e_2, Je_2, \ldots, e_n, Je_n$ are orthonormal vectors in UM_x orthogonal to v, then we can consider $\{e_2, Je_2, \ldots, e_n, Je_n\}$ as an orthonormal basis of $T_v(UM_x)$. We remark that $\{v = e_1, Je_1, e_2, Je_2 \ldots, e_n, Je_n\}$ is an orthonormal basis of $T_x M$. We denote the Laplacian of $UM_x \cong S^{2n-1}$ by Δ.

Since M is a complex Kaehler submanifold of $CP_{n+p}(c)$, then from the Gauss equation we get

$$S(v, w) = \frac{n+1}{2} c < v, w > - \sum_{i=1}^{2n} < A_{h(v,e_i)} e_i, w >, \tag{9}$$

$$\rho = n(n+1)c - |h|^2. \tag{10}$$

Also, the sectional curvature K of M determined by orthonormal vectors X and Y is given by

$$K(X, Y) = \frac{c}{4}\{1 + 3g(X, JY)^2\} + < h(X, X), h(Y, Y) > - < h(X, Y), h(X, Y) > .$$

In particular, the holomorphic sectional curvature H of M determined by a unit vector X is given by

$$H(X) = c - 2|h(X,X)|^2. \tag{11}$$

Define a function f on $UM_x, x \in M$, by

$$f(v) = < (\nabla h)(v,v,v), h(v,v) > .$$

Using the minimality of M we can prove that

$$
\begin{aligned}
(\Delta f)(v) &= \sum_{i=1}^{2n} \nabla_i \nabla_i f(v) \\
&= -5(2n+3)f(v) \\
&\quad +12 \sum_{i=1}^{2n} < (\nabla h)(v,e_i), h(e_i,v) > .
\end{aligned} \tag{12}
$$

From the assumption of Theorem 2 we have

$$S = \frac{n+1}{2} cI - \sum_{i=1}^{2n} A_{h(v,e_i)} e_i = \frac{\rho}{2n} I, \tag{13}$$

where I denotes the identity transformation. Combining (12) with (13), we have

$$(\Delta f)(v) = -5(2n+3)f(v).$$

If $f(v) = < (\nabla h)(v,v,v), h(v,v) > \geq 0$ (resp. ≤ 0), then $(\Delta f)(v) \leq 0$ (resp. ≥ 0). Since M is compact, we have

$$(\nabla f)(v) = 0,$$

i.e.,

$$f(v) = 0.$$

So, if e_i is a unit vector at x orthogonal to v, we have using the symmetry of ∇h

$$0 = 3 < (\nabla h)(e_i,v,v), h(v,v) > +2 < (\nabla h)(v,v,v), h(v,e_i) >= 0.$$

Interchanging Jv and v, we get

$$< (\nabla h)(v,v,v), h(v,e_i) > = 0.$$

By the same argument we obtain

$$< (\nabla h)(v,v,v), h(e_i,e_j) >= 0.$$

From the assumption of $\dfrac{n(n+1)}{2}$ M is parallel (See [2], [11] and [13]). By the result of Nakagawa and Takagi [10] we get the conclusion.

ACKNOWLEDGMENTS

The author wishes to thank Prof. Dr. Kenan Tas for invitation to visit Cankaya University, for providing nice working conditions and for his helpful comments. The author also wishes to thank Prof. Dr. Dumitru Baleanu, Cankaya University, for his helpful comments and friendship.

REFERENCES

1. Chen, B. Y and Ogiue, K, *On totally real submanifolds*, Trans. Amer. Math. Soc., **193**,(1974), 257-266.
2. Maeda, S, *Isotropic immersions*, Can. J. Math., **6**, (1983), 416-430.
3. Matsuyama, Y, *On a 2-dimensional Einstein Kaehler submanifolds of a complex space form*, Proc. Amer. Math. Soc., **95**, (1985), 595-603.
4. Matsuyama, Y, *On some pinchings of minimal submanifolds*, Proceedings of the Workshop on Geometry and its applications in Honor of Morio Obata, November 1991, edited by Tadashi Nagano et al., World Scieentific, Singapore, (1993), 121-134.
5. Matsuyama, Y, *On curvature pinching for totally real submanifolds of $CP^n(c)$*, J. Ramanujan Math. Soc., **9**, (1994), 13-24.
6. Matsuyama, Y", *Curvature pinching for totally real submanifolds of a complex projective space*, J. Math. Soc. Japan, **52**, (2000), 51-62.
7. Matsuyama, Y, *On totally real submanifolds of a complex projective space*, Nihonkai Math. J., **13**, (2002), 153-157.
8. Montiel, S, Ros, A and Urbano, F, *Curvature pinching and eigenvalue rigidity for minimal submanifolds*, Math. Z., **191**, (1986), 537-548.
9. Naitoh, H, *Totally real parallel submanifolds of $P^n(C)$*, Tokyo J. Math., **4**, (1981), 279-306.
10. Nakagawa, H and Takagi, R, *On locally symmetric Kaehler submanifolds in a complex projective space*, J. Math. Soc. Japan, **28**, (1976), 638-667.
11. Ogiue, K, *Differential geometry of Kaehler submanifolds*, Adv. Math., **13**, (1974), 73-114.
12. Ogiue, K, *Recent topics of differential geometry*, Mathematics: the puplication of the Mathematical Society of Japan, **39**, (1987), 305-319.
13. O'Neill, B, *Isotropic and Kaehler immersions*, Can. J. Math., (1965), 907-915.
14. Tsukada, K, *Parallel Kaehler submanifolds of Hermitian symmetric spaces*, Math. Z., **190**, (1985), 129-150.
15. Umehara, M, *Kaehler submanifolds of complex space forms*, Tokyo J. Math., **10**, (1987), 203-214.
16. Xia, C, *Minimal submanifolds with bounded second fundamental form*, Math. Z., **208**, (1991), 537-543.
17. Xia, C, *On the minimal submanifolds in $CP^m(c)$ and $S^N(1)$*, Kodai Math. J., **15**, (1992), 143-153.

Generalized connections on affine bundles

Tom Mestdag

Department of Mathematical Physics and Astronomy, Ghent University
Krijgslaan 281 - S9, B-9000 Gent, Belgium
email: Tom.Mestdag@UGent.be

Abstract. Connections are among the most important tools to study qualitative features of dynamical systems. In this paper, we discuss generalized connections on affine bundles and show how they appear in the context of Lagrangian systems on affine Lie algebroids.
Key words. Affine bundle, Lie algebroid, Lagrangian system, connection.
PACS. 02.40.-k.
MSC. 37J99, 53C05, 53C15, 70H03.

1. INTRODUCTION

Let $\pi : E \to M$ be an affine bundle, modelled on a vector bundle $\overline{\pi} : \overline{E} \to M$. Coordinates on M will be denoted by (x^I); adapted coordinates on E by (x^I, y^α). Suppose that also an affine map $\rho : E \to TM$ (with corresponding linear map $\overline{\rho} : \overline{E} \to TM$) is given. We will refer to ρ as the anchor map and use ρ_α^I and ρ_0^I for its coordinate functions. This paper concerns the geometric study of dynamical systems on E of the form

$$
\begin{aligned}
\dot{x}^I &= \rho_\alpha^I(x)y^\alpha + \rho_0^I(x), \\
\dot{y}^\alpha &= f^\alpha(x,y).
\end{aligned}
\tag{1}
$$

Lagrangian systems on so-called 'affine' Lie algebroids motivate our interest in such systems, as we will explain in the next section (see also [6, 7]). Dynamical systems of the form (1) are called 'pseudo-SODEs', where SODE stands for 'second-order ordinary differential equations'. In the special case where the base manifold M is fibred over $\mathbf{R}$, $E = J^1M$ is its first jet extension $((x^I, y^\alpha) = (t, x^i, \dot{x}^i))$ and ρ is the canonical injection $J^1M \to TM$, pseudo-SODEs become indeed a system of second-order differential equations on M

$$
\ddot{x}^i = f^i(t,x,\dot{x}).
\tag{2}
$$

In the following we will often refer to this situation as the 'time-dependent SODE-case'. Two important tools are extensively used in the geometrical analysis of (time-dependent) SODEs (see [2, 4, 5]):

1. Every SODE gives rise to a *non-linear connection* on $\pi_M : J^1M \to M$.

CP729, *Global Analysis and Applied Mathematics: International Workshop on Global Analysis,*
edited by K. Taş, D. Krupka, O. Krupková, and D. Baleanu

2. This non-linear connection can be linearized to a linear connection, the so-called *Berwald-type connection.*

This contribution is centred around the following question: *Can we generalize the above mentioned tools in such a way that they can be used to study similar questions for the dynamical systems (1)?* In the time-dependent SODE-case ($E = J^1M$ and ρ the canonical injection) the newly found tools should of course correspond to the known situation.

In the next section we discuss pseudo-SODEs of Lagrangian type on affine Lie algebroids. In Section 3 we relate to any pseudo-SODE a section of a prolonged Lie algebroid and we generalize the notion of a connection in such a way that it becomes compatible with the affine anchor map ρ. We further show how any pseudo-SODE on an affine Lie algebroid can generate such a generalized connection. In the last section we discuss affine generalized connections and, finally, we define generalized Berwald-type connections. This article is an overview of joint work with W. Sarlet and E. Martínez. Proofs and detailed calculations have been omitted, they can be found in [9, 11, 13, 14, 15] and the references therein.

2. GENERALIZATIONS OF LAGRANGIAN SYSTEMS

We will first investigate the special subclass of pseudo-SODEs of the form

$$\dot{x}^J = \rho_0^J + \rho_\alpha^J y^\alpha,$$

$$\frac{d}{dt}\left(\frac{\partial L}{\partial y^\alpha}\right) = \rho_\alpha^J \frac{\partial L}{\partial x^J} + (C_{0\alpha}^\beta + C_{\gamma\alpha}^\beta y^\gamma)\frac{\partial L}{\partial y^\beta}, \tag{3}$$

where $L \in C^\infty(E)$ and the matrix $\left(\dfrac{\partial^2 L}{\partial y^\alpha y^\beta}\right)$ is supposed to have maximal rank. Obviously, in the time-dependent SODE-case, (3) corresponds to the equations of Lagrangian mechanics. Also here, there is a variational problem at work. Consider curves $\gamma : [t_0, t_1] \to E, t \mapsto (x(t), y(t))$ whose projection on M have fixed endpoints and which satisfy the constraint $\dot{x}^J = \rho_0^J + \rho_\alpha^J y^\alpha$ (i.e. 'admissible' curves). Making use of admissible variations, the equations (3) can be derived from extremizing the functional $J(\gamma) = \int_{t_0}^{t_1} L(\gamma(t))dt$, provided the functions ρ_α^J, ρ_0^J, $C_{\gamma\alpha}^\beta$ and $C_{0\alpha}^\beta$ satisfy the following relations

$$\rho_\alpha^J \frac{\partial \rho_\beta^J}{\partial x^J} - \rho_\beta^J \frac{\partial \rho_\alpha^J}{\partial x^J} = C_{\alpha\beta}^\gamma \rho_\gamma^J \quad \text{and} \quad \rho_0^J \frac{\partial \rho_\beta^J}{\partial x^J} - \rho_\beta^J \frac{\partial \rho_0^J}{\partial x^J} = C_{0\beta}^\alpha \rho_\alpha^J. \tag{4}$$

These equations are closely related to some of the structure equations of a 'Lie algebroid'. A Lie algebroid is a vector bundle $\tau : V \to Q$, which comes equipped with a real Lie bracket on the set of its sections and a linear bundle map $\lambda : V \to TQ$ (and its extension $\lambda : \text{Sec}(\tau) \to \mathscr{X}(Q)$). Moreover, the bracket and the map λ should satisfy, for all $s, r \in \text{Sec}(\tau)$, $f \in C^\infty(Q)$, the Leibnitz rule $[s, fr] = f[s, r] + \lambda(s)(f)r$.

To understand better the structure (4), we have to extend the notion of a Lie algebroid to affine bundles. Although there are several ways to do so, we will mention here only one. Let E_m be a fibre of the affine bundle π. It is an affine space, modelled on a vector space $\overline{E}_m$. We will denote by $E_m^\dagger = \mathrm{Aff}(E_m, \mathbf{R})$ the space of affine functions from E_m to $\mathbf{R}$. This is a vector space, so we can look at its dual $\tilde{E}_m = (E_m^\dagger)^*$ which is again a vector space, called the bi-dual of E_m. The collection of all bi-duals gives rise to a vector bundle $\tilde{\pi} : \tilde{E} \to M$, called bi-dual bundle of π. For example, if $M \to \mathbf{R}$ and π is $\pi_M : J^1M \to M$, then $\tilde{\pi}$ is the tangent bundle $\tau_M : TM \to M$.

$\tilde{E}$ contains both a copy of E and $\overline{E}$. Indeed, there exists a canonical injection $\iota : E \to \tilde{E}$, given by $\iota_m(e)(\phi) = \phi(e)$ ($e \in E_m$, $\phi \in \mathrm{Aff}(E_m, \mathbf{R})$). ι is an affine map and we will use $\overline{\iota} : \overline{E} \to \tilde{E}$ for its underlying linear map. The above observations lead to a decomposition of elements in $\tilde{E}_m$: For a fixed $e_0 \in E_m$ and for an arbitrary $\tilde{e} \in \tilde{E}_m$, there exist $r \in \mathbf{R}$ and $\overline{e} \in \overline{E}_m$ such that $\tilde{e} = r\iota_m(e_0) + \overline{\iota}_m(\overline{e})$. This property can be used to extend the maps ρ and $\overline{\rho}$ to a linear map $\tilde{\rho} : \tilde{E} \to TM$: $\tilde{\rho}_m(r\iota(a_0) + \overline{\iota}(\overline{a})) = r\rho_m(a_0) + \overline{\rho}_m(\overline{a})$. If $(o, \{\overline{e}_\alpha\})$ is a frame for $\mathrm{Sec}(\pi)$, then $\{e_0 = \iota(o), e_\alpha = \overline{\iota}(\overline{e}_\alpha)\}$ is a basis for $\mathrm{Sec}(\tilde{\pi})$. In coordinates, $\tilde{\rho}$ is the map $(x^i, y^0, y^\alpha) \mapsto \left(\rho_0^i(x)y^0 + \rho_\alpha^i(x)y^\alpha\right)\dfrac{\partial}{\partial x^i}$.

There is no need to define structures *directly* on the affine bundle π, since we can conveniently make use of the vector bundle structure of $\tilde{\pi}$.

Definition 1 *An affine Lie algebroid on π is a Lie algebroid on the vector bundle $\tilde{\pi} : \tilde{E} \to M$ with anchor map $\tilde{\rho} : \tilde{E} \to TM$ and with a bracket such that, $\forall \zeta, \eta \in \mathrm{Sec}(\pi)$, $[\iota(\zeta), \iota(\eta)] \in \mathrm{Im}\,\overline{\iota}$.*

Locally, the bracket must be of the form

$$[e_0, e_0] = 0 \qquad [e_0, e_\beta] = C_{0\beta}^\gamma e_\gamma \qquad [e_\alpha, e_\beta] = C_{\alpha\beta}^\gamma e_\gamma.$$

$\{e^0, e^\alpha\}$ will denote the basis, dual to $\{e_0, e_\alpha\}$. It is easy to see that e^0 (given by $e^0(\zeta^0 e_0 + \zeta^\alpha e_\alpha) = \zeta^0$) is in fact a globally defined 1-form on $\mathrm{Sec}(\tilde{\pi})$. On any Lie algebroid, one can define an exterior derivative (see [8]). It is completely determined by its action on functions and 1-forms. For an affine Lie algebroid, one finds that

$$dx^i = \rho_0^i e^0 + \rho_\alpha^i e^\alpha, \qquad de_0 = 0 \qquad \text{and} \qquad de^\alpha = -C_{0\beta}^\alpha e^0 \wedge e^\beta - \frac{1}{2}C_{\beta\gamma}^\alpha e^\beta \wedge e^\gamma.$$

Proposition 1 *A Lie algebroid on $\tilde{\pi}$ with anchor $\tilde{\rho}$ is affine if and only if $de^0 = 0$.*

3. PSEUDO-SODES AND GENERALIZED CONNECTIONS

We will show next that a Lie algebroid can be 'lifted' to a Lie algebroid structure on a 'prolonged' bundle $\pi^1 : T^{\tilde{\rho}}E \to E$ and that it is more convenient to look at (not necessarily Lagrangian) pseudo-SODEs as sections of this bundle, rather than as vector fields on E. As announced in the introduction we will generalize the concept of a 'connection' to the current set-up.

The bundle π^1 has been visualized in the diagram. It is a vector bundle whose total manifold is the pullback bundle $\tilde{\rho}^*TE$ (i.e. $(\tilde{e}, X_e)$ is an element of $T^{\tilde{\rho}}E$ if $\tilde{\rho}(\tilde{e}) = T\pi(X_e)$), but whose bundle projection is given by $\pi^1(\tilde{e}, X_e) = \tau_E(X_e) = e$. Further, π^2

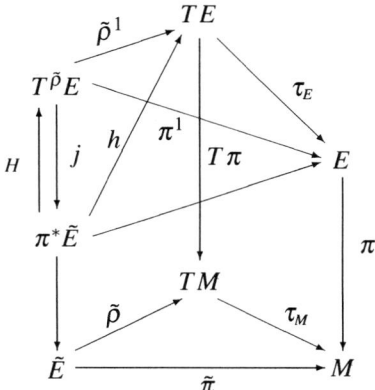

FIGURE 1. The prolonged bundle π^1.

maps the element $(\tilde{e}, X_e) \in T^{\tilde{\rho}}E$ onto $\tilde{e} \in \tilde{E}$, while $\tilde{\rho}^1$ projects the same element of $T^{\tilde{\rho}}E$ onto X_e. Finally, there is also a canonical map $j: T^{\tilde{\rho}}E \to \pi^*\tilde{E}, (\tilde{e}, X_e) \mapsto (e, \tilde{e})$. A frame $(o, \{\bar{e}_\alpha\})$ for $\text{Sec}(\pi)$ induces a basis $\{\mathcal{X}_0, \mathcal{X}_\alpha, \mathcal{V}_\alpha\}$ for $\text{Sec}(\pi^1)$, where (with $A = 0, \alpha$)

$$\mathcal{X}_A(e) = \left(e_A(\pi(e)), \rho'_A(x) \frac{\partial}{\partial x^l} \Big|_e \right) \qquad \text{and} \qquad \mathcal{V}_\alpha(e) = \left(o(\pi(e)), \frac{\partial}{\partial y^\alpha} \Big|_e \right).$$

Remark that $\tilde{\rho}^1(\zeta^0 \mathcal{X}_0 + \zeta^\alpha \mathcal{X}_\alpha + Z^\alpha \mathcal{V}_\alpha) = (\rho'_0 \zeta^0 + \rho'_\alpha \zeta^\alpha) \frac{\partial}{\partial x^l} + Z^\alpha \frac{\partial}{\partial y^\alpha}$.

It is well-known that a connection on a vector bundle $\tau: V \to M$ selects a (horizontal) distribution which is complementary to the set of vertical elements in TV. For the bundle π^1 too, we can identify a set of 'vertical' vectors: they are those elements of $T^{\tilde{\rho}}E$ whose projection on $\tilde{E}$ (by means of π^2) vanishes. The set of all such verticals is denoted by $Ver(\pi^1)$. A section $\mathcal{Z} \in \text{Sec}(\pi^1)$ is vertical if its image lies in $Ver(\pi^1)$, or, locally, if it lies in the span of the sections $\mathcal{V}_\alpha$. The required extension of the concept 'connection' will generate a direct complement of $Ver(\pi^1)$ within $T^{\tilde{\rho}}E$. As usual, the definition can be cast in terms of a horizontal lift.

Definition 2 *A (generalized) $\tilde{\rho}$-connection on π is a map $^H: \pi^*\tilde{E} \to T^{\tilde{\rho}}E$ for which $j \circ {}^H = id_{\pi^*\tilde{E}}$ holds. Then $T^{\tilde{\rho}}E = Im^H \oplus Ver(\pi^1)$.*

In what follows we will usually refer to a $\tilde{\rho}$-connection in terms of its associated map $h := \tilde{\rho}^1 \circ {}^H : \pi^*\tilde{E} \to TE$. Locally, a generalized connection is completely determined by a set of connection coefficients $\Gamma_0^\beta, \Gamma_\alpha^\beta \in C^\infty(E)$. If $\tilde{X} = X^0 e_0 + X^\alpha e_\alpha \in \text{Sec}(\pi^*\tilde{\pi})$, then $\tilde{X}^H = X^0 \mathcal{H}_0 + X^\alpha \mathcal{H}_\alpha \in \text{Sec}(\pi^1)$ with $\mathcal{H}_0 = e_0{}^H = \mathcal{X}_0 - \Gamma_0^\beta \mathcal{V}_\beta$ and $\mathcal{H}_\alpha = e_\alpha{}^H = \mathcal{X}_\alpha - \Gamma_\alpha^\beta \mathcal{V}_\beta$.

Let us now come back to the dynamical systems (1). To any pseudo-SODE we can associate a section Γ of π^1 with the property that $\pi^2 \circ \Gamma = \iota$. Indeed, such sections are locally of the form $\Gamma = \mathcal{X}_0 + y^\alpha \mathcal{X}_\alpha + f^\alpha \mathcal{V}_\alpha$ and can thus represent pseudo-SODEs in this framework. Remark that $\tilde{\rho}^1(\Gamma)$ is then exactly the vector field whose integral curves are solutions of (1).

Proposition 2 *Every pseudo-*SODE $\Gamma = \mathscr{X}_0 + y^\alpha \mathscr{X}_\alpha + f^\alpha \mathscr{V}_\alpha$ *on a affine Lie algebroid gives rise to a $\tilde{\rho}$-connection with connection coefficients*

$$\Gamma_\beta^\alpha = -\frac{1}{2}\left(\frac{\partial f^\alpha}{\partial y^\beta} + y^\gamma C_{\gamma\beta}^\alpha + C_{0\beta}^\alpha\right) \qquad and \qquad \Gamma_0^\alpha = -f^\alpha - y^\beta \Gamma_\beta^\alpha.$$

The above theorem generalizes the well-known non-linear connection of a SODE which has been mentioned in the introduction (see e.g. [3]). We can also give a more intrinsic formulation of this proposition. Let $\{\mathscr{X}^0, \mathscr{X}^\alpha, \mathscr{V}^\alpha\}$ be the basis of $\mathrm{Sec}((\pi^1)^*)$ which is dual to $\{\mathscr{X}_0, \mathscr{X}_\alpha, \mathscr{V}_\alpha\}$. Then, $\mathscr{X}^0$ is a global 1-form on $\mathrm{Sec}(\pi^1)$ and $S = (\mathscr{X}^\alpha - y^\alpha \mathscr{X}^0) \otimes \mathscr{V}_\alpha$, the vertical endomorphism, a globally defined vector-valued 1-form on $\mathrm{Sec}(\pi^1)$. Moreover, it is possible to prolong the Lie algebroid structure of π to the bundle π^1.

Proposition 3 *An (affine) Lie algebroid on $\tilde{\pi}$ with anchor $\tilde{\rho}$ can be extended to an (affine) Lie algebroid on π^1 with anchor $\tilde{\rho}^1$ and Lie bracket* $[\mathscr{X}_\alpha, \mathscr{X}_\beta] = C_{\alpha\beta}^\gamma \mathscr{X}_\gamma$, $[\mathscr{X}_0, \mathscr{X}_\beta] = C_{0\beta}^\gamma \mathscr{X}_\gamma$, $[\mathscr{V}_\alpha, \mathscr{X}_\beta] = 0$, $[\mathscr{X}_0, \mathscr{V}_\beta] = 0$ *and* $[\mathscr{V}_\alpha, \mathscr{V}_\beta] = 0$.

If d is the exterior derivative of this extended Lie algebroid (which acts on forms of $\mathrm{Sec}(\pi^1)$), then $d_\Gamma = [i_\Gamma, d]$ is a degree zero derivation, which extends to vector-valued forms in the usual way. The horizontal part of a $\mathscr{X} \in \mathrm{Sec}(\pi^1)$ for the connection in Prop. 2 is then given by $\frac{1}{2}\left(I - d_\Gamma S + \mathscr{X}^0 \otimes \Gamma\right)(\mathscr{X})$.

Remark that $\Gamma \in \mathrm{Sec}(\pi^1)$ is a pseudo-SODE if and only if $S(\Gamma) = 0$ and $\langle \Gamma, \mathscr{X}^0 \rangle = 1$. In particular, if d is again the exterior derivative of the lifted Lie algebroid, we can define for a function $L \in C^\infty(E)$ a 1-form $\theta_L = S(dL) + L\mathscr{X}^0$ on $\mathrm{Sec}(\pi^1)$. The pseudo-SODE, associated to the Lagrangian system (3), is then a solution of the equation $i_\Gamma d\theta_L = 0$.

4. AFFINE $\tilde{\rho}$-CONNECTIONS AND BERWALD-TYPE CONNECTIONS

A special subclass of $\tilde{\rho}$-connections on an affine bundle is formed by those for which the connection coefficients are affine functions, i.e. $\Gamma_0^\alpha(x,y) = \Gamma_{00}^\alpha(x) + \Gamma_{0\beta}^\alpha(x)y^\beta$ and $\Gamma_\mu^\alpha(x,y) = \Gamma_{\mu 0}^\alpha(x) + \Gamma_{\mu\beta}^\alpha(x)y^\beta$. Before we can characterize such connections, we need to say a few words about 'linear' generalized connections (see also [1]).

It is not difficult to introduce also 'generalized' connections for the vector bundle $\bar{\pi}$ on which π is modelled. Indeed, by replacing the affine bundle π with the vector bundle $\bar{\pi}$ in the diagram, we arrive at a new prolonged vector bundle $\bar{\pi}^1 : T^{\tilde{\rho}}\bar{E} \to E$. Using the corresponding $\bar{j} : T^{\tilde{\rho}}\bar{E} \to \bar{\pi}^*\tilde{E}$, it is easy to give a well-defined meaning to the concept of a $\tilde{\rho}$-connection $\bar{h} : \bar{\pi}^*\tilde{E} \to T\bar{E}$ on $\bar{\pi}$. When in addition the connection coefficients are linear functions, we will call the connection 'linear'. Putting $\bar{\Sigma}(\bar{e}_1, \bar{e}_2) = \bar{e}_1 + \bar{e}_2$, for $\bar{e}_1, \bar{e}_2 \in \bar{E}_m$, a more intrinsic formulation of this property is the following: a $\tilde{\rho}$-connection

$\bar{h}$ on $\bar{\pi}$ is linear if and only if

$$\bar{h}(\bar{e}_1 + \bar{e}_2, \tilde{e}) = T_{(\bar{e}_1, \bar{e}_2)} \bar{\Sigma} \left(\bar{h}(\bar{e}_1, \tilde{e}), \bar{h}(\bar{e}_2, \tilde{e}) \right),$$

It is easy to find an equivalent characterization in terms of a covariant derivative operator. One can check that the existence of a linear $\bar{h}$ is equivalent with the existence of an operator $\bar{\nabla} : \mathrm{Sec}(\tilde{\pi}) \times \mathrm{Sec}(\bar{\pi}) \to \mathrm{Sec}(\bar{\pi}) : (\tilde{\zeta}, \bar{\eta}) \mapsto \bar{\nabla}_{\tilde{\zeta}} \bar{\eta}$ which is $\mathbf{R}$-linear in both its arguments and satisfies

$$\bar{\nabla}_{f\tilde{\zeta}} \bar{\eta} = f \bar{\nabla}_{\tilde{\zeta}} \bar{\eta} \quad \text{and} \quad \bar{\nabla}_{\tilde{\zeta}}(f\bar{\eta}) = f \bar{\nabla}_{\tilde{\zeta}} \bar{\eta} + \tilde{\rho}(\tilde{\zeta})(f)\bar{\eta}, \quad (f \in C^{\infty}(M)).$$

Generalized connections on π with affine coefficients have properties which are similar to those of linear connections. Let $\Sigma(e, \bar{e}) = e + \bar{e}$, for $e \in E_m$ and $\bar{e} \in \bar{E}_m$.

Proposition 4 *A $\tilde{\rho}$-connection h on π is affine, if there exists a linear $\tilde{\rho}$-connection $\bar{h}$ on $\bar{\pi}$, which is related to h in such a way that*

$$h(e + \bar{e}, \tilde{e}) = T_{(e, \bar{e})} \Sigma \left(h(e, \tilde{e}), \bar{h}(\bar{e}, \tilde{e}) \right).$$

Equivalently, a $\tilde{\rho}$-connection on π is affine if there exists an operator $\nabla : \mathrm{Sec}(\tilde{\pi}) \times \mathrm{Sec}(\pi) \to \mathrm{Sec}(\bar{\pi}) : (\tilde{\zeta}, \eta) \mapsto \nabla_{\tilde{\zeta}} \eta$ which is $\mathbf{R}$-linear in its first argument, satisfies $\nabla_{f\tilde{\zeta}} \sigma = f \nabla_{\tilde{\zeta}} \sigma$ and is related to a covariant derivative operator $\bar{\nabla}$ of a linear $\tilde{\rho}$-connection on $\bar{\pi}$ by means of

$$\nabla_{\tilde{\zeta}}(\sigma + f\bar{\eta}) = \nabla_{\tilde{\zeta}} \sigma + f \bar{\nabla}_{\tilde{\zeta}} \bar{\eta} + \tilde{\rho}(\tilde{\zeta})(f)\bar{\eta}.$$

It is possible to give an explicit relation between the horizontal lift h and the operator ∇. First, we need to define the vertical lift: it is the map $v : \mathrm{Sec}(\pi^* \tilde{\pi}) \to \mathscr{X}(E)$ given by $v(X^0 e_0 + X^\alpha e_\alpha) = (X^\alpha - y^\alpha X^0)\frac{\partial}{\partial y^\alpha}$. There is also an inverse map for vertical vector fields: $\left(Y^\alpha \frac{\partial}{\partial y^\alpha} \right)_v = Y^\alpha e_\alpha \in \mathrm{Sec}(\pi^* \tilde{\pi})$. It is not difficult to see that the brackets $[h(\tilde{\zeta}), v(\bar{\sigma})]$ and $[h(\tilde{\zeta}), v(\sigma)]$ of vector fields on E are vertical. $\tilde{\zeta} \in \mathrm{Sec}(\tilde{\pi})$, $\bar{\sigma} \in \mathrm{Sec}(\bar{\pi})$ and $\sigma \in \mathrm{Sec}(\pi)$ have here been identified with 'basic' sections of $\mathrm{Sec}(\pi^* \tilde{\pi})$. A section $\tilde{X}$ of $\pi^* \tilde{\pi}$ is basic if it is of the form $e \mapsto (e, \tilde{\zeta})$ for a certain $\tilde{\zeta} \in \mathrm{Sec}(\tilde{\pi})$. If the connection is affine, then also $[h\tilde{\zeta}, v(\bar{\sigma})]_v$ and $[h\tilde{\zeta}, v(\sigma)]_v$ are basic sections and one can prove that

$$\bar{\nabla}_{\tilde{\zeta}} \bar{\sigma} = [h\tilde{\zeta}, v(\bar{\sigma})]_v \quad \text{and} \quad \nabla_{\tilde{\zeta}} \sigma = [h\tilde{\zeta}, v(\sigma)]_v. \tag{5}$$

The connection coefficients of the connection $\bar{h}$ in Prop. 4 can be found in the linear part of those of h, i.e. $\bar{\Gamma}_0^\alpha(x, w) = \Gamma_{0\beta}^\alpha(x)w^\beta$ and $\bar{\Gamma}_\mu^\alpha(x, w) = \Gamma_{\mu\beta}^\alpha(x)w^\beta$. An other associated connection is the one whose coefficients are $\tilde{\Gamma}_0^\alpha(x, y^0, y) = \Gamma_{00}^\alpha(x)y^0 + \Gamma_{0\beta}^\alpha(x)y^\beta$ and $\tilde{\Gamma}_\mu^\alpha(x, y^0, y) = \Gamma_{\mu 0}^\alpha(x)y^0 + \Gamma_{\mu\beta}^\alpha(x)y^\beta$. This linear $\tilde{\rho}$-connection on $\tilde{\pi}$ is related to h as follows:

Proposition 5 *h is affine, if there exists a linear $\tilde{\rho}$-connection $\tilde{h} : \tilde{\pi}^* \tilde{E} \to T\tilde{E}$ on $\tilde{\pi} : \tilde{E} \to M$ such that,*

$$\tilde{h} \circ \iota = T\iota \circ h.$$

There is, of course, also a corresponding derivative operator $\tilde{\nabla} : \mathrm{Sec}(\tilde{\pi}) \times \mathrm{Sec}(\tilde{\pi}) \rightarrow \mathrm{Sec}(\tilde{\pi})$.

Proposition 6 *A linear $\tilde{\rho}$-connection on $\tilde{\pi}$ is associated with an affine $\tilde{\rho}$-connection on π if and only if e^0 is parallel w.r.t. $\tilde{\nabla}$.*

We are now ready to define Berwald-type connections. In the time-dependent SODE-case, they are linearized versions of an originally non-linear connection on π_M. The price to pay is that the bundle on which they live is a little bit more complicated: usually Berwald-type connections are defined on the (pull-back) vector bundle $\pi_M^* \tau_M$ as a covariant derivative operator. However, it can already be noticed in the time-dependent case that this operator is in fact of the type $\tilde{\nabla}$ and thus associated to an affine connection ∇ on the affine bundle $\pi_M^* \tau_M$. In the general set-up, Berwald-type connections will therefore be affine $\tilde{\rho}^1$-connections on the affine bundle $\pi^* \pi : \pi^* E \rightarrow E$. We will explain below that the affine bundle does not need to have the structure of an affine Lie algebroid for this purpose.

We briefly explain two ways to define them. First, it is easy to see that an affine connection $D : \mathrm{Sec}(\pi^1) \times \mathrm{Sec}(\pi^* \pi) \rightarrow \mathrm{Sec}(\pi^* \pi)$ is completely determined by its action on horizontal and vertical lifts of *basic* sections in its first argument and its action on *basic* sections in its second argument. In the particular case that the $\tilde{\rho}$-connection one starts from is already affine, the Berwald-type connections should essentially reproduce a copy of themselves. Therefore, we can find inspiration in the relations (5) and put

$$D_{\tilde{\zeta}^H} \sigma = [h\tilde{\zeta}, v\sigma]_v, \quad \overline{D}_{\tilde{\zeta}^H} \overline{\sigma} = [h\tilde{\zeta}, v\overline{\sigma}]_v, \quad D_{\overline{\eta}^v} \sigma = \overline{D}_{\overline{\eta}^v} \overline{\sigma} = 0.$$

One can easily check that this definition is consistent with the module structure over $C^\infty(M)$ and that the above operator can be extended to arbitrary sections in the obvious way.

The second method uses the fact that the connection is completely determined by specifying the rule of parallel transport along two specific classes of admissible curves in $T^{\tilde{\rho}} E$. A detailed analysis in [13] revealed that, for the subclass of vertical curves, there are in fact two 'natural' ways to fix such a rule, leading thus to two different Berwald-type connections. We will not go deeper into this matter here; we will only give the defining relations for this second Berwald-type connection $\hat{D}$.

$$\hat{D}_{\tilde{\zeta}^H} \sigma = [h\tilde{\zeta}, v\sigma]_v, \quad \overline{\hat{D}}_{\tilde{\zeta}^H} \overline{\sigma} = [h\tilde{\zeta}, v\overline{\sigma}]_v, \quad \hat{D}_{\overline{\eta}^v} \sigma = -\overline{\eta}, \quad \overline{\hat{D}}_{\overline{\eta}^v} \overline{\sigma} = 0.$$

A similar behaviour was detected before in the time-dependent SODE-case: the first connection corresponds with the one in [4], while the second one is given in [12] and is basically the same one as in [10]. We now list the coordinate expression (for the associated operator $\tilde{D}$). Both connections share the expressions (with $A = 0, \alpha$),

$$\tilde{D}_{\mathcal{H}_A} e_0 = \left(\Gamma_A^\gamma - y^\beta \frac{\partial \Gamma_A^\gamma}{\partial y^\beta} \right) e_\gamma, \quad \tilde{D}_{\mathcal{H}_A} e_\beta = \frac{\partial \Gamma_A^\gamma}{\partial y^\beta} e_\gamma, \quad \tilde{D}_{\mathcal{V}_\alpha} e_\beta = 0.$$

while for the first $\tilde{D}_{\mathcal{V}_\alpha} e_0 = 0$ and for the second $\tilde{\hat{D}}_{\mathcal{V}_\alpha} e_0 = -e_\alpha$. Unlike in the time-dependent SODE-case, there are no direct defining formulas for the covariant derivatives.

Such formulas become available, however, when the affine bundle $\pi : E \to M$ has the structure of an affine Lie algebroid. Remark that it is clear from the coordinate expressions that such an additional structure is not required to define Berwald-type connections.

Some final comments: Berwald-type connections have been successfully applied in a number of applications concerning qualitative feature of SODEs. A few examples of such problems are mentioned below (see also [2, 4, 5]):

- Does a coordinate transformation $x^i = x^i(t, \bar{x})$ exist, such that the system (2) becomes linear (i.e. of the form $\ddot{\bar{x}}^j = A^j_k(t)\,\bar{x}^k + B^j_k(t)\,\dot{\bar{x}}^k + C^j(t)$)?
- Can we construct a Lagrangian for this system?
- What are the conditions for the existence of a coordinate transformation in which (2) completely decouples?

It is our belief that the above 'generalized' Berwald-type connections will prove to be equally useful to answer similar questions for pseudo-SODEs.

REFERENCES

1. F. Cantrijn and B. Langerock, Generalised connections over a bundle map, *Diff. Geom. and its Appl.* **18** (2003), 295–317.
2. F. Cantrijn, W. Sarlet, A. Vandecasteele and E. Martínez, Complete separability of time-dependent second-order ordinary differential equations, *Acta Appl. Math.* **42** (1996), 309–334.
3. M. Crampin, *Jet bundle techniques in analytical mechanics*, Quaderni del consiglio nazionale delle ricerche, gruppo nazionale di fisica matematica **47** (1995).
4. M. Crampin, E. Martínez and W. Sarlet, Linear connections for systems of second-order ordinary differential equations, *Ann. Inst. H. Poincaré, Phys. Théor.* **65** (1996), 223–249.
5. M. Crampin, W. Sarlet, E. Martínez, G.B. Byrnes and G.E. Prince, Towards a geometrical understanding of Douglas's solution of the inverse problem of the calculus of variations, *Inverse problems* **10** (1994), 245–260.
6. K. Grabowska, J. Grabowski and P. Urbański, Lie brackets on affine bundles, *Ann. Glob. Anal. Geom.* **24** (2003), 101–130.
7. K. Grabowska, J. Grabowski and P. Urbański, AV-differential geometry: Poisson and Jacobi structures, preprint 2004 (math.DG/0402435).
8. K. Mackenzie, *Lie groupoids and Lie algebroids in differential geometry*, London Math. Soc. Lect. Notes Series 124 (Cambridge Univ. Press, 1987).
9. E. Martínez, T. Mestdag and W. Sarlet, Lie algebroid structures and Lagrangian systems on affine bundles, *J. Geom. Phys.* **44** (2002), 70–95.
10. E. Massa and E. Pagani, Jet bundle geometry, dynamical connections, and the inverse problem of Lagrangian mechanics, *Ann. Inst. H. Poincaré Phys. Théor.* **61** (1994), 17–62.
11. T. Mestdag, Generalized connections on affine Lie algebroids, *Rep. on Math. Phys.* **51** (2003), 297–305.
12. T. Mestdag and W. Sarlet, The Berwald-type connection associated to time-dependent second-order differential equations, *Houston J. Math.*, **27** (2001), 763–797.
13. T. Mestdag and W. Sarlet, The Berwald-type linearization of generalized connections, *J. Phys. A* **36** (2003), 8049–8069.
14. T. Mestdag, W. Sarlet and E. Martínez, Note on generalized connections and affine bundles, *J. Phys. A* **35** (2002), 9843–9856.
15. W. Sarlet, T. Mestdag and E. Martínez, Lie algebroid structures on a class of affine bundles, *J. Math. Phys.* **43** (2002), 5654–5674.

On the path integral of constrained systems

Sami I. Muslih

Physics Department, Al-Azhar University, Gaza, Palestine

Abstract. Constrained Hamiltonian systems are investigated by using Güler's method. Integration of a set of equations of motion and the action function is discussed. It is shown that the canonical path integral quantization is obtained directly as an integration over the canonical phase-space coordinates without any need to enlarge the initial phase-space by introducing extra- unphysical variables as in the Batalin-Fradkin-Tyutin (BFT) method. The abelian Proca model is analyzed by the two methods.
Key words. Hamiltonian and Lagrangian approaches, Hamilton-Jacobi approach, Integrable systems, Field theories.
PACS. 11.10.Ef, 11.10.z, 02.30.Ik.
MSC. 70H20, 70H06, 81S40, 81T70.

1. INTRODUCTION

The generalized Hamiltonian dynamics describing systems with constraints is initiated by Dirac [6]. Dirac's theory of constrained systems plays an important role in theoretical physics, and is widely used in investigating theoretical models in a contemporary elementary particle physics [9]. In this method, the Poisson brackets in a second class constraint systems are converted into Dirac brackets to attain self- consistency. However, whenever we adopt the Dirac method, we frequently meet the problem of the operating ordering ambiguity. In order to avoid this problem, Batalin, Fradkin and Tyutin (BFT) have developed a method [4] by enlarging the phase space with some extra variables, such that the second -class constraints became converted into first- class ones.

Recently the canonical path integral quantization was initiated in [15] and it is based on the Güler's method [10] to investigate singular systems. In Güler's method the equations of motion are obtained as total differential equations in many variables. If the system is integrable then one can obtain the canonical reduced phase-space coordinates and the path integral as an integration over the canonical phase-space coordinates. The advantage of using the canonical path integral approach [15], is that we have no difference between first and second class constraints and we do not need gauge fixing term, because the gauge variables are separated in the process of constructing an integrable system of total differential equation. Besides, this approach has been employed successfully in different physical systems (see references [14-17]).

On the other hand, Güler and Baleanu [1] have used the Güler's and the chain *(gauge un − fixing)* [12] approaches, to obtain the involutive Hamiltonian without any

CP729, *Global Analysis and Applied Mathematics: International Workshop on Global Analysis,*
edited by K. Taş, D. Krupka, O. Krupková, and D. Baleanu
© 2004 American Institute of Physics 0-7354-0209-4/04/$22.00

need to introduce extra un-physical variables. Besides, Stückelberg [19] approach based on the Lagrangian formulation, leads us to convert second class theories into first class by extending the initial configuration space. A similar (Lagrangian) functional approach has been used by Faddeev and Shatashvili [7] by introducing Wess-Zumino scalars [20] to interpret anomalous gauge theories as true (i.e. first class) gauge systems. Recently there have been suggestions [8, 11] that the extra fields introduced in the Hamiltonian formalism may be identified with the Stückelberg scalars or the Wess-Zumino fields.

The aim of this paper is to compare the results obtained by using two different Hamiltonian formulations, firstly by using the Hamilton-Jacobi method and secondly by using the Batalin, Fradkin and Tyutin (BFT) method.

2. HAMILTON-JACOBI METHOD

In the Hamilton-Jacobi method if we start with a singular Lagrangian $L = L(q_i, \dot{q}_i, t)$, $i = 1, 2, ..., n$, with Hessian matrix of rank $(n - r)$, $r < n$, then the generalized momenta can be written as

$$p_a = \frac{\partial L}{\partial \dot{q}_a}, \quad a = 1, 2, ..., n - r, \tag{1}$$

$$p_\mu = \frac{\partial L}{\partial \dot{t}_\mu}, \quad \mu = n - r + 1, ..., n, \tag{2}$$

where q_i are divided into two sets, q_a and t_μ. Since the rank of the Hessian matrix is $(n - r)$, one can solve the expressible velocities from (1) and after substituting in (2), we get

$$p_\mu = -H_\mu(q_i, \dot{t}_\mu, p_a; t). \tag{3}$$

The canonical Hamiltonian H_0 reads

$$H_0 = p_a \dot{q}_a + p_\mu \dot{t}_\mu|_{p_\nu = -H_\nu} - L(t, q_i, \dot{t}_\nu, \dot{q}_a), \quad \mu, \nu = n - r + 1, ..., n. \tag{4}$$

The set of Hamilton-Jacobi partial differential equations [HJPDE] is expressed as [10]

$$H'_\alpha \left(t_\beta, q_a, \frac{\partial S}{\partial q_a}, \frac{\partial S}{\partial t_\alpha} \right) = p_\alpha + H_\alpha = 0, \quad \alpha, \beta = 0, n - r + 1, ..., n, \tag{5}$$

where $p_\beta = \partial S[q_a; t_\alpha]/\partial t_\beta$ and $p_a = \partial S[q_a; t_\alpha]/\partial q_a$ with $t_0 = t$ and S being the action. The equations of motion are obtained as total differential equations in many variables as follows [10]:

$$dq_a = \frac{\partial H'_\alpha}{\partial p_a} dt_\alpha, \, dp_a = -\frac{\partial H'_\alpha}{\partial q_a} dt_\alpha, \, dp_\beta = -\frac{\partial H'_\alpha}{\partial t_\beta} dt_\alpha. \tag{6}$$

$$dz = \left(-H_\alpha + p_a \frac{\partial H'_\alpha}{\partial p_a} \right) dt_\alpha; \tag{7}$$

$$\alpha, \beta = 0, n - r + 1, ..., n, a = 1, ..., n - r$$

241

where $z = S(t_\alpha; q_a)$.

As was clarified, that the equations (6,7) are obtained as total differential equations in many variables, which require the investigation of integrability conditions. To achieve this goal we define the linear operators X_α which correspond to the total differential equations (6,7) as [13, 5]

$$
\begin{aligned}
X_\alpha f(t_\beta, q_a, p_a, z) \quad &= \frac{\partial f}{\partial t_\alpha} + \frac{\partial H'_\alpha}{\partial p_a}\frac{\partial f}{\partial q_a} - \frac{\partial H'_\alpha}{\partial q_a}\frac{\partial f}{\partial p_a} \\
&+(-H_\alpha + p_a\frac{\partial H'_\alpha}{\partial p_a})\frac{\partial f}{\partial z}, \\
&= [H'_\alpha, f] - \frac{\partial f}{\partial z}H'_\alpha, \\
&\alpha, \beta = 0, n-r+1, ..., n, a = 1, ..., n-r,
\end{aligned}
\tag{8}
$$

where the commutator $[,]$ is the square bracket (for details, see the appendix).

Lemma. A system of total differential equations (6,7) is integrable if and only if [13, 18]

$$
\{H'_\alpha, H'_\beta\} = 0, \quad \forall \, \alpha, \, \beta,
\tag{9}
$$

where the commutator $\{,\}$ is the Poisson bracket. If the set of equations (6, 7) is integrable, then one can obtain the canonical action function (7) in terms of the canonical coordinates. In this case, the path integral representation may be written as [15]

$$
K(q'_a, t'_\alpha; q_a, t_\alpha) = \int_{q_a}^{q'_a} Dq^a \, Dp^a \times
$$

$$
\exp i\left\{ \int_{t_\alpha}^{t'_\alpha} \left[-H_\alpha + p_a\frac{\partial H'_\alpha}{\partial p_a}\right] dt_\alpha \right\},
$$

$$
a = 1, ..., n-r, \quad \alpha = 0, n-r+1, ..., n.
\tag{10}
$$

3. BATALIN-FRANDKIN-TYUTIN FORMALISM

In this section, we shall briefly review the abelian conversion of a second class system into a first class as developed by Batalin and Tyutin [4, 3].

Let us assume that the canonical variables, (q_i, p_i), $i = 1, ..., n$ and the Grassman parity $\varepsilon(q_i) = \varepsilon(p_i) = \varepsilon_i$ define the initial phase space of a dynamical system. It is supposed that the system possesses only second class constraints. Denoting them by T_a, with $a = 1, ..., m < 2n$, so that the matrix

$$
\{T_a, T_b\} = \Delta_{ab},
\tag{11}
$$

has a non-vanishing determinant.

We now convert the second class system into a first class system by introducing for each second class constraint auxiliary variables ξ^a which satisfy

$$\{\xi^a, \xi^b\} = \omega^{ab}, \tag{12}$$

where ω^{ab} is a constant non-degenerate matrix (i.e., $\det(\omega^{ab}) \neq 0$). The obtained first class constraints are denoted as $\tilde{T}_a$

$$\tilde{T}_a = \tilde{T}_a(q, p; \xi), \tag{13}$$

and are supposed to satisfy the boundary condition

$$\tilde{T}_a(q, p; 0) = \tilde{T}_a(q, p), \quad T_a^{(0)} = T_a. \tag{14}$$

Then the abelian conversion of Batalin-Frandkin-Tyutin formalism implies that these constraints are strongly involutive,

$$\{\tilde{T}_a, \tilde{T}_a\} = 0. \tag{15}$$

The involutive Hamiltonian $\tilde{H}$ is defined in the extended phase space as

$$\tilde{H} = \tilde{H}(q, p, \xi), \tag{16}$$

and we then have

$$\{\tilde{T}_a, \tilde{H}\} = 0, \tag{17}$$

subject to the boundary condition

$$\tilde{H}(q, p, 0) = H_0(q, p), \quad H^{(0)} = H_0, \tag{18}$$

where $H_0(q, p)$ is the canonical Hamiltonian.

4. THE ABELIAN PROCA MODEL

In this section we shall analyze the abelian Proca model by using the Hamilton-Jacobi and the Batalin-Frandkin-Tyutin methods.

-A) Hamilton-Jacobi method:

The Lagrangian density for abelian Proca model in $(3+1)$ dimensions is given by [2, 3, 15]

$$L = -\frac{1}{4}F_{\mu\nu}F^{\mu\nu} + \frac{m^2}{2}A_\mu A^\mu, \tag{19}$$

where $F_{\mu\nu} = \partial_\mu A_\nu - \partial_\nu A_\mu$, and $g_{\mu\nu} = diag(+, -, -, -)$.

The momenta π_μ conjugated to the fields A_μ are given by

$$\pi_i = -F_{0i}, \quad \pi_0 = 0. \tag{20}$$

The canonical Hamiltonian H_c is calculated as follows

$$H_c = \int d^3x \left(\frac{1}{2}\pi_i\pi_i + \frac{1}{4}F_{ij}F^{ij} - \frac{m^2}{2}(A_0^2 - A_i^2) + A_0(\partial_i\pi_i) \right). \tag{21}$$

Following the Hamilton-Jacobi method we obtain the set of Hamilton-Jacobi partial differential equations as

$$H'_0 = p_0 + \int d^3x(\frac{1}{2}\pi_i\pi_i + \frac{1}{4}F_{ij}F^{ij} - \frac{m^2}{2}(A^20 - A_i^2) + A_0(\partial_i\pi_i)), \tag{22}$$

$$H'_1 = \int d^3x\,\pi_0 = 0. \tag{23}$$

The equations of motion and the action function are obtained as total differential equations in many variables as follows:

$$dA_i = (\pi_i - \partial_iA_0)dt, \tag{24}$$

$$d\pi_i = (m^2A_i - \partial_jF_{ji})dt, \tag{25}$$

$$d\pi_0 = (-\partial_i\pi_i + m^2A_0)dt, \tag{26}$$

$$dS = (\pi_i\dot{A}^i - H_c)dt. \tag{27}$$

To check whether this system is integrable or not, we take the total variations of the constraints. In fact, the total variation of H'_1 gives the following condition [2, 15]

$$H'_2 = (-\partial_i\pi_i + m^2A_0). \tag{28}$$

Making use of the integrability condition (28), the integrable set of equations of motion and the action function is calculated as follows:

$$dA_i = (\pi_i - \frac{1}{m^2}\partial_i\partial_i\pi_i)dt, \tag{29}$$

$$d\pi_i = (m^2A_i - \partial_jF_{ji})dt, \tag{30}$$

$$d\pi_0 = 0, \tag{31}$$

$$dS = (\pi_i\dot{A}^i - \frac{1}{2}\pi_i\pi_i - \frac{1}{4}F_{ij}F^{ij} - \frac{m^2}{2}A_i^2 - \frac{(\partial_i\pi_i)^2}{2m^2})dt. \tag{32}$$

The path integral quantization for the abelian Proca model is given by

$$K = \int dA^i\,d\pi^i\exp i \int d^3x \left(\pi_i\dot{A}^i - \frac{1}{2}\pi_i\pi_i - \frac{1}{4}F_{ij}F^{ij} - \frac{m^2}{2}A_i^2 - \frac{(\partial_i\pi_i)^2}{2m^2} \right)dt. \tag{33}$$

-B) Batalin Fradkin Tyutin method:

In the BFT analysis of the Proca model, we have two second-class constraints [3, 4]

$$H'_1 = \pi_o \approx 0, \tag{34}$$

$$H'_2 = (-\partial_i\pi_i + m^2A_0) \approx 0. \tag{35}$$

The matrix Δ of equation (11) is given by

$$\Delta = \begin{pmatrix} 0 & -m^2 \\ m^2 & 0 \end{pmatrix}. \tag{36}$$

To convert H'_1 and H'_2 into first class constraints, we introduce two extra canonical pairs of fields (ρ, π_ρ) such that, $\{\rho(x), \pi_\rho(y)\} = \delta(x-y)$.

Following the BFT method [3, 4], we calculate the following first class constraints

$$\tilde{H}'_1 = H'_1 + m^2\rho, \tag{37}$$

$$\tilde{H}'_2 = H'_2 + \pi_\rho. \tag{38}$$

The corresponding gauge invariant Hamiltonian is

$$\tilde{H}_{BFT} = H_c + \int d^3x \left(\frac{\pi_\rho^2}{2m^2} + \frac{m^2}{2}(\partial_i\rho)^2 - m^2\rho\partial_i A_i \right). \tag{39}$$

5. CONCLUSION

In this work we have studied the constrained Hamiltonian systems by using two different methods: The fist method is the Güler's method [10] and the second method is the Batalin, Fradkin and Tyutin (BFT) [3, 4] method. Following the prescriptions of the Güler's method, we obtain the set of Hamilton-Jacobi partial differential equations which has the multi- Hamiltonians H'_0 and H'_μ. The integrability conditions lead us to obtain the canonical reduced phase-space coordinates (q^a, p^a) and the integrable action function in terms of these canonical variables. The path integral is obtained directly as an integration over the canonical phase space coordinates . For the Batalin, Fradkin and Tyutin (BFT) [3, 4] formalism, in order to obtain the quantization procedure for singular systems, one should enlarge the initial phase space by introducing un-physical auxiliary field variables, such that the second class constraints became converted into first class constraints.

The Abelian Proca model was investigated by using the two methods and it was observed that the Güler's method leads us to obtain the gauge invariant Hamiltonian and the integrable canonical action without any need to enlarge the initial phase space as in the Batalin, Fradkin and Tyutin (BFT) [3, 4] formalism.

The advantage of using the canonical path integral approach [15] is that, it is simpler, more economical and we have no difference between first and second class constraints, as well as we do not need to enlarge the initial phase-space by introducing unphysical auxiliary fields (see refs. [3, 4] and the references therein) . All that, is needed the set of Hamilton-Jacobi partial differential equations and the set of equations of motion and the action function. If the system is integrable, then one can obtain the path integral directly as an integration over the canonical reduced phase-space coordinates q_a and p_a. In other words, unlike the Batalin, Fradkin and Tyutin (BFT) method, in order to get the quantization procedure by using the Güler's method, one need not to look outside the original system, where the canonical *integrable* action function is present within the original system it self.

6. SQUARE BRACKETS, POISSON BRACKETS

In this appendix we shall give a brief review on two kinds of commutators: the square and the Poisson brackets. The square bracket is defined as

$$[F,G]_{q_i,p_i,z} = \frac{\partial F}{\partial p_i}\frac{\partial G}{\partial q_i} - \frac{\partial G}{\partial p_i}\frac{\partial F}{\partial q_i} + \frac{\partial F}{\partial p_i}(p_i\frac{\partial G}{\partial z}) - \frac{\partial G}{\partial p_i}(p_i\frac{\partial F}{\partial z}). \tag{40}$$

The Poisson bracket is defined as

$$\{f,g\}_{q_i,p_i} = \frac{\partial f}{\partial p_i}\frac{\partial g}{\partial q_i} - \frac{\partial g}{\partial p_i}\frac{\partial f}{\partial q_i}. \tag{41}$$

According to above definitions, the following relation holds

$$[H'_\alpha, H'_\beta] = \{H'_\alpha, H'_\beta\}. \tag{42}$$

ACKNOWLEDGMENTS

I would like to thank the organizing committee and Çankaya university for their support and hospitality during the workshop. Also, I would like to thank the referee for his valuable comments. My sincere thanks to Prof. Dr. Yurdahan Güler for his continuous encouragement during all my research studies.

REFERENCES

1. D. Baleanu and Y. Güler, *Chain and Hamilton-Jacobi appoaches for systems with purely second class constarints, Nuovo Cimento B* **118**, 615-623 (2003), hep-th/ 0310250.
2. D. Baleanu and Y. Güler, *Hamilton-Jacobi formalism of abelian Proca's model revisited, Nuovo Cimento B* **117**, 353-357 (2002).
3. N. Banerjee, S. Ghosh and R. Banerjee, *Quantization of O(N) invariant nonlinear sigma model in the Batalin-Tyutin formalism, Nucl. Phys.* **B 417** , 257-266 (1994), hep-th/9310043; *Batalin-Tyutin quantization of the CP(N-1) model, Phys.Rev.* **D49**, 1996-2000 (1994); *Quantization of second class systems in the Batalin-Tyutin foramlism, Annals Phys.* **241**, 237-257 (1995); M. I. Park and Y. J. park, *Nonabelin Proca model based on the improved BFT formalism, Int. J. Mod. Phys.* **A 13**, 2179-2199 (1998); Y. W. Kim, M. I. Park and Y. J. Park, *BRST quantization of the Proca model based on the BFT and the BFV formalism, Int. J. Mod. Phys.* **A 12**, 4217-4239 (1997); Y. W. Kim and K. D. Rothe, *BFT Hamiltonian embeding of nonabelian selfdaul model, Nucl. Phys.* **B 511**, 510-520 (1998); S. T. Hong, Y.W. Kim, Y. J. Park and K. D. Rothe, *Symplectic embedding and Hamilton-Jacobi analysis of Proca model, Mod. Phys. Lett.* **A 17**, 435-452 (2002).
4. I. A. Batalin and I. V. Tyutin, *Existance theorem for the effective gauge algebra in the generalized canonical formalism with abelian conversion of second class constraints, Int. J. Mod. Phys.* **A6**, 3255-3282 (1991); I. A. Batalin and E. S. Fradkin, *Operator quantization of dynamical systems with irreducible first- and second- class constraints, Phys. Lett.* **B180**, 157-162 (1986).
5. C. Caratheodory, *Calculus of Variation and Partial Differential Equations of First Order*, Part II (Holden-Day, 1976).
17. P. A. M. Dirac, *Lectures on Quantum Mechanics*, Belfer Graduate School of Science (A cademic Press, 1964); *Generalized Hamiltonian dynamics, Can. J. Math.* **2**, 129-148 (1950).
7. L. D. Faddeev and S. Shatashvili, *Realization of the Schwinger term in the gauss law and the possibilty of correct quantization of a theory with anomalies, Phys. Lett.* **B 167**, 225-228 (1986).

8. T. Fujiwara, Y. Igarashi and J. Kubo, *Anomalous gauge theories and subcritical strings based on the Batalin-Fradkin fornalism*, *Nucl. Phys.* **B 341**, 695-713 (1990).

9. D.M. Gitman and I. V. Tyutin, *Quantization of Fields with Constraints*, Spinger (1980); K. Sundermeyer, *Constrained Dynamics*, Lectures Notes in Physics Vol. 169 (Springer, New York, 1982); M. Henneaux and C. Teitelboim, *Quantization of Guage Systems* (Princeton Univ. Press, 1992); A. Hanson, T. Regge and C. Teitelboim, *Constrained Hamiltonian Systems* (Academia Nazionle dei Lincei, Rome, 1976).

10. Y. Güler, *Integration of singular systems*, *Nuovo Cimento B* **107**, 1143-1149 (1992); *Canonical formulation of constrained systems*, *Nuovo Cimento B* **107**, 1389-1395 (1992); *Hamilton-Jacobi theory of continuous systems*, *Nuovo Cimento B* **100**, 251-266 (1987); *On the dynamics of singular continuous systems*, *J. Math. Phys.* **30**, 785-788 (1989).

11. Y.-W. Kim, S.-K. kim, W. Kim, Y. J. Park, K. Y. Kim and Y. Kim, *The Chiral schwinger model based on the Batalin-Fradkin-Vilkovisky formalism*, *Phys. Rev.* **D46**, 4574-4579 (1992); R. Banerjee, H. Rothe and K. D. Rothe, *Batalin-Fradkin quantization of anomalous Chiral gauge theories* , *Phys. Rev.* **D 49**, 5438-5445 (1994).

12. P. Mitra and R. Rrajaraman, *New results on systems with second class constraints*, *Ann. Phys. (N. Y.)* **203**, 137-156 (1990).

13. S. I. Muslih, *Completely and partially integrable systems of total differential equations*, *Nuovo Cimento B***118**, 505-511 (2003); *Integrability and action function in multi-Hamiltonian systems*, *Mod. Phys. Lett* **A 19**, 863-870 (2004).

14. S. I. Muslih, *Hamilton-Jacobi quantization of two-dimensional gravity with torsion*, *Mod. Phys. Lett.* **A 19**, 151-157 (2004).

15. S. I. Muslih, *The canonical path integral quantization of Chiral Schwinger model*, *Mod. Phys. Lett.* **A 18**, 1187-1196 (2003); S. I. Muslih, *Quantization of parametrization invariant theories*, *Nuovo Cimento B* **115**, 1-6 (2000); *Path integral quantization of electromagnetic theory*, **115**, 7-12 (2000); S. I. Muslih , H. A. El-Zalan and F. El-Sabaa , *Quantization of Yang-Mills theory*, *Int. J. Theor. Phys.* **39**, 2495-2502 (2000); S. I. Muslih, *Canonical path integral quantization of Einstein's gravitational field*, *Gen. Rel. Grav.* **34**, 1059-1069 (2002); S. I. Muslih, *Quantization of singular systems with second-order Lagrangians*, *Mod. Phys. Lett.* **A 17**, 2383-2391 (2002); S. I. Muslih, *Path integral formulation of constrained systems*, *Hadronic J.* **23**, 203-218 (2000).

16. S. I. Muslih, *The canonical path integral quantization of Cp(1) model*, *Czech. J. Phys.* **52**, 919-925 (2002).

17. S. I. Muslih, *The Hamilton-Jacobi treatment for an abelian Chern-Symons system*, *Nucl. Phys.* **B106**, (Proc. Supp.), 879-881 (2002).

18. E. Rabei and Y. Güler, *Hamilton-Jacobi treatment of second-class constraints*, *Phys. Rev. A* **46**, 3513-3515 (1992); S. I. Muslih and Y. Güler, *On the integrability conditions of constrained systems*, *Nuovo Cimento B***110**, 307-315 (1995).

19. E. G. Stückelberg, *Théorie de la radiation de photons de mass arbitrairement petite, (Radiation theory of photons of arbitrarily small mass)*, *Helv. Phys. Act.* **30**, 209-215 (1957); R. Banerjee and J. Barcelos-Neto, *Hamiltonian embedding of the massive Yang-Mills theory and the generalized Stückelberg formalism*, *Nucl. Phys.* **499**, 453-478 (1997); C. G. Knetter and R. Kogerler, *Unitary gauge, Stückelberg formalism, and gauge-invariant models for effective Lagrangians*, *Phys. Rev.* **48 D**, 2865-2876 (1993).

20. J. Wess and B. Zumino, *Consequences of anomalous ward identities*, *Phys. Lett.* **B 37**, 95-97 (1971).

247

An Extension of the Mazur-Ulam Theorem

Constantin P. Niculescu

University of Craiova, Department of Mathematics, Street A. I. Cuza 13, Craiova 200585, Romania

Abstract. One proves that the Mazur-Ulam theorem can be extended in the framework of metric spaces as long as a well behaved concept of midpoint is available. This leads to the new concept of Mazur-Ulam space. Besides the classical case of real normed spaces, other examples such as $\mathrm{Sym}^{++}(n, \mathbb{R})$, the space of all $n \times n$ dimensional positive definite matrices, appear as symmetric cones attached to Euclidean Jordan algebras. It turns out that the Mazur-Ulam spaces provide a framework for new generalizations of the concept of convex function.
Key words. Metric space, isometry, midpoint.
PACS. Primary 02.40.Ft; Secondary 02.40.Ky, 02.40.Ma.
MSC. Primary 54E40; Secondary 46B20, 52A41.

1. INTRODUCTION

The Mazur-Ulam theorem asserts that every bijective isometry $T : E \to E$ acting on a real normed space is an affine map, that is,

$$T(\lambda x + (1 - \lambda)y) = \lambda T(x) + (1 - \lambda)T(y) \tag{1}$$

for all $x, y \in E$ and all $\lambda \in \mathbb{R}$. See [8], [2], [12] and [13] for details and extensions within the framework of linear spaces. The essence of this result is the property of T to preserve midpoints of line segments, that is,

$$T\left(\frac{x+y}{2}\right) = \frac{T(x) + T(y)}{2} \tag{2}$$

for all $x, y \in E$. In fact, the condition (2) implies (1) for dyadic affine combinations, and thus for all convex combinations (since every isometry is a continuous map). Finally, it is routine to pass from convex combinations to general affine combinations in (1).

A careful inspection of a recent argument given by J. Väisälä [12] to the Mazur-Ulam theorem, makes clear that the linear structure of E is needed only to support the notion of midpoint. In other words, a property like (2), of midpoint preservation, works in the framework of metric spaces as long as a well behaved concept of midpoint is available. We are thus led to rephrase the Mazur-Ulam theorem as a result saying that the midpoints of line segments in a real normed space have a nice behavior.

Definition 1 *A Mazur-Ulam space is any metric space $M = (M, d)$ on which there is given a pairing $\sharp : M \times M \to M$ with the following four properties:*

CP729, *Global Analysis and Applied Mathematics: International Workshop on Global Analysis*,
edited by K. Taş, D. Krupka, O. Krupková, and D. Baleanu
© 2004 American Institute of Physics 0-7354-0209-4/04/$22.00

(*The idempotent property*) $x \sharp x = x$ for all $x \in M$;
(*The commutative property*) $x \sharp y = y \sharp x$ for all $x, y \in M$;
(*The midpoint property*) $d(x, y) = 2d(x, x \sharp y) = 2d(y, x \sharp y)$ for all $x, y \in M$;
(*The transformation property*) $T(x \sharp y) = T(x) \sharp T(y)$, for all $x, y \in M$ and all bijective isometries $T : M \to M$.

A Mazur-Ulam space should be viewed as a triplet $(M, d, \sharp)$. In this context, the point $x \sharp y$ is called a *midpoint* between x and y.

In a real normed space, the midpoint has the classical definition,

$$x \sharp y = \frac{x + y}{2},$$

and the Mazur-Ulam theorem is equivalent to the assertion that *every real normed space is a Mazur-Ulam space*. It is this way we want to extend the Mazur-Ulam theorem, by investigating other classes of Mazur-Ulam spaces. And they are plenty.

In the above example, $\sharp$ coincides with the arithmetic mean, A. The simplest example of a Mazur-Ulam space where the midpoint is associated to the geometric mean is $\mathbb{R}_+^\star = (0, \infty)$, endowed with the metric

$$\delta(x, y) = \left| \log \frac{x}{y} \right|,$$

and the midpoint pairing $x \sharp y = G(x, y) = \sqrt{xy}$.

Notice that no discrete metric space M (of cardinality greater than 1) supports a midpoint pairing $\sharp$.

In Section 1 we give a new proof of the classical Mazur-Ulam theorem by taking into account the existence of sufficiently many reflections in a real normed space. This leads to new examples of Mazur-Ulam spaces such as $\text{Sym}^{++}(n, \mathbb{R})$, the space of all $n \times n$ dimensional positive definite matrices with real coefficients (which provides a higher dimensional generalization of $(\mathbb{R}_+^\star, \delta)$). In this case, the midpoint is precisely the geometric mean in the operator-theoretical sense.

In Section 2 we briefly discuss the class of Bruhat-Tits spaces, which also includes $\text{Sym}^{++}(n, \mathbb{R})$ (and other symmetric cones attached to suitable Euclidean Jordan algebras). Our exposition follows the approach by J. D. Lawson and Y. Lim [6]. See also [5].

In Section 3 we make an attempt to built up a theory of generalized convexity in the framework of Mazur-Ulam spaces. The idea is to replace the arithmetic mean by a midpoint combination. Precisely, if $M' = (M', d', \sharp')$ and $M'' = (M'', d'', \sharp'')$ are Mazur-Ulam spaces, with M'' a subinterval of $\mathbb{R}$, it is natural to say that a continuous function $f : M' \to M''$ is *convex* (more precisely, ($\sharp', \sharp''$)-*convex*) if

$$f(x \sharp' y) \leq f(x) \sharp'' f(y) \quad \text{for all } x, y \in M'.$$

This theory encompasses a large variety of functions such as the usual convex functions, the log-convex functions, the multiplicatively convex functions etc. The case of multiplicatively convex functions corresponds to the choice $M' = M'' = (\mathbb{R}_+^\star, \delta, G)$ and refers to all functions $f : \mathbb{R}_+^\star \to \mathbb{R}_+^\star$ such that

$$f(\sqrt{xy}) \leq \sqrt{f(x)f(y)} \quad \text{for all } x, y > 0.$$

It turns out that this property is shared by a large variety of special functions like Γ (the gamma function), Si (the integral sine), Li (the logarithmic integral) etc. See our paper [9] for details. While the entire theory of convex functions has a companion within multiplicative convexity, things become unclear in the case of $\mathrm{Sym}^{++}(n, \mathbb{R})$, and we end this paper with a number of open problems.

2. MIDPOINTS AS FIXED POINTS

The classical Mazur-Ulam theorem can be easily proved by noticing the presence of sufficiently many reflections on any normed vector space. This idea can be considerably extended.

Theorem 1 *Suppose that $M = (M, d)$ is a metric space such that for every pair (a, b) of points of M there exists a bijective isometry $G_{(a,b)} : M \to M$ with the following two properties:*
(MU1) $G_{(a,b)}a = b$ and $G_{(a,b)}b = a$;
(MU2) $G_{(a,b)}$ has a unique fixed point z (denoted $a \sharp b$) and

$$d(G_{(a,b)}x, x) = 2d(x, z) \quad \text{for all } x \in E.$$

Then M is a Mazur-Ulam space.

The geometrical framework of Theorem 1 is illustrated in Figure 1, while its proof will constitute the objective of Lemma 3 below.

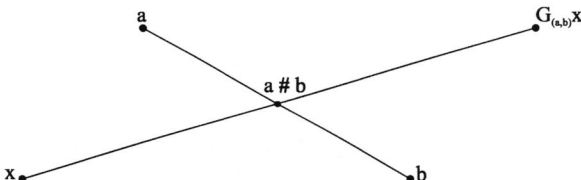

FIGURE 1. The geometrical framework of Theorem 1

Every normed vector space verifies the hypotheses of Theorem 1. In fact, in that case the maps $G_{(a,b)}$ are precisely the reflections

$$G_{(a,b)}x = a + b - x.$$

The unique fixed point of $G_{(a,b)}$ is the midpoint of the line segment $[a, b]$, that is,

$$a \sharp b = \frac{a+b}{2}.$$

250

An example of a different nature is offered by $\mathbb{R}^{\star}_{+} = (0, \infty)$, endowed with the metric

$$\delta(a,b) = \left| \log \frac{a}{b} \right|.$$

The hypotheses of Theorem 1 are fulfilled by the family of isometries

$$G_{(a,b)}x = \frac{ab}{x}$$

and the fixed point of $G_{(a,b)}$ is precisely the geometric mean $\sqrt{ab}$, of a and b. A higher dimensional generalization of this example is provided by the space $\mathrm{Sym}^{++}(n, \mathbb{R})$, of all $n \times n$ positive definite matrices, with real coefficients. We consider on $\mathrm{Sym}^{++}(n, \mathbb{R})$, the *trace metric*,

$$d_{\mathrm{trace}}(A,B) = \left(\sum_{k=1}^{n} \log^2 \lambda_k \right)^{1/2}, \tag{3}$$

where $\lambda_1, ..., \lambda_n$ are the eigenvalues of AB^{-1}. Since similarities preserve eigenvalues, this metric is invariant under similarities, that is,

$$d_{\mathrm{trace}}(A,B) = d_{\mathrm{trace}}(C^{-1}AC, C^{-1}BC) \quad \text{for every } C \in \mathrm{GL}(n, \mathbb{R}).$$

Notice that AB^{-1} is similar with

$$A^{-1/2} \left(AB^{-1} \right) A^{1/2} = A^{1/2}B^{-1/2} \left(A^{1/2}B^{-1/2} \right)^{\star} > 0$$

and this fact assures the positivity of the eigenvalues of AB^{-1}.

The proof that $\mathrm{Sym}^{++}(n, \mathbb{R})$ admits a midpoint pairing follows from Theorem 1. We shall need the following technical result:

Lemma 2 *Given two matrices A and B in $\mathrm{Sym}^{++}(n, \mathbb{R})$, their geometric mean*

$$A \sharp B = A^{1/2}(A^{-1/2}BA^{-1/2})^{1/2}A^{1/2}$$

is the unique matrix C in $\mathrm{Sym}^{++}(n, \mathbb{R})$ such that

$$d_{\mathrm{trace}}(A,C) = d_{\mathrm{trace}}(B,C) = \frac{1}{2} d_{\mathrm{trace}}(A,B).$$

The geometric mean $A \sharp B$ of two positive definite matrices A and B was introduced by Pusz and Woronowicz [11]. It is the unique solution of the equation

$$XA^{-1}X = B.$$

and this fact has a number of useful consequences such as:

$$A \sharp B = (AB)^{1/2} \quad \text{if } A \text{ and } B \text{ commute}$$
$$A \sharp B = B \sharp A$$
$$(C^{\star}AC) \sharp (C^{\star}BC) = C^{\star}(A \sharp B)C \quad \text{for all } C \in \mathrm{GL}(n, \mathbb{R})$$

251

as well as the fact that the maps

$$G_{(A,B)}X = (A\sharp B)X^{-1}(A\sharp B)$$

verify the condition $(MU1)$ in Theorem 1 above. As concerns the condition $(MU2)$, let us check first the fixed points of $G_{(A,B)}$. Clearly, $A\sharp B$ is a fixed point. It is the only fixed point because any solution $X \in \mathrm{Sym}^{++}(n,\mathbb{R})$ of the equation

$$CX^{-1}C = X,$$

with $C \in \mathrm{Sym}^{++}(n,\mathbb{R})$, verifies the relation

$$\left(X^{-1/2}CX^{-1/2}\right)\left(X^{-1/2}CX^{-1/2}\right) = I.$$

Since the square root is unique, we get $X^{-1/2}CX^{-1/2} = I$, that is, $X = C$. The second part of the condition $(MU2)$ asks for

$$d_{\mathrm{trace}}\left(G_{(A,B)}X,X\right) = 2d\left(X,A\sharp B\right),$$

that is,

$$d_{\mathrm{trace}}\left((A\sharp B)X^{-1}(A\sharp B),X\right) = 2d_{\mathrm{trace}}\left(X,A\sharp B\right),$$

for every $X \in \mathrm{Sym}^{++}(n,\mathbb{R})$. This follows directly from the definition (3) of the trace metric. Notice that $\sigma(C^2) = \left\{\lambda^2 \,|\, \lambda \in \sigma(C)\right\}$ for all $C \in \mathrm{Sym}^{++}(n,\mathbb{R})$.

Lemma 3 *Suppose that $M_1 = (M_1,d_1)$ and $M_2 = (M_2,d_2)$ are two metric spaces which verify the conditions $(MU1)$ and $(MU2)$ of Theorem 1. Then*

$$T\left(x\sharp_{1/2}y\right) = Tx\sharp_{1/2}Ty$$

for all bijective isometries $T : M_1 \to M_2$.

Proof. For $x,y \in M_1$ arbitrarily fixed, consider the set $\mathscr{G}_{(x,y)}$ of all bijective isometries $G : M_1 \to M_1$ such that $Gx = x$ and $Gy = y$. Notice that the identity of M_1 belongs to $\mathscr{G}_{(x,y)}$. Put

$$\alpha = \sup_{G \in \mathscr{G}_{(x,y)}} d(Gz,z),$$

where $z = x\sharp y$. Since

$$\begin{aligned}
d(Gz,z) &\leq d(Gz,x) + d(x,z) \\
&= d(Gz,Ga) + d(x,z) \\
&= 2d(x,z)
\end{aligned}$$

we infer that $\alpha < \infty$. If $G \in \mathscr{G}_{(x,y)}$, so is

$$G' = G_{x,y}G^{-1}G_{x,y}G,$$

which yields
$$d(G_{(x,y)}G^{-1}G_{(x,y)}Gz,z) \leq \alpha.$$

Then
$$\begin{aligned}
d(G'z,z) &= d(G_{(x,y)}G^{-1}G_{(x,y)}Gz,z) = d(G_{(x,y)}G^{-1}G_{(x,y)}Gz,G_{x,y}z) \\
&= d(G^{-1}G_{(x,y)}Gz,z) \\
&= d(G_{(x,y)}Gz,Gz) \\
&= 2d(Gz,z)
\end{aligned}$$

and thus $d(Gz,z) \leq \alpha/2$ for all G. Consequently $\alpha = 0$ and this yields
$$G(z) = z$$

for all $G \in \mathscr{G}_{(x,y)}$.

Now, for $T : M_1 \to M_2$ an arbitrary bijective isometry, we want to show that $Tz = z'$, where $z' = Tx \sharp Ty$. In fact, $G_{(x,y)}T^{-1}G_{(Tx,Ty)}T$ is a bijective isometry in $\mathscr{G}_{(x,y)}$, so
$$G_{(x,y)}T^{-1}G_{(Tx,Ty)}Tz = z.$$

This implies
$$G_{(Tx,Ty)}Tz = Tz.$$

Since z' is the only fixed point of $G_{(Tx,Ty)}$, we conclude that $Tz = z'$. ∎

As noticed A. Vogt [13], the Mazur-Ulam theorem can be extended to all surjective maps $T : E \to F$ (acting on real normed spaces of dimension ≥ 2) which preserve equality of distances,
$$\|x - y\| = \|u - v\| \text{ implies } \|Tx - Ty\| = \|Tu - Tv\|.$$

It is open whether this result remains valid in the more general framework provided by Theorem 1.

3. MIDPOINTS WITHIN BRUHAT-TITS SPACES

The presence of $\mathrm{Sym}^{++}(n,\mathbb{R})$ among the Mazur-Ulam spaces is just the peak of the iceberg. In fact, many other symmetric cones play the same property due to the presence of a special metric structure.

A *Bruhat-Tits space* is a complete metric space $M = (M,d)$ which verifies the semiparallelogram law, that is, for every pair of points x and y in M there is a point z such that
$$d(x,y)^2 + 4d(z,w)^2 \leq 2d(x,w)^2 + 2d(y,w)^2 \text{ for all } w \in M.$$

As in the particular case of Hilbert spaces (when equality occurs), the point z appearing in the semiparallelogram law is the unique point in M satisfying
$$d(x,z) = d(y,z) = \frac{1}{2}d(x,y).$$

We shall call w the *midpoint* between u and v.

A basic source of Bruhat-Tits spaces (which are also Mazur-Ulam spaces) is as follows:

Theorem 4 *Let $M = (M,d)$ be a metric space such that:*

i) For each pair of points x and y of M there is an isometry $G : M \to M$ with $Gx = y$;

ii) There exists a Hilbert space H and and a continuous bijection $\exp : H \to M$ such that $\|a - b\| \le d(\exp a, \exp b)$ for all $a, b \in H$, and $\exp$ restricted to any line $\mathbb{R}c$ is an isometry;

iii) There is a pairing $\sharp : M \times M \to M$ such that any equality of the form

$$G(x \sharp y) = \exp(0)$$

for $G : M \to M$ an isometry, is possible only when $\log Gx = -\log Gy$, where $\log : X \to M$ is the inverse of $\exp : H \to M$.

Then M is a Bruhat-Tits space and $x \sharp y$ is the midpoint of x and y for all $x, y \in M$. Furthermore, each isometry $G : M \to M$ preserves midpoints and $\sharp$ is the only pairing satisfying $\exp((a+b)/2) = \exp a \sharp \exp b$ whenever $a, b \in \mathbb{R}c$, and $c \in H$.

The details are covered by Proposition 4.2 in [6].

One can easily prove that $M = \mathrm{Sym}^{++}(n, \mathbb{R})$ verifies the hypotheses of Theorem 2 above. In fact, in this case we may choose as H the space $\mathrm{Sym}(n, \mathbb{R})$, of all $n \times n$ dimensional symmetric matrices with real coefficients, endowed with the inner product

$$\langle A, B \rangle = \mathrm{trace}(AB).$$

The corresponding exponential map $\exp : H \to M$ is given by the familiar formula,

$$\exp(A) = \sum_{n=0}^{\infty} A^n / n!.$$

It is worth to notice that $\mathrm{Sym}^{++}(n, \mathbb{R})$ is actually a Cartan-Hadamard manifold. These manifolds are complete simply connected Riemannian manifolds with seminegative curvature. See [5], Ch. XI. In their case the Riemannian metric provides a structure of Bruhat-Tits space. Examples are the symmetric cones $\Omega = \exp(V)$ attached to Euclidean Jordan algebras V. See [6] and references therein.

4. GENERALIZED CONVEXITY IN MAZUR-ULAM SPACES

As mentioned in the Introduction, the Mazur-Ulam spaces constitute a natural framework for a generalized theory of convexity, where the role of the arithmetic mean is played by a suitable midpoint pairing.

Suppose that $M' = (M', d', \sharp')$ and $M'' = (M'', d'', \sharp'')$ are two Mazur-Ulam spaces, with M'' a subinterval of $\mathbb{R}$. A continuous function $f : M' \to M''$ is called *convex* (more precisely, $(\sharp', \sharp'')$-*convex*) if

$$f(x \sharp' y) \le f(x) \sharp'' f(y) \quad \text{for all } x, y \in M' \tag{4}$$

and *concave* if the opposite inequality holds. If

$$f\left(x\sharp'y\right) = f(x)\sharp''f(y) \quad \text{for all } x,y \in M' \tag{5}$$

then the function f is called *affine*.

Since every subinterval of $\mathbb{R}$ (endowed with the pairing associated to the arithmetic mean) is a Mazur-Ulam space, the above framework provides a generalization of the usual notion of convex function.

When $M' = M'' = (\mathbb{R}_+^\star, \delta, G)$, the convex functions in the sense of (4) are precisely the functions $f : \mathbb{R}_+^\star \to \mathbb{R}_+^\star$ such that

$$f\left(\sqrt{xy}\right) \leq \sqrt{f(x)f(y)} \quad \text{for all } x,y > 0. \tag{6}$$

They are known as the *multiplicatively convex* functions. Their theory can be easily deduced from the general theory of usual convex functions by a change of variable and function. In fact, $f : \mathbb{R}_+^\star \to \mathbb{R}_+^\star$ is multiplicatively convex if and only if $\log \circ f \circ \exp : \mathbb{R} \to \mathbb{R}$ is convex in the usual sense. See [9] or [10] for details.

When $M' = (\mathbb{R}_+^\star, \delta, G)$ and $M'' = \mathbb{R}$, we get the class of log-*convex* functions, that is, of those functions $f : \mathbb{R}_+^\star \to \mathbb{R}$ such that

$$f\left(\sqrt{xy}\right) \leq \frac{f(x) + f(y)}{2} \quad \text{for all } x,y > 0. \tag{7}$$

Due to the arithmetic-geometric mean inequality, this class contains all nondecreasing convex functions $f : \mathbb{R}_+^\star \to \mathbb{R}$, and also all multiplicatively convex functions.

Things become considerably more complicated in the case where M' is the cone $\mathrm{Sym}^{++}(n, \mathbb{R})$. A notable example of an affine function $f : \mathrm{Sym}^{++}(n, \mathbb{R}) \to \mathbb{R}$ is $f = \log \det$, but few is known on the corresponding class of convex functions.

In fact, an important feature of the usual theory of convex functions is the possibility to extend the basic inequality (4) to all convex combinations of finitely many points, and then to random variables attached to probability fields.

The analogue of $(1 - \lambda)x + \lambda y$ in the context of $\mathrm{Sym}^{++}(n, \mathbb{R})$ is

$$A\sharp_\lambda B = A^{1/2}(A^{-1/2}BA^{-1/2})^\lambda A^{1/2},$$

an this fact was investigated by F. Kubo and T. Ando [4] from the point of view of noncommutative means. As concerns the noncommutative analogue of $A\sharp_\lambda B$ for three (or finitely many) positive matrices, the theory is only at the beginning. An interesting approach was recently proposed by C.-K. Li and R. Mathias [7].

Besides the question on the generality of Jensen inequality (as well as of all other basic inequalities) within the above theory of convexity associated to midpoints, many others important questions remain open. We shall mention here the problem of an analogue of the gamma function,

$$\Gamma : (0, \infty) \to \mathbb{R}, \quad \Gamma(x) = \int_0^\infty t^{x-1}e^{-t}\, dt \quad \text{for } x > 0$$

in the case of $\mathrm{Sym}^{++}(n, \mathbb{R})$, $n \geq 2$.

The famous Bohr-Mollerup theorem asserts that Γ is the only function $f : (0, \infty) \to \mathbb{R}$ which verifies the following three conditions:

$\Gamma 1$) the functional equation $f(x+1) = xf(x)$;

$\Gamma 2$) the normalization condition $f(1) = 1$;

$\Gamma 3$) the condition of log-convexity.

See [1], or [10], pp. 50-52. On the other hand, D. Gronau and J. Matkowski [3] proved a characterization of Γ within the class of multiplicatively convex functions, by replacing $\Gamma 3$) with the following condition:

$\Gamma 3'$) f is multiplicatively convex on an interval (a, ∞), for some $a > 0$.

The problem whether these results have an analogue when $(0, \infty)$ is replaced by $\text{Sym}^{++}(n, \mathbb{R})$, with $n \geq 2$, is open.

REFERENCES

1. Artin, E., *The Gamma Function*, Holt, Rinehart and Winston, New York, 1964. English translation of German original, *Einführung in die Theorie der Gammafunktion*, Teubner, 1931.
2. Day, M. M., *Normed Linear Spaces*, 3rd Edition, Springer-Verlag, Berlin, 1973.
3. Gronau, D., and Matkowski, J., *Geometrical convexity and generalizations of the Bohr-Mollerup theorem on the Gamma function*, Mathematica Panonica, **4**, 153-160 (1993).
4. Kubo, F., and Ando, T., *Means of Positive Linear Operators*, Math. Ann., **246**, 205-224 (1980).
5. Lang, S., *Fundamentals of differential geometry*, Graduate Texts in Math., Springer-Verlag, Berlin, 1999.
6. Lawson, J. D., and Lim, Y., *The Geometric Mean, Matrices, Metrics and More*, Amer. Math. Month., **108**, 797-812 (2001).
7. Li, Ch.-K., and Mathias, R., *Geometric Means*. Preprint, 2003.
8. Mazur, S., and Ulam, S., *Sur les transformations isométriques d'espaces vectoriels normés*, C. R. Acad. Sci. Paris, **194**, 946-948 (1932).
9. Niculescu, C. P., *Convexity according to the geometric mean*, Math. Inequal. Appl., **3**, 155-167 (2000).
10. Niculescu, C. P., and Persson, L.-E., *Convex Functions and their Applications. An Introduction*, Universitaria Press, Craiova, 2003.
11. Pusz, W., and Woronowicz, S. L., *Functional calculus for sesquilinear forms and the purification map*, Rep. Math. Phys., 8, 159-170 (1975).
12. Väisälä, J., *A Proof of the Mazur-Ulam Theorem*, Amer. Math. Month., **110**, 633-635 (2003).
13. Vogt, A., *Maps which preserve equality of distance*, Studia Math., **XLV**, 43-48 (1973).

The Types of Centro-affine Curves

Ömer Pekşen

Department of Mathematics, Karadeniz Technical University, 61080, Trabzon, Turkey
E-mail: peksen@ktu.edu.tr

Abstract. Let $GL(n,R)$ be the general linear group of $n \times n$ real matrices. Definitions of $GL(n,R)$-equivalence and the centro-affine type of curves are introduced. All possible centro-affine types are founded. For every centro-affine type all invariant parametrizations of a curve are described. The problem of $GL(n,R)$-equivalence of curves is reduced to that of paths.
Key words. Centro-affine geometry, curve, path, centro-affine equivalence of curves, centro-affine type of a curve, differential invariants of a path.
PACS. 02.40.Ma.
MSC. 53A15, 53A55.

1. INTRODUCTION

The fundamental theorem of curves in n-dimensional centro-affine geometry are obtained by Gardner and Wilkens [5]. In the paper they used Cartan's method [4] of moving frames in order to find the formulation of the local rigidity theorem for curves that is amenable to direct application to problems in control theory. A discussion of centro-affine plane curves, as well as a very brief discussion of centro-affine space curves, can be found in ([15],[13],[17]). A very detailed discussion of the centro-affine theory of plane curves can be found in Laugwitz [11]. The first comprehensive treatment of affine geometry is given in the seminal work of Blaschke [3]. For further developments of the subject, we refer the reader to [14], and the more modern texts [20], [12], commentaries [16], [17], and survey papers [19], [2], [18]. Equi-affine invariants of 3-dimensional space curves are investigated by Izumiya and Sano [7]. For curvatures of curves in n-dimensional equi-affine geometry see ([6], $pp.170-172$, [13]). But in all these works, equivalence of curves is investigated locally. The global $SL(n)$-equivalence of paths in R^n and C^n is considered by Khadjiev [8] and Suhtaeva [21]. Complete systems of global equi-affine invariants for plane and space paths are obtained by Angelis, Moons, Van Gool and Verstraelen [1]. The complete system of global differential and integral invariants for curves in n-dimensional equi-affine geometry is obtained by Khadjiev and Pekşen [9].

This paper is concerned with the problem of the global equivalence of centro-affine curves. A centro-affine type of a curve is introduced. The centro-affine type of a curve coincides with the centro-affine arclength if it is finite. But curves with infinite centro-affine arclength have three different centro-affine types. All possible invariant

CP729, *Global Analysis and Applied Mathematics: International Workshop on Global Analysis,*
edited by K. Taş, D. Krupka, O. Krupková, and D. Baleanu
© 2004 American Institute of Physics 0-7354-0209-4/04/$22.00

parametrizations of a curve are described for every centro-affine type.

2. THE CENTRO-AFFINE TYPE OF A CURVE

Let R be the field of real numbers and $I = (a,b)$ be an open interval of R.

Definition 1. *A C^∞-map $x : I \to R^n$ will be called an I-path (shortly, a path) in R^n.*

Definition 2. *An I_1-path $x(t)$ and an I_2-path $y(r)$ in R^n will be called D-equivalent if there exists a C^∞-diffeomorphism $\varphi : I_2 \to I_1$ such that $\varphi'(r) > 0$ and $y(r) = x(\varphi(r))$ for all $r \in I_2$. A class of D-equivalent paths in R^n will be called a curve in R^n, ([9], p.9). A path $x \in \alpha$ will be called a parametrization of a curve α.*

Remark 1. *There exist different definitions of a curve ([6], p.2, [8]).*

Let $G = GL(n,R)$ be the general linear group of $n \times n$ regular matrices. G acts by $(g,x) \to gx$ on R^n, where gx is the multiplication of a matrix g and a column vector $x \in R^n$.

If $x(t)$ is an I-path in R^n then $gx(t)$ is an I-path in R^n for any $g \in G$.

Definition 3. *Two I-paths x and y in R^n will be called G-equivalent and written $x \overset{G}{\sim} y$ if there exists $g \in G$ such that $y(t) = gx(t)$.*

Let α be a curve in R^n, that is, $\alpha = \{h_\tau, \tau \in Q\}$, where h_τ is a parametrization of α. Then $g\alpha = \{gh_\tau, \tau \in Q\}$ is a curve in R^n for any $g \in G$.

Definition 4. *Two curves α and β in R^n will be called G-equivalent (or G-congruent) and written $\alpha \overset{G}{\sim} \beta$ if $\beta = g\alpha$ for some $g \in G$.*

Remark 2. *This definition is essentially different from the definition ([6], p.21) of a congruence of curves for the group of euclidean motions. By the definition ([6], p.21), two curves with different lengths may be congruent.*

Let x be an I-path in R^n and $x'(t)$ be the derivative of $x(t)$. Put $x^{(0)} = x$, $x^{(n)} = (x^{(n-1)})'$. For $a_k \in R^n$, $k = 1, ..., n$, the determinant $\det(a_{ij})$ (where a_{ki} are coordinates of a_k) will be denoted by $[a_1 a_2 ... a_n]$. So $\left[x(t) x'(t) ... x^{(n-1)}(t) \right]$ is the determinant of the vectors $x(t)$, $x'(t)$, ..., $x^{(n-1)}(t)$. For $I = (a,b)$, $q, p \in I$, put

$$l_x(q,p) = \int_q^p \left| \frac{\left[x^{(n)}(t) x'(t) ... x^{(n-1)}(t) \right]}{\left[x(t) x'(t) ... x^{(n-1)}(t) \right]} \right|^{\frac{1}{n}} dt$$

and $l_x(a,p) = \lim\limits_{q \to a} l_x(q,p)$, $l_x(q,b) = \lim\limits_{p \to b} l_x(q,p)$. There are only four possible cases:

(i) $l_x(a,p) < +\infty$, $l_x(q,b) < +\infty$; (ii) $l_x(a,p) < +\infty$, $l_x(q,b) = +\infty$;
(iii) $l_x(a,p) = +\infty$, $l_x(q,b) < +\infty$; (iv) $l_x(a,p) = +\infty$, $l_x(q,b) = +\infty$.

Suppose that the case (i) or (ii) holds for some q, $p \in I$. Then the number $l = l_x(a,p) + l_x(q,b) - l_x(q,p)$, where $0 \le l \le +\infty$, does not depend on q, p. In this case, it is said that x belongs to the centro-affine type of $(0,l)$. The cases (iii) and (iv) do not depend on q, p. In these cases, it is said that x belongs to the centro-affine types of $(-\infty,0)$ and $(-\infty,+\infty)$, respectively. There exist paths of all types $(0,l)$ (where $0 \le l \le +\infty$), $(-\infty,0)$ and $(-\infty,+\infty)$. The centro-affine type of a path x will be denoted by $L(x)$.

Proposition 1. (i) *If* $x \overset{G}{\sim} y$ *then* $L(x) = L(y)$;
(ii) *Let* α *be a curve and* $x,y \in \alpha$. *Then* $L(x) = L(y)$.

The centro-affine type of a path $x \in \alpha$ will be called the centro-affine type of the curve α and denoted by $L(\alpha)$. According to Proposition 1, $L(\alpha)$ is a G-invariant of a curve α.

Definition 5. [5]*An* I-*path* $x(t)$ *in* R^n *will be called substantial if both*

$$\left[x(t)x'(t)...x^{(n-1)}(t) \right] \ne 0 \ and \ \left[x^{(n)}(t)x'(t)...x^{(n-1)}(t) \right] \ne 0$$

for all $t \in I$. *A curve will be called substantial if it contains a substantial path.*

Now an invariant parametrization of a substantial curve in R^n can be defined.
Let $I = (a,b)$ and $x(t)$ be a substantial I-path in R^n. The centro-affine arc length function $s_x(t)$ for each centro-affine type can be defined as follows. Put $s_x(t) = l_x(a,t)$ for the case $L(x) = (0,l)$, where $0 < l \le +\infty$, and $s_x(t) = -l_x(t,b)$ for the case $L(x) = (-\infty,0)$. Let $L(x) = (-\infty,+\infty)$. Choose a fixed point in every interval $I = (a,b)$ of R and denote it by a_I. Let $a_I = 0$ for $I = (-\infty,+\infty)$. Set $s_x(t) = l_x(a_I,t)$.
Since $s'_x(t) > 0$ for all $t \in I$, the inverse function of $s_x(t)$ exists. Let us denote it by $t_x(s)$. The domain of $t_x(s)$ is $L(x)$ and $t'_x(s) > 0$ for all $s \in L(x)$.

Proposition 2. *Let* $I = (a,b)$ *and* x *be a substantial* I-*path in* R^n. *Then*

(i) $s_{gx}(t) = s_x(t)$ *and* $t_{gx}(s) = t_x(s)$ *for all* $g \in G$;
(ii) *the equalities* $s_{x(\varphi)}(r) = s_x(\varphi(r)) + s_0$ *and* $\varphi(t_{x(\varphi)}(s+s_0)) = t_x(s)$ *hold for any* C^∞-*diffeomorphism* $\varphi : J = (c,d) \to I$ *such that* $\varphi'(r) > 0$ *for all* $r \in J$, *where* $s_0 = 0$ *for* $L(x) \ne (-\infty,+\infty)$ *and* $s_0 = l_x(\varphi(a_J),a_I)$ *for* $L(x) = (-\infty,+\infty)$.

Let α be a substantial curve and $x \in \alpha$. Then $x(t_x(s))$ is a parametrization of α.

Definition 6. *The parametrization* $x(t_x(s))$ *of a substantial curve* α *will be called an invariant parametrization of* α.

The set of all invariant parametrizations of α will be denoted by ϕ_α. Every $y \in \phi_\alpha$ is an I-path, where $I = L(\alpha)$.

Proposition 3. *Let* α *be a substantial curve,* $x \in \alpha$ *and* x *be an* I-*path, where* $I = L(\alpha)$. *Then the following conditions are equivalent:*

(i) x *is an invariant parametrization of* α;

(ii) $\left| \dfrac{\left[x^{(n)}(s)x'(t)...x^{(n-1)}(s) \right]}{\left[x(s)x'(s)...x^{(n-1)}(s) \right]} \right| = 1$ *for all* $s \in L(\alpha)$;

(iii) $s_x(s) = s$ *for all* $s \in L(\alpha)$.

Proposition 4. *Let α be a substantial curve and $L(\alpha) \neq (-\infty, +\infty)$. Then the invariant parametrization of α is unique.*

Let α be a substantial curve and $L(\alpha) = (-\infty, +\infty)$. Then it is easy to see that the set ϕ_α is not countable.

Proposition 5. *Let α be a substantial curve, $L(\alpha) = (-\infty, +\infty)$ and $x \in \phi_\alpha$. Then $\phi_\alpha = \{y : y(s) = x\left(s + s'\right), s' \in (-\infty, +\infty)\}.$*

Theorem 1. *Let α, β be substantial curves and $x \in \phi_\alpha$, $y \in \phi_\beta$. Then,*

(i) *for $L(\alpha) = L(\beta) \neq (-\infty, +\infty)$, $\alpha \overset{G}{\sim} \beta$ if and only if $x(s) \overset{G}{\sim} y(s)$;*

(ii) *for $L(\alpha) = L(\beta) = (-\infty, +\infty)$, $\alpha \overset{G}{\sim} \beta$ if and only if $x(s) \overset{G}{\sim} y\left(s + s'\right)$ for some $s' \in (-\infty, +\infty)$.*

Theorem 1 reduces the problem of the G-equivalence of substantial curves to that of paths.

REFERENCES

1. E. De Angelis, T. Moons, L. Van Gool, P. Verstraelen, *Complete systems of affine semi-differential invariants for plane and space curves*, In: Dillen, F.(ed.) et al., *Geometry and topology of submanifolds*, VIII. Proceedings of the international meeting on geometry of submanifolds, Brussels, Belgium, July 13-14, 1995 and Nordfjordeid, Norway, July 18-August 7, 1995. Singapore: World Scientific, (1996), 85-94.
2. W. Barthel, *Zur affinen Differentialgeometrie -Kurventheorie in der allgemeinen Affingeometrie*, Proceedings of the Congress of Geometry, Thessaloniki, (1987), 5-19.
3. W. Blaschke, *Affine Differentialgeometrie*, Berlin, (1923).
4. É. Cartan, *La théorie des groupes finis et continus et la géométrie différentielle*, Gauthier-Villars, Paris, (1951).
5. R.B. Gardner and G.R. Wilkens, *The fundamental theorems of curves and hypersurfaces in centro-affine geometry*, Bull. Belg. Math. Soc., **4**, (1997), 379-401.
6. H.W. Guggenheimer, *Differential Geometry*, McGraw-Hill, New York, (1963).
7. S. Izumiya and T. Sano, *Generic affine differential geometry of space curves*, Proceedings of the Royal Society of Edinburg, **128**A, (1998), 301-314.
8. Dj. Khadjiev, *The Application of Invariant Theory to Differential Geometry of Curves*, Fan Publ., Tashkent, (1988).
9. Dj. Khadjiev and Ö. Pekşen, *The complete system of global differential and integral invariants for equi-affine curves*, Differential Geometry and its Applications, **20**, (2004), 167-175.
10. W. Klingenberg, *A Course in Differential Geometry*, Springer-Verlag, New York, (1978).
11. D. Laugwitz, *Differentialgeometrie in Vectorraumen*, Friedr. Vieweg & Sohn, Braunschweig, (1965).
12. K. Nomizu and T. Sasaki, *Affine Differential Geometry*, Cambridge Univ. Press, (1994).
13. H.P. Paukowitsch, *Begleitfiguren und Invariantensystem minimaler Differentiationsordnung von Kurven im reellen n-dimensionalen affinen Raum*, Mh. Math., **85**, no.2, (1978), 137-148.
14. E. Salkowski, *Affine Differentialgeometrie*, W. de Gruyter, Berlin, (1934).
15. P.A. Schirokow and A.P. Schirokow, *Affine Differentialgeometrie*, Teubner, Leipzig, (1962).
16. U. Simon and W. Burau, *Blaschkes Beitrage zur affinen Differentialgeometrie*, in: W. Blaschke (Ed.), Gesammelte Werke, Vol. **IV**, (1985), 11-34.
17. U. Simon, *Entwicklung der affinen Differentialgeometrie nach Blaschkes*, in: W. Blaschke (Ed.), Gesammelte Werke, Vol. **IV**, (1985), 35-88.

18. U. Simon, *Recent developments in affine differential geometry*, Diff. Geom. and its Applications, Proc. Conf. Dubrovnik/Yugosl. 1988, (1989), 327-347.
19. U. Simon, H.L. Liu, M. Magid and Ch. Scharlach, *Recent developments in affine differential geometry*, In: Geometry and Topology of Submanifolds VIII, World Scientific, Singapore (1966), 1-15 and 293-408.
20. B. Su, *Affine Differential Geometry*, Science Press, Beijing, Gordon and Breach, New York, (1983).
21. A.M. Suhtaeva, *On the equivalence of curves in C^n with respect to the action of groups $SL(n,C)$ and $GL(n,C)$*, Dokl. Akad. Nauk of SSRUz, N6, (1987), 11-13.

Sprays and Cartan projective connections

D. J. Saunders

Faculty of Mathematics and Computing, The Open University, Walton Hall, Milton Keynes, MK7 6AA, UK ,
Address for correspondence: 30 Little Horwood Road, Great Horwood, Milton Keynes, MK17 0QE, UK

Abstract. Around 80 years ago, several authors (for instance H. Weyl, T.Y. Thomas, J. Douglas and J.H.C. Whitehead) studied the projective geometry of paths, using the methods of tensor calculus. The principal object of study was a spray, namely a homogeneous second-order differential equation, or more generally a projective equivalence class of sprays. At around the same time, E. Cartan studied the same topic from a different point of view, by imagining a projective space attached to a manifold, or, more generally, attached to a 'manifold of elements'; the infinitesimal 'glue' may be interpreted in modern language as a Cartan projective connection on a principal bundle. This paper describes the geometrical relationship between these two points of view.
Key words. Cartan connection, projective equivalence, spray.
PACS. 02.40.Hw, 02.40.Ma.
MSC. 53A20, 53B10, 53C13.

1. INTRODUCTION

This paper is a report of some joint work with Mike Crampin [3, 4] where we investigate the relationship between two different theories of the projective geometry of paths: that is, the theories consider sets of curves in a manifold without regard to their parametrization, where there is a unique curve through each point in any given direction. In both cases there is a restricted version, the affine case, and also a more general version: in the affine case, the paths are the (unparametrized) geodesics of a symmetric affine connection. The two approaches to the subject are quite different in philosophy, but we have been able to describe in some detail the relationships between the two theories.

The first two sections below describe the two theories. The remaining sections describe our results about the relationships between the two theories in both the affine case and the general case.

2. SPRAYS AND PATH SPACES

The study of path spaces in terms of symmetric affine connections, or more generally of sprays, is due to several authors, in particular H. Weyl [10], T.Y. Thomas [8, 9], J.

CP729, *Global Analysis and Applied Mathematics: International Workshop on Global Analysis,*
edited by K. Taş, D. Krupka, O. Krupková, and D. Baleanu
© 2004 American Institute of Physics 0-7354-0209-4/04/$22.00

Douglas [5] and J.H.C. Whitehead [11].

2.1. The affine case

Let M be a manifold M (with $\dim M = m$), and let ∇ be a symmetric affine connection on M. Write Γ for the geodesic spray Γ defined by ∇. We recall that a spray is a vector field on the tangent bundle TM satisfying

$$S(\Gamma) = \Delta, \qquad [\Delta, \Gamma] = \Gamma$$

where S and Δ are, respectively, the almost-tangent structure and the dilation field on TM. If we take coordinates (x^a) on M, so that $(x^a, \dot{x}^a)$ are the induced coordinates on TM, then

$$\Gamma = \dot{x}^a \frac{\partial}{\partial x^a} - \Gamma^a \frac{\partial}{\partial \dot{x}^a}$$

where the functions Γ^a are quadratic in fibre coordinates $\dot{x}^a$,

$$\Gamma^a = \Gamma^a_{bc}(x)\dot{x}^b \dot{x}^c.$$

A curve γ in M is geodesic of the connection if the lifted curve $\tilde{\gamma}$ in TM is an integral curve of Γ.

2.2. Projective equivalence

Two connections ∇, $\tilde{\nabla}$ are said to be *projectively equivalent* if they have the same geometric geodesics (although their parametrizations may be different). The corresponding geodesic sprays are related by

$$\Gamma - \hat{\Gamma} = \alpha\Delta$$

where the function α satisfies $\alpha = \alpha_a(x)\dot{x}^a$.

The fundamental descriptive invariant (Douglas) of a connection ∇ is given by

$$\Pi^a_{bc} = \Gamma^a_{bc} - \frac{1}{m+1}(\Gamma^d_{bd}\delta^a_c + \Gamma^d_{dc}\delta^a_b),$$

and depends only on the equivalence class of connections. Such an equivalence class is called a *restricted path space*.

2.3. Generalized path spaces

In general, a spray Γ on the slit tangent bundle $T^\circ M$ (excluding the zero section) has coefficients Γ^a satisfying

$$\Delta(\Gamma^a) = 2\Gamma^a,$$

so that they are homogeneous of degree 2 in the fibre coordinates $\dot{x}^a$ but need not be quadratic. Such a spray gives rise to a connection ∇ on the pull-back vector bundle

$$\tau_M^{\circ *}(TM) \to T^\circ M$$

called the *Berwald connection* of the spray. The connection coefficients of ∇ are given by

$$\Gamma_{bc}^a(x,\dot{x}) = \tfrac{1}{2} \frac{\partial^2 \Gamma^a}{\partial \dot{x}^b \partial \dot{x}^c} .$$

Once again, a curve γ in M is a geodesic of the connection if the lifted curve $\widetilde{\gamma}$ in $T^\circ M$ is an integral curve of the spray Γ.

In this general case, projectively equivalent sprays are related by

$$\Gamma - \widehat{\Gamma} = \alpha \Delta$$

where α satisfies $\Delta(\alpha) = \alpha$, and so is homogeneous of degree 1 in the fibre coordinates, but now need not be linear. The fundamental descriptive invariant is now

$$\Pi_{bc}^a = \Gamma_{bc}^a - \frac{1}{m+1}\left(\Gamma_{bd}^d \delta_c^a + \Gamma_{dc}^d \delta_b^a + \dot{x}^a \frac{\partial \Gamma_{bc}^d}{\partial \dot{x}^d}\right)$$

with an additional term which vanishes in the affine case.

3. CARTAN'S APPROACH

In 1924, Cartan [2] studied a manifold M with projective spaces at each point, 'infinitesimally connected': a curve in M could be 'developed' into one of these spaces, using the connection. A curve is a geodesic if its development is a straight line in one of these projective spaces. Each class of torsion-free connections giving the same geodesics has a distinguished representative, called the 'normal' connection.

In the same work, Cartan also considered the general second order differential equation (in the case $m = 2$). He used the 'manifold of elements' — a point, and a direction at that point — or, in modern terminology, the projective tangent bundle PTM.

3.1. Cartan connections — a modern interpretation

Sharpe [7] compares Ehresmann connections and Cartan connections on a principal H-bundle $P \to M$. For an *Ehresmann connection*, the connection form ω takes values in $\mathfrak{h}$, the Lie algebra of H. The kernel of $\omega : TP \to \mathfrak{h}$ contains the horizontal vectors of the connection. On the other hand, a *Cartan connection* involves a larger group G, where $\dim G = \dim P$ and $H \subset G$. The connection form ω now takes values in $\mathfrak{g}$, the Lie algebra of G, rather than in $\mathfrak{h}$, and $\omega : TP \to \mathfrak{g}$ is isomorphism on each fibre

A *Cartan projective connection* is a Cartan connection where $G = \mathrm{PGL}(m+1)$ and where H is the stabilizer of $[1,0,\ldots,0] \in P^m$. The curvature of the connection is $\Omega =$

$d\omega + \frac{1}{2}[\omega,\omega]$, and the torsion is $\kappa(\Omega)$ where $\kappa\colon \mathfrak{g} \to \mathfrak{g}/\mathfrak{h}$. The canonical example of a Cartan projective connection arises where $P = G$ and where M is the homogeneous space $G/H = P^m$. The connection form ω is then the Maurer-Cartan form of G.

4. SOME RESULTS

In [3, 4] we show that associated with every manifold M there are two principal bundles. In each case the group of the bundle is a subgroup of $\mathrm{PGL}(m+1)$, the quotient of $\mathrm{GL}(m+1)$ by non-zero multiples of the identity matrix. If m is even this group may be identified with $\mathrm{SL}(m+1)$, whereas if m is odd then it may be identified with the group of $+/-$ equivalence classes of matrices having determinant ± 1.

In the affine case, the *Cartan bundle* $\mathscr{C}M \to M$ has group $\mathrm{H}_{m+1} \subset \mathrm{PGL}(m+1)$; if $h \in \mathrm{H}_{m+1}$ then

$$
h = \varepsilon_m \begin{pmatrix} h_0^0 & h_1^0 & \cdots & h_m^0 \\ 0 & h_1^1 & \cdots & h_m^1 \\ \vdots & \vdots & \ddots & \vdots \\ 0 & h_1^m & \cdots & h_m^m \end{pmatrix} \qquad \begin{cases} m \text{ even:} & \varepsilon_m = 1, \quad \det h = 1 \\ m \text{ odd:} & \varepsilon_m = \pm 1, \quad \det h = \pm 1 \end{cases}
$$

A projective equivalence class of symmetric affine connections on M gives rise to a family of torsion-free Cartan connections on $\mathscr{C}M \to M$ with the same geodesics; one of these is the distinguished normal Cartan connection ω.

In the general case, the *projective Cartan bundle* $\mathscr{P}\mathscr{C}M \to PTM$ has group $\mathrm{K}_{m+1} \subset \mathrm{PGL}(m+1)$; if $k \in \mathrm{K}_{m+1}$ then

$$
k = \varepsilon_m \begin{pmatrix} k_0^0 & k_1^0 & k_2^0 & \cdots & k_m^0 \\ 0 & k_1^1 & k_2^1 & \cdots & k_m^1 \\ 0 & 0 & k_2^2 & \cdots & k_m^2 \\ \vdots & \vdots & \vdots & \ddots & \vdots \\ 0 & 0 & k_2^m & \cdots & k_m^m \end{pmatrix} \qquad \begin{cases} m \text{ even:} & \varepsilon_m = 1, \quad \det k = 1 \\ m \text{ odd:} & \varepsilon_m = \pm 1, \quad \det k = \pm 1 \end{cases}
$$

A projective equivalence class of sprays on $T^\circ M$ gives rise to a family of Cartan connections on $\mathscr{P}\mathscr{C}M \to PTM$ with the same geodesics; again there is a distinguished normal Cartan connection ω.

Thus, in the affine case, we have constructed two bundles

$$
\mathscr{C}M \to M, \qquad \mathscr{P}\mathscr{C}M \to PTM;
$$

it turns out, however, that the total spaces may be identified:

$$
\mathscr{C}M \cong \mathscr{P}\mathscr{C}M.
$$

>From a symmetric affine connection ∇ on M and its geodesic spray Γ restricted to $T^\circ M$, we have therefore constructed two normal Cartan connection forms

$$
\omega\colon T(\mathscr{C}M) \to \mathfrak{sl}(m+1), \qquad \omega\colon T(\mathscr{P}\mathscr{C}M) \to \mathfrak{sl}(m+1),
$$

but after identifying $\mathscr{C}M$ and $\mathscr{PC}M$ we find, as the notation suggests, that the two forms are identical.

The rest of this paper gives an outline of the methods used to obtain these results.

5. THE CONSTRUCTION IN THE AFFINE CASE

The clues to the affine construction come from the following observations:

- a quadratic spray on TM (indeed, any spray on $T^{\circ}M$) defines a 1-dimensional distribution on PTM,
- the fibres of PTM are projective spaces of dimension $m-1$, and
- there are no distinguished points in the fibres of PTM;

whereas

- Cartan used projective spaces of dimension m,
- each space has a distinguished point (where it is 'attached' to M), and
- the group G for a Cartan connection is $\mathrm{PGL}(m+1)$ with Lie algebra $\mathfrak{sl}(m+1)$.

It is clear from these observations that the construction should involve a manifold with an additional dimension.

5.1. The volume bundle

An extra dimension was used by T. Y. Thomas, and by J. H. C. Whitehead; also, more recently, by T. N. Bailey, M. G. Eastwood & A. R. Gover [1] and by C. Roberts [6]. We use the (unoriented) *volume bundle*

$$\mathscr{V}M = \{\pm\theta : \theta \in \textstyle\bigwedge^m M, \theta \neq 0\}.$$

This is a principal $\mathbf{R}_+$-bundle with action $\mu_s : \mathscr{V}M \to \mathscr{V}M$ given by

$$[\pm\theta] \quad \mapsto \quad s^{1/(m+1)}[\pm\theta]$$

(note that this is slightly different from the formulation given by Roberts). We take coordinates x^α (with $\alpha = 0, 1, \ldots, m$) on $\mathscr{V}M$, where

$$[\pm\theta] = \left(x^0[\pm\theta]\right)^{m+1} [\pm dx^1 \wedge \ldots \wedge dx^m].$$

The volume bundle has as a fundamental vector field

$$\Psi = x^0 \frac{\partial}{\partial x^0};$$

it also has a canonical odd scalar density

$$\Theta = \pm dx^0 \wedge dx^1 \wedge \ldots \wedge dx^m.$$

5.2. The Cartan algebroid

The action μ_s of $\mathbf{R}_+$ on $\mathscr{V}M$ gives rise to the derived action $\mu_{s*} \colon T\mathscr{V}M \to T\mathscr{V}M$; the *Cartan algebroid* is the quotient manifold

$$(\mathscr{W}M = T\mathscr{V}M/\mu_{s*}) \to M$$

with coordinates

$$x^a, \quad w = (x^0)^{-1}\dot{x}^0, \quad \dot{x}^a.$$

The Lie algebroid structure of $\mathscr{W}M$ (which we do not use) is derived from that of $T\mathscr{V}M$.

The Cartan algebroid has a distinguished global section

$$e_0 = \frac{\partial}{\partial w}$$

which is the projection of the fundamental vector field Ψ on $\mathscr{V}M$.

The projectivized Cartan algebroid

$$\mathrm{P}\mathscr{W}M = \mathscr{W}^\circ M/(\zeta \sim \lambda\zeta, \lambda \neq 0)$$

has m-dimensional projective spaces as fibres, each with a distinguished point $[e_0]$; these are the projective spaces used by Cartan.

5.3. The Cartan bundle

The *simplex bundle* of the Cartan algebroid, $\mathscr{S}_\mathscr{W}M \to M$, is defined by

$$\mathscr{S}_\mathscr{W}M = \{[\xi_0, \xi_1, \dots, \xi_m]\} \quad \text{where } \{\xi_0, \xi_1, \dots, \xi_m\} \text{ is a basis for } \mathscr{W}M.$$

Each element is therefore a reference simplex for the corresponding fibre of $\mathrm{P}\mathscr{W}M \to M$. There is a standard right action of $\mathrm{PGL}(m+1)$ on the fibres of the simplex bundle $\mathscr{S}_\mathscr{W}M \to M$, given as usual by matrix multiplication.

The *Cartan bundle* $\mathscr{C}M \to M$, with $\mathscr{C}M \subset \mathscr{S}_\mathscr{W}M$, is defined by

$$\mathscr{C}M = \{[\xi_0, \xi_1, \dots, \xi_m]\} \quad \text{where} \quad \begin{cases} \{\xi_0, \xi_1, \dots, \xi_m\} \text{ is a basis for } \mathscr{W}_pM \\ \xi_0 = \lambda e_0 \end{cases}$$

Each element of the Cartan bundle is therefore a reference simplex for the corresponding fibre of $\mathrm{P}\mathscr{W}M \to M$ with the additional property that its first point is the distinguished point $[e_0]$. The standard right action of $\mathrm{PGL}(m+1)$ on the fibres of the simplex bundle restricts to a right action of H_{m+1} on the fibres of the Cartan bundle $\mathscr{C}M \to M$.

5.4. *TW*-connections

Roberts [6] defined a *TW* ('Thomas-Whitehead') connection $\widetilde{\nabla}$ to be a symmetric affine connection on $\mathscr{V}M$ with the additional properties

$$\widetilde{\nabla}\Psi = \mathrm{id}, \qquad [\Psi, \widetilde{\nabla}_X Y] = \widetilde{\nabla}_{[\Psi,X]}Y + \widetilde{\nabla}_X[\Psi,Y]$$

267

(the exact definition given in [6] is slightly different as he uses a different definition of the volume bundle). Given this, each Ehresmann connection ϑ on $v: \mathcal{V}M \to M$ maps a symmetric TW-connection $\widetilde{\nabla}$ on $\mathcal{V}M$ to a symmetric connection ∇^ϑ on M by

$$\nabla^\vartheta_X Y = v_* \left(\widetilde{\nabla}_{X^h} Y^h \right)$$

(where X^h, Y^h are the horizontal lifts of X, Y by ϑ). Then different connections ϑ, ϑ' map $\widetilde{\nabla}$ to projectively equivalent affine connections ∇^ϑ, $\nabla^{\vartheta'}$ on M, and every affine connection in the projective equivalence class may be obtained in this way.

There is a unique TW-connection $\widetilde{\nabla}$ mapping to the projective equivalence class of ∇ satisfying the additional conditions

$$\mathrm{tr}(\widetilde{\nabla}) = 0, \qquad \mathrm{Ric}(\widetilde{\nabla}) = 0$$

(where the trace is taken with respect to the odd scalar density Θ); this is the *normal TW*-connection.

5.5. From *TW*-connection to Cartan connection

The normal TW-connection $\widetilde{\nabla}$ has, in the usual way, an Ehresmann connection form $\widetilde{\omega}$ on the frame bundle $\mathcal{F}(\mathcal{V}M) \to \mathcal{V}M$,

$$\widetilde{\omega}: T(\mathcal{F}(\mathcal{V}M)) \to \mathfrak{gl}(m+1).$$

The connection form $\widetilde{\omega}$ gives rise to another Ehresmann connection form $\widehat{\omega}$ on the simplex bundle $\mathcal{S}_\mathscr{W}M \to M$,

$$\widehat{\omega}: T(\mathcal{S}_\mathscr{W}M) \to \mathfrak{sl}(m+1),$$

by

$$\widehat{\omega}_\xi(v) = p_*(\widetilde{\omega}_\zeta(w))$$

where

$$\zeta \in \mathcal{F}(\mathcal{V}M) \text{ maps to } \xi \in \mathcal{S}_\mathscr{W}M$$
$$w \in T_\zeta \mathcal{F}(\mathcal{V}M) \text{ maps to } v \in T_\xi \mathcal{S}_\mathscr{W}M$$
$$p: \mathrm{GL}(m+1) \to \mathrm{PGL}(m+1)$$
$$p_*: \mathfrak{gl}(m+1) \to \mathfrak{sl}(m+1);$$

the conditions on the TW-connection ensure that $\widehat{\omega}$ is well-defined. We also find that $\ker \widehat{\omega} \cap T(\mathscr{C}M) \subset T(\mathcal{S}_\mathscr{W}M)$ contains only zero vectors, and so the restriction

$$\omega = \widehat{\omega}|_{T(\mathscr{C}M)}$$

is a Cartan connection (Sharpe [7]); this is the *normal Cartan projective connection* corresponding to the projective equivalence class of sprays.

6. THE CONSTRUCTION IN THE GENERAL CASE

A general spray on the slit tangent bundle $T°M$ is associated with a Berwald connection ∇ on the pull-back bundle $\tau_M^{\circ} TM$. This suggests consideration of the pull-back bundles

$$\pi_M^*(\mathscr{W}M) \to PTM, \qquad \pi_M^*(P\mathscr{W}M) \to PTM$$

where $\pi_M \colon PTM \to M$. Elements of $\pi_M^*(P\mathscr{W}M)$ are pairs $([v], [\xi])$ where $v \in T_pM$ and $\xi \in \mathscr{W}_pM$. Then each fibre of $\pi_M^*(P\mathscr{W}M) \to PTM$ has a distinguished point

$$([v], [e_0])$$

as before, but now it also has a distinguished line

$$\{([v], [\xi])\}, \qquad \rho(\xi) = \mu v$$

where $\rho \colon \mathscr{W}M \to TM$. Note that $\rho(e_0) = 0$, so that the distinguished point lies on the distinguished line. This line is the inverse image of the diagonal of $\pi_M^*(PTM)$, and is therefore a geometric representation of the total time derivative operator

$$d/dt \colon TM \to \tau_M^*(TM), \qquad v \mapsto (v, v).$$

6.1. The projective Cartan bundle

Following the same philosophy, we consider the pull-back of the simplex bundle $\pi_M^*(\mathscr{S}_{\mathscr{W}}M) \to PTM$. This has elements

$$([v], [\xi_0, \xi_1, \ldots, \xi_m]) \qquad \text{where } v \in T_pM \text{ and } \{\xi_0, \xi_1, \ldots, \xi_m\} \text{ is a basis for } \mathscr{W}_pM,$$

and as before there is a right action of $\mathrm{PGL}(m+1)$ on its fibres, given by matrix multiplication.

The projective Cartan bundle $\mathscr{PC}M \to PTM$ is the sub-bundle $\mathscr{PC}M \subset \pi_M^*(\mathscr{S}_{\mathscr{W}}M)$ given by

$$\mathscr{PC}M = \{([v], [\xi_0, \xi_1, \ldots, \xi_m])\}$$

where

$$\begin{cases} v \in T_pM \\ \{\xi_0, \xi_1, \ldots, \xi_m\} \text{ is a basis for } \mathscr{W}_pM \\ \xi_0 = \lambda e_0 \\ \rho(\xi_1) = \mu v \end{cases}$$

Here, each element is a reference simplex for a fibre of $\pi_M^*(P\mathscr{W}M) \to PTM$, where the first point is the distinguished point $[e_0]$, and where the second point is some other point in the distinguished line. The standard right action of $\mathrm{PGL}(m+1)$ on the fibres of $\pi_M^*(\mathscr{S}_{\mathscr{W}}M) \to PTM$ restricts to a right action of K_{m+1} on the fibres of the projective Cartan bundle $\mathscr{PC}M \to PTM$.

6.2. *BTW*-connections

A spray $\widetilde{\Gamma}$ on $T^\circ \mathscr{V}M$ defines a Berwald connection $\widetilde{\nabla}$ on the pull-back vector bundle

$$\tau_{\mathscr{V}M}^{\circ*}(T\mathscr{V}M) \to T^\circ \mathscr{V}M.$$

We say that $\widetilde{\nabla}$ is a *BTW*-connection if $\widetilde{\nabla}(\Psi \circ \tau_{\mathscr{V}M}^\circ) = \tau_{\mathscr{V}M*}^\circ$ and in addition

$$
\begin{aligned}
[\Psi^c, \widetilde{\nabla}_X Z] &= \widetilde{\nabla}_{[\Psi^c, X]} Z + \widetilde{\nabla}_X [\Psi^c, Z], \\
[\Psi^v, \widetilde{\nabla}_X Z] &= \widetilde{\nabla}_{[\Psi^v, X]} Z + \widetilde{\nabla}_X [\Psi^v, Z].
\end{aligned}
$$

Here,

$$X \in \mathfrak{X}(T^\circ \mathscr{V}M), \quad Z \in \mathfrak{X}(\tau_{\mathscr{V}M}^\circ),$$

Ψ^c, Ψ^v are the complete and vertical lifts of Ψ,

and the brackets $[\Psi^c, Z]$, $[\Psi^v, Z]$ are defined as commutators, using the fact that Ψ^c and Ψ^v are projectable to $\mathscr{V}M$. The spray $\widetilde{\Gamma}$ is a *BTW*-spray if its Berwald connection $\widetilde{\nabla}$ is a *BTW*-connection; its coordinate representation is

$$\widetilde{\Gamma} = \dot{x}^\alpha \frac{\partial}{\partial x^\alpha} - \left(G^a + (x^0)^{-1}\dot{x}^0\dot{x}^a\right)\frac{\partial}{\partial \dot{x}^a} - x^0 G^0 \frac{\partial}{\partial \dot{x}^0}$$

where G^a, G^0 are functions pulled back from $T^\circ M$.

6.3. *BTW*-connections and equivalence classes of Berwald connections

The relationship between *TW*-connections and equivalence classes of symmetric affine connections may be generalized in the following way, using the maps

$$T^\circ \mathscr{V}M \to \mathscr{W}^\circ M \xrightarrow{\rho} T^\circ M.$$

First, any *BTW*-spray $\widetilde{\Gamma}$ on $T^\circ \mathscr{V}M$ is projectable to a vector field $\bar{\Gamma}$ on $\mathscr{W}^\circ M$, represented in coordinates by

$$\bar{\Gamma} = \dot{x}^a \frac{\partial}{\partial x^a} - (G^a + w\dot{x}^a)\frac{\partial}{\partial \dot{x}^a} - (G^0 + w^2)\frac{\partial}{\partial w}.$$

Next, we consider sections $\sigma: T^\circ M \to \mathscr{W}^\circ M$ of ρ. Such a section is homogeneous if

$$\sigma(\lambda v) = \lambda \sigma(v);$$

an example of such a homogeneous section would be the linear section defined by a connection ϑ on $\mathscr{V}M \to M$. Then any homogeneous section σ maps a *BTW*-spray $\widetilde{\Gamma}$ to a spray Γ^σ on $T^\circ M$ by

$$\Gamma^\sigma = \rho_*\left(\bar{\Gamma}|_\sigma\right).$$

270

Different sections σ, σ' map $\widetilde{\Gamma}$ to projectively equivalent sprays Γ^{σ}, $\Gamma^{\sigma'}$, and there is a unique BTW-spray $\widetilde{\Gamma}$ mapping to the projective equivalence class of Γ satisfying

$$\operatorname{tr}(\widetilde{\nabla}) = 0, \qquad \operatorname{Ric}(\widetilde{\nabla}) = 0$$

(where the trace is taken with respect to the odd scalar density Θ); this is the *normal* BTW-connection.

6.4. *BTW*-connection to Cartan connection

The normal BTW-connection $\widetilde{\nabla}$ has an Ehresmann connection form $\widetilde{\omega}$ on the pull-back of the frame bundle $\tau^{\circ*}_{\mathscr{V}M}(\mathscr{F}(\mathscr{V}M)) \to T^{\circ}\mathscr{V}M$,

$$\widetilde{\omega} \colon T\left(\tau^{\circ*}_{\mathscr{V}M}(\mathscr{F}(\mathscr{V}M))\right) \to \mathfrak{gl}(m+1)\,.$$

This connection form $\widetilde{\omega}$ gives rise to an Ehresmann connection form $\widehat{\omega}$ on the pull-back of the simplex bundle $\pi^{*}_{M}(\mathscr{S}_{\mathscr{W}}M) \to PTM$,

$$\widehat{\omega} \colon T\left(\pi^{*}_{M}(\mathscr{S}_{\mathscr{W}}M)\right) \to \mathfrak{sl}(m+1)\,,$$

by

$$\widehat{\omega}_{\xi}(v) = p_{*}(\widetilde{\omega}_{\zeta}(w))\,,$$

where

$$\zeta \in \tau^{\circ*}_{\mathscr{V}M}(\mathscr{F}(\mathscr{V}M)) \text{ maps to } \xi \in \pi^{*}_{M}(\mathscr{S}_{\mathscr{W}}M)$$
$$w \in T_{\zeta}\tau^{\circ*}_{\mathscr{V}M}(\mathscr{F}(\mathscr{V}M)) \text{ maps to } v \in T_{\xi}\pi^{*}_{M}(\mathscr{S}_{\mathscr{W}}M)$$
$$p \colon GL(m+1) \to PGL(m+1)$$
$$p_{*} \colon \mathfrak{gl}(m+1) \to \mathfrak{sl}(m+1)\,;$$

the conditions on the BTW-connection ensure that $\widehat{\omega}$ is well-defined. We also find that $\ker \widehat{\omega} \cap T(\mathscr{P}\mathscr{C}M) \subset T(\pi^{*}_{M}(\mathscr{S}_{\mathscr{W}}M))$ contains only zero vectors, and so the restriction

$$\omega = \widehat{\omega}|_{T(\mathscr{P}\mathscr{C}M)}$$

is a Cartan connection (Sharpe [7]).

6.5. The Cartan bundle and the Projective Cartan bundle

Recall that the Cartan bundle $\mathscr{C}M \to M$ has

$$
\begin{aligned}
\text{base dimension} &= \dim M = m \\
\text{fibre dimension} &= \dim H_{m+1} = (m+1)^2 - m \\
\text{so that } \dim \mathscr{C}M &= (m+1)^2\,,
\end{aligned}
$$

whereas the projective Cartan bundle $\mathscr{PC}M \to PTM$ has

$$
\begin{aligned}
\text{base dimension} &= \dim PTM = 2m - 1 \\
\text{fibre dimension} &= \dim K_{m+1} = (m+1)^2 - (2m-1) \\
\text{so that } \dim \mathscr{PC}M &= (m+1)^2.
\end{aligned}
$$

As $\mathscr{C}M = \{[\xi_0, \xi_1, \ldots, \xi_m]\}$ where

$$
\begin{cases}
\{\xi_0, \xi_1, \ldots, \xi_m\} \text{ is a basis for } \mathscr{W}_p M \\
\xi_0 = \lambda e_0
\end{cases}
$$

whereas $\mathscr{PC}M = \{([v], [\xi_0, \xi_1, \ldots, \xi_m])\}$ where

$$
\begin{cases}
v \in T_p M \\
\{\xi_0, \xi_1, \ldots, \xi_m\} \text{ is a basis for } \mathscr{W}_p M \\
\xi_0 = \lambda e_0 \\
\rho(\xi_1) = \mu v
\end{cases}
$$

we may identify $\mathscr{C}M \leftrightarrow \mathscr{PC}M$ by

$$
[\xi_0, \xi_1, \ldots, \xi_m] \leftrightarrow ([\rho(\xi_1)], [\xi_0, \xi_1, \ldots, \xi_m]).
$$

6.6. The connection forms in a gauge

For a general equivalence class of sprays on $T^\circ M$, we use the projective Cartan bundle $\mathscr{PC}M \to PTM$. A gauge is local section of this bundle: with a suitable choice, ω is represented (in coordinates x^a, y^i on PTM, $i = 2, \ldots, m$, $y^i = (\dot{x}^1)^{-1}\dot{x}^i$) as the matrix

$$
\begin{pmatrix}
0 & \left(\alpha_{1c} + y^k \alpha_{kc}\right) dx^c & \alpha_{jc} dx^c \\
dx^1 & \left(\Pi^1_{1c} + y^k \Pi^1_{kc}\right) dx^c & \Pi^1_{jc} dx^c \\
dx^i - y^i dx^1 & \omega^i_1 & \left(\Pi^i_{jc} - y^i \Pi^1_{jc}\right) dx^c
\end{pmatrix}
$$

where $\omega^i_1 = \left(\Pi^i_{1c} - y^i \Pi^1_{1c} + y^k (\Pi^i_{kc} - y^i \Pi^1_{kc})\right) dx^c + dy^i$, and α_{bc} is related to the curvature of the corresponding Berwald connection.

If Γ affine, we use instead the coarser fibration $\mathscr{C}M \to M$: with a suitable choice of gauge, ω is now represented as the matrix

$$
\begin{pmatrix}
0 & \alpha_{bc} dx^c \\
dx^a & \Pi^a_{bc} dx^c
\end{pmatrix}
$$

These connections, in gauged form, are essentially the same as those given by Cartan [2].

ACKNOWLEDGMENTS

I should like to record my thanks to the organisers of the International Workshop on Global Analysis at Çankaya University for providing the opportunity to give the talk upon which this paper is based.

REFERENCES

1. T. N. Bailey, M. G. Eastwood and A. R. Gover, *Thomas's structure bundle for conformal, projective and related structures*, Rocky Mountain J. Math., **24**, (1994), 1191-1217.
2. E. Cartan, *Sur les variétés à connexion projective*, Bull. Soc. Math. France, **52**, (1924), 205-241.
3. M. Crampin and D. J. Saunders, *On projective connections: the affine case*, Preprint, Ghent University, (2004).
4. M. Crampin and D. J. Saunders, *On projective connections: the general case*, Preprint, Ghent University, (2004).
5. J. Douglas, *The general geometry of paths*, "Ann. Math., **29**, (1928), 143-168.
6. C. Roberts, *The projective connections of T.Y. Thomas and J.H.C. Whitehead applied to invariant connections*, Diff. Geom. Appl., **5**, (1995), 237-255.
7. R. W. Sharpe, *Differential Geometry: Cartan's Generalization of Klein's Erlangen Program*, Berlin: Springer, (1997).
8. T. Y. Thomas, *On the projective and equi-projective geometries of paths*, Proc. Nat. Acad. Sci., **11**, (1925), 199-203.
9. T. Y. Thomas, *A projective theory of affinely connected manifolds*, Math. Zeit., **25**, (1926), 723-733.
10. H. Weyl, *Zur Infinitesimalgeometrie; Einordnung der projektiven und der konformen Auffassung*, Gött. Nach., (1921), 99-112.
11. J. H. C. Whitehead, *The representation of projective spaces*, Ann. Math., **32**, (1931), 327-360.

Some properties of the Cauchy-type integral for the Laplace vector fields theory

Baruch Schneider* and Michael Shapiro[†]

*Department of Mathematics, Izmir University of Economics, 35330, Balcova, Izmir, Turkey
[†]Departamento de Matemáticas, Escuela Superior de Física y Mathemáticas, 07300 México, D.F.,
México

Abstract. We study the analog of the Cauchy-type integral for the Laplace vector fields theory in case of a piece-wise Liapunov surface of integration and we prove the Sokhotski-Plemelj theorem for it as well as the necessary and sufficient condition for the possibility to extend a given Hölder function from such a surface up to a Laplace vector field. Formula for the square of the singular Cauchy-type integral is given. The proofs of all these facts are based on intimate relations between Laplace vector field and some versions of quaternionic analysis.
Key words. Cauchy-type integral, Laplace vector fields theory, quaternionic analysis.
PACS. 11.10.-z.
MSC. 30G35, 37C10.

1. INTRODUCTION

In the present paper we follow the approach presented in paper [6] in which we studied the analog of the Cauchy-type integral for the theory of time-harmonic electromagnetic fields in case of a piece-wise Liapunov surface of integration. The paper is organized as follows. In Section 2 there is formulated a series of theorems which cover basic properties of the Cauchy-type integral for the Laplace vector fields theory in case of the piece-wise Liapunov surface of integration. The proofs of all of them one can find in Section 4, and they are given in the form of more or less direct corollaries of the corresponding facts valid for hyperholomorphic function theory which is being developed in Section 3 and Reference [7].

2. LAPLACE VECTOR FIELDS THEORY AND THE CAUCHY-LAPLACE INTEGRAL

2.1 Let Ω denote a domain in $\mathbb{R}^3$, $\Gamma := \partial\Omega$ be its boundary, and let $\vec{f} : \Omega \subset \mathbb{R}^3 \to \mathbb{R}^3$. We'll consider the following system:

$$\begin{cases} div \, \vec{f} = 0, \\ rot \, \vec{f} = 0. \end{cases} \tag{2.1}$$

CP729, Global Analysis and Applied Mathematics: International Workshop on Global Analysis,
edited by K. Taş, D. Krupka, O. Krupková, and D. Baleanu

If $\vec{f}$ is a solution of the system (2.1) in Ω, then $\vec{f}$ is called a **Laplace vector field**. The integral

$$C_\Gamma(x, \vec{f}(\tau)) := -\frac{1}{4\pi} \int_\Gamma (grad\frac{1}{|\tau - x|} \langle \vec{n}(\tau), \vec{f}(\tau) \rangle + [grad\frac{1}{|\tau - x|}, [\vec{n}(\tau), \vec{f}(\tau)]]) ds, \ x \notin \Gamma,$$

plays the role of the Cauchy-type integral in the theory of three-dimensional Laplace vector fields with $\vec{f} : \Gamma \to \mathbb{R}^3$ (see [11]), where ds is the differential form of the two-dimensional surface area Γ in $\mathbb{R}^3$, $\vec{n}(\tau) = (n_1(\tau), n_2(\tau), n_3(\tau))$ is the outward unit normal to surface Γ at τ. We shall call it the Cauchy-Laplace-type integral.

2.2 Let $H_\mu(\Gamma, \mathbb{R}^3)$ denote the class of functions satisfying the Hölder condition with the exponent $0 < \mu \leq 1$. Let Γ be a surface in $\mathbb{R}^3$ which contains a finite number of conical points and a finite number of non-intersecting edges such that none of the edges contain any of conical points. If the complement (in Γ) of the union of conical points and edges, is a Liapunov surface, then we shall refer to Γ as a piece-wise Liapunov surface in $\mathbb{R}^3$.

2.3 We shall use the following notation:

$$\mathfrak{W} := \left\{ \vec{f} : \Gamma \to \mathbb{R}^3 \ \middle| \ \int_\Gamma \langle grad\frac{1}{|\tau - x|}, [\vec{n}(\tau), \vec{f}(\tau)] \rangle ds = 0, \ x \notin \Gamma \right\},$$

$$\widetilde{\mathfrak{W}} := \left\{ \vec{f} : \Gamma \to \mathbb{R}^3 \ \middle| \ \int_\Gamma \langle grad\frac{1}{|\tau - t|}, [\vec{n}(\tau), \vec{f}(\tau)] \rangle ds = 0, \ \forall t \in \Gamma \right\},$$

note that $\widetilde{\mathfrak{W}}$ can be described in purely physical terms [11].

2.4 Theorem *(Sokhotski-Plemelj formulas for the Cauchy-Laplace-type integral with the piece-wise Liapunov surface of integration). Let Ω be a bounded domain in $\mathbb{R}^3$ with the piece-wise Liapunov boundary. Let $\vec{f} \in H_\mu(\Gamma, \mathbb{R}^3) \cap \mathfrak{W}$. Then the following limits exist:*

$$\lim_{\Omega^\pm \ni x \to t \in \Gamma} C_\Gamma(x, \vec{f}) =: C_\Gamma^\pm(t, \vec{f}),$$

moreover, the following identities hold:

$$C_\Gamma^+(t, \vec{f}) = (1 - \frac{\gamma(t)}{4\pi})\vec{f}(t) + C_\Gamma(t, \vec{f}) =: (1 - \frac{\gamma(t)}{4\pi})\vec{f}(t) + \frac{1}{2} S_{C_\Gamma}[\vec{f}](t),$$

$$(2.2)$$

$$C_\Gamma^-(t, \vec{f}) = -\frac{\gamma(t)}{4\pi}\vec{f}(t) + C_\Gamma(t, \vec{f}) =: -\frac{\gamma(t)}{4\pi}\vec{f}(t) + \frac{1}{2} S_{C_\Gamma}[\vec{f}](t),$$

for all $t \in \Gamma$, where $S_{C_\Gamma}[\vec{f}](t) := 2C_\Gamma(t, \vec{f})$, the integrals being understood in the sense of the Cauchy principal value, $\gamma(t)$ is the measure of a solid angle of the tangential conical surface at the point t or is the solid measure of the tangential dihedral angle at the point t.

2.5 We shall call the operator S_{C_Γ} the singular Cauchy-Laplace integral operator. It's appeared that many properties which are of interest for us, can be expressed better in terms of another operator.

Set

$$\check{S}_{C_\Gamma}[\vec{f}](t) := \frac{2\pi - \gamma(t)}{2\pi}\vec{f}(t) + S_{C_\Gamma}[\vec{f}](t),$$

for any $t \in \Gamma$. We shall call $\check{S}_{C_\Gamma}$ the modified singular Cauchy-Laplace integral operator.

2.6 Theorem (*Plemelj-Privalov's type theorem for the Laplace vector fields*). *Let Ω be a bounded domain in $\mathbb{R}^3$ with piece-wise Liapunov boundary. Then for $0 < \mu < 1$*

$$\vec{f} \in H_\mu(\Gamma, \mathbb{R}^3) \cap \mathfrak{W} \Rightarrow \check{S}_{C_\Gamma}[\vec{f}] \in H_\mu(\Gamma, \mathbb{R}^3) \cap \mathfrak{W}. \tag{2.3}$$

2.7 Theorem (*extension of a given on Γ Hölder function up to a Laplace vector field*). *Let Ω be a bounded domain in $\mathbb{R}^3$ with the piece-wise Liapunov boundary.*

1. *In order to a function $\vec{f} \in H_\mu(\Gamma, \mathbb{R}^3) \cap \mathfrak{W}$ be the boundary value of a vector field $\vec{\tilde{f}}$ which is a Laplace vector field in Ω^+ and is continuous in $\overline{\Omega^+}$, it is necessary and sufficient that*

$$\vec{f}(t) = \check{S}_{C_\Gamma}[\vec{f}](t), \ \forall t \in \Gamma. \tag{2.4}$$

2. *In order to a function $\vec{f} \in H_\mu(\Gamma, \mathbb{R}^3) \cap \mathfrak{W}$ be a boundary value of a vector field $\vec{\tilde{f}}$ which is a Laplace vector field in Ω^- and is continuous in $\overline{\Omega^-}$ and vanishes at infinity, it is necessary and sufficient that*

$$\vec{f}(t) = -\check{S}_{C_\Gamma}[\vec{f}](t), \ \forall t \in \Gamma.$$

2.8 Theorem (*On the square of the operators S_{C_Γ} and $\check{S}_{C_\Gamma}$*). *If Γ is a piece-wise Liapunov surface, then for $\vec{f} \in H_\mu(\Gamma, \mathbb{R}^3) \cap \mathfrak{W}$, $0 < \mu < 1$, we have the following formulas:*

$$S_{C_\Gamma}^2[\vec{f}](t) = a_1(t)\vec{f}(t) + a_2(t)S_{C_\Gamma}[\vec{f}](t) + S_{C_\Gamma}[a_3\vec{f}](t), \tag{2.5}$$

$$\check{S}_{C_\Gamma}^2[\vec{f}](t) = \vec{f}(t), \tag{2.6}$$

for $\forall t \in \Gamma$, i.e., the modified singular Cauchy-Laplace integral operator $\check{S}_{C_\Gamma}$ is an involution on $H_\mu(\Gamma, \mathbb{R}^3) \cap \mathfrak{W}$, $0 < \mu < 1$,

$$\check{S}_{C_\Gamma}^2 = I,$$

where

$$a_1(t) := \frac{\gamma(t)}{\pi} - \frac{\gamma^2(t)}{4\pi^2}, \ a_2(t) := \frac{\gamma(t)}{2\pi} - 2, \ a_3(t) := \frac{\gamma(t)}{2\pi}.$$

The proofs of these theorems can be found in Section 4. Note that theorems 2.4 and 2.7 were proved in [11] under more restrictive conditions on densities and another method.

3. HYPERHOLOMORPHIC FUNCTION THEORY: GENERAL INFORMATION

In this section, we provide some background on quaternionic analysis needed in this paper. For more information, we refer the reader to [1], [2], [3], [9], [10].

3.1 Let $\mathbb{H}$ be the set of real quaternions, it means that each quaternion a is represented in the form $a = \sum_{k=0}^{3} a_k i_k$ with the standard basis $\{i_0 := 1, i_1, i_2, i_3\}$, where $\{a_k \mid k \in \mathbb{N}_3^0 := \mathbb{N}_3 \cup \{0\}; \ \mathbb{N}_3 := \{1, 2, 3\}\} \subset \mathbb{R}$. We use the Euclidean norm $|a|$ in $\mathbb{H}$, defined by $|a| := \sqrt{\sum_{k=0}^{3} |a_k|^2}$. Let $a = \sum_{k=0}^{3} a_k i_k \in \mathbb{H}$, then $a_0 =: Sc(a); \ \vec{a} := \sum_{k=1}^{3} a_k \cdot i_k =: Vect(a)$ are called, respectively, the scalar and the vector part of a quaternion. We write $a = a_0 + \vec{a}$. In vector terms, the multiplication of two arbitrary real quaternions a, b can be rewritten as follows:

$$a \cdot b = (a_0 + \vec{a}) \cdot (b_0 + \vec{b}) = a_0 \cdot b_0 - \langle \vec{a}, \vec{b} \rangle + a_0 \vec{b} + b_0 \vec{a} + [\vec{a}, \vec{b}], \tag{3.1}$$

where $\langle \cdot, \cdot \rangle$ and $[\cdot, \cdot]$ denote the usual scalar and vector products of three-dimensional vectors. In particular if $a_0 = b_0 = 0$ then we have

$$a \cdot b = -\langle \vec{a}, \vec{b} \rangle + [\vec{a}, \vec{b}]. \tag{3.2}$$

3.2 We shall consider functions ranged in $\mathbb{H}$ and defined in a domain $\Omega \subset \mathbb{R}^3$. Notations $C^p(\Omega, \mathbb{H})$, $p \in \mathbb{N} \cup \{0\}$, have the usual component-wise meaning. A function f is called left-hyperholomorphic if

$$D[f] := \sum_{k=1}^{3} i_k \cdot \frac{\partial f}{\partial x_k} =: \sum_{k=1}^{3} i_k \partial_k[f] = 0$$

holds in Ω. The vector representation of a quaternion (see Subsection 3.1) gives rise to the following representation of the operator D: for any $f \in C^1(\Omega, \mathbb{H})$, $f = f_0 + \vec{f}$,

$$
\begin{aligned}
D[f] &:= \sum_{k=1}^{3} i_k \frac{\partial f}{\partial x_k} = \sum_{k=1}^{3} i_k \frac{\partial}{\partial x_k} (f_0 + \vec{f}) = \sum_{k=1}^{3} i_k \frac{\partial f_0}{\partial x_k} - \sum_{k=1}^{3} \langle i_k, \frac{\partial \vec{f}}{\partial x_k} \rangle + \sum_{k=1}^{3} [i_k, \frac{\partial \vec{f}}{\partial x_k}] = \\
&= \operatorname{grad} f_0 - \operatorname{div} \vec{f} + \operatorname{rot} \vec{f}.
\end{aligned}
\tag{3.3}
$$

Let $\theta = -\frac{1}{4\pi} \frac{1}{|x|}$ be the fundamental solution of the Laplace operator, then the fundamental solution to the operator D, $\mathbf{K}$, is given by the formula (see [9]):

$$\mathbf{K}(x) := -D[\theta](x) = \frac{1}{4\pi} \sum_{k=1}^{3} \bar{i}_k \frac{x_k}{|x|^3} = \frac{1}{4\pi} \sum_{k=1}^{3} i_k \frac{\partial}{\partial x_k} \left(\frac{1}{|x|} \right) = \frac{1}{4\pi} \operatorname{grad} \frac{1}{|x|}, \tag{3.4}$$

where $\operatorname{grad} = \sum_{k=1}^{3} i_k \frac{\partial}{\partial x_k}$ is a quaternion-valued operator. Let $\Omega = \Omega^+$ be a domain in $\mathbb{R}^3$ with the boundary Γ which is assumed to be a piece-wise Liapunov surface; denote $\Omega^- := \mathbb{R}^3 \setminus (\Omega^+ \cup \Gamma)$. If f is a Hölder function then its left-hyperholomorphic Cauchy-type integral is defined:

$$K_\Gamma[f](x) := \int_\Gamma \mathbf{K}(\tau - x) \cdot \vec{n}(\tau) \cdot f(\tau) ds, \ x \in \Omega^{\pm}.$$

For more information about hyperholomorphic functions we refer to [1], [2], [3], [9], [10], see also [5].

4. PROOFS OF THE THEOREMS FROM SECTION 2

4.1 In this Section we'll prove all theorems from Section 2 using the relations between the Laplace vector theory and the theory of hyperholomorphic functions. We start this Section with a brief description of the relations between the Laplace vector theory and the theory of hyperholomorphic functions. One can find more about this in [3], [4], [8]. If $f =: \vec{f}$ is a purely imaginary quaternion-valued function (i.e. $Sc\, f = 0$), then by (3.3) we have

$$D[\vec{f}] = -div\,\vec{f} + rot\,\vec{f}.$$

Thus

$$D[\vec{f}] = 0 \Longleftrightarrow \left\{ \begin{array}{l} div\,\vec{f} = 0, \\ rot\,\vec{f} = 0, \end{array} \right.$$

i.e., the class of Laplace vector fields one can identify with the set of purely imaginary hyperholomorphic functions.

4.2 By the equality (3.4) for a purely imaginary quaternionic function $\vec{f}$ we have

$$K_\Gamma[\vec{f}](x) = \frac{1}{4\pi} \int_\Gamma grad \frac{1}{|\tau - x|} \cdot (-\langle \vec{n}(\tau), \vec{f}(\tau) \rangle + [\vec{n}(\tau), \vec{f}(\tau)]) ds =$$

$$= \frac{1}{4\pi} \int_\Gamma (grad \frac{1}{|\tau - x|} \cdot (-\langle \vec{n}(\tau), \vec{f}(\tau) \rangle) + grad \frac{1}{|\tau - x|} \cdot [\vec{n}(\tau), \vec{f}(\tau)]) ds =$$

$$= \frac{1}{4\pi} \int_\Gamma (grad \frac{1}{|\tau - x|} (-\langle \vec{n}(\tau), \vec{f}(\tau) \rangle) - \langle grad \frac{1}{|\tau - x|}, [\vec{n}(\tau), \vec{f}(\tau)] \rangle +$$

$$+ \ [grad \frac{1}{|\tau - x|}, [\vec{n}(\tau), \vec{f}(\tau)]]) ds = -\frac{1}{4\pi} \int_\Gamma \langle grad \frac{1}{|\tau - x|}, [\vec{n}(\tau), \vec{f}(\tau)] \rangle ds -$$

$$- \ \frac{1}{4\pi} \int_\Gamma (grad \frac{1}{|\tau - x|} \langle \vec{n}(\tau), \vec{f}(\tau) \rangle + [grad \frac{1}{|\tau - x|}, [\vec{n}(\tau), \vec{f}(\tau)]]) ds.$$

Thus we expressed $K_\Gamma[\vec{f}](x)$ in vectorial terms and separating the scalar and the vector components we have:

$$Sc(K_\Gamma[\vec{f}](x)) = -\frac{1}{4\pi} \int_\Gamma \langle grad \frac{1}{|\tau - x|}, [\vec{n}(\tau), \vec{f}(\tau)] \rangle ds,$$

$$Vect(K_\Gamma[\vec{f}](x)) = -\frac{1}{4\pi} \int_\Gamma (grad \frac{1}{|\tau - x|} \langle \vec{n}(\tau), \vec{f}(\tau) \rangle + [grad \frac{1}{|\tau - x|}, [\vec{n}(\tau), \vec{f}(\tau)]]) ds.$$

So, the integral $K_\Gamma[\vec{f}](x)$ for $\vec{f} \in \mathfrak{W}$ coincides with $C_\Gamma(x, \vec{f})$. In the same way

$$S_\Gamma[\vec{f}](t) := 2K_\Gamma[\vec{f}](t) = -\frac{1}{2\pi} \int_\Gamma \langle grad \frac{1}{|\tau - t|}, [\vec{n}(\tau), \vec{f}(\tau)]\rangle ds -$$

$$-\frac{1}{2\pi} \int_\Gamma (grad \frac{1}{|\tau - t|} \langle \vec{n}(\tau), \vec{f}(\tau)\rangle + [grad \frac{1}{|\tau - t|}, [\vec{n}(\tau), \vec{f}(\tau)]])ds, \ \forall t \in \Gamma,$$

so, the singular integral $S_\Gamma[\vec{f}]$ for $\vec{f} \in H_\mu(\Gamma, \mathbb{R}^3) \cap \tilde{\mathfrak{W}}$ coincides with $S_{C_\Gamma}[\vec{f}]$, also we one can find more details and certain physical reasonings in Reference [11, Part 2]. Let us consider the relation between $H_\mu(\Gamma, \mathbb{R}^3) \cap \mathfrak{W}$ and $H_\mu(\Gamma, \mathbb{R}^3) \cap \tilde{\mathfrak{W}}$. Recalling definitions of $\mathfrak{W}$ and $\tilde{\mathfrak{W}}$ we have:

$$H_\mu(\Gamma, \mathbb{R}^3) \cap \mathfrak{W} = \left\{\vec{f} \in H_\mu(\Gamma, \mathbb{R}^3), \ Sc(K_\Gamma[\vec{f}](x)) = 0, \ x \in \Omega^\pm\right\}, \quad (4.1)$$

$$H_\mu(\Gamma, \mathbb{R}^3) \cap \tilde{\mathfrak{W}} = \left\{\vec{f} \in H_\mu(\Gamma, \mathbb{R}^3), \ Sc(S_\Gamma[\vec{f}](t)) = 0, \ t \in \Gamma\right\}. \quad (4.2)$$

The structure of formula (2.2) show that

$$\lim_{\Omega^\pm \ni x \to t \in \Gamma} Sc(K_\Gamma[\vec{f}](x)) = Sc(K_\Gamma[\vec{f}]^\pm(t)) = 2Sc(S_\Gamma[\vec{f}](t))$$

which implies the inclusion:

$$\vec{f} \in H_\mu(\Gamma, \mathbb{R}^3) \cap \mathfrak{W} \subset \vec{f} \in H_\mu(\Gamma, \mathbb{R}^3) \cap \tilde{\mathfrak{W}}.$$

4.3 Proof of Theorem 2.4 Let $\vec{f} \in H_\mu(\Gamma, \mathbb{R}^3) \cap \mathfrak{W}$, consider $C_\Gamma(x, \vec{f})$. It was proved that

$$C_\Gamma(x, \vec{f}) = K_\Gamma[\vec{f}](x).$$

By Reference [7, Theorem 2.1 for $\alpha = 0$], see also [1] and [2], there exist $K_\Gamma[\vec{f}]^\pm(t)$ and

$$K_\Gamma[\vec{f}]^+(t) = (1 - \frac{\gamma(t)}{4\pi})\vec{f}(t) + K_\Gamma[\vec{f}](t) =: (1 - \frac{\gamma(t)}{4\pi})\vec{f}(t) + \frac{1}{2}S_\Gamma[\vec{f}](t),$$

$$K_\Gamma[\vec{f}]^-(t) = -\frac{\gamma(t)}{4\pi}\vec{f}(t) + K_\Gamma[\vec{f}](t) =: -\frac{\gamma(t)}{4\pi}\vec{f}(t) + \frac{1}{2}S_\Gamma[\vec{f}](t).$$

Hence $\exists C_\Gamma^\pm(t, \vec{f})$, and after not complicated computation we obtain the result. $\square$

Set

$$\check{S}_\Gamma[\vec{f}](t) := \frac{2\pi - \gamma(t)}{2\pi}\vec{f}(t) + S_\Gamma[\vec{f}](t),$$

for any $t \in \Gamma$.

4.4 Proof of Theorem 2.6 Let $\vec{f} \in H_\mu(\Gamma, \mathbb{R}^3) \cap \mathfrak{W}$, consider $C_\Gamma(x, \vec{f}(\tau))$. By Theorem 2.4 $\exists C_\Gamma^\pm(t, \vec{f})$ and

$$C_\Gamma^+(t, \vec{f}) = \frac{1}{2}[\vec{f}(t) + \check{S}_{C_\Gamma}[\vec{f}](t)],$$

$$C_\Gamma^-(t, \vec{f}) = \frac{1}{2}[-\vec{f}(t) + \check{S}_{C_\Gamma}[\vec{f}](t)],$$

where $\check{S}_{C_\Gamma}$ was defined in Subsection 2.5. By Subsection 4.2, $\vec{f} \in H_\mu(\Gamma, \mathbb{R}^3) \cap \tilde{\mathfrak{W}}$, hence on Γ, $\check{S}_{C_\Gamma}[\vec{f}] = \check{S}_\Gamma[\vec{f}]$. In Reference [7, Subsection 2.2 for $\alpha = 0$] it was proved that $\check{S}_\Gamma$ satisfy the Hölder condition. So, recalling the relationship between the operators $\check{S}_\Gamma$ and $\check{S}_{C_\Gamma}$ we have that $\check{S}_{C_\Gamma}[\vec{f}] \in H_\mu(\Gamma, \mathbb{R}^3) \cap \mathfrak{W}$. $\square$

4.5 Proof of Theorem 2.7 This follows from Reference [7, Theorem 2.3 for $\alpha = 0$] taking into account the above relation between the class of Laplace vector fields and the set of purely imaginary hyperholomorphic functions. $\square$

4.6 Proof of Theorem 2.8 Let $\vec{f} \in H_\mu(\Gamma, \mathbb{R}^3) \cap \mathfrak{W}$ and consider $S_{C_\Gamma}[\vec{f}]$. In Subsection 4.2 it was proved:

$$\vec{f} \in H_\mu(\Gamma, \mathbb{R}^3) \cap \mathfrak{W} \Longrightarrow \vec{f} \in H_\mu(\Gamma, \mathbb{R}^3) \cap \tilde{\mathfrak{W}}.$$

So, we obtain (2.5) after taking into account Theorem 2.4 [7, for $\alpha = 0$] see also [2] combined with a straightforward calculation. Using the definition of the modified singular operator $\check{S}_{C_\Gamma}$ we obtain (2.6). $\square$

ACKNOWLEDGMENTS

Michael Shapiro was partially supported by CONACYT projects as well as by Instituto Politéchnico National in the framework of COFAA and CGPI programs.

REFERENCES

1. K. Gürlebeck and W. Sprössig, *Quaternionic and Clifford Calculus for Physicists and Engineers*. John Wiley & Sons, England 1997, 371 pp.
2. K. Gürlebeck and W. Sprössig, *Quaternionic Analysis and Elliptic Boundary Value Problems*. Math. Res. 56, Akademie-Verlog, Berlin 1989, 253 pp.
3. V. Kravchenko and M. Shapiro, *Integral representations for spatial models of mathematical physics*. Addison Wesley Longman, Pitman Research Notes in Mathematics Series 351, 1996, 247 pp.
4. I. Mitelman and M. Shapiro, *Formulae of changing of integration order and of inversion for some multidimensional singular integrals and hypercomplex analysis*. Journal of Natural Geometry, 1994, v. 5, pp. 11–27.
5. M. Mitrea, *Clifford Wavelets, Singular Integrals, and Hardy Spaces*. Lecture Notes in Mathematics, No. 1575, Springer-Verlag, New York 1994.
6. B. Schneider and M. Shapiro, *Some properties of the Cauchy-type integral for the time-harmonic Maxwell equations*. Integral Equations and Operator Theory 44(1), pp. 93 – 126, 2002.
7. B. Schneider and M. Shapiro, *Some properties of quaternionic Cauchy-type integral for a piece-wise Liapunov surface of integration*. IMCP, vol. 16, pp. 243–260, 2004.
8. M. Shapiro, *Structure of the quaternionic modules and some properties of the involutive operators*. Journal of Natural Geometry, 1992, v. 1, pp. 9–37.
9. M. Shapiro and N. Vasilevski, *Quaternionic ψ-hyperholomorphic functions, singular integral operators and boundary value problems. I. ψ-hyperholomorphic function theory*. Complex variables, theory and applications, 1995, v. 27, pp. 17–46.
10. M. Shapiro and N. Vasilevski, *Quaternionic ψ-hyperholomorphic functions, singular integral operators and boundary value problems. II. Algebras of singular integral operators and Riemann type boundary value problems*. Complex variables, theory and applications, 1995, v. 27, pp. 67–96.
11. M. Zhdanov, *Integral transforms in geophysics*. Springer-Verlag, Heidelberg, 1988.

A generalization of Lepage forms in mechanics

J. Šeděnková

Department of Mathematics, Tomas Bata University, Zlín, Czech Republic

Abstract. In this paper we generalize the concept of a Lepage form, introduced by Krupka, to forms of arbitrary degree in mechanics. These forms allow us to find a suitable representation of the classes of forms, appearing in variational sequences in mechanics. The structure of Lepage 2-forms is discussed in detail. The Lepage equivalents of the dynamical forms are mentioned.
Key words. Lepage form, Euler-Lagrange form, variational sequence.
PACS. 02.30.Xx, 02.40.Vh, 02.40.Yy.
MSC. 35A15, 58A10, 58A20, 70G75.

1. INTRODUCTION

In this paper, a construction is introduced, allowing us to generalize the concept of a Lepage form (Krupka [9, 10]) to forms of arbitrary degree in the higher order variational sequences on fibered manifolds over one-dimensional bases (i.e., in mechanics).

The r-th order variational sequence is by definition the quotient sequence of the De Rham sequence on the r-jet prolongation of a fibered manifold, factored through its contact subsequence (Krupka [12]). Basic general properties of the sequence, and in particular, of the variational terms (lagrangians, Euler-Lagrange forms and Helmholtz-Sonin forms) have been studied by several authors. A complete local representation of the r-th order variational sequence in mechanics was found by Štefánek [19]. Another representation of all classes in the first order variational sequence was given by Krupka [11]. Musilová and Krbek [16] found a representation of the variational terms in higher order variational sequence in mechanics. Kašparová [6] found a representation of classes of n-forms, $(n+1)$-forms and $(n+2)$-forms of the variational sequence in the first order field theory. Her results were extended to the general order by Krbek, Musilová and Kašparová [8]. The representation of all terms in the r-th order field theory were found by Krbek, Musilová [7] by the use of a finite version of Anderson's interior Euler operator [1].

Francaviglia, Palese and Vitolo discussed, among others, such questions as the correspondence of variational sequences and bicomplexes, and their relations to spectral sequences ([4, 5, 20]).

The need of global concepts in higher order variational theory led to the introduction of the so called Lepage n-forms in field theory, and Lepage equivalents of lagrangians. The main idea, going back to Lepage and Dedecker, was that there should exist a

CP729, *Global Analysis and Applied Mathematics: International Workshop on Global Analysis,*
edited by K. Taş, D. Krupka, O. Krupková, and D. Baleanu
© 2004 American Institute of Physics 0-7354-0209-4/04/$22.00

close connection between the Euler-Lagrange mapping and the exterior derivative of forms (Krupka [10]). Later, the concept of the Lepage form was extended to 2-forms in mechanics and to $(n+1)$-forms in field theory (Krupková [14, 13]); the Lepage forms have been introduced as closed counterparts of the Euler-Lagrange forms. In [14], the Lepage forms have been applied to the inverse problem in higher order mechanics, and to the order reducibility problem.

In our generalization of Lepage forms we use a slight (finite order) modification of an operator $\mathscr{I}$, acting on forms on jet manifolds, given by Anderson [1] for the case of the variational bicomplex and called by Anderson the interior Euler operator. This operator was already used, and denoted by different symbols, by Kuperschmidt [15], Dedecker and Tulczyjew [3], and Bauderon [2].

2. VARIATIONAL SEQUENCE

Let $\pi : Y \to X$ be a fibered manifold with fibered coordinate systems (V, ψ), $\psi = (t, q^\sigma)$, on Y and (U, φ), $\varphi = (t)$ on X, $\dim X = 1$, $\dim Y = m + 1$. Denote by $\pi^r :$ $J^r Y \to X$ or just $J^r Y$ the r-jet prolongation of the fibered manifold $\pi : Y \to X$, the coordinate system is (V^r, ψ^r), $\psi^r = (t, q^\sigma, q_1^\sigma, \cdots, q_r^\sigma)$ on $J^r Y$. For small r we denote $q_0^\sigma = q^\sigma, q_1^\sigma = \dot{q}^\sigma, q_2^\sigma = \ddot{q}^\sigma$. The canonical jet projections are $\pi^{r,s} : J^r Y \to J^s Y$, $r > s$ and $\pi^{r,0} : J^r Y \to Y$.

A differential k-form ρ on $J^r Y$ is called *contact*, if it vanishes along the r-jet prolongation $J^r \gamma$ of every section γ of π.

If (V, ψ), $\psi = (t, q^\sigma)$, is a fibered chart on Y, then we often use the *contact basis* $dt, \omega^\sigma, \omega_1^\sigma, \ldots, \omega_r^\sigma, dq_{r+1}^\sigma$ on $V^{r+1} = (\pi^{r+1,0})^{-1} V$ given by the forms

$$\omega_j^\sigma = dq_j^\sigma - q_{j+1}^\sigma dt, \qquad 0 \leq j \leq r. \tag{1}$$

Recall that a form which contains exactly k expressions (1) is called k-contact. Every form ρ on $J^r Y$ can be uniquely decomposed, after the lifting to $J^{r+1} Y$, as the sum of the k-contact components $p_k \rho$.

Let Ω_k^r be the direct image of the sheaf of smooth k-forms over $J^r Y$ by the jet projection $\pi^{r,0}$, where $k \geq 0$. Denote

$$\Omega_{0,c}^r = \{0\}, \quad \Omega_{k,c}^r = \ker p_{k-1}, \quad \Theta_k^r = \Omega_{k,c}^r + d\Omega_{k-1,c}^r, \tag{2}$$

where $k \geq 1$, and $d\Omega_{k-1,c}^r$ is the image sheaf of $\Omega_{k-1,c}^r$ by d. Then for every open set $V \subset Y$, $\Omega_k^r V$ (resp. $\Omega_{k,c}^r V$) is the Abelian group of k-forms (resp. k-contact k-forms) on $V^r = (\pi^{r,0})^{-1}(V)$, $d\Omega_{k-1,c}^r V$ is the Abelian group of forms which can be locally expressed as differentials of $(k-1)$-contact $(k-1)$-forms on V^r, and $\Theta_k^r V$ is a subgroup of $\Omega_k^r V$. We get a sequence

$$0 \to \Theta_1^r \to \Theta_2^r \to \Theta_3^r \to \ldots \to \Theta_M^r \to 0, \tag{3}$$

in which all arrows denote the exterior differentiation d, and $M = mr + 1$. Sequence (3) is a subsequence of the De Rham sequence

$$0 \to \mathbb{R}_Y \to \Omega_0^r \to \Omega_1^r \to \Omega_2^r \to \ldots \to \Omega_{N-1}^r \to \Omega_N^r \to 0, \tag{4}$$

where $N = \dim J^r Y = 1 + m(r+1)$. The quotient sequence

$$
\begin{aligned}
0 \to \mathbb{R}_Y \to \Omega_0^r \to \Omega_1^r/\Theta_1^r \to \Omega_2^r/\Theta_2^r \to \ldots \\
\ldots \to \Omega_M^r/\Theta_M^r \to \Omega_{M+1}^r \to \ldots \to \Omega_{N-1}^r \to \Omega_N^r \to 0
\end{aligned}
\tag{5}
$$

is also exact. Sequence (5) is called the *r-th order variational sequence*. The class of a differential form $\rho \in \Omega_k^r V$ in the variational sequence (5) is denoted by $[\rho]$.

The quotient mapping $E : \Omega_k^r/\Theta_k^r \to \Omega_{k+1}^r/\Theta_{k+1}^r$ is defined by

$$
E([\rho]) = [d\rho].
\tag{6}
$$

This mapping satisfies the condition $E^2 = 0$. The quotient mapping $E : \Omega_1^r/\Theta_1^r \to \Omega_2^r/\Theta_2^r$ is called the *Euler-Lagrange mapping*. The quotient mapping $E : \Omega_2^r/\Theta_2^r \to \Omega_3^r/\Theta_3^r$ is called the *Helmholtz-Sonin mapping*.

A *lagrangian* of order r is a π^r-horizontal n-form λ. In coordinates, the following can be written

$$
\lambda = L dt,
\tag{7}
$$

where L is a function on $J^r Y$ called *Lagrange function*.

Let ρ be a 1-form on $J^r Y$. A form ρ is called a *Lepage 1-form* if $p_1 d\rho$ is a $\pi^{r+1,0}$-horizontal 2-form. A Lepage form ρ is called a *Lepage equivalent* of a lagrangian λ if $h\rho = \lambda$. It is known that in higher order mechanics, Lepage equivalents are uniquely determined by lagrangians. We denote by θ_λ the Lepage equivalent of a lagrangian λ. If $r = 1$, θ_λ is the well known *Poincaré-Cartan form*, if $r > 1$, we have the *generalized Poincaré-Cartan form*. If in a fibered chart $\lambda = L dt$, then

$$
p_1 d\theta_\lambda = E_\sigma(L)\omega^\sigma \wedge dt,
\tag{8}
$$

where

$$
E_\sigma(L) = \sum_{l=0}^{r} (-1)^l \frac{d^l}{dt^l} \frac{\partial L}{\partial q_l^\sigma}.
\tag{9}
$$

The form (8) is called the *Euler-Lagrange form* and it is denoted by E_λ. The components (9) are called the *Euler-Lagrange expressions*.

3. THE INTERIOR EULER-LAGRANGE OPERATOR

We recall basic properties of the interior Euler-Lagrange operator rewritten in the form presented in Šeděnková [18].

Let (V, ψ), $\psi = (t, q^\sigma)$, be a fibered chart on Y and let (V^{2r+1}, ψ^{2r+1}), $\psi^{2r-1} = (t, q^\sigma, q_1^\sigma, \ldots, q_{2r+1}^\sigma)$, be the associated fibered chart on $J^{2r+1}Y$. We set

$$
\Xi = \frac{\partial}{\partial t} + \sum_{j=0}^{2r} q_{j+1}^\sigma \frac{\partial}{\partial q_j^\sigma},
\tag{10}
$$

Ξ is a vector field on V^{2r+1}. If $\rho \in \Omega^r_{k+1}V$, $k \geq 1$, we define a form on V^{2r+1} by

$$\mathscr{I}_{(V,\psi)}(\rho) = \frac{1}{k}\omega^\alpha \wedge \sum_{j=0}^{r}(-1)^j \partial^j_\Xi i_{\frac{\partial}{\partial q^\alpha_j}} p_k\rho, \tag{11}$$

where ∂_Ξ is the Lie derivative with respect to the vector field Ξ, ∂^j_Ξ is the j-th power of ∂_Ξ, and $i_{\partial/\partial q^\alpha_j}$ denotes the contraction by the vector field $\partial/\partial q^\alpha_j$. For $k=0$ and $\rho \in \Omega^r_1 V$, we define

$$\mathscr{I}_{(V,\psi)}(\rho) = h\rho. \tag{12}$$

Note that the form $\mathscr{I}_{(V,\psi)}(\rho)$ depends only on the k-contact $(k+1)$-form $p_k\rho$. The following two lemmas and Theorem 1 can be proved in fibered coordinates.

Lemma 1. *Let $\rho \in \Omega^r_{k+1}V$, $k \geq 1$. Then the following equation is satisfied*

$$\frac{1}{k}\omega^\alpha \wedge \sum_{j=0}^{r}(-1)^j \partial^j_\Xi i_{\frac{\partial}{\partial q^\alpha_j}} p_k\rho = p_k\rho + \frac{1}{k}\sum_{j=1}^{r}\sum_{l=1}^{j}(-1)^l \binom{j}{l} \partial^l_\Xi(\omega^\alpha_{j-l} \wedge i_{\frac{\partial}{\partial q^\alpha_j}} p_k\rho). \tag{13}$$

Lemma 2. *Let (V,ψ), $\psi = (t,q^\sigma)$, $(\bar{V},\bar{\psi})$, $\bar{\psi} = (\bar{t},\bar{q}^\sigma)$ be two fibered charts on Y such that $V \cap \bar{V} \neq 0$. Then for every $\rho \in \Omega^r_{k+1}(V \cap \bar{V})$, $k \geq 0$,*

$$\mathscr{I}_{(V,\psi)}(\rho) = \mathscr{I}_{(\bar{V},\bar{\psi})}(\rho). \tag{14}$$

It follows from Lemma 2 that equations (11) define a global operator $\mathscr{I} : \Omega^r_{k+1} \to \Omega^{2r+1}_{k+1}$. $\mathscr{I}$ is called the *interior Euler-Lagrange operator*. The differential form $\mathscr{I}(\rho)$ is called the *canonical representative* of a differential form ρ. The operator $\mathscr{I}$ generates new sequence

$$0 \to \mathbb{R}_Y \to \Omega^r_0 \to \mathscr{I}\Omega^r_1 \to \mathscr{I}\Omega^r_2 \to \ldots \to \mathscr{I}\Omega^r_{N-1} \to \mathscr{I}\Omega^r_N \to 0, \tag{15}$$

which is isomorfic with the variational sequence (5).

The following theorem characterizes properties of $\mathscr{I}$. In particular, it turns out that the kernels of $\mathscr{I}$ coincide with the spaces in the subsequence (3) of the De Rham sequence (4).

Theorem 1. *Let $\pi : Y \to X$ be a fibered manifold over one-dimensional base X. Let $k \geq 0$.*

(a) *For every open set $V \subset Y$ and every $\rho \in \Omega^r_{k+1}V$, $\mathscr{I}(\rho)$ lies in the same class as $(\pi^{2r+1,r})^*\rho$.*

(b) *The operator $\mathscr{I}$ satisfies $\mathscr{I}^2 = \mathscr{I}$ (up to the canonical projection).*

(c) *For every open set $V \subset Y$, the kernels of $\mathscr{I}$ coincide with $\Theta^r_{k+1}V$.*

4. LEPAGE FORMS

Let $k \geq 0$. A form $\rho \in \Omega_{k+1}^r V$ is called a *Lepage form*, if

$$p_{k+1}d\rho = \mathscr{I}(d\rho). \tag{16}$$

For $k = 0$, this definition reduces to the original one (Krupka [9, 10]; for more details we refer to Šeděnková [18]). If ρ is a Lepage form, then the forms $d\rho$ and $\rho + d\eta$, where η is arbitrary, are trivially also Lepage forms. The meaning of Lepage forms consists in a generalization of formulas (8), (9); if $k = 1$, then $p_2 d\rho$ is the *Helmholtz-Sonin form* (compare with Krupka [11] for the first order case).

We now analyze the structure of Lepage 2-forms in higher order mechanics. Because of the lack of space, we restrict ourselves to preliminary results; more details as well as proofs will be given elsewhere.

Theorem 2. *Let $\rho \in \Omega_2^1 V$, let in a fibered chart*

$$\rho = a_\sigma \omega^\sigma \wedge dt + b_\sigma d\dot{q}^\sigma \wedge dt + c_{\sigma v}\omega^\sigma \wedge \omega^v + d_{\sigma v}d\dot{q}^\sigma \wedge \omega^v + e_{\sigma v}d\dot{q}^\sigma \wedge d\dot{q}^v, \tag{17}$$

the coefficients $c_{\sigma v}$, $e_{\sigma v}$ are antisymmetric in σ, v. The following three conditions are equivalent:

(a) *ρ is a Lepage form*

(b) *ρ satisfies*

$$\frac{\partial e_{\sigma v}}{\partial \dot{q}^\lambda} + \frac{\partial e_{v\lambda}}{\partial \dot{q}^\sigma} + \frac{\partial e_{\lambda\sigma}}{\partial \dot{q}^v} = 0, \tag{18}$$

$$d_{\sigma v} - d_{v\sigma} - \frac{\partial b_\sigma}{\partial \dot{q}^v} + \frac{\partial b_v}{\partial \dot{q}^\sigma} + 2\frac{\partial e_{\sigma v}}{\partial t} + 2\frac{\partial e_{\sigma v}}{\partial q^\lambda}\dot{q}^\lambda = 0, \tag{19}$$

$$\frac{\partial d_{\sigma v}}{\partial \dot{q}^\lambda} - \frac{\partial d_{v\sigma}}{\partial \dot{q}^\lambda} + \frac{\partial d_{\lambda\sigma}}{\partial \dot{q}^v} - \frac{\partial d_{\lambda v}}{\partial \dot{q}^\sigma} + 2\frac{\partial e_{\lambda\sigma}}{\partial q^v} - 2\frac{\partial e_{\lambda v}}{\partial q^\sigma} = 0, \tag{20}$$

$$\frac{\partial a_v}{\partial \dot{q}^\sigma} - \frac{\partial a_\sigma}{\partial \dot{q}^v} + \frac{\partial b_v}{\partial q^\sigma} - \frac{\partial b_\sigma}{\partial q^v} + \frac{\partial d_{\sigma v}}{\partial t} - \frac{\partial d_{v\sigma}}{\partial t}$$

$$+ \frac{\partial d_{\sigma v}}{\partial q^\lambda}\dot{q}^\lambda - \frac{\partial d_{v\sigma}}{\partial q^\lambda}\dot{q}^\lambda + 4c_{\sigma v} = 0. \tag{21}$$

(c) *There exist functions A_σ, and a 1-form η such that*

$$\rho = A_\sigma \omega^\sigma \wedge dt + \frac{1}{4}\left(\frac{\partial A_\sigma}{\partial \dot{q}^v} - \frac{\partial A_v}{\partial \dot{q}^\sigma}\right)\omega^\sigma \wedge \omega^v + d\eta. \tag{22}$$

Now we consider second order Lepage 2-forms. We have the following result.

Theorem 3. *Let $\rho \in \Omega_2^2 V$, let in a fibered chart*

$$\begin{aligned}\rho &= a_\sigma \omega^\sigma \wedge dt + b_\sigma \dot{\omega}^\sigma \wedge dt + c_\sigma d\ddot{q}^\sigma \wedge dt + d_{\sigma v}\omega^\sigma \wedge \omega^v + e_{\sigma v}\dot{\omega}^\sigma \wedge \omega^v \\ &+ f_{\sigma v}\dot{\omega}^\sigma \wedge \dot{\omega}^v + g_{\sigma v}d\ddot{q}^\sigma \wedge \omega^v + h_{\sigma v}d\ddot{q}^\sigma \wedge \dot{\omega}^v + i_{\sigma v}d\ddot{q}^\sigma \wedge d\ddot{q}^v\end{aligned} \tag{23}$$

and

$$
\begin{aligned}
p_2 d\rho &= P_{\sigma\nu}\omega^\sigma \wedge \omega^\nu \wedge dt + Q_{\sigma\nu}\dot\omega^\sigma \wedge \omega^\nu \wedge dt + R_{\sigma\nu}\ddot\omega^\sigma \wedge \omega^\nu \wedge dt \\
&+ S_{\sigma\nu}\dot\omega^\sigma \wedge \dot\omega^\nu \wedge dt + T_{\sigma\nu}\ddot\omega^\sigma \wedge \dot\omega^\nu \wedge dt + U_{\sigma\nu}\ddot\omega^\sigma \wedge \ddot\omega^\nu \wedge dt,
\end{aligned}
\tag{24}
$$

the coefficients $d_{\sigma\nu}, f_{\sigma\nu}, i_{\sigma\nu}, P_{\sigma\nu}, S_{\sigma\nu}, U_{\sigma\nu}$ *are antisymmetric in* σ, ν. *The following three conditions are equivalent:*

(a) ρ *is a Lepage form*

(b) *The components of* $p_2 d\rho$ *satisfy*

$$
\begin{aligned}
U_{\sigma\nu} - U_{\nu\sigma} = 0, \qquad T_{\sigma\nu} = 0, \qquad S_{\sigma\nu} - S_{\nu\sigma} = 0, \\
R_{\sigma\nu} + R_{\nu\sigma} = 0, \qquad Q_{\sigma\nu} - Q_{\nu\sigma} - 2\tfrac{d}{dt}R_{\sigma\nu} = 0.
\end{aligned}
\tag{25}
$$

(c) *There exist functions* A_σ *satisfying*

$$
\frac{\partial}{\partial \ddot{q}^\tau}\left(\frac{\partial a_\sigma}{\partial \ddot{q}^\nu} - \frac{\partial a_\nu}{\partial \ddot{q}^\sigma} \right) = 0
\tag{26}
$$

and a 1-forn η *such that*

$$
\begin{aligned}
\rho &= a_\sigma \omega^\sigma \wedge dt + \frac{1}{4}\left(\frac{\partial a_\sigma}{\partial \dot{q}^\nu} - \frac{\partial a_\nu}{\partial \dot{q}^\sigma} - \frac{d}{dt}\left(\frac{\partial a_\sigma}{\partial \ddot{q}^\nu} - \frac{\partial a_\nu}{\partial \ddot{q}^\sigma} \right) \right) \omega^\sigma \wedge \omega^\nu \\
&- \frac{1}{2}\left(\frac{\partial a_\sigma}{\partial \ddot{q}^\nu} + \frac{\partial a_\nu}{\partial \ddot{q}^\sigma} \right) \dot\omega^\sigma \wedge \omega^\nu + d\eta,
\end{aligned}
\tag{27}
$$

We note that the coefficients in (24) can be expressed in terms of the coefficients in (23); then (25) become conditions for the coefficients of ρ.

Analogous results can also be given for r-th order 2-forms.

Theorem 4. *Let* $\rho \in \Omega_2^r V$, *let in a fibered chart*

$$
\begin{aligned}
\rho &= \sum_{i=0}^{r-1} a_\sigma^i \omega_i^\sigma \wedge dt + b_\sigma^r dq_r^\sigma \wedge dt \\
&+ \sum_{i,j=0}^{r-1} c_{\sigma\nu}^{ij}\omega_i^\sigma \wedge \omega_j^\nu + \sum_{j=0}^{r-1} d_{\sigma\nu}^{rj}dq_r^\sigma \wedge \omega_j^\nu + e_{\sigma\nu}^{rr}dq_r^\sigma \wedge dq_r^\nu
\end{aligned}
\tag{28}
$$

and

$$
p_2 d\rho = \sum_{j=0}^{r} H_{\sigma\nu}^{0j}\omega^\sigma \wedge \omega_j^\nu \wedge dt + \sum_{i,j=1}^{r} H_{\sigma\nu}^{ij}\omega_i^\sigma \wedge \omega_j^\nu \wedge dt.
\tag{29}
$$

the coefficients $e_{\sigma\nu}^{rr}$ *are antisymmetric in* σ, ν, *the coefficients* $c_{\sigma\nu}^{ij}, H_{\sigma\nu}^{ij}$ *are antisymmetric in pairs* $\binom{i}{\sigma}, \binom{j}{\nu}$. *Then* ρ *is Lepage if and only if*

$$
\begin{aligned}
H_{\sigma\nu}^{ij} - H_{\nu\sigma}^{ji} = 0, \qquad 1 \le i, j \le r, \\
H_{\sigma\nu}^{0j} + (-1)^j H_{\nu\sigma}^{0j} + \Sigma_{l=j+1}^r (-1)^l \binom{l}{j}\frac{d^{l-j}}{dt^{l-j}}H_{\nu\sigma}^{0l} = 0, \qquad 1 \le j \le r.
\end{aligned}
\tag{30}
$$

Finally, we define Lepage equivalents of the canonical representatives of differential forms. Let $\beta \in \Omega_{k+1}^s / \Theta_{k+1}^s$ be a class, i.e., let $\beta = \mathscr{I}(\eta)$ for some $\eta \in \Omega_{k+1}^s V$. A form $\rho \in \Omega_{k+1}^r V$ is said to be a *Lepage equivalent* of β, if ρ is a Lepage form, and

$$p_k \rho = \beta. \tag{31}$$

In particular, this definition includes Lepage equivalents of *dynamical forms* (i.e., the canonical representatives of 2-forms). In particular, let $E = E_\sigma \omega^\sigma \wedge dt$ be the second order dynamical form with the functions E_σ linear in coordinates $\ddot{q}^\nu$. Then Lepage equivalent ρ_E of the dynamical form E has the form

$$
\begin{aligned}
\rho_E \;=\;& E_\sigma \omega^\sigma \wedge dt + \frac{1}{4}\left(\frac{\partial E_\sigma}{\partial \dot{q}^\nu} - \frac{\partial E_\nu}{\partial \dot{q}^\sigma} - \frac{d}{dt}\left(\frac{\partial E_\sigma}{\partial \ddot{q}^\nu} - \frac{\partial E_\nu}{\partial \ddot{q}^\sigma} \right) \right) \omega^\sigma \wedge \omega^\nu \\
&- \frac{1}{2}\left(\frac{\partial E_\sigma}{\partial \ddot{q}^\nu} + \frac{\partial E_\nu}{\partial \ddot{q}^\sigma} \right) \dot{\omega}^\sigma \wedge \omega^\nu
\end{aligned}
\tag{32}
$$

(compare with second order Lepage form (27)).

ACKNOWLEDGMENTS

I would like to thank Prof. Demeter Krupka for his remarks and comments. I am obliged to Cankaya University for its scholarship which allowed me to present my talk in their conference. I would also like to acknowledge the financial support of the Czech Grant Agency (grant no. 201/03/0512).

REFERENCES

1. I. M. Anderson, *The Variational Bicomplex*, Utah State University, Technical Reports, 1989.
2. M. Bauderon, *Le probleme inverse du calcul des variations*, Ann. Inst. Henri Poincaré, A **36**, (1982), 159-179.
3. P. Dedecker, W. M. Tulczyjew, *Spectral sequences and the inverse problem of the calculus of variations*, Diff. Geom. Methods in Math. Phys., Proc. Conf. Aix-en-Provence and Salamanca 1979, Lect. Notes Math., **836**, (1980), 498-503.
4. M. Francaviglia, M. Palese, *Second order variations in variational sequences*, Coll. on Diff. Geom. Proc. Conf., Debrecen, Hungary, July 2000, 119-130.
5. M. Francaviglia, M. Palese, R. Vitolo, *Symmetries in finite order variational sequences*, Czech. Math. J., **52**, 127, (2002), 197-213.
6. J. Kašparová (Krpcová), *A representation of the 1st-order variational sequence in field theory*, Diff. Geom. Appl., Satellite Conference of ICM in Berlin, Aug. 10-14, 1998, Brno, Masaryk University in Brno (1999), 493-502.
7. M. Krbek, J. Musilová, *Representation of the variational sequence*, Rep. Math. Phys., Torun, 2002, **51**, (2003), 251-258.
8. M. Krbek, J. Musilová, J. Kašparová, *Representation of the variational sequence in field theory*, Steps in Diff. Geom., L. Kozma, P. T. Nagy and L. Tamássy, eds., Proc. Coll. on Diff. Geom., Debrecen, Hungary, July 2000 (Univ. Debrecen, Debrecen, 2001), 147-160.

9. D. Krupka, *Lepage forms in higher order variational theory*, Modern developments in analytical mechanics, Vol. **I**, Geometrical Dynamics, Proc. IUTAM-ISIMM Symp., Torino/Italy 1982, (1983), 197-238.

10. D. Krupka, *Some Geometric Aspects of Variational Problems in Fibered Manifolds*, Folia Fac. Sci. Nat. Univ. Purk. Brunensis, Physica, XIV, Brno, Czechoslovakia, 1973, pp. 65., arXiv: math-ph/0110005.

11. D. Krupka, *Variational sequences in mechanics*, Calc. Var., **5**, (1997), 557-583.

12. D. Krupka, *Variational sequences on finite order jet spaces*, Diff. Geom. Appl., Proc. Conf., Brno (Czechoslovakia), J. Janyška and D. Krupka, eds., August 1989; World Scientific, Singapore (1990), 236-254.

13. O. Krupková, *Hamiltonian field theory revisited: A geometric approach to regularity*, L. Kozma, (ed.) et al., Steps in Diff. Geom., Proc. Coll. on Diff. Geom., Debrecen, Hungary, July 25-30, 2000, Debrecen: Univ. Debrecen, Institute of Mathematics and Informatics (2001), 187-207.

14. O. Krupková, *Lepage 2-forms in higher order Hamiltonian mechanics I. Regularity*, Arch. Math., **22**, No. 2, Brno (1986), 97-120.

15. B. A. Kuperschmidt, *Geometry of jet bundles and the structure of Lagrangian and Hamiltonian formalisms*, Geometric Methods in Math. Phys., Proc. NSF-CBMS Conf., Lowell/Mass. 1979, Lect. Notes Math. 775 (1980), 162-218.

16. J. Musilová, M. Krbek, *A note to the representation of the variational sequence in mechanics*, Diff. Geom. Appl., I. Kolář, O. Kowalski, D. Krupka. J. Slovák, eds., Proc. Conf., Brno, Czech Republic, August 1998 (Masaryk University, Brno, 1999) 511-523.

17. J. Šeděnková, *On the invariant variational sequences in mechanics*, The Proceedings of the 22th Winter School Geometry and Physics, Srni, January 2002; Rend. Circ. Mat. Palermo, Ser. II, **71**, (2003), 185-190.

18. J. Šeděnková, *Representations of variational sequences and Lepage forms*, Ph.D. Thesis, Palacky University, Olomouc, 2004.

19. J. Štefánek, *A representation of the variational sequence in higher order mechanics*, J. Janyska, (ed.) et al., Diff. Geom. Appl., Proc. of the 6th Int. Conf., Brno, Czech Republic, 1995, (Masaryk University, Brno, 1996), 469-478.

20. R. Vitolo, *On different geometric formulations of Lagrangian formalism*, Diff. Geom. Appl., **10**, No. 3, (1999), 225-255.

On regularization of second order Lagrangians

Dana Smetanová

Department of Algebra and Geometry, Faculty of Science, Palacky University, Tomkova 40, 779 01 Olomouc, Czech Republic

Abstract. The aim of the paper is to apply some recent results on regularization of Lagrangians to the case of second order Lagrangians corresponding to 3rd order Euler–Lagrange equations.
Key words. Hamilton extremals, Dedecker–Hamilton extremals, Hamilton equations, Lagrangian, Lepagean equivalents, Poincaré–Cartan form, regular and strongly regular systems.
PACS. 02.40.Ma, 02.40.Vh, 11.10.Ef.
MSC. 35A15, 49N60, 58Z05.

1. INTRODUCTION

In this paper we consider an extension of the classical Hamilton–Cartan variational theory on fibered manifolds, recently proposed by Krupková [8], [9], [10]. In the generalized Hamiltonian field theory, to a variational problem represented by a Lagrangian one can associate different Hamilton equations corresponding to different Lepagean equivalents of the Euler–Lagrange form. This concept is a generalization to $(n+1)$-forms of Krupka's Lepagean equivalent of Lagrangian [3], [4]. The arising Hamilton equations and regularity conditions depend not only on a Lagrangian, but also on some "free" functions, which correspond to the choice of a concrete Lepagean equivalent. Within this setting, a proper choice of a Lepagean equivalent can lead to a "regularization" of a Lagrangian. A regularization of some interesting traditionally singular physical fields, the Dirac field, the Electromagnetic field and Scalar Curvature Lagrangian is studied in [1], [7] and [11], [12], some second order Lagrangians are discussed also in [15].

In this paper we are interested in second order Lagrangians which give rise to Euler–Lagrange equations of the 3rd order. All these Lagrangians are singular in the standard Hamilton–De Donder theory. However, in the generalized setting, the question on existence of regular Hamilton equations has sense. We apply to this case regularity conditions found in [9], [10] and find their explicit expression for the above mentioned type of Lagrangians. The regularization procedure is then illustrated on a concrete example.

Throughout the paper all manifolds and mappings are smooth and summation convention is used. We consider a fibered manifold (i.e., surjective submersion) $\pi : Y \to X$, dim $X = n$, dim $Y = n + m$, its r-jet prolongation $\pi_r : J^r Y \to X$, $r \geq 1$ and canonical jet projections $\pi_{r,k} : J^r Y \to J^k Y$, $0 \leq k < r$ (with an obvious notations $J^0 Y = Y$). A fibered

CP729, *Global Analysis and Applied Mathematics: International Workshop on Global Analysis,*
edited by K. Taş, D. Krupka, O. Krupková, and D. Baleanu
© 2004 American Institute of Physics 0-7354-0209-4/04/$22.00

char on Y (resp. associated fibered chart on $J^r Y$) is denoted by (V, ψ), $\psi = (x^i, y^\sigma)$ (resp. (V_r, ψ_r), $\psi_r = (x^i, y^\sigma, y^\sigma_i, \ldots, y^\sigma_{i_1 \ldots i_r})$).

A vector field ξ on $J^r Y$ is called π_r-vertical (resp. $\pi_{r,k}$-vertical) if it projects onto the zero vector field on X (resp. on $J^k Y$).

Recall that every q-form η on $J^r Y$ admits a unique (canonical) decomposition into a sum of q-forms on $J^{r+1} Y$ as follows:

$$\pi^*_{r+1,r} \eta = h\eta + \sum_{k=1}^{q} p_k \eta,$$

where $h\eta$ is a horizontal form, called the *horizontal part of* η, and $p_k \eta$, $1 \le k \le q$, is a *k-contact part of* η (see [3], [4]).

We use the following notations:

$$\omega_0 = dx^1 \wedge dx^2 \wedge \ldots \wedge dx^n, \quad \omega_i = i_{\frac{\partial}{\partial x^i}} \omega_0, \quad \omega_{ij} = i_{\frac{\partial}{\partial x^j}} \omega_i,$$

and

$$\omega^\sigma = dy^\sigma - y^\sigma_j dx^j, \quad \ldots, \quad \omega^\sigma_{i_1 i_2 \ldots i_k} = dy^\sigma_{i_1 i_2 \ldots i_k} - y^\sigma_{i_1 i_2 \ldots i_k j} dx^j$$

For more details on fibered manifolds and the corresponding geometric structures we refer e.g. to [13].

2. LEPAGEAN EQUIVALENTS AND REGULAR HAMILTONIAN SYSTEMS

In this section we briefly recall basic concepts on Lepagean equivalents of Lagrangians, due to Krupka [3], [4], and on Lepagean equivalents of Euler–Lagrange forms and generalized Hamiltonian field theory, due to Krupková [8], [9], [10].

By an *r-th order Lagrangian* we shall mean a horizontal n-form λ on $J^r Y$.

A n-form ρ is called a *Lepagean equivalent of a Lagrangian* λ if (up to a projection) $h\rho = \lambda$, and $p_1 d\rho$ is a $\pi_{r+1,0}$-horizontal form.

For an r-th order Lagrangian we have all its Lepagean equivalents of order $(2r - 1)$ characterized by the following formula

$$\rho = \Theta + \mu, \tag{1}$$

where Θ is a (global) Poincaré–Cartan form associated to λ and μ is an arbitrary n-form of order of contactness ≥ 2, i.e., such that $h\mu = p_1 \mu = 0$. Recall that for a Lagrangian of order 1, $\Theta = \Theta_\lambda$ where Θ_λ is the classical Poincaré–Cartan form of λ. If $r \ge 2$, Θ is no more unique, however, there is an *non-invariant* decomposition

$$\Theta = \Theta_\lambda + p_1 dv, \tag{2}$$

where

$$\Theta_\lambda = L\omega_0 + \sum_{k=0}^{r-1} \left(\sum_{l=0}^{r-k-1} (-1)^l d_{p_1} d_{p_2} \ldots d_{p_l} \frac{\partial L}{\partial y^\sigma_{j_1 \ldots j_k p_1 \ldots p_l i}} \right) \omega^\sigma_{j_1 \ldots j_k} \wedge \omega_i, \tag{3}$$

and v is an arbitrary at least 1-contact $(n-1)$-form.

A *closed* $(n+1)$-form α is called a *Lepagean equivalent of an Euler–Lagrange form* $E = E_\sigma \omega^\sigma \wedge \omega_0$ if $p_1\alpha = E$.

Recall that the Euler–Lagrange form corresponding to an r-th order $\lambda = L\omega_0$ is the following $(n+1)$-form of order $\leq 2r$

$$E = \left(\frac{\partial L}{\partial y^\sigma} - \sum_{l=1}^{r}(-1)^l d_{p_1} d_{p_2} \ldots d_{p_l} \frac{\partial L}{\partial y^\sigma_{p_1 \ldots p_l}} \right) \omega^\sigma \wedge \omega_0. \tag{4}$$

By definition of a Lepagean equivalent of E, one can find using Poincaré lemma local forms ρ, such that $\alpha = d\rho$, and ρ is an Lepagean equivalent of a Lagrangian for E. The family of Lepagean equivalents of E is also called a *Lagrangian system*, and denoted by $[\alpha]$. The corresponding Euler–Lagrange equations now take the form

$$J^s \gamma^* i_{J^s\xi} \alpha = 0 \quad \text{for every } \pi - \text{vertical vector field } \xi \text{ on Y,} \tag{5}$$

where α is any representative of order s of the class $[\alpha]$. A (single) Lepagean equivalent α of E on J^sY is also called a *Hamiltonian system of order s* and the equations

$$\delta^* i_\xi \alpha = 0 \quad \text{for every } \pi_s - \text{vertical vector field } \xi \text{ on } J^sY \tag{6}$$

are called *Hamilton equations*. They represent equations for integral sections δ (called *Hamilton extremals*) of the *Hamiltonian ideal*, generated by the system $\mathscr{D}^s_\alpha$ of n-forms $i_\xi\alpha$, where ξ runs over π_s-vertical vector fields on J^sY. Also, considering π_{s+1}-vertical vector fields on $J^{s+1}Y$, one has the ideal $\mathscr{D}^{s+1}_{\hat\alpha}$ of n-forms $i_\xi\hat\alpha$ on $J^{s+1}Y$, where $\hat\alpha$ denotes the at most 2-contact part of α. Its integral sections which moreover annihilate all at least 2-contact forms, are called *Dedecker–Hamilton extremals*. It holds that if γ is an extremal then its s-prolongation (resp. $(s+1)$-prolongation) is a Hamilton (resp. Dedecker–Hamilton) extremal, and (up to projection) every Dedecker-Hamilton extremal is a Hamilton extremal.

Denote by r_0 the minimal order of Lagrangians corresponding to E. A Hamiltonian system α on J^sY, $s \geq 1$, associated with E is called *regular* [10] if the system of local generators of $\mathscr{D}^{s+1}_{\hat\alpha}$ contains all the $n-forms$

$$\omega^\sigma \wedge \omega_i, \ \omega^\sigma_{(j_1} \wedge \omega_{i)}, \ \ldots, \ \omega^\sigma_{(j_1 \ldots j_{r_0-1}} \wedge \omega_{i)}, \tag{7}$$

where $(\ldots)$ denotes symmetrization in the indicated indices. If α is regular then every Dedecker–Hamilton extremal is holonomic up to the order r_0, and its projection is an extremal. (In case of first order Hamiltonian systems there is an bijection between extremals and Dedecker–Hamilton extremals). α is called *strongly regular* if the above correspondence holds between extremals and Hamilton extremals. It can be proved that every strongly regular Hamiltonian system is regular, and it is clear that if α is regular and such that $\alpha = \hat\alpha$ then it is strongly regular. A Lagrangian system is called *regular* (resp. *strongly regular*) if it has a regular (resp. strongly regular) associated Hamiltonian system.

Theorem 1. [10] *Let α be Hamiltonian system of order $s \leq 2r_0 - 1$, where r_0 denote the minimal order of the corresponding Lagrangians, let*

$$\hat{\alpha} = E_\sigma \omega^\sigma \wedge \omega_0 + \sum_{|J|, |P|=0}^{s} F_{\sigma\nu}^{P,J,i} \omega_J^\sigma \wedge \omega_P^\nu \wedge \omega_i \tag{8}$$

be the principal part of α. Assume that $F_{\sigma\nu}^{P,J,i} = 0$, $|J| + |P| \geq s + 1$, and

$$F_{\sigma\nu}^{P,J,i} = F_{\sigma\nu}^{P,(J,i)}, \quad 1 \leq |J| \leq r_0 - 1, \ s - r_0 + 1 \leq |P| \leq s - 1, \tag{9}$$

and

$$\text{rank} \left(F_{\sigma\nu}^{p_1 \cdots p_s, 0, i} \right) = mn, \tag{10}$$

$$\text{rank} \left(F_{\sigma\nu}^{p_1 \cdots p_{s-1}, j_1, i} \right) = m \binom{n+1}{2},$$

$$\cdots$$

$$\text{rank} \left(F_{\sigma\nu}^{p_1 \cdots p_{s-r_0+1}, j_1 \cdots j_{r_0-1}, i} \right) = m \binom{n+r_0-1}{r_0},$$

where in the above matrices, the (ν, P) label rows, and the (σ, J, i) label columns. Then α is regular.

Remark. As pointed out in [8], in case that $\alpha = d\theta_\lambda$ (resp. $\alpha = d\Theta$), the above generalized Hamilton equations reduce to standard higher-order local (respectively global) Hamilton–De Donder equations. In this case, however, the concepts of a Hamilton extremal and Dedecker-Hamilton extremal, as well as those of regularity and strong regularity coincide. Moreover, for Lagrangians of order r such that $d\Theta$ is actually of order $2r - 1$ (i.e., their Euler–Lagrange equations are nontrivially of order $2r$) the regularity conditions (10) for $\hat{\alpha} = d\Theta$ take the form of Shadwick's conditions for Lagrangians ([14], see also [2], [5] and [6] for regular higher-order Hamilton–De Donder equations).

3. REGULAR SECOND ORDER LAGRANGIANS FOR 3RD ORDER EQUATIONS

In general, a second order Lagrangian gives rise to an Euler–Lagrange form on J^4Y. We shall consider second order Lagrangians corresponding to Euler-Lagrange forms of order 3, i.e. satisfying the conditions

$$\left(\frac{\partial^2 L}{\partial y_{ij}^\sigma \partial y_{kl}^\nu} \right)_{Sym(ijkl)} = 0, \tag{11}$$

where $Sym(ijkl)$ means symmetrization in the indicated indices. In what follows, we shall study Hamiltonian systems corresponding to a special choice of a Lepagean equivalent of such Lagrangians, namely, α of order 3, $\alpha = d\rho$, where

$$\rho = L\omega_0 + \left(\frac{\partial L}{\partial y_j^\sigma} - d_k\frac{\partial L}{\partial y_{jk}^\sigma}\right)\omega^\sigma \wedge \omega_j + \frac{\partial L}{\partial y_{ij}^\sigma}\omega_i^\sigma \wedge \omega_j + \bar{\mu} \tag{12}$$

$$+ \quad a_{\sigma v}^{ij}\omega^\sigma \wedge \omega^v \wedge \omega_{ij} + b_{\sigma v}^{kij}\omega^\sigma \wedge \omega_k^v \wedge \omega_{ij}$$

$$+ \quad c_{\sigma v}^{klij}\omega^\sigma \wedge \omega_{kl}^v \wedge \omega_{ij} + d_{\sigma v}^{klij}\omega_k^\sigma \wedge \omega_l^v \wedge \omega_{ij},$$

with an arbitrary at least 3-contact n-form $\bar{\mu}$, and functions $a_{\sigma v}^{ij}$, $b_{\sigma v}^{kij}$, $c_{\sigma v}^{klij}$, $d_{\sigma v}^{klij}$ dependent on variables x^k, y^κ, y_k^κ, y_{kl}^κ and satisfying the conditions

$$a_{\sigma v}^{ij} = -a_{\sigma v}^{ji}, \quad a_{\sigma v}^{ij} = -a_{v\sigma}^{ij}; \quad b_{\sigma v}^{kij} = -b_{\sigma v}^{kji}; \tag{13}$$

$$c_{\sigma v}^{klij} = c_{\sigma v}^{lkij}, \quad c_{\sigma v}^{klij} = -c_{\sigma v}^{klji}; \quad d_{\sigma v}^{klij} = -d_{v\sigma}^{lkij}, \quad d_{\sigma v}^{klij} = -d_{\sigma v}^{klji}.$$

In the following proposition, regularity conditions (10) are transcribed to the case $\alpha = d\rho$ with the above ρ, and conditions from strong regularity are found

Proposition *Let* $\dim X \geq 2$. *Let* $\lambda = L\omega_0$ *be a second order Lagrangian with the Euler–Lagrange form (nontrivially) of order 3, and* $\alpha = d\rho$ *with* ρ *of the form (12), (13), be its Lepagean equivalent. Assume that*

$$d_{\sigma v}^{ijkl} = d_{\sigma v}^{ilkj} \tag{14}$$

and the matrix

$$P_{\sigma v}^{ijkl} = \left(\frac{\partial^2 L}{\partial y_{ij}^v \partial y_{kl}^\sigma} + 2\,c_{v\sigma}^{klij}\right)_{Sym(jkl)}, \tag{15}$$

with mn^3 *rows (resp. mn columns) labelled by* σjkl *(resp. vi) has rank mn and matrix*

$$Q_{\sigma v}^{ijkl} = \left(\frac{\partial^2 L}{\partial y_{ij}^\sigma \partial y_{kl}^v} - 2c_{\sigma v}^{klij} - 4d_{\sigma v}^{jkil}\right), \tag{16}$$

with mn^2 *rows (resp. mn^2 columns) labelled by* σij *(resp. vkl) has rank* $mn(n+1)/2$.

Then the Hamiltonian system $\alpha = d\rho$ *is regular. If moreover* $\bar{\mu}$ *is closed and functions* $d_{\sigma v}^{klij}$ *do not depend on variables* y_{kl}^κ *then* α *is strongly regular.*

Proof. The prove regularity, we shall show that the above regularity conditions coincide with those of Theorem 1. Since $s = 3$ and $r_0 = 2$, we have $2r_0 - 1 = 3 = s$, as required. Explicit computation $\alpha = d\rho$ gives:

$$\pi_{4,3}^*\alpha = E_\sigma\omega^\sigma \wedge \omega_0 + \left(\frac{\partial^2 L}{\partial y_j^\sigma \partial y^v} - \frac{\partial}{\partial y^v}d_j\frac{\partial L}{\partial y_{ij}^\sigma} - 2d_k a_{\sigma v}^{ij}\right)\omega^v \wedge \omega^\sigma \wedge \omega_i \tag{17}$$

$$+ \left(\frac{\partial^2 L}{\partial y_i^\sigma \partial y_k^v} - \frac{\partial^2 L}{\partial y^\sigma \partial y_{ik}^v} - \frac{\partial}{\partial y_k^v}d_j\frac{\partial L}{\partial y_{ij}^\sigma} + 4a_{v\sigma}^{ik} - 2d_j b_{\sigma v}^{kij}\right)\omega_k^v \wedge \omega^\sigma \wedge \omega_i$$

$$+ \left(\frac{\partial^2 L}{\partial y_i^\sigma \partial y_{kl}^\nu} - \frac{\partial}{\partial y_{kl}^\nu} d_j \frac{\partial L}{\partial y_{ij}^\sigma} - 2(b_{\sigma\nu}^{kil})_{Sym(kl)} - 2d_j c_{\sigma\nu}^{klij} \right) \omega_{kl}^\nu \wedge \omega^\sigma \wedge \omega_i$$

$$- \left(\frac{\partial^2 L}{\partial y_{ij}^\sigma \partial y_{kl}^\nu} + 2c_{\sigma\nu}^{klij} \right)_{Sym(jkl)} \omega_{jkl}^\nu \wedge \omega^\sigma \wedge \omega_i$$

$$+ \left(\frac{\partial^2 L}{\partial y_{ij}^\sigma \partial y_k^\nu} - 4(b_{\sigma\nu}^{kij})_{Alt((\sigma j)(\nu k))} + 2d_p d_{\nu\sigma}^{kjip} \right) \omega_k^\nu \wedge \omega_j^\sigma \wedge \omega_i$$

$$+ \left(\frac{\partial^2 L}{\partial y_{ij}^\sigma \partial y_{kl}^\nu} - 2c_{\sigma\nu}^{klij} - 4d_{\sigma\nu}^{jkil} \right) \omega_{kl}^\nu \wedge \omega_j^\sigma \wedge \omega_i + \left(\frac{\partial a_{\sigma\nu}^{ij}}{\partial y^\kappa} \right)_{Alt(\kappa\sigma\nu)}$$

$$\omega^\kappa \wedge \omega^\sigma \wedge \omega^\nu \wedge \omega_{ij} + \left(\frac{\partial a_{\sigma\nu}^{ij}}{\partial y_p^\kappa} + \frac{\partial b_{\nu\kappa}^{pij}}{\partial y^\sigma} \right)_{Alt(\sigma\nu)} \omega_p^\kappa \wedge \omega^\sigma \wedge \omega^\nu \wedge \omega_{ij}$$

$$+ \left(\left(\frac{\partial a_{\sigma\nu}^{ij}}{\partial y_{pq}^\kappa} \right)_{Sym(pq)} + \left(\frac{\partial c_{\nu\kappa}^{pqij}}{\partial y_{pq}^\sigma} \right)_{Alt(\sigma\nu)} \right) \omega_{pq}^\kappa \wedge \omega^\sigma \wedge \omega^\nu \wedge \omega_{ij}$$

$$+ \left(\frac{\partial b_{\sigma\nu}^{qij}}{\partial y_p^\kappa} + \frac{\partial d_{\nu\kappa}^{qpij}}{\partial y^\sigma} \right)_{Alt((\kappa p)(\nu q))} \omega^\sigma \wedge \omega_q^\nu \wedge \omega_p^\kappa \wedge \omega_{ij} + \left(\frac{\partial b_{\sigma\nu}^{kij}}{\partial y_{pq}^\kappa} - \frac{\partial c_{\nu\kappa}^{pqij}}{\partial y_k^\nu} \right)_{Sym(pq)}$$

$$\omega^\sigma \wedge \omega_k^\nu \wedge \omega_{pq}^\kappa \wedge \omega_{ij} - \left(\frac{\partial c_{\sigma\nu}^{klij}}{\partial y_{pq}^\kappa} \right)_{Alt((\kappa pq)(\nu kl))} \omega^\sigma \wedge \omega_{pq}^\kappa \wedge \omega_{kl}^\nu \wedge \omega_{ij}$$

$$+ \left(\frac{\partial d_{\sigma\nu}^{klij}}{\partial y_p^\kappa} \right)_{Alt((\kappa p)(\sigma k)(\nu l))} \omega_p^\kappa \wedge \omega_k^\sigma \wedge \omega_l^\nu \wedge \omega_{ij}$$

$$+ \left(\frac{\partial d_{\sigma\nu}^{klij}}{\partial y_{pq}^\kappa} \right) \omega_{pq}^\kappa \wedge \omega_k^\sigma \wedge \omega_l^\nu \wedge \omega_{ij} + d\bar{\mu}.$$

Considering $\hat{\alpha} = p_2\alpha$, we can see that $F_{\nu\sigma}^{jkl,0,i} = P_{\sigma\nu}^{ijkl} = \left(\frac{\partial^2 L}{\partial y_{ij}^\nu \partial y_{kl}^\nu} + 2c_{\nu\sigma}^{klij} \right)_{Sym(jkl)}$, $F_{\sigma\nu}^{kl,j,i} = Q_{\sigma\nu}^{ijkl} = \left(\frac{\partial^2 L}{\partial y_{ij}^\sigma \partial y_{kl}^\nu} - 2c_{\sigma\nu}^{klij} - 4d_{\sigma\nu}^{jkil} \right)$. The condition that functions $F_{\sigma\nu}^{kl,j,i}$ are symmetric in the indices j, i, gives the symmetry condition (14) on the functions $d_{\sigma\nu}^{ijkl}$, and the rank conditions on the matrices $P_{\sigma\nu}^{ijkl}$ and $Q_{\sigma\nu}^{ijkl}$ coincide with (10).

Let us prove strong regularity: We have to show that under our assumptions, for every section δ satisfying Hamilton equations, one has $\pi_{3,2} \circ \delta = J^2\gamma$, where γ is a solution of the Euler–Lagrange equations of the Lagrangian λ. Assuming $d\bar{\mu} = 0$, we obtain: $\delta^*(i_{\partial/\partial y_{jkl}^\sigma}\alpha) = \delta^*(P_{\sigma\nu}^{ijkl}\omega^\nu \wedge \omega_i) = 0$, i.e. $\delta^*\omega^\nu = 0$ by the rank condition on $P_{\sigma\nu}^{ijkl}$. Hence, $\delta^*(i_{\partial/\partial y_{kl}^\nu}\alpha) = \delta^*\left(Q_{\sigma\nu}^{ijkl}\omega_j^\sigma \wedge \omega_i + \partial d_{\sigma\kappa}^{pqij}/\partial y_{kl}^\nu \ \omega_p^\sigma \wedge \omega_q^\kappa \wedge \omega_{ij} \right) = 0$.

If $d_{\sigma\kappa}^{pqij}$ do not depend on y_{kl}^ν then, due to the rank condition on $Q_{\sigma\nu}^{ijkl}$, $\delta^*\omega_j^\sigma = 0$. The above obtained conditions on δ mean that every solution of Hamilton equations is holonomic up to the second order, i.e. we can write $\pi_{3,2} \circ \delta = J^2\gamma$, where γ is a section of π. Now, the equations $J^3(\pi_{3,0} \circ \delta)^*(i_{\partial/\partial y_k^\sigma}\alpha) = 0$ are satisfied identically, and the last set of Hamilton equations, i.e. $J^3(\pi_{3,0} \circ \delta)^*(i_{\partial/\partial y^\sigma}\alpha) = 0$ take the form $E_\sigma \circ J^3\gamma = 0$, proving that γ is an extremal of λ. This completes the proof.

Let us recall the following definitions [9]:

Let $W \subset J^2Y$ be an open set. A Lagrangian λ is called *regularizable* (resp. *strongly regularizable*) on W if its Euler–Lagrange form has a regular (resp. strongly regular) Lepagean equivalent defined on W. If $W = J^2Y$ we say that λ is *globally regularizable* (resp. *globally strongly regularizable*). We say that λ is *locally regularizable* (resp. *locally strongly regularizable*) if it is regularizable (resp. strongly regularizable) in a neighbourhood of every point in J^2Y. The corresponding Lepagean equivalent α is then called a *global* (resp. *local*) *regularization* (resp. *strong regularization*) of λ.

The above results can by directly applied to concrete Lagrangians. Let us consider the following example as an illustration of the regularization procedure, which means to find to a given Lagrangian a suitable (local or global) regularization.

Example. *Regularization by constants.*
Let us consider $X = R^2, Y = R^2 \times R^2$ (i.e., $n = 2, m = 2$) and the Lagrangian corresponding to Euler–Lagrange form of order 3, $L = y_{11}^1 y_{22}^2 - y_{22}^1 y_{11}^2$.

Denote (V, ψ), $\psi = (x^i, y^\sigma)$ a fibered chart on $R^2 \times R^2$. In view of the above consideration we take a Lepagean equivalent α of E in the following form: $\alpha = d\rho$,

$$\rho = \theta_\lambda + p_1 dv + a_{\sigma v}^{ij}\omega^\sigma \wedge \omega^v \wedge \omega_{ij} + b_{\sigma v}^{kij}\omega^\sigma \wedge \omega_k^v \wedge \omega_{ij}$$
$$+ \quad 4\,\omega^1 \wedge \omega_{12}^1 \wedge \omega_{12} + 4\,\omega^2 \wedge \omega_{12}^2 \wedge \omega_{12},$$

where v is an arbitrary $(n-1)$-form and $a_{\sigma v}^{ij}$, $b_{\sigma v}^{kij}$ are arbitrary functions satisfying (13). Note that we have only 8 non zero constant $c_{\sigma v}^{klij}$. The matrices (15) and (16) take the forms:

$$(P_{\sigma v}^{ijkl})^T = \frac{1}{3}\begin{pmatrix} 0 & 0 & 0 & 4 & 0 & 4 & 4 & 0 & 0 & 0 & 0 & 1 & 0 & 1 & 1 & 0 \\ 0 & -4 & -4 & 0 & -4 & 0 & 0 & 0 & 0 & -1 & -1 & 0 & -1 & 0 & 0 & 0 \\ 0 & 0 & 0 & -1 & 0 & -1 & -1 & 0 & 0 & 0 & 0 & 4 & 0 & 4 & 4 & 0 \\ 0 & 1 & 1 & 0 & 1 & 0 & 0 & 0 & 0 & -4 & -4 & 0 & -4 & 0 & 0 & 0 \end{pmatrix},$$

and

$$Q_{\sigma v}^{ijkl} = \begin{pmatrix} 0 & 0 & 0 & 0 & 0 & 0 & 0 & 1 \\ 0 & -2 & -2 & 0 & 0 & 0 & 0 & 0 \\ 0 & 2 & 2 & 0 & 0 & 0 & 0 & 0 \\ 0 & 0 & 0 & 0 & -1 & 0 & 0 & 0 \\ 0 & 0 & 0 & -1 & 0 & 0 & 0 & 0 \\ 0 & 0 & 0 & 0 & 0 & -2 & -2 & 0 \\ 0 & 0 & 0 & 0 & 0 & 2 & 2 & 0 \\ 1 & 0 & 0 & 0 & 0 & 0 & 0 & 0 \end{pmatrix}.$$

We can easily see that $\text{rank}(P_{\sigma v}^{ijkl}) = 4$ and $\text{rank}(P_{\sigma v}^{ijkl}) = 6$. The form $\alpha = d\rho$ is strongly regular.

ACKNOWLEDGMENTS

Research supported by Grants 201/04/P186 and 201/03/0512 of the Czech Grant Agency, and MSM 153100011 of the Czech Ministry of Education, Youth and Sports.

The author wishes to thank Prof. Olga Krupková for her valuable stimulation and discussions in course of the present work.

The author also wishes to thank organizers for support and kind hospitality during the International Workshop on Global Analysis in Ankara.

REFERENCES

1. P. Dedecker, *On the generalization of symplectic geometry to multiple integrals in the calculus of variations*, in: Lecture Notes in Math. 570 (Springer, Berlin, 1977) 395–456.
2. M. J. Gotay, *A multisymplectic framework for classical field theory and the calculus of variations, I. covariant Hamiltonian formalism*, in: M. Francaviglia, D. D. Holm (Eds.), Mechanics, Analysis and Geometry: 200 Years After Lagrange, North-Holland, Amsterdam, 1990, pp. 203–235.
3. D. Krupka, *Some geometric aspects of variational problems in fibered manifolds*, Folia Fac. Sci. Nat. UJEP Brunensis 14 (1973) 1–65.
4. D. Krupka, *Lepagean forms in higher order variational theory*, in: Modern Developments in Analytical Mechanics I: Geometrical Dynamics, Proc. IUTAM-ISIM Symp., Torino, Italy 1982, (S. Benenti, M. Francaviglia and A. Lichnerowitz, eds.), Accad. delle Scienze di Torino, Torino, 1983, 197–238.
5. D. Krupka, *On the higher order Hamilton theory in fibered spaces*, in: D. Krupka (Ed.), Geometrical Methods in Physics, Poceedings of the Conference on Differential Geometry and Applications, Nové Město na Moravě, 1983, J. E. Purkyně University, Brno, Czechoslovakia, 1984, pp. 167–183.
6. D. Krupka, *Regular Lagrangians and Lepagean forms*, in: D. Krupka, A. Švec (Eds.), Proceedings of the Conference on Differential Geometry and its Applications, Brno Czechoslovakia, 1986, Reidel, Dordrecht,1986, pp. 111–148.
7. D. Krupka and O. Štěpánková, On the *Hamilton form in second order calculus of variations*, in: Geometry and Physics, Proc. Int. Meeting, Florence, Italy, 1982, M. Modugno, ed. (Pitagora Ed., Bologna, 1983) 85–101
8. O. Krupková, *Hamiltonian field theory*, J. Geom. Phys. 43 (2002), 93–132.
9. O. Krupková, *Hamiltonian field theory revisited: A geometric approach to regularity*, in: Steps in Differential Geometry, Proc. of the Coll. on Diff. Geom., Debrecen 2000 (University of Debrecen, Debrecen, 2001) 187–207.
10. O. Krupková, *Higher-order Hamiltonian field theory*, Paper in preparation.
11. O. Krupková and D. Smetanová, *On regularization of variational problems in first-order field theory*, Proceedings of the 20th Winter School "Geometry and Physics" (Srní, 2000). Rend. Circ. Mat. Palermo (2) Suppl. No. 66, (2001), 133–140.
12. O. Krupková and D. Smetanová, *Legendre transformation for regularizable Lagrangians in field theory*, Letters in Math. Phys. 58 (2001) 189–204.
13. D. J. Saunders, *The Geometry of Jets Bundles*, Cambridge University Press, Cambridge, 1989.
14. W. F. Shadwick, *The Hamiltonian formulation of regular r-th order Lagrangian field theories*, Letters in Math. Phys. 6 (1982) 409–416.
15. D. Smetanová, *On Hamilton p_2-equations in second-order field theory*, in: Steps in Differential Geometry, Proc. of the Coll. on Diff. Geom., Debrecen 2000 (University of Debrecen, Debrecen, 2001) 329–341.

On the nonholonomic variational principle

M. Swaczyna

*Department of Mathematics, Faculty of Science, University of Ostrava, 30. dubna 22, 701 03
Ostrava, Czech Republic, e-mail: Martin.Swaczyna@osu.cz*

Abstract. A variational principle for the first order mechanical systems subjected to general non-holonomic constraints is presented, providing the (reduced) nonholonomic motion equations in a form of constraint Euler–Lagrange equations.
Key words. Lagrangian system, nonholonomic constraints, canonical distribution, constraint ideal, constraint horizontalization and contactization, constraint Lepage equivalent, constraint first variational formula, constraint Euler-Lagrange equations.
PACS. 45.20.Jj, 45.90.+t.
MSC. 58A10, 58A20, 58A30, 58F05, 70F25, 70H05.

1. INTRODUCTION

In classical mechanics one naturally encounters different kinds of constraints on the motion, mostly *holonomic* and *linear non-holonomic constraints*. While mechanics of holonomic systems is well-established both from the physical and geometric point of view, in non-holonomic mechanics still many questions remain open. The main problems studied concern namely geometric structures arising due to presence of constraints, a direct geometric characterization of constrained systems (without help of Lagrange multipliers), variationality and regularity of constrained systems, the concept of virtual displacements, Principle of virtual works, etc., all for the case of *general non-holonomic constraints* (which are not supposed linear in the first derivatives).

These aspects of the geometry of Lagrangian systems with non-holonomic constraints have been of interest in many recent papers (see e.g. [1], [2], [3], [6]—[19]). The aim of this contribution is to announce, in a preliminary form, some results on a variational formulation of nonholonomic mechanics recently obtained in the collaboration with O. Krupková and to be published elsewhere ([11], [12]). We present a *variational principle* formulated directly on a constraint submanifold, the corresponding *constraint first variational formula* and *"constraint Euler–Lagrange equations"*. The obtained equations for "constraint extremals" are a particular kind of *reduced Chetaev-type equations* (i.e. with "eliminated" Lagrange multipliers), as found by Krupková [6]. They coincide with the *variational reduced equations* characterized by means of "constraint Helmholtz conditions", found by Krupková and Musilová [9], [10]. It should be stressed that our "constraint Euler-Lagrange equations" need not be variational in the standard sense, i.e. they need not come from a *Lagrangian unconstrained system*. If this happens to be the

CP729, *Global Analysis and Applied Mathematics: International Workshop on Global Analysis,*
edited by K. Taş, D. Krupka, O. Krupková, and D. Baleanu

case, then our equations coincide with the reduced equations for constrained Lagrangian systems, found by Sarlet [17].

2. LAGRANGIAN SYSTEMS ON FIBERED MANIFOLDS

In this section we recall basic structures and concepts used in the calculus of variations on fibered manifolds, for more details we refer to [4], [5] and [6]. Throughout the paper we consider a fibered manifold $\pi : Y \to X$ with $\dim X = 1$, $\dim Y = m + 1$, its jet prolongations $\pi_1 : J^1Y \to X$ and $\pi_2 : J^2Y \to X$ and the jet projections $\pi_{1,0} : J^1Y \to Y$ and $\pi_{2,1} : J^2Y \to J^1Y$. We denote (t, q^σ), $1 \leq \sigma \leq m$, fibered coordinates on Y, $(t, q^\sigma, \dot{q}^\sigma)$ associated coordinates on J^1Y, and by

$$\omega^\sigma = dq^\sigma - \dot{q}^\sigma \, dt, \quad 1 \leq \sigma \leq m \tag{1}$$

canonical contact 1-forms annihilating the *Cartan distribution* on J^1Y. Whenever possible, the summation convention is used. If $f(t, q^\sigma, \dot{q}^\sigma)$ is a function we write

$$\frac{df}{dt} = \frac{\partial f}{\partial t} + \frac{\partial f}{\partial q^\sigma} \dot{q}^\sigma + \frac{\partial f}{\partial \dot{q}^\sigma} \ddot{q}^\sigma, \quad \frac{\bar{d}f}{dt} = \frac{\partial f}{\partial t} + \frac{\partial f}{\partial q^\sigma} \dot{q}^\sigma. \tag{2}$$

A (local) section δ of π_1 is called *holonomic* if $\delta = J^1\gamma$ for a section γ of π.

A vector field ξ is called π_1-*vertical* (or simply *vertical*) if $T\pi_1 \cdot \xi = 0$. Similarly, ξ is called $\pi_{1,0}$- *vertical* if $T\pi_{1,0} \cdot \xi = 0$. A differential form ρ is called *contact* if $J^1\gamma^*\rho = 0$ for every section γ of π. A differential form ρ is called *horizontal* if $i_\xi \rho = 0$ for every vertical vector field ξ. We denote by h the operator assigning to ρ its horizontal part. Every 2-form on J^1Y is contact and admits a *unique decomposition* $\pi_{2,1}^* \rho = \rho_1 + \rho_2$, where ρ_1 is a 1-*contact* form on J^2Y (i.e. for every vertical vector field ξ, $i_\xi \rho_1$ is a horizontal form), and ρ_2 is a 2-*contact* form (i.e. for every vertical vector field ξ, $i_\xi \rho_2$ is a 1-contact form). We denote by p_1, and p_2 operators assigning to ρ its 1-contact and 2-contact part, respectively.

By a *distribution* on J^1Y we mean a mapping D assigning to every point $z \in J^1Y$ a vector subspace $D(z)$ of the vector space $T_z J^1Y$. A distribution is said to be of a *constant rank* if $\dim D(z)$ does not depend on z. A distribution can be spanned by a system of (local) vector fields. If D is a distribution, we denote by D^0 its annihilator, i.e., the set of all 1-forms η_κ on J^1Y such that $i_{\xi_\iota} \eta_\kappa = 0$ for every vector field ξ_ι belonging to D. Recall that a section δ of π_1 is called an *integral section* of D if $\delta^*\eta = 0$ for every 1-form η belonging to D^0.

If λ is a Lagrangian on J^1Y, we denote by θ_λ its *Lepage equivalent* or *Cartan form*, and E_λ its *Euler-Lagrange form*. Recall that $E_\lambda = p_1 d\theta_\lambda$. In fibered coordinates where $\lambda = Ldt$ we have

$$\theta_\lambda = Ldt + \frac{\partial L}{\partial \dot{q}^\sigma} \omega^\sigma, \tag{3}$$

and $E_\lambda = E_\sigma(L)dq^\sigma \wedge dt$, where

$$E_\sigma(L) = \frac{\partial L}{\partial q^\sigma} - \frac{d}{dt} \frac{\partial L}{\partial \dot{q}^\sigma}. \tag{4}$$

Euler-Lagrange expressions are *affine in the second derivatives*, i.e. are of the form

$$E_\sigma = A_\sigma(t, q^\nu, \dot{q}^\nu) + B_{\sigma\rho}(t, q^\nu, \dot{q}^\nu)\ddot{q}^\rho, \tag{5}$$

where

$$B_{\sigma\nu} = -\frac{\partial^2 L}{\partial \dot{q}^\sigma \partial \dot{q}^\nu}, \quad A_\sigma = \frac{\partial L}{\partial q^\sigma} - \frac{\partial^2 L}{\partial t \partial \dot{q}^\sigma} - \frac{\partial^2 L}{\partial q^\nu \partial \dot{q}^\sigma}\dot{q}^\nu. \tag{6}$$

The corresponding Euler–Lagrange equations $E_\sigma \circ J^2\gamma = 0$ can be written in an intrinsic form as follows

$$J^1\gamma^* i_\xi d\theta_\lambda = 0, \tag{7}$$

where ξ is a π_1-vertical vector field on J^1Y, or quite equivalently in the form

$$J^1\gamma^* i_\xi \alpha = 0, \tag{8}$$

where α is any 2-form on J^1Y such that $p_1\alpha = E_\lambda$, apparently, $\alpha = d\theta_\lambda + F$, where F runs over all $\pi_{1,0}$-horizontal 2-contact 2-forms. Recall from [6] that the family of all such (local) 2-form α is called a *Lagrangian system*, and is denoted by $[\alpha]$.

3. THE NONHOLONOMIC CONSTRAINT STRUCTURE

By a *constraint submanifold* in J^1Y we mean a fibered submanifold $\pi_{1,0}|_Q : Q \to Y$ of the fibered manifold $\pi_{1,0} : J^1Y \to Y$. We denote by ι the canonical embedding of Q into J^1Y, and suppose $\operatorname{codim} Q = k < m$. (cf. for example [6], [7], [14], [17], [18], [19]). Locally, Q can be given by equations

$$f^i = 0, \quad 1 \le i \le k, \quad \text{where} \quad \operatorname{rank}\left(\frac{\partial f^i}{\partial \dot{q}^\sigma}\right) = k, \tag{9}$$

or, equivalently in an explicit form

$$\dot{q}^{m-k+i} = g^i(t, q^\sigma, \dot{q}^1, \dot{q}^2, \ldots, \dot{q}^{m-k}), \quad 1 \le i \le k, \tag{10}$$

called a *system of k nonholonomic constraints*.

The presence of a constraint submanifold in J^1Y gives rise to a concept of a *constrained section* as a local section $\bar{\delta}$ of the fibered manifold π_1 such that for every $x \in \operatorname{dom} \bar{\delta} : \bar{\delta}(x) \in Q$, and a Q-admissible section as a section $\bar{\gamma}$ of the fibered manifold π such that $J^1\bar{\gamma}(x) \in Q$ for every $x \in \operatorname{dom} \bar{\gamma}$. The set of all Q-admissible sections $\bar{\gamma}$ of π will be denoted by $\bar{\Gamma}^Q(\pi)$.

The submanifold Q is naturally endowed with a distribution, called the *canonical distribution* [6], or *Chetaev bundle* [14], and denoted by C. It is annihilated by a system of k linearly independent (local) 1-forms

$$\varphi^i = \iota^*\phi^i, \quad \text{where} \quad \phi^i = f^i dt + \frac{\partial f^i}{\partial \dot{q}^\sigma}\omega^\sigma, \quad 1 \le i \le k, \tag{11}$$

i.e.

$$\varphi^i = -\sum_{l=1}^{m-k} \frac{\partial g^i}{\partial \dot{q}^l}\, \omega^l + \iota^* \omega^{m-k+i}, \quad 1 \le i \le k, \tag{12}$$

called *canonical constraint 1-forms*. The ideal in the exterior algebra of forms on Q generated by the annihilator of C is called the *constraint ideal*, and denoted by $I(C^0)$, or simply I; its elements are called *constraint forms*. The pair (Q,C) is then called a *(nonholonomic) constraint structure* on the fibered manifold π [6], [7].

Remark 1. Notice that the holonomic integral sections of the canonical distribution C coincide with the holonomic constrained sections of π_1, i.e. have the form $\bar{\delta} = J^1\bar{\gamma}$, where $\bar{\gamma}$ is a Q-admissible section of π.

Let $\tilde{Q}$ be the *lift* of Q in J^2Y, i.e. the manifold of all points $J_x^2\gamma \in J^2Y$ such that $J_x^1\gamma \in Q$. If Q is given by (10) then equations of $\tilde{Q}$ are

$$\dot{q}^{m-k+i} = g^i(t,q^\sigma,\dot{q}^1,\dot{q}^2,\dots,\dot{q}^{m-k}), \quad \ddot{q}^{m-k+i} = \frac{dg^i}{dt}. \tag{13}$$

We denote by $\rho : \tilde{Q} \to Q$ the corresponding jet projection (i.e. $\rho = \pi_{2,1}|_{\tilde{Q}}$). The distribution $\tilde{C}$ on $\tilde{Q}$, such that for every $y \in \tilde{Q}$

$$T_y\rho\left(\tilde{C}(y)\right) = C\left(\rho(y)\right) \tag{14}$$

is called the *lift of the canonical distribution C* [11]. Since $\operatorname{rank}\tilde{C} = 3m+1-3k$, and $\dim\tilde{Q} = 3m+1-2k$ one has $\operatorname{corank}\tilde{C} = \operatorname{corank}C = k$. The annihilator of $\tilde{C}$ is locally spanned by 1-forms $\tilde{\varphi}^i = \rho^*\varphi^i$, $1 \le i \le k$. But since canonical constraint 1-forms are $\pi_{1,0}$-horizontal, $\rho^*\varphi^i = \varphi^i$, and thus $\tilde{C}^0 = C^0 = \operatorname{span}\{\varphi^i\}$. We denote by $\tilde{I}$ the ideal on $\tilde{Q}$, generated by $\tilde{C}^0$.

4. CONSTRAINT HORIZONTAL AND CONTACT FORMS

If Q is a constraint in J^1Y, we denote by $\Lambda^q(Q)$, resp. $\Lambda^q(\tilde{Q})$ the module of q-forms on Q, resp. $\tilde{Q}$. The concepts of a π_1-horizontal and contact form directly transfer to forms on Q: a form η is called π_1-*horizontal* if $i_\xi\eta = 0$ for every π_1-vertical vector field ξ on Q, η is called *contact* if $J^1\bar{\gamma}^*\eta = 0$ for every Q-admissible section $\bar{\gamma}$ of π. The same definitions as on J^1Y are used for 1-contact, 2-contact and q-contact forms on Q, where contractions are provided by π_1-vertical vector fields on Q. Notice, that contact 1-forms on Q are locally generated by 1-forms $\iota^*\omega^\sigma$, $1 \le \sigma \le m$, i.e. $dq^l - \dot{q}^l dt, 1 \le l \le m-k$, $dq^{m-k+i} - g^i dt$, $1 \le i \le k$. We denote by $\Lambda_X(Q)$ the submodule of π_1-horizontal 1-forms on Q, by $\Omega^1(Q)$ the submodule of contact 1-forms on Q, by $\Omega_1^2(Q)$ and $\Omega_2^2(Q)$ the submodule of 1-contact and 2-contact 2-forms on Q. Similar definitions and notations are used for $\tilde{Q}$. Mappings h and p for forms on Q are defined in a similar way as in the unconstrained case, making use of the projection $\rho : \tilde{Q} \to Q$. For a form η on Q, $h\eta$ and $p\eta$ are defined on $\tilde{Q}$. If I is the constraint ideal, we denote by $\Lambda^q(I)$ the submodule of constraint q-forms. Similar notations are used if $\tilde{I}$, the constraint ideal on $\tilde{Q}$ is considered.

Constraint ideal I induced by the canonical constraint structure enables the construction of the quotient modules $\Lambda^q(Q)/\Lambda^q(I)$, resp. $\Lambda^q(\tilde{Q})/\Lambda^q(\tilde{I})$, which elements are equivalence classes $[\eta]_{\Lambda^q(I)}$, resp. $[\eta]_{\Lambda^1(\tilde{I})}$ of q-forms modulo constraint q-forms. The corresponding module operations, as well as the wedge product of classes and contraction of a class by a vector field from the canonical distribution C are defined as usual.

Let us recall some basic facts on the "constraint calculus", as introduced in [11], [12]. A 1-form η on Q is called *constraint horizontal* if $i_\xi \eta = 0$ for every π_1-vertical vector field $\xi \in C$. Constraint horizontal 1-forms take the form $\eta = \eta_0 + \varphi$, where η_0 is a horizontal form and φ is a constraint form. If $h : \Lambda^1(Q) \to \Lambda^1_X(\tilde{Q})$ and $p : \Lambda^1(Q) \to \Omega^1(\tilde{Q})$ are the "unconstrained" horizontalization and contactization mappings restricted to the module of 1-forms, we can define corresponding mappings between quotient modules

$$\bar{h} : \Lambda^1(Q)/\Lambda^1(I) \to (\Lambda^1_X(\tilde{Q}) \oplus \Lambda^1(\tilde{I}))/\Lambda^1(\tilde{I}), \quad \bar{h}[\eta]_{\Lambda^1(I)} = [h\eta]_{\Lambda^1(\tilde{I})} = h\eta + \varphi, \quad (15)$$

$$\bar{p} : \Lambda^1(Q)/\Lambda^1(I) \to \Omega^1(\tilde{Q})/\Lambda^1(\tilde{I}), \quad \bar{p}[\eta]_{\Lambda^1(I)} = [p\eta]_{\Lambda^1(\tilde{I})} = p\eta + \varphi, \quad (16)$$

which are defined on equivalence classes modulo constraint 1-forms (φ is an arbitrary constraint 1-form defined on $\tilde{Q}$). The mappings $\bar{h}$ and $\bar{p}$ will be called *constraint horizontalization* and *constraint contactization*.

Proposition 1. *Every equivalence class of* 1*-forms on* Q *admits a unique decomposition*

$$\rho^*[\eta]_{\Lambda^1(I)} = \bar{h}[\eta]_{\Lambda^1(I)} + \bar{p}[\eta]_{\Lambda^1(I)}. \quad (17)$$

In particular, for $\eta = df$ (17) gives a unique decomposition into a constraint-horizontal differential $\bar{h}[df]_{\Lambda^1(I)}$ and constraint-contact differential $\bar{p}[df]_{\Lambda^1(I)}$. It holds

$$\bar{h}[df]_{\Lambda^1(I)} = \frac{d_C f}{dt} dt + \varphi, \quad \bar{p}[df]_{\Lambda^1(I)} = \sum_{l=1}^{m-k} \frac{\partial_C f}{\partial q^l} \omega^l + \sum_{l=1}^{m-k} \frac{\partial f}{\partial \dot{q}^l} \dot{\omega}^l + \varphi, \quad (18)$$

where φ runs over constraint 1-forms on Q, and the operators

$$\frac{d_C f}{dt} = \frac{\partial f}{\partial t} + \sum_{l=1}^{m-k} \frac{\partial f}{\partial q^l} \dot{q}^l + \sum_{i=1}^{k} \frac{\partial f}{\partial q^{m-k+i}} g^i + \frac{\partial f}{\partial \dot{q}^l} \ddot{q}^l, \quad (19)$$

$$\frac{\partial_C f}{\partial q^l} = \frac{\partial f}{\partial q^l} + \sum_{i=1}^{k} \frac{\partial f}{\partial q^{m-k+i}} \frac{\partial g^i}{\partial \dot{q}^l} \quad (20)$$

are called *constraint total derivative* and *constraint partial derivative* respectively.

If $p_1 : \Lambda^2(Q) \to \Omega^2_1(\tilde{Q})$ and $p_2 : \Lambda^2(Q) \to \Omega^2_2(\tilde{Q})$ are standard "unconstrained" 1-contactization and 2-contactization mappings we can define corresponding mappings between quotient modules

$$\bar{p}_1 : \Lambda^2(Q)/\Lambda^2(I) \to (\Omega^2_1(\tilde{Q}) + \Lambda^2(\tilde{I}))/\Lambda^2(\tilde{I}), \quad \bar{p}_1[\eta]_{\Lambda^2(I)} = [p_1\eta]_{\Lambda^2(\tilde{I})} = p_1\eta + \varphi, \quad (21)$$

301

$$\bar{p}_2 : \Lambda^2(Q)/\Lambda^2(I) \to (\Omega_2^2(\tilde{Q}) + \Lambda^2(\tilde{I}))/\Lambda^2(\tilde{I}), \; \bar{p}_2[\eta]_{\Lambda^2(I)} = [p_2\eta]_{\Lambda^2(\tilde{I})} = p_2\eta + \varphi, \quad (22)$$

where φ runs over constraint 2-forms on $\tilde{Q}$. Mappings $\bar{p}_1$ and $\bar{p}_2$ are called *constraint 1-contactization* and *constraint 2-contactization*.

Proposition 2. *Every class of 2-forms on Q admits a decomposition*

$$\rho^*[\eta]_{\Lambda^2(I)} = \bar{p}_1[\eta]_{\Lambda^2(I)} + \bar{p}_2[\eta]_{\Lambda^2(I)}. \quad (23)$$

Corollary 1. *Every class of the exterior derivative of a 1-form η is decomposable in the form :*

$$[d(\rho^*\eta)]_{\Lambda^2(\tilde{I})} = [\rho^*(d\eta)]_{\Lambda^2(\tilde{I})} = \bar{p}_1[d\eta]_{\Lambda^2(I)} + \bar{p}_2[d\eta]_{\Lambda^2(I)}. \quad (24)$$

Note that the contraction of forms restricted to vector fields belonging to the canonical distribution transfers to a well-defined operation of classes modulo constraint q-forms: For $\Xi \in C$ we put

$$i_\Xi[\eta]_{\Lambda^q(I)} = [i_\Xi\eta]_{\Lambda^{q-1}(I)}. \quad (25)$$

5. CONSTRAINT EULER–LAGRANGE EQUATIONS

Variational aspects of nonholonomic mechanical systems are studied in many recent papers e.g. [2], [8]–[11], [15]. We approach this problem within the framework of Lepage forms introduced in [4] and non-holonomic structures proposed in [6]–[8]. The concepts and results are explained in more details in our paper [11].

Consider a 1-form η on the constraint submanifold Q. Let Ω be a piece of X, i.e. a compact submanifold of X with boundary $\partial\Omega$ (without loss of generality we can assume $\Omega = [a, b] \subset \mathbb{R}$). Denote by $\bar{\Gamma}_\Omega^Q(\pi)$ the set of all Q-admissible sections $\bar{\gamma}$ of π such that dom $\bar{\gamma} \supset \Omega$. The function

$$\bar{\Gamma}_\Omega^Q(\pi) \ni \bar{\gamma} \longmapsto \int_\Omega J^1\bar{\gamma}^*\eta \in \mathbb{R}. \quad (26)$$

is called the *action function of the 1-form η over Ω*.

Remark 2. Notice that the action function does not change if one takes instead of η the 1-form $\eta + \varphi$, where φ is an arbitrary constraint 1-form, since $J^1\bar{\gamma}$ are integral sections of the canonical distribution C. This enables one to define the *action function of the class* $[\eta]_{\Lambda^1(I)}$

$$\bar{\Gamma}_\Omega^Q(\pi) \ni \bar{\gamma} \longmapsto \int_\Omega J^1\bar{\gamma}^*[\eta]_{\Lambda^1(I)} \in \mathbb{R}. \quad (27)$$

Let Ξ be a π_1-projectable vector field on Q belonging to the canonical distribution C, and ξ its π_1-projection. Denote ϕ_u (resp. ϕ_{0u}) the local one-parameter group of Ξ (resp.

ξ). Let $\bar{\gamma}$ be a Q-admissible section of π. We define the *constraint deformation* of the section $\bar{\gamma}$ induced by Ξ as the one-parameter family of sections $\{\bar{\delta}_u\} = \{\phi_u \circ J^1\bar{\gamma} \circ \phi_{0u}^{-1}\}$. Notice that deformed sections $\bar{\delta}_u$ remain in the set of constrained sections, but need not be the first jet prolongations of Q-admissible sections.

Consider the real valued function :

$$(-\varepsilon, \varepsilon) \ni u \longmapsto \int_{\phi_{0u}(\Omega)} \bar{\delta}_u^* \eta \in \mathbb{R}, \tag{28}$$

where ε is a suitable number. This function is differentiable. Differentiating it with respect to u at $u = 0$ we obtain

$$\left(\frac{d}{du} \int_{\phi_{0u}(\Omega)} \bar{\delta}_u^* \eta \right)_{u=0} = \int_{\Omega} J^1\bar{\gamma}^* \partial_\Xi \eta. \tag{29}$$

The arising action function of the 1-form $\partial_\Xi \eta$ over Ω will be called *the constraint first variation of the action function* (26), induced by Ξ. In view of Remark 2 the action function of $\partial_\Xi \eta$ does not change if one takes instead of $\partial_\Xi \eta$ a form $\partial_\Xi \eta + \varphi$, where φ is any constraint 1-form. This enables us better to consider the function (29) defined on the class $[\partial_\Xi \eta]_{\Lambda^1(I)}$. Now we obtain:

Theorem 1. *Let η be a 1-form on Q. The following conditions are equivalent:*

1. $[d\eta]_{\Lambda^2(I)}$ is decomposable in the form :

$$\rho^*[d\eta]_{\Lambda^2(I)} = [E]_{\Lambda^2(\tilde{I})} + [F]_{\Lambda^2(\tilde{I})}, \tag{30}$$

where $[E]_{\Lambda^2(\tilde{I})} = \bar{p}_1[d\eta]_{\Lambda^2(I)}$ is horizontal with respect to the projection $\tilde{Q} \to Y$.

2. The constraint 1-contact part $\bar{p}_1[d\eta]_{\Lambda^2(I)}$ is the equivalence class of a dynamical form on $\tilde{Q}$.

3. For every π_1-vertical vector field Ξ on Q belonging to the canonical distribution C, the coordinate representation of the class $\bar{h}i_\Xi[d\eta]_{\Lambda^2(I)}$ does not depend on components $\tilde{\Xi}^l$ of the vector field Ξ.

4. In every fibered chart,

$$\eta = \bar{L}dt + \frac{\partial \bar{L}}{\partial \dot{q}^l} \omega^l + \bar{L}_i \varphi^i, \tag{31}$$

where $\bar{L}$ and $\bar{L}_i$ are (local) functions on Q, and φ^i are canonical constraint 1-forms.

We call a form η satisfying any of the equivalent conditions above a *constraint Lepage form*. The (non-invariant) constraint horizontal part of η, i.e. the local form

$$\lambda_C = \bar{L}dt + \bar{L}_i \varphi^i \tag{32}$$

is then said to be a *constraint Lagrangian*. Note that contrary to the unconstrained variational calculus, where a Lagrangian is determined by a single function, here a constraint Lagrangian is determined rather by $(k+1)$ functions $\bar{L}, \bar{L}_i$.

The meaning of the constraint Lepage form is similar as in the unconstrained case; they provide the invariant decomposition of the Lie derivative into a "constraint Euler–Lagrange term" and a boundary term. Namely, if η is a constraint Lepage form, then for every π_1-vertical vector field on Q belonging to the canonical distribution C,

$$\bar{h}[\partial_\Xi \eta]_{\Lambda^1(I)} = \bar{h}\left(i_\Xi[d\eta]_{\Lambda^2(I)}\right) + \bar{h}d\left(i_\Xi[\eta]_{\Lambda^1(I)}\right), \tag{33}$$

and for every Q-admissible section $\bar{\gamma}$ of π

$$\int_a^b J^1\bar{\gamma}^*[\partial_\Xi \bar{\lambda}]_{\Lambda^1(I)} = \int_a^b J^1\bar{\gamma}^*\left(i_\Xi[d\eta]_{\Lambda^2(I)}\right) + \tag{34}$$

$$+ J^1\bar{\gamma}^*\left(i_\Xi[\eta]_{\Lambda^1(I)}\right)(b) - J^1\bar{\gamma}^*\left(i_\Xi[\eta]_{\Lambda^1(I)}\right)(a).$$

The formulas above are an infinitesimal and integral form of the *constraint first variation formula*, respectively. Indeed, computing the first term on the right-hand side of (33) we obtain

$$\bar{p}_1[d\eta]_{\Lambda^2(I)} = [E]_{\Lambda^2(I)}, \tag{35}$$

where

$$E = E_l(\bar{L}, \bar{L}_i)\omega^l \wedge dt, \quad 1 \leq l \leq m - k, \tag{36}$$

take the form of the "constraint Euler–Lagrange expressions", as follows :

$$E_l = \frac{\partial_C \bar{L}}{\partial q^l} - \frac{d_C}{dt}\left(\frac{\partial \bar{L}}{\partial \dot{q}^l}\right) + \bar{L}_j\left[\frac{d_C}{dt}\left(\frac{\partial g^j}{\partial \dot{q}^l}\right) - \frac{\partial_C g^j}{\partial q^l}\right], \tag{37}$$

where operators d_C/dt and $\partial_C/\partial q^l$ are defined by (19) and (20), respectively.

In the analogy with the unconstrained variational calculus we can now define the concept of a *constraint extremal of η on Ω* as a Q-admissible section such that the constraint first variation over Ω is zero for every deformation vanishing over the boundary of Ω (a "fixed end-point deformation"), and the concept of a *constraint extremal of η* as a Q-admissible section the restriction of which to any piece $\Omega \subset X$ is a constraint extremal of η on Ω.

Consequently, the following theorem holds:

Theorem 2. *Let η be a constraint Lepage form on Q, let $\bar{\gamma}$ be a Q-admissible section of π. The following conditions are equivalent:*

1. $\bar{\gamma}$ is a constraint extremal of η.

2. For every vector field Ξ on Q belonging to the canonical distribution C,

$$J^1\bar{\gamma}^*i_\Xi\alpha = 0 \tag{38}$$

for every element $\alpha \in [d\eta]_{\Lambda^2(I)}$.

3. The constraint Euler-Lagrange expressions vanish along $J^2\bar{\gamma}$, i.e.

$$E_l \circ J^2\bar{\gamma} = 0, \quad 1 \leq l \leq m - k, \tag{39}$$

304

where E_l are given by (37).

Remark 3. If λ is a Lagrangian on J^1Y then one can find the corresponding constrained motion equations of λ on Q ([17], [6], [8]). It is clear now, that these equations arise as a constraint Euler–Lagrange equations from the *constraint variational principle* defined by the *constraint Lepage* 1-*form*

$$\eta = \iota^* \theta_\lambda = (L \circ \iota) dt + \left(\frac{\partial (L \circ \iota)}{\partial \dot{q}^l} \right) \omega^l + \left(\frac{\partial L}{\partial \dot{q}^{m-k+i}} \circ \iota \right) \iota^* \omega^{m-k+i}, \qquad (40)$$

i.e., within the notations in (31),

$$\bar{L} = (L \circ \iota), \quad \bar{L}_i = \left(\frac{\partial L}{\partial \dot{q}^{m-k+i}} \circ \iota \right), \qquad (41)$$

or, alternatively, by a (local) *constraint Lagrangian*

$$\lambda_C = (L \circ \iota) dt + \left(\frac{\partial L}{\partial \dot{q}^{m-k+i}} \circ \iota \right) \varphi^i. \qquad (42)$$

ACKNOWLEDGMENTS

Research supported by the Grants GAČR 201/03/0512 of the Czech Grant Agency and IGA PřF OU 8/2004 of the University of Ostrava. The author is indebted to Professor Olga Krupková for her continual interest. Numerous consultations with her have contributed significantly to the progress of this work. The author is also grateful to the Cankaya University for a scholarship to cover living expenses during the International Workshop on Global Analysis, as well as to the Department of Mathematics, Faculty of Science, University of Ostrava for the payment of travel expenses.

REFERENCES

1. J. F. Cariñena and M. F. Rañada, Lagrangian systems with constraints: a geometric approach to the method of Lagrange multipliers, *J. Phys. A: Math. Gen.* 26 (1993) 1335–1351.
2. F. Cardin and M.Favreti, On nonholonomic and vakonomic dynamics of mechanical systems with nonintegrable constraints, *J. Geom. Phys.* 18 (1996) 295–325.
3. G. Giachetta, Jet methods in nonholonomic mechanics, *J. Math. Phys.*33 (1992) 1652–1655.
4. D. Krupka, Some geometric aspects of variational problems in fibered manifolds, *Folia Fac. Sci. Nat. UJEP Brunensis* 14 (1973) 1–65.
5. O. Krupková, *The Geometry of Ordinary Variational Equations*, Lecture Notes in Mathematics 1678, Springer, Berlin, 1997.
6. O. Krupková, Mechanical systems with nonholonomic constraints, *J. Math. Phys.* 38 (1997) 5098–5126.
7. O. Krupková, On the geometry of non-holonomic mechanical systems in: Proc. Conf. Diff. Geom. Appl., Brno, August 1998, edited by O. Kowalski, I. Kolář, D. Krupka and J. Slovák, (Masaryk University, Brno, 1999) 533–546.

8. O. Krupková, Recent results in the geometry of constrained systems, *Rep. Math. Phys.* 49 (2002) 269–278.
9. O. Krupková and J. Musilová, Constraint Helmholtz conditions, Preprint 6/2002, Inst. Theor. Phys. and Astrophys., Masaryk University, Brno, (2002) 10 pp.
10. O. Krupková and J. Musilová, Non-holonomic variational systems, Poster, 36 Symp. on Math. Phys., Toruň, Poland, June 9–12, 2004, 8 pp.
11. O. Krupková and M.Swaczyna, The non-holonomic variational principle, Preprint 8/2002, Inst. Theor. Phys. and Astrophys., Masaryk University, Brno, (2002) 34 pp.; Paper in preparation
12. O. Krupková and M.Swaczyna, Horizontal and contact forms on constraint manifolds, submitted.
13. M. de León, J.C. Marrero and D.M. de Diego, Non-holonomic Lagrangian systems in jet manifolds, *J. Phys. A: Math. Gen.* 30 (1997) 1167–1190.
14. E. Massa and E. Pagani, A new look at classical mechanics of constrained systems, *Ann. Inst. Henri Poincaré* 66 (1997) 1–36.
15. P. Morando and S. Vignolo, A geometric approach to constrained mechanical systems, symmetries and inverse problems, *J. Phys. A.: Math. Gen.* 31 (1998) 8233–8245.
16. M. F. Rañada, Time-dependent Lagrangian systems: A geometric approach to the theory of systems with constraints, *J. Math. Phys.* 35 (1994) 748–758.
17. W. Sarlet, A direct geometrical construction of the dynamics of non-holonomic Lagrangian systems, *Extracta Mathematicae* 11 (1996) 202–212.
18. W. Sarlet, F. Cantrijn and D.J. Saunders, A geometrical framework for the study of non-holonomic Lagrangian systems, *J. Phys. A: Math. Gen.* 28 (1995) 3253–3268.
19. D.J. Saunders, W. Sarlet and F. Cantrijn A geometrical framework for the study of non-holonomic Lagrangian systems: II, *J. Phys. A: Math. Gen.* 29 (1996) 4265–4274.

Applied Mathematics

Fredholm Joint Spectrum for Families of Operators

Arzu Akgül* and Sadi Bayramov*

*Department of Mathematics, University of Kocaeli, Ataturk Bulvari, 41300, Kocaeli, Turkey

Abstract. In this work, Fredholm joint spectrum of noncommuting operator family which generates nilpotent Lie algebra is defined and investigated some properties of it. The joint spectrum of a is defined to be the set $\sigma_F(a)$ of those $\lambda \in C^n$ for which the complex $Kos(a - \lambda, X)$ whose cohomology spaces are not of finite dimension. It is proved that $\sigma_F(a)$ is a compact set in C^n. Polynomial spectral mapping theorem with respect to polynomials in noncommuting variables is proved such that $P(\sigma_F(a)) \subset \sigma_F(P(a))$.
Key words. General theory of linear operators, Spectrum, Resolvent, Cochain complexes.
PACS. 02., 02.30.-f, 02.30.Sa, 02.30.Tb.
MSC. 47A10, 55U15.

1. INTRODUCTION

The problem of joint spectrum has turned into a modern problem by the work of Taylor (1970) and Atkinson (1968). There are several definitions for commutative family of operators. The most important of these definitions was introduced by Taylor for a commutative family of operators in Banach spaces, by using homological algebra [5]. The functional calculus is satisfied for Taylor spectrum. For noncommuting family of operators there have been different views but, a spectral theory satisfying all features of functional calculus has't been constructed, so far. The most convenient version of Taylor joint spectrum for tuples of operators generating nilpotent Lie algebra was constructed by Feinshtein, in 1993 [3]. In 2002, Dosiyev proved more general spectral mapping theorem for representations of a finite dimensional Lie algebra [2]. In this work, we introduce a joint spectrum for families of noncommuting operators by using Taylor methods and prove a version of spectral mapping theorem with respect to polynomials, in noncommuting variables. We use the homological methods so, we shall consider the space of cochain complexes.

CP729, *Global Analysis and Applied Mathematics: International Workshop on Global Analysis*, edited by K. Taş, D. Krupka, O. Krupková, and D. Baleanu

2. THE SPACE OF COCHAIN COMPLEXES

In particular, let $\{X_i\}_{i=0,...,N}$ be Banach spaces and $\{T_i\}_{i=0,...,N-1}$ be bounded linear operators. In that case,

$$CC(X,N) = \left\{(X,T) : X = \{X_i\}_{i=0,...,N}, T = \{T_i\}_{i=0,...,N-1}\right\} \tag{1}$$

is defined as the category of cochain complexes having length N on the Banach spaces. The subcategories of $CC(X,N)$ are exact complexes $EC(X,N)$ and Fredholm complexes which are defined as

$$FC(X,N) = \left\{(X,T) \in CC(X,N) : \dim(H^i(X,T)) < \infty\right\}. \tag{2}$$

If we are given the Tychonoff topology on the product set $\prod_{i=0}^{n-1} L(X_i, X_{i+1})$ than the subspace topology on the $CC(X,N)$ is given.

Theorem 1.a) $CC(X,N)$ is a subspace which is closed and nowhere dense in the product space $\prod_{i=0}^{n-1} L(X_i, X_{i+1})$.

b) The subspaces $EC(X,N)$ and $FC(X,N)$ are open sets in the product space $\prod_{i=0}^{n-1} L(X_i, X_{i+1})$.

This theorem is going to be used to prove the property of compactness.

3. THE JOINT SPECTRUM FOR FAMILIES OF OPERATORS

For Banach spaces X, let $L(X)$ be a set of all bounded linear operators. Let E be a complex Lie algebra whose generators $u_1,...,u_n$ are noncommuting and $\rho : E \to L(X)$ be representation of E on X. Denote by ΛE the exterior algebra generated by E, and by $\Lambda^p E$ the $p-th$ exterior space of E. The generators of $\Lambda^p E$ are represented by $\underline{u} = u_1 \Lambda ... \Lambda u_p \in \Lambda^p E$. The chain Kozsul complex $Kos(E,\rho,X)$ or simply $Kos(E,X)$ is defined as follows by Feinshtein (1993) [2,5]:

$$0 \leftarrow X \leftarrow X \otimes E \leftarrow ... \leftarrow X \otimes \Lambda^P E \leftarrow ... \tag{3}$$

$$\alpha(x \otimes \underline{u}) = \sum_{i=1}^{p}(-1)^{i-1}\rho(u_i)x \otimes \widehat{\underline{u}}^i + \sum_{i<j}(-1)^{i+j-1}x \otimes [u_i,u_j] \wedge \widehat{\underline{u}}^{i,j}.$$

The notation of $\widehat{\underline{u}}^i$ means to omit u_i. In particular when generators of E are commutative, than the complex (3) turns into the Taylor complex. Let $a = (a_1,...,a_n)$ be a family of noncommuting operators on the Banach space X, $\lambda = (\lambda_1,...,\lambda_n) \in C^n$, $a - \lambda = (a_1 - \lambda_1,...,a_n - \lambda_n)$, $E(a)$ the nilpotent Lie algebra generated by a which is in the algebra $L(X)$. We use the notation $Kos(a - \lambda,X) = Kos(E(a-\lambda),X)$ for the Kozsul complex generated by $E(a-\lambda)$ module X. We consider instead of complex (3) the following cochain complex:

$$... \leftarrow X \otimes \Lambda^P E \leftarrow ... \leftarrow X \otimes E \leftarrow X \leftarrow 0 \tag{4}$$

$$\alpha\,(x \otimes \underline{u}) = \sum_{i=1}^{p}(-1)^{i-1}\rho(u_i)x \otimes u_i \wedge \underline{u} + \sum_{i<j}(-1)^{i+j-1}x \otimes [u_i,u_j] \wedge \underline{u}.$$

In this study we define the Fredholm joint spectrum as following:

Definition 1. The Fredholm joint spectrum of a is the set $\sigma_F(a)$ of those $\lambda \in C^n$ for which the cochain complex $Kos(a - \lambda, X)$ whose cohomology spaces are not of finite dimension.

It is obvious that $\sigma_F(a) \subset \sigma(a)$. The next theorem is proved by using this containment and Theorem 1.

Theorem 2. For each family a , the set $\sigma_F(a)$ is a compact subset of C^n.

Let $p = (p_1, ..., p_m)$ be a family of polynomials in noncommuting variables. For n tuples of noncommuting operators a we denote $p(a) = (p_1(a_1), ..., p_m(a_n)) = (b_1, ..., b_m) = b$. The Lie algebra generated by b is denoted by $F(b)$. Denote the Kozsul complexes generated by a and b $Kos((a,X),\alpha)$ and $Kos((b,X),\beta)$ respectively. Since $F(b)$ is an ideal of $E(a)$, b_i can be represented as following:

$$b_i = \sum_{j=1}^{n} c_{ij}a_j \qquad i = 1, ..., m \qquad (5)$$

with $C = (c_{ij})$ is the coefficient matrix . To prove the spectral mapping theorem, it is necessary to define the following homomorphism.

$$\delta_*^p : H^p(Kos((a,X),\alpha)) \to H^p(Kos((b,X),\beta)). \qquad (6)$$

For this, firstly we define the operator

$$\delta^{p,q} : X \otimes \Lambda^P E \to X \otimes \Lambda^{p-q} E \otimes \Lambda^q F \qquad , \qquad for\ q > 0$$

$$\delta^{p,q}(x \otimes u_{j_1} \wedge ... \wedge u_{j_p}) = \sum_{1 \le i_1 < ... < i_q \le q} \sum_{1 \le k_1 < ... < k_q \le q} (-1)^{\frac{q(q-1)}{2}+|k|}$$

$$\Delta_{i_1,...,i_q,j_{k_1},...,j_{k_q}} x \otimes u_{j_1} \wedge ... \wedge \hat{u}_{j_{k_1}} \wedge \hat{...} \wedge \hat{u}_{j_{k_q}} \wedge ... \wedge u_{j_p} \wedge v_{i_1} \wedge ... \wedge v_{i_q}. \qquad (7)$$

If we set $q = 0$ than $\delta^{p,0} = I$.

Here $\Delta_{i_1,...,i_q,j_{k_1},...,j_{k_q}}$ is the minor of $i_1,...,i_q$ rows and $j_{k_1},...,.j_{k_q}$ columns of C, $q \le \min(m,n)$ and $|k| = k_1 + ... + k_q$.

Lemma 1. For all $p, q > 0$

$$\delta^{p+1,q}\alpha^p + (-1)^{q-1}\alpha^{p-q}\delta^{p,q} = \beta^{q-1}\delta^{p,q-1}. \qquad (8)$$

If $p = q$, then

$$\delta^{p,p} : X \otimes \Lambda^P E \to X \otimes \Lambda^P F$$

is a morphism of cochain complexes.

Proof: The proof of this lemma is omitted here even though its proof is complicated and not difficult to prove. Since

$$\{\delta^p := \delta^{p,p} : X \otimes \Lambda^P E \to X \otimes \Lambda^P F\} : Kos((a,X),\alpha) \to (Kos((b,X),\beta)$$

is a morphism of cochain complexes, we can define the operator δ_*^p . But, the operator δ_*^p depends on c_{ij} coefficients. We prove this theorem by using following bicomplex:

$$
\begin{array}{ccccccccc}
 & & 0 & & & & 0 & & 0 \\
 & & \downarrow & & & & \downarrow & & \downarrow \\
\dots \leftarrow & & X \otimes \Lambda^P E & \leftarrow & \dots & \leftarrow & X \otimes E & \leftarrow & X & \leftarrow 0 \\
 & & \downarrow & & & & \downarrow & & \downarrow \\
\dots \leftarrow & & X \otimes \Lambda^P E \otimes F & \leftarrow & \dots & \leftarrow & X \otimes E \otimes F & \leftarrow & X \otimes F & \leftarrow 0 \\
 & & \downarrow & & & & \downarrow & & \downarrow \\
 & & \vdots & & & & \vdots & & \vdots \\
 & & \downarrow & & & & \downarrow & & \downarrow \\
\dots \leftarrow & & X \otimes \Lambda^P E \otimes \Lambda^q F & \leftarrow & \dots & \leftarrow & X \otimes E \otimes \Lambda^q F & \leftarrow & X \otimes \Lambda^q F & \leftarrow 0 \\
 & & \downarrow & & & & \downarrow & & \downarrow \\
 & & \vdots & & & & \vdots & & \vdots \\
\end{array}
$$

Theorem 3. If $H^k(Kos((a,X),\alpha) = 0$, $k = 0,...,p-1$ then the operators δ_*^p do not depend on coefficients c_{ij} .

Proof: We give two different statements b_i with respect to a_j:

$$b_i = \sum_{j=1}^{n} c_{ij} a_j = \sum_{j=1}^{n} d_{ij} a_j ; \quad i = 1,...,m. \tag{9}$$

312

Let δ^p and γ^p be operators corresponding to (c_{ij}) and (d_{ij}) respectively. For $\forall [x] \in H^p(Kos((a,X),\alpha)$, $x \in Ker\alpha^p \subset X \otimes \Lambda^P E$, to prove the theorem, we must show that there exists an element $y \in X \otimes \Lambda^{P-1}F$ such that the equality

$$\delta^p(x) - \gamma^p(x) = \beta^{p-1}(y) \tag{10}$$

is valid. By using Lemma1 for δ^p and γ^p, and applying x we obtain

$$(-1)^{q-1}\alpha^{p-q,q}(\delta^{p,q} - \gamma^{p,q})(x) = \beta^{p-q+1,q-1}(\delta^{p,q-1} - \gamma^{p,q-1})(x) \tag{11}$$

and for $q = 0,...,p-1,p$

$$(-1)^{p-1}\alpha^{0,p}(\delta^p - \gamma^p)(x) = \beta^{1,p-1}(\delta^{p,p-1} - \gamma^{p,p-1})(x) \tag{12}$$
$$(-1)^{p-2}\alpha^{1,p-1}(\delta^{p,p-1} - \gamma^{p,p-1})(x) = \beta^{2,p-2}(\delta^{p,p-2} - \gamma^{p,p-2})(x) \tag{13}$$

$$\cdots$$

$$\alpha^{p-2,2}(\delta^{p,2} - \gamma^{p,2})(x) = \beta^{p-1,1}(\delta^{p,1} - \gamma^{p,1})(x) \tag{14}$$
$$\alpha^{p-1,1}(\delta^{p,1} - \gamma^{p,1})(x) = \beta^{p,0}(\delta^{p,0} - \gamma^{p,0})(x). \tag{15}$$

For $[x] \in \ker\alpha^p / Im\alpha^{p-1}$, using the commutativity of diagrams in the bicomplex, triviality of cohomology spaces $H^k(Kos((a,X),\alpha)$, for $k = 0,...,p-1$ and equalities (12-15) we construct the sequence of elements

$$y_q \in X \otimes \Lambda^{p-q+1}E \otimes \Lambda^q F, \qquad q = 0,...,p-1 \tag{16}$$

for which the following equalities are valid:

$$(\delta^{p,1} - \gamma^{p,1})(x) = \alpha^{p-2,1}(y_1) \tag{17}$$
$$(\delta^{p,2} - \gamma^{p,2})(x) = \alpha^{p-3,2}(y_2) - \beta^{p-2,1}(y_1) \tag{18}$$

$$\cdots$$

$$(\delta^{p,p-1} - \gamma^{p,p-1})(x) = \alpha^{0,p-1}(y_{p-1}) + (-1)^{p-1}\beta^{1,p-2}(y_{p-2}) \tag{19}$$
$$(\delta^p - \gamma^p)(x) = (-1)^{p-1}\beta^{0,p-1}(y_{p-1}). \tag{20}$$

Thus, we obtain $y = (-1)^{p-1}y_{p-1} \in X \otimes \Lambda^{P-1}F$ such that the equality (10) holds. That is, $(\delta^p - \gamma^p)(x) \in Im\beta^{0,p-1} \subset X \otimes \Lambda^P F$ and $[(\delta^p - \gamma^p)(x)] \in H^p(Kos((b,X),\beta)$. So, $\delta_*^p = \gamma_*^p$.

Theorem 4. If $H^k(Kos((b,X),\beta) = 0, k = 0,...,p-1$ then $H^k(Kos((a,X),\alpha) = 0$ too, $k = 0,...,p-1$, and the operator δ_*^p is one-to-one.

313

Proof: Let $H^k(Kos((b,X),\beta)), k = 0, ..., p-1$ be trivial. To prove that δ_*^p is one-to-one operator, it is sufficient to show that $ker\,\delta_*^p$ is zero.

$$ker\,\delta_*^p = \{[x] \in H^p(Kos((a,X),\alpha) \ : \ x \in ker\,\alpha^p, \ \delta^p(x) \in ker\,\beta^p \} \tag{21}$$

To prove that $ker\,\delta_*^p$ is zero, we must show that there exists an element $y \in X \otimes \Lambda^{p-1}E$ such that

$$\alpha^{p-1}(y) = x, \ x \in ker\,\alpha^p \tag{22}$$

holds. For $[x] \in H^p(Kos((a,X),\alpha)$, by using Lemma1 for $q = p, ..., 1$ and triviality of diagrams in bicomplex we obtain

$$\beta^{1,p-1}\delta^{p,p-1}(x) = (-1)^{p-1}\alpha^{0,p}\delta^p(x) \tag{23}$$
$$\beta^{2,p-2}\delta^{p,p-2}(x) = (-1)^{p-2}\alpha^{1,p-1}\delta^{p,p-1}(x) \tag{24}$$

$$\ldots$$

$$\beta^{p-1,1}\delta^{p,1}(x) = -\alpha^{p-2,2}\delta^{p,2}(x) \tag{25}$$
$$\beta^{p,0}\delta^{p,0}(x) = \alpha^{p-1,1}\delta^{p,1}(x). \tag{26}$$

Now, we use the triviality of $H^k(Kos((b,X),\beta)$, $k = 0, ..., p-1$, commutativity of diagrams in bicomplex and the equalities (23-26). We firstly use (23)and then (24-26). Thus, we obtain the sequence of elements

$$y_q \in X \otimes \Lambda^{p-q-1}E \otimes \Lambda^q F, \tag{27}$$

such that following equalities are valid:

$$\beta^{p-1}(y_{p-1}) = \delta^p(x) \tag{28}$$
$$\beta^{1,p-2}(y_{p-2}) = \delta^{p,p-1}(x) - (-1)^{p-1}\alpha^{0,p-1}(y_{p-1}) \tag{29}$$

$$\ldots$$

$$\beta^{p-2,1}(y_1) = \delta^{p,2}(x) - \alpha^{p-3,2}(y_2) \tag{30}$$
$$\beta^{p-1,0}(y_0) = \delta^{p,1}(x) - \alpha^{p-2,1}(y_1). \tag{31}$$

We consider the equation (26). We know that

$$\delta^{p,0}(x) = x \tag{32}$$

from definition of $\delta^{p,0}$. We use $\delta^{p,1}(x)$ in the equation in (31) instead of $\delta^{p,1}(x)$ in the equation (26). Now we use the triviality of $H^p(Kos(b,X),\beta) = 0$ thus, we obtain that

314

$y \in X \otimes \wedge^{p-1}E$ such that the equality (22) holds. So, ker δ_*^p is zero.

Theorem 5. If $H^k(X,b)$ is finite dimensional and $H^k(X,a) = 0, k = 0, ..., p-1$, then δ_*^p is one-to-one operator.

Proof: Firstly, we show that $H^j(H^k(X,b),a) = 0$. For this, it is sufficient to show that

$$H^0(H^k(X,b),a) = 0 \tag{33}$$

holds by [1]. We give the proof by induction:

$$H^0(H^0(X,b),a) = 0 \tag{34}$$

is valid, because for $k = 0$,

$$H^0(H^0(X,b),a) = \ker\alpha^{0,0} \cap \ker\beta^{0,0}. \tag{35}$$

We assume that (33) holds for $i < k$. Let us show that for $\forall k$,

$$H^0(H^k(X,b),a) = 0 \tag{36}$$

$$H^0(H^k(X,b),a) = \ker\alpha_*^{0,k}. \tag{37}$$

For $\forall [x] \in \ker\alpha_*^{0,k}$, to prove the theorem, it is sufficient to show that $x \in Im\beta^{k-1}$. We obtain that, there exist elements $y_i, z_i \in X \otimes \wedge^{k-i}E \otimes \wedge F^i$ such that

$$\beta^{k-i+1,i-1}(y_{i-1}) = \alpha^{k-i,i}(z_i - y_i), \ \beta^{k-i,i}(z_i) = 0 \tag{38}$$

is valid. For this, firstly, we set $z_k = x$ and $y_k = 0$.

Thus, there exists an element $y_{k-1} \in X \otimes E \otimes \wedge^{k-1}F$ such that

$$\beta^{1,k-1}(y_{k-1}) = \alpha^{0,k}(x) \tag{39}$$

holds. That is, (33) is valid.

To prove that δ_*^p is one-to-one operator, we must show that ker δ_*^p is zero. That is, for $x \in X \otimes \wedge^p E$, $x \in Im\alpha^{p-1}$. By using Lemma 3.2 and triviality of $H^j(H^k(X,b),a)$, we obtain that there exists $u_{p-q}, v_{p-q} \in X \otimes \wedge^{q-1} \otimes \wedge^{p-q}F$ such that for $q = 0,.,p$

$$\delta^{p,p-q}(x) = \beta^{q,p-q-1}(u_{p-q-1}) + (-1)^{p-q-1}\alpha^{q-1,p-q}(u_{p-q} + v_{p-q}) \tag{40}$$

holds. Finally,

$$x = \alpha^{p-1}(u_0 + y_0). \tag{41}$$

We obtain the following consequence by using these theorems:

Consequence 6.

$$p(\sigma_F(a)) \subset \sigma(p_F(a)). \tag{42}$$

REFERENCES

1. Albrethch, E., and Frunza, S., *Nonanalytic functional calculus in several variables*, Manuscripta Math., (1976), 327-336.
2. Dosiyev, A., *Spectra of infinite parametrized Banach complexes* , J. Operator Theory, (2002), 585-614.
3. Feinshtein, A. S., *Taylor joint spectrum for families of operators generating nilpotent Lie algebras*, J. Operator Theory, (1993), 3-27.
4. Müller, V., *Spectral Theory of Linear Operators and spectral systems in Banach Algebras*, Springer Verlag, London, (2003).
5. Taylor, Y. L., *A joint spectrum for several commuting operators*, J.Funct.Anal., (1970), 172-191.

Oscillation Criteria for Second Order Impulsive Delay Differential Equation

J. Alzabut* and A. Zafer†

* *Department of Mathematics and Computer Science, Çankaya University, 06530 Ankara, Turkey*
e-mail: jehad@cankaya.edu.tr
† *Department of Mathematics, Middle East Technical University, 06531 Ankara, Turkey*

Abstract. A necessary and sufficient condition is obtained for oscillation of bounded solutions of second order impulsive delay differential equations of the form

$$(r(t)x'(t))' + p(t)f(x(\tau(t))) = 0, \quad t \neq \theta_i,$$
$$\Delta(r(\theta_i)x'(\theta_i)) + b_i g(x(\sigma(\theta_i))) = 0, \quad i \in Z,$$
$$\Delta x(\theta_i) = 0.$$

An example is also inserted to illustrate the effect of impulses on the oscillatory behavior of the solutions.

Key words. Impulse, delay, oscillation, nonoscillation.
PACS. 02.30.Hq.
MSC. 34A37, 34K11.

1. INTRODUCTION

In recent years there has been intensive studies on the theory of impulsive differential equations, see for instance the monographs [7, 4, 1]. Although there exists also a well developed oscillation theory for delay differential equations [3, 5], the oscillation of solutions of impulsive delay differential equations seems to have been rarely considered [6, 2]. It is worth mentioning that the impulsive delay differential equations are adequate mathematical models of many physical phenomena in which the changes depend not only on the present but also on the past states. The magnitude of the impulses may greatly alter the state variable, causing oscillation and non oscillation of the solutions.

In this paper, our aim is to derive a new oscillation criterion for solutions of second order impulsive delay differential equations of the form

$$(r(t)x'(t))' + p(t)f(x(\tau(t))) = 0, \quad t \neq \theta_i, \quad t \geq t_0$$
$$\Delta(r(\theta_i)x'(\theta_i)) + b_i g(x(\sigma(\theta_i))) = 0, \quad i \in Z, \qquad (1)$$
$$\Delta x(\theta_i) = 0,$$

where $f, g \in C[R,R], r, p \in C[R,R^+], \tau, \sigma \in C[R^+,R^+], R^+ = (0,\infty)$, such that

CP729, Global Analysis and Applied Mathematics: International Workshop on Global Analysis,
edited by K. Taş, D. Krupka, O. Krupková, and D. Baleanu
© 2004 American Institute of Physics 0-7354-0209-4/04/$22.00

$\lim_{t\to\infty} \tau(t) = \lim_{t\to\infty} \sigma(t) = \infty$ and $\{b_i\}$ and $\{\theta_i\}$ are sequences of real numbers such that $\theta_i < \theta_{i+1}$ for all $i \in Z$ and $\lim_{i\to\infty} \theta_i = \infty$. We also assume that the functions f and g are nondecreasing. By $\Delta x(t)$ we mean the difference $x(t^+) - x(t^-)$ where $x(t^+) = \lim_{r\to t^+} x(r)$ and $x(t^-) = \lim_{r\to t^-} x(r)$. A continuous function $x(t)$ is called a solution of (1) if it satisfies (1) for $t \in [t_0, \infty)$. We assume that $x'(t)$ is continuous from the left at θ_i, that is, $x'(\theta_i^-) = x'(\theta_i)$. As usual, such a solution of (1) is called oscillatory if it has arbitrarily large zeros, otherwise, it is called nonoscillatory. It is clear that a nonoscillatory solution is eventually either positive or negative.

2. THE MAIN RESULTS

In this section, we give a necessary and sufficient condition for all bounded solutions of (1) to be oscillatory. We start with a lemma that we will rely on later. The proof can be accomplished simply by expanding and re-grouping the terms.

Lemma 1 *Let $A_i(s)$ be continuous on $[a,b]$, then*

$$\int_a^b \sum_{s\le\theta_i<b} A_i(s)ds = \sum_{a\le\theta_i<b} \int_a^{\theta_i} A_i(s)ds.$$

In what follows we denote

$$R(t,u) = \int_u^t \frac{1}{r(s)}ds, \tag{2}$$

and assume that

$$\lim_{t\to\infty} R(t,u) = \infty. \tag{3}$$

We first establish a sufficient condition for oscillation of bounded solutions of (1).

Theorem 1 *Suppose that $xf(x) > 0$ and $xg(x) > 0$ for $x \ne 0, r(t) > 0, p(t) \ge 0$ is not identically zero on any interval of the form $[t_*, \infty), t_* \ge t_0$, the sequence $\{b_i\}$ is nonnegative with infinitely many positive terms. If for every $T > t_0$,*

$$\int^\infty R(u,T)p(u)du + \sum^\infty R(\theta_i,T)b_i = \infty, \tag{4}$$

then every bounded solution $x(t)$ of (1) is oscillatory.

Proof. Suppose that there exists a bounded nonoscillatory solution $x(t)$ of (1). We may assume that $x(t) > 0$ and $x(\tau(t)) > 0$ for $t \ge t_1 \ge t_0$. The case $x(t) < 0$ is similar. In fact, in view of the sign conditions on f and g, it suffices to replace x by $-x$. It is clear from the differential equation in (1) that $(r(t)x'(t))' < 0$ if $t \ne \theta_i$, that is, $r(t)x'(t)$ is nonincreasing on each interval (θ_i, θ_{i+1}). Without loss of generality, we may assume that $\theta_1 > t_1$, then, since $r(\theta_i^+)x'(\theta_i^+) - r(\theta_i)x'(\theta_i) = -b_i g(x(\sigma(\theta_i))) < 0$, we see at once that $r(t)x'(t)$ is nonincreasing for all $t \ge t_1$. We claim that $r(t)x'(t) > 0$ for $t \ge t_1$. If $r(t)x'(t) \le 0$ then since $r(t)x'(t)$ is nonincreasing, there exists $t_2 > t_1$ such that $r(t)x'(t) \le r(t_2)x'(t_2) < 0$

for all $t > t_2$. Integrating this inequality divided by $r(t)$ from t_2 to t and then taking limit as $t \to \infty$ in the resulting inequality, we obtain

$$x(t) - x(t_2) \leq r(t_2)x'(t_2)R(t,t_2),$$

which in view of (3) implies that $x(t)$ is eventually negative. This is a contradiction. Therefore we must have $r(t)x'(t) > 0$ eventually. Obviously, $x'(t) > 0$ eventually as well.

Integrating (1) from s to t, we have

$$r(t)x'(t) - r(s)x'(s) + \int_s^t p(u)f(x(\tau(u)))du + \sum_{s \leq \theta_i < t} b_i g(x(\sigma(\theta_i))) = 0,$$

and so

$$r(s)x'(s) > \int_s^t p(u)f(x(\tau(u)))du + \sum_{s \leq \theta_i < t} b_i g(x(\sigma(\theta_i))). \tag{5}$$

Dividing the above inequality by $r(s)$ and integrating again from t_1 to t lead to

$$x(t) - x(t_1) > \int_{t_1}^t \left[\frac{1}{r(s)} \int_s^t p(u)f(x(\tau(u)))du + \frac{1}{r(s)} \sum_{s \leq \theta_i < t} b_i g(x(\sigma(\theta_i))) \right] ds,$$

from which by changing the order of integration and using Lemma 1, we get

$$x(t) - x(t_1) > \int_{t_1}^t R(u,t_1)p(u)f(x(\tau(u)))du + \sum_{t_1 \leq \theta_i < t} R(\theta_i,t_1)b_i g(x(\sigma(\theta_i))).$$

Since $x(t)$ is eventually positive and increasing, there exists a positive real number c such that $x(\tau(t)) > c$ for all $t > t_1$ and $x(\sigma(\theta_i)) > c$ for all $\theta_i \geq t_1$. Since f and g are nondecreasing, we also have $f(x(\tau(t))) > f(c)$ and $g(x(\sigma(\theta_i))) > g(c)$. Thus in view of the last inequality we obtain

$$\int_{t_1}^t R(u,t_1)p(u)du + \sum_{t_1 \leq \theta_i < t} R(\theta_i,t_1)b_i < K[x(t) - x(t_1)], \tag{6}$$

where $\frac{1}{K} = \min\{f(c), g(c)\}$. Letting $t \to \infty$ in (6) and using (4), we see that $x(t) \to \infty$ as $t \to \infty$ which contradicts the boundedness of $x(t)$. Thus, every bounded solution must be oscillatory.

If condition (4) is strengthen, then the conclusion of the theorem remains valid not only for bounded but also for all solutions. Indeed, we have the following theorem generalizing Theorem 2 in [3].

Theorem 2 *If condition (4) of the previous theorem is replaced by*

$$\int^\infty p(u)du + \sum^\infty b_i = \infty,$$

then every solution $x(t)$ of (1) is oscillatory.

Proof. We proceed as in the proof of the previous theorem until (5). Setting $s = t_1$ there leads to

$$\int_{t_1}^t p(u)du + \sum_{t_1 \le \theta_i < t} b_i < Kr(t_1)x'(t_1).$$

Letting $t \to \infty$ in this inequality contradicts our assumption. Therefore, every solution is oscillatory.

In the next theorem we show that condition (4) is not only sufficient but also necessary for oscillation of the bounded solutions.

Theorem 3 *Suppose that for some $T \ge t_0$,*

$$\int_T^\infty |R(u,T)p(u)|du + \sum_{T \le \theta_i} |R(\theta_i,T)b_i| < \infty \tag{7}$$

then (1) has a bounded nonoscillatory solution.

Proof. Let $T_1 \ge T$ be sufficiently large so that

$$\int_{T_1}^\infty |R(u,T_1)p(u)|du + \sum_{T_1 \le \theta_i} |R(\theta_i,T_1)b_i| < \frac{a}{4\max\{|f(2a)|,|g(2a)|\}},$$

where a is positive real number to be specified. Define $T_0 = \inf_{t \ge T_1} \tau(t)$ and let X be the space of bounded continuous functions on $[T_0,\infty)$ with the supremum norm $\|x\| = \sup\{|x(t)| : t \ge T_0\}$. Let $Y \subset X$ be defined by $Y = \{x \in X : a \le \|x\| \le 2a\}$. Clearly, Y is bounded convex closed subset of X. Define an operator ϕ on Y by

$$
\begin{aligned}
(\phi x)(t) &= \frac{3a}{2} + \int_t^\infty R(t,u)p(u)f(x(\tau(u)))du \\
&\quad + \sum_{t \le \theta_i} R(t,\theta_i)b_ig(x(\sigma(\theta_i))), \quad t \ge T_1 \\
(\phi x)(t) &= \frac{3a}{2} + \int_{T_1}^\infty R(t,u)p(u)f(x(\tau(u)))du \\
&\quad + \sum_{T_1 \le \theta_i} R(t,\theta_i)b_ig(x(\sigma(\theta_i))), \quad t \in [T_0,T_1].
\end{aligned}
$$

We will show that ϕ maps Y to Y, is continuous and ϕY is relatively compact.
(i) ϕ *maps Y into itself.* It is clear that

$$|(\phi x)(t)| \le \frac{3a}{2} + \max\{|f(2a)|,|g(2a)|\}\left[\int_{T_1}^\infty |R(t,u)p(u)|du + \sum_{T_1 \le \theta_i} |R(t,\theta_i)b_i|\right].$$

Since

$$|R(t,u)| \le 2|R(u,T_1)| \quad \text{for all } u \ge t \ge T_1,$$

320

it follows that

$$|(\phi x)(t)| \;\leq\; \frac{3a}{2} + 2\max\{|f(2a)|,|g(2a)|\}\left[\int_{T_1}^{\infty}|R(u,T_1)p(u)|\,du + \sum_{T_1\leq\theta_i}|R(\theta_i,T_1)b_i|\right]$$

$$\leq\; 2a,$$

and

$$|(\phi x)(t)| \;\geq\; \frac{3a}{2} - |\int_{t}^{\infty}R(t,u)p(u)f(x(\tau(u)))\,du + \sum_{t\leq\theta_i}R(t,\theta_i)b_ig(x(\theta_i))|$$

$$\geq\; \frac{3a}{2} - 2\max\{|f(2a)|,|g(2a)|\}\left[\int_{T_1}^{\infty}|R(u,T_1)p(u)|\,du + \sum_{T_1\leq\theta_i}|R(\theta_i,T_1)b_i|\right]$$

$$\geq\; a.$$

Therefore, $\phi x \in Y$.

  (ii) ϕ *is continuous.* Let $\{x_n\}$ be a Cauchy sequence in Y such that

$$\lim_{n\to\infty}\|x_n - x\| = 0.$$

Because Y is closed, $x \in Y$. To prove the continuity of ϕ, we first see that

$$|(\phi x_n)(t) - (\phi x)(t)| \;\leq\; 2\int_{t}^{\infty}R(u,T_1)p(u)|f(x_n(\tau(u))) - f(x(\tau(u)))|\,du$$

$$+\; 2\sum_{t\leq\theta_i}R(\theta_i,T_1)b_i|g(x_n(\theta_i)) - g(x(\theta_i))|.$$

If we set

$$G_n(u) = R(u,T_1)p(u)|f(x_n(\tau(u))) - f(x(\tau(u)))|,$$

and

$$H_n(\theta_i) = R(\theta_i,T_1)b_i|g(x_n(\theta_i)) - g(x(\theta_i))|,$$

then we may write that

$$|\phi x_n - \phi x| \leq 2\int_{T_1}^{\infty}G_n(u)\,du + 2\sum_{T_1\leq\theta_i<\infty}H_n(\theta_i).$$

Notice that

$$G_n(u) \leq 2|f(2a)||R(u,T_1)p(u)|,$$

and

$$H_n(\theta_i) \leq 2|g(2a)||R(\theta_i,T_1)b_i|.$$

Furthermore $\lim_{n\to\infty}G_n(u) = \lim_{n\to\infty}H_n(\theta_i) = 0$. Since $|R(u,T_1)p(u)|$ is integrable over $[T,\infty)$ and the series with general term $|R(\theta_i,T)b_i|$ is convergent, applying *Lebesque Convergence theorem* on integral and the uniform convergence on the sum, we have

$$\lim_{n\to\infty}\|\phi x_n - \phi x\| = 0.$$

Hence ϕ is continuous.

(iii) ϕ *is precompact.* It is clear that $\phi(x)$, $x \in Y$ is uniformly bounded. Therefore, we need to prove that ϕY is an equicontinuous family of functions on $[T_0, \infty)$. Let $x \in Y$ and $t_2 > t_1$, we have

$$
\begin{aligned}
|\phi x(t_2) - \phi x(t_1)| &= |\int_{t_2}^{\infty} R(t_2, u)p(u)f(x(\tau(u)))du + \sum_{t_2 \leq \theta_i} R(t_2, \theta_i)b_i g(x(\sigma(\theta_i))) \\
&\quad - \int_{t_1}^{\infty} R(t_1, u)p(u)f(x(\tau(u)))du - \sum_{t_1 \leq \theta_i} R(t_1, \theta_i)b_i g(x(\sigma(\theta_i)))| \\
&\leq 2|f(2a)| \int_{t_1}^{\infty} |R(t, u)p(u)|du + 2|g(2a)| \sum_{t_1 \leq \theta_i} |R(t, \theta_i)b_i| \\
&\leq 4|f(2a)| \int_{t_1}^{\infty} |R(u, T_1)p(u)|du + 4|g(2a)| \sum_{t_1 \leq \theta_i} |R(\theta_i, T_1)b_i|.
\end{aligned}
$$

For a given $\varepsilon > 0$, there exists $T^* > T_1$ such that

$$
\int_{T^*}^{\infty} R(u, T_1)p(u)du + \sum_{T^* \leq \theta_i < \infty} R(\theta_i, T_1)b_i < \varepsilon.
$$

So if $T^* \leq t_1 < t_2$, we have

$$
|\phi x(t_2) - \phi(x_1)| < \varepsilon \quad \text{for all } x \in Y.
$$

Now, suppose that $T_1 \leq t_1 < t_2 \leq T^*$. Clearly,

$$
\begin{aligned}
|\phi x(t_2) - \phi x(t_1)| &= |\int_{t_2}^{\infty} (R(t_2, u) - R(t_1, u))p(u)f(x(\tau(u)))du \\
&\quad + \sum_{t_2 \leq \theta_i} (R(t_2, \theta_i) - R(t_1, \theta_i))b_i g(x(\theta_i)) \\
&\quad - \int_{t_1}^{t_2} R(t_1, u)p(u)f(x(\tau(u)))du - \sum_{t_1 \leq \theta_i < t_2} R(t_1, \theta_i)b_i g(x(\theta_i))|.
\end{aligned}
$$

Using the fact $R(t_2, u) - R(t_1, u) = R(t_2, t_1)$, we have

$$
\begin{aligned}
|\phi x(t_2) - \phi x(t_1)| &\leq |f(2a)|R(t_2, t_1) \int_{t_2}^{\infty} |p(s)|ds + |g(2a)|R(t_2, t_1) \sum_{t_2 \leq \theta_i} |b_i| \\
&\quad + |f(2a)| \int_{t_1}^{t_2} |R(t_1, u)p(u)|du + |g(2a)| \sum_{t_1 \leq \theta_i < t_2} |R(t_1, \theta_i)b_i|,
\end{aligned}
$$

or

$$
\begin{aligned}
|\phi x(t_2) - \phi x(t_1)| &\leq \max\{|f(2a)|, |g(2a)|\}R(t_2, t_1)\left[\int_{T_1}^{\infty} |p(u)|du + \sum_{T_1 \leq \theta_i} |b_i|\right] \\
&\quad + 2|f(2a)| \int_{t_1}^{t_2} |R(u, T_1)p(u)|du + 2|g(2a)| \sum_{t_1 \leq \theta_i < t_2} |R(\theta_i, T_1)b_i|.
\end{aligned}
$$

322

Clearly,

$$\int_{T_1}^{\infty} |p(u)| du + \sum_{T_1 \le \theta_i} |b_i| < \int_{T_1}^{\infty} |R(u,T_1)p(u)| du + \sum_{T_1 \le \theta_i} |R(\theta_i,T_1)b_i|$$

$$< \frac{a}{4 \max\{|f(2a)|, |g(2a)|\}}.$$

Hence, for any given $\varepsilon > 0$, there exists a δ such that

$$|\phi x(t_2) - \phi x(t_1)| < \varepsilon, \quad |t_2 - t_1| < \delta \quad \text{for all } x \in Y.$$

That is, the interval $[T_0, \infty)$ can be divided into two subintervals $[T_0, T^*]$ and $[T^*, \infty)$ on which every $\phi x(t)$, $x \in Y$, has variation less than ε. Therefore ϕY is an equicontinuous family on $[T_0, \infty)$, and hence ϕY is a compact subset of Y. According to the *Schauder fixed point theorem* there exists $x \in Y$ such that $x = \phi x$. This x is a bounded nonoscillatory solution of (1). The proof is complete.

Combining Theorem 1 and Theorem 3, we obtain the following necessary and sufficient condition for every bounded solution of (1) to be oscillatory.

Theorem 4 *Suppose that $xf(x) > 0$ and $xg(x) > 0$ for $x \ne 0$, $r(t) > 0$, $p(t) \ge 0$ is not identically zero on any interval of the form $[t_*, \infty)$, $t_* \ge t_0$, the sequence $\{b_i\}$ is nonnegative with infinitely many positive terms. Then every bounded solution $x(t)$ of (1) is oscillatory if and only if*

$$\int^{\infty} R(u,T)p(u) du + \sum^{\infty} R(\theta_i,T)b_i = \infty.$$

Example 1 Consider

$$x''(t) + \frac{1}{t^3} x(t-\tau) = 0, \quad t \ne i, \quad t \ge 1,$$

$$\Delta x'(i) + \frac{1}{i^\beta} x(i-\sigma) = 0, \quad i = 2, 3, \ldots$$

$$\Delta x(i) = 0,$$

where τ, β, σ are fixed positive real numbers. Comparing with (1), we see that $p(t) = \frac{1}{t^3}$, $\tau(t) = t - \tau$, $\sigma(\theta_i) = i - \sigma$, $\theta_i = i$, $b_i = \frac{1}{i^\beta}$ and $r(t) = 1$. It is clear that if $\beta \le 2$ then

$$\int^{\infty} \frac{1}{u^2} du + \sum^{\infty} \frac{1}{i^{\beta-1}} = \infty.$$

Therefore, by Theorem 1, every bounded solution of the above impulsive delay equation is oscillatory. We should note that if the solution is not subject to any impulse effect, that is, if we consider the equation

$$x''(t) + \frac{1}{t^3} x(t-\tau) = 0,$$

then, since

$$\int^{\infty} \frac{1}{u^2} du < \infty,$$

it follows from Theorem 4.6.1 in [3] that there is a nonoscillatory solution.

As observed in this example, we may deduce that the presence of impulses may in general change the oscillatory and/or nonoscillatory behavior of the solutions.

REFERENCES

1. D. Bainov and P. Simeonov, *Oscillation Theory of Impulsive Differential Equations*, International Publications, (1998).
2. L. Berzansky and E. Braverman, *Oscillation of a Linear Delay Impulsive Differential Equations*, Comm. Appl. Nonlinear Anal. 3, (1996), 61-77.
3. G. S. Ladde, V. Lakshmikantham and B. G. Zhang, *Oscillation Theory of Differential Equations with Deviating Arguments*, Marcel Dekker, Inc., New York, (1998).
4. V. Lakshmikantham, D.D. Bainov and P.S. Simeonov, *Theory of Impulsive Differential Equations*, World Scientific, Singapor, (1989).
5. I. Gyori and G. Ladas, *Oscillation Theory of Delay Differential Equations*, Clarendon Press, Oxford, (1991).
6. K. Gopalsamy and B. G. Zhang, *On Delay Differential Equations with Impulses*, J. Math. Anal. App. 139, (1989), 110-122.
7. A.M. Samoilenko and N.A. Perestyuk, *Impulsive Differential Equations*, World Scientific, (1995).

Nonlocal Boundary-Value Problems for PDE:Well-Posedness

Allaberen Ashyralyev

Department of Mathematics, Fatih University, Istanbul, Turkey
and
International Turkmen-Turkish University, Ashgabat, Turkmenistan

Abstract. The role played by coercive inequalities in the study of local boundary-value problems for parabolic and elliptic differential equations is well known (see, e.g.,[14],[21]). In present paper we consider the nonlocal boundary-value problems for parabolic and elliptic differential equations. The coercive inequalities for the solution of these nonlocal boundary-value problems are presented.
Key words. Parabolic equation, Elliptic equation, Nonlocal boundary-value problem, Well-Posedness.
PACS. 02.30.Jr.
MSC. 35K90, 35J67.

1. PARTIAL DIFFERENTIAL EQUATIONS OF PARABOLIC TYPE

It is known (see, e.g.,[2],[12],[13]) that various nonlocal boundary-value problems for the parabolic equations can be reduced to the nonlocal boundary-value problem for differential equation

$$v'(t) + Av(t) = f(t) \quad (0 \le t \le 1), v(0) = v(1) + \mu \tag{1}$$

in an arbitrary Banach space with linear (unbounded) operator A.
A function $v(t)$ is called a solution of the problem (1) if the following conditions are satisfied:

  i. $v(t)$ is continuously differentiable on the segment $[0,1]$.

 ii. The element $v(t)$ belongs to $D(A)$ for all $t \in [0,1]$, and the function $Av(t)$ is continuous on $[0,1]$.

iii. $v(t)$ satisfies the equation and the nonlocal boundary condition (1).

A solution of problem (1) defined in this manner will from now on be referred to as a solution of problem (1) in the space $C(E) = C([0,1],E)$ of all continuous functions $\varphi(t)$ defined on $[0,1]$ with values in E equipped with the norm

$$\|\varphi\|_{C(E)} = \max_{0 \le t \le 1} \|\varphi(t)\|_E.$$

CP729, *Global Analysis and Applied Mathematics: International Workshop on Global Analysis*,
edited by K. Taş, D. Krupka, O. Krupková, and D. Baleanu

It is said that the problem (1) is well posed in $C(E)$ if the following conditions are satisfied:

1. Problem (1) is uniquely solvable for any $f(t) \in C(E)$ and any $\mu \in D(A)$. This means that an additive and homogeneous operator $v(t) = v(t; f(t), \mu)$ is defined which acts from $C(E) \times D(A)$ to $C(E)$ and gives the solution of problem (1) in $C(E)$.

2. $v(t; f(t), \mu)$, regarded as an operator from $C(E) \times D(A)$ to $C(E)$, is continuous. Here $C(E) \times D(A)$ is understood as the normed space of the pairs $(f(t), \mu)$, $f(t) \in C(E)$ and $\mu \in D(A)$, equipped with the norm

$$\|(f(t), \mu)\|_{C(E) \times D(A)} = \|f\|_{C(E)} + \|\mu\|_{D(A)}.$$

By Banach's theorem in $C(E)$ and these properties one has coercive inequality

$$\|v'\|_{C(E)} + \|Av\|_{C(E)} \le M_C[\|f\|_{C(E)} + \|\mu\|_{D(A)}],$$

where M_C $(1 \le M_C < +\infty)$ does not depend on μ and $f(t)$.

The coercivity inequality implies the analyticity of the semigroup $\exp\{-tA\}(t \ge 0)$. Thus, the analyticity of the semigroup $\exp\{-tA\}(t \ge 0)$ is a necessary for the well-posedness of problem (1) in $C(E)$. Unfortunately, the analyticity of the semigroup $\exp\{-tA\}$ $(t \ge 0)$ is not a sufficient for the well-posedness of problem (1) in $C(E)$.

In [5] the well-posedness of problem (1) was established in the spaces $C_0^{\beta,\gamma}(E)(0 \le \gamma \le \beta, 0 < \beta < 1)$ and $C_0^{\beta,\gamma}(E_{\alpha-\beta})(0 \le \gamma \le \beta \le \alpha, 0 < \alpha < 1)$ under the assumption that the operator $-A$ generates an analytic semigroup $\exp\{-tA\}(t \ge 0)$ with exponentially decreasing norm, when $t \longrightarrow +\infty$.

Now, let us study the nonlocal boundary-value problem (1) in the spaces $L_p(E) = L_p([0,1], E)$ $(1 \le p < \infty)$ of all strongly measurable E−valued functions $v(t)$ on $[0,1]$ for which the norm

$$\| v \|_{L_p(E)} = \left(\int_0^1 \| v(t) \|_E^p \, dt \right)^{\frac{1}{p}}$$

is finite.

A function $v(t)$ is said to be absolutely continuous if it has a derivative $v'(t)$ for almost every t such that $v'(t) \in L_1(E)$, and if the Newton-Leibniz formula

$$v(t) - v(\tau) = \int_\tau^t v'(s) ds$$

holds for all t, $\tau \in [0, 1]$. Here the integral is understood in the sense of Bochner.

A function $v(t)$ is said to be a solution of the problem (1) in $L_p(E)$ if it is absolutely continuous, the functions $v'(t)$ and $Av(t)$ belong to $L_p(E)$, equation (1) is satisfied for almost every t, and $v(0) = v(1) + \mu$. From this definition it follows that a necessary condition for the solvability of problem (1) in $L_p(E)$ is that $f(t) \in L_p(E)$. It will be shown that in certain cases this condition is also sufficient for the solvability of problem

(1). As concerns the boundary elements, in contrast to the situation considered earlier, from the solvability of problem (1) in $L_p(E)$ it follows only that $\mu \in E$. In the case of an unbounded operator A this does not allow us to prove the solvability of problem (1).

From the unique solvability of (1) it follows that the operator $v(t; f(t), \mu)$ is bounded in $L_p(E)$ and one has coercive inequality

$$\|v'\|_{L_p(E)} + \|Av\|_{L_p(E)} \leq M_C[\|f\|_{L_p(E)} + \|A\mu\|_E]\,,$$

where M_C $(1 \leq M_C < +\infty)$ does not depend on μ and $f(t)$.

Theorem 1 *Suppose problem (1) is well posed in $L_{p_0}(E)$ for some p_0, $1 < p_0 < \infty$. Then it is well posed in $L_p(E)$ for any $p, 1 < p < \infty$ and the coercivity inequality holds:*

$$\|v'\|_{L_p(E)} + \|Av\|_{L_p(E)} + \|v\|_{C(E_{1-1/p,p})} \leq \frac{M(p_0)p^2}{p-1}\|f\|_{L_p(E)} + M\|\mu\|_{E_{1-1/p,p}},$$

where $M(p_0)$ and M does not depend on p, μ and $f(t)$. Here, the Banach space $E_{1-1/p,p} = E_{1-1/p,p}(E,A)(0 < \alpha < 1)$ consists of those $v \in E$ for which the norm

$$\| v \|_{E_{1-1/p,p}} = (\int\limits_0^1 \| A exp\{-zA\}v \|_E^p \, dz)^{\frac{1}{p}}, 1 \leq p < \infty$$

is finite.

The proof of this theorem is based on the well-posedness of Cauchy problem

$$v'(t) + Av(t) = f(t)(0 \leq t \leq 1), \quad v(0) = v_0 \tag{2}$$

in $L_p(E)$ [16] and the estimate

$$\|v_0\|_{E_{1-1/p,p}} \leq \frac{M(p_0)p^2}{p-1}\|f\|_{L_p(E)} + M\|\mu\|_{E_{1-1/p,p}(E,A)}.$$

Introduce the Banach space $E_{\alpha,q} = E_{\alpha,q}(E,A)(0 < \alpha < 1)$ consists of those $v \in E$ for which the norm

$$\| v \|_{E_{\alpha,q}} = (\int\limits_0^\infty [z^{1-\alpha} \| A exp\{-zA\}v \|_E]^q \frac{dz}{z})^{\frac{1}{q}}, \ 1 \leq q < \infty,$$

$$\| v \|_{E_{\alpha,\infty}} = \sup_{0<z} z^{1-\alpha} \| A exp\{-zA\}v \|_E, q = \infty$$

is finite.

Theorem 2 *Let $1 \leq p \leq \infty$ and $0 < \alpha < 1$. Then problem (1) is well posed in $L_p(E_{\alpha,p})$ and the coercivity inequality holds:*

$$\|v'\|_{L_p(E_{\alpha,p})} + \|Av\|_{L_p(E_{\alpha,p})} \leq \frac{M}{\alpha(1-\alpha)}\|f\|_{L_p(E_{\alpha,p})} + M\|\mu\|_{E_{1-1/p,p}(E_{\alpha,p},A)},$$

where M does not depend on α, p, μ and $f(t)$.

The proof of this theorem is based on the well-posedness of Cauchy problem (2) in $L_p(E_{\alpha,p})$ [9],[10],[18] and the estimate

$$\|v_0\|_{E_{1-1/p,p}(E_{\alpha,p},A)} \leq \frac{M}{\alpha(1-\alpha)}\|f\|_{L_p(E_{\alpha,p})} + M\|\mu\|_{E_{1-1/p,p}(E_{\alpha,p},A)}.$$

From Theorems 1 and 2 it follows that

Theorem 3 *Let* $1 < p,\ q < \infty$ *. Then problem (1) is well posed in* $L_p(E_{\alpha,q})$ *and the coercivity inequality holds:*

$$\|v'\|_{L_p(E_{\alpha,q})} + \|Av\|_{L_p(E_{\alpha,q})} \leq \frac{M(q)}{\alpha(1-\alpha)}\|f\|_{L_p(E_{\alpha,q})} + M\|\mu\|_{E_{1-1/p,p}(E_{\alpha,q},A)},$$

where $M(q)$ *and* M *does not depend on* $\alpha,\ p,\ \mu$ *and* $f(t)$.

Theorems 1-3 and theorem on the structure of the fractional spaces $E_{\alpha,q}(L_q(\mathbf{R}^n),A^x)$ [7],[20] permit us to establish the coercive inequalities in $L_p(L_2(\mathbf{R}^n))\,(1 < p < \infty)$ and in $L_p\left(W_q^{2m\alpha}(\mathbf{R}^n)\right)\,(1 < p,\ q < \infty, 0 < \alpha < \frac{1}{2m})$for the solution of the nonlocal boundary-value problem on the range $\{0 \leq t \leq 1,\ x \in \mathbf{R}^n\}$ for parabolic equation

$$\frac{\partial u}{\partial t} + \sum_{|r|=2m} a_r(x)\frac{\partial^{|\tau|}u}{\partial x_1^{r_1}...\partial x_n^{r_n}} + \delta u(t,x) = f(t,x), 0 \leq t \leq 1,\ x,r \in \mathbf{R}^n,$$

$$|r| = r_1 + \cdots + r_n, u(0,x) = u(1,x) + \varphi(x), f(0,x) = f(1,x), x \in \mathbf{R}^n,$$

where $a_r(x)$ and $f(t,x)$, $\varphi(x)$ are given sufficiently smooth functions and $\delta > 0$ is the sufficiently large number. Here Banach space $W_q^\beta(\mathbf{R}^n)$ is the space of the all integrable functions $f(x)$ defined on $\mathbf{R}^n$, equipped with the norm

$$\|f\|_{W_q^\beta(\mathbf{R}^n)} = \{ \int\limits_{x\in\mathbf{R}^n} \int\limits_{y\in\mathbf{R}^n} \frac{|f(x)-f(x+y)|^q}{|y|^{n+\beta q}}dxdy + \|f\|_{L_q(\mathbf{R}^n)}\}^{\frac{1}{q}}, 0 < \beta < 1, 1 \leq q \leq \infty.$$

where $L_q(\mathbf{R}^n)$ is the space of the all integrable functions defined on $\mathbf{R}^n$, equipped with the norm

$$\|f\|_{L_q(\mathbf{R}^n)} = \{ \int\limits_{x\in\mathbf{R}^n} |f(x)|^q dx\}^{\frac{1}{q}}.$$

2. PARTIAL DIFFERENTIAL EQUATIONS OF ELLIPTIC TYPE

Now, let us consider the nonlocal boundary- value problem

$$-v''(t) + Av(t) = f(t) \quad (0 \leq t \leq 1), \qquad v(0) = v(1),\ v'(0) = v'(1) \qquad (3)$$

in an arbitrary Banach space with positive operator A. It is known (see, for example [1],[11],[15]) that various nonlocal-boundary-value problems for the elliptic equations can be reduced to the boundary-value problem (3).

A function $v(t)$ is called a solution of the problem (3) if the following conditions are satisfied:

i. $v(t)$ is twice continuously differentiable function on the segment [0,1].
ii. The element $v(t)$ belongs to $D(A)$ for all $t \in [0,1]$, and the function $Av(t)$ is continuous on the segment [0,1].
iii. $v(t)$ satisfies the equation and boundary conditions (3).

A solution of problem (3) defined in this manner will from now on be referred to as a solution of problem (3) in the space $C(E)$. The well-posedness in $C(E)$ of the boundary-value problem (3) means that coercive inequality

$$\|v''\|_{C(E)} + \|Av\|_{C(E)} \le M\|f\|_{C(E)}$$

is true for its solution $v(t) \in C(E)$ with some M, does not depend on $f(t) \in C(E)$.

It is known that from this coercive inequality the positivity of the operator A in the Banach space E follows under the assumption that the operator has bounded in E inverse $(I\lambda + A)^{-1}$ for any $\lambda \ge 0$, and estimate

$$\|(\lambda I + A)^{-1}\|_{E \to E} \le M(1+\lambda)^{-1}$$

holds for some $1 \le M < \infty$. It turns out that this positivity property of the operator A in E is necessary condition of well-posedness of the boundary-value problem (3) in $C(E)$. Unfortunately, the positivity property of the operator A is not a sufficient for the well-posedness of problem (3) in $C(E)$ [6]. It is known (see, for example [17]) that the operator $A^{1/2}$ has better spectral properties than the positive operator A. In particular, the operator $\lambda I + A^{1/2}$ has a bounded inverse for any complex number λ with $\text{Re}\lambda \ge 0$, and the estimate

$$\|(\lambda I + A^{1/2})^{-1}\|_{E \to E} \le M(|\lambda| + 1)^{-1}$$

is true for some $M \ge 1$. This means that $B = A^{\frac{1}{2}}$ is a strongly positive operator in a Banach space E . In [4] the well-posedness of problem (3) was established in the spaces $C_{01}^{\beta,\gamma}(E)(0 \le \gamma \le \beta, 0 < \beta < 1)$ and $C_{01}^{\beta,\gamma}(E_{\alpha-\beta})(0 \le \gamma \le \beta \le \alpha, 0 < \alpha < 1)$ under the assumption that A is the positive operator.

Now, let us study the boundary value problem in the spaces $L_p(E) = L_p([0,1],E)$ $(1 \le p < \infty)$.

A function $v(t)$ is said to be a solution of the problem (3) in $L_p(E)$ if it is absolutely continuous, the functions $v''(t)$ and $Av(t)$ belong to $L_p(E)$, equation (3) is satisfied for almost every t, and $v(0) = v(1)$, $v'(0) = v'(1)$.

From the unique solvability of (3) it follows that the operator $v(t; f(t))$ is bounded in $L_p(E)$ and one has coercive inequality

$$\|v''\|_{L_p(E)} + \|Av\|_{L_p(E)} \le M_C \|f\|_{L_p(E)} \ ,$$

where M_C $(1 \le M_C < +\infty)$ does not depend on $f(t)$. From that the positivity of A can be obtained under the stronger assumption that the operator A^{-1} is compact in E.

Theorem 4 *Let A is the positive operator in a Banach space E. Suppose problem (3) is well posed in $L_{p_0}(E)$ for some p_0, $1 < p_0 < \infty$. Then it is well posed in $L_p(E)$ for any*

$p, 1 < p < \infty$ and the coercivity inequality holds:

$$\|v''\|_{L_p(E)} + \|Av\|_{L_p(E)} + \|v\|_{C(\ddot{E}_{1-1/p,p})} \leq \frac{M(p_0)p^2}{p-1}\|f\|_{L_p(E)},$$

where $M(p_0)$ does not depend on p and $f(t)$. Here $\ddot{E}_{1-1/p,p} = E_{1-1/p,p}(D(B), B)$.

The proof of this theorem is based on the well-posedness of the boundary value problem

$$-v''(t) + Av(t) = f(t)(0 \leq t \leq 1), \quad v(0) = v_0, v(1) = v_1 \qquad (4)$$

in $L_p(E)$ [19] and on the estimate

$$\|v_0\|_{\ddot{E}_{1-1/p,p}} \leq \frac{M(p_0)p^2}{p-1}\|f\|_{L_p(E)}, \|v_1\|_{\ddot{E}_{1-1/p,p}} \leq \frac{M(p_0)p^2}{p-1}\|f\|_{L_p(E)}.$$

Theorem 5 Let $1 \leq p \leq \infty$ and $0 < \alpha < 1$. Suppose that A is the positive operator in a Banach space E. Then problem (3) is well posed in $L_p(E_{\alpha,p})$ and the coercivity inequality holds:

$$\|v''\|_{L_p(E_{\alpha,p})} + \|Av\|_{L_p(E_{\alpha,p})} \leq \frac{M}{\alpha(1-\alpha)}\|f\|_{L_p(E_{\alpha,p})},$$

where M does not depend on α, p and $f(t)$. Here $E_{\alpha,p} = E_{\alpha,p}(E, B)(0 < \alpha < 1, 1 \leq p \leq \infty)$.

The proof of this theorem is based on the well-posedness of the boundary value problem (4) in $L_p(E_{\alpha,p}(E, B))$ [11] and the estimate

$$\|v_0\|_{\ddot{E}_{1-1/p,p}(E_{\alpha,p}(E,B))}, \|v_1\|_{\ddot{E}_{1-1/p,p}(E_{\alpha,p}(E,B))} \leq \frac{M}{\alpha(1-\alpha)}\|f\|_{L_p(E_{\alpha,p})}.$$

From Theorems 4 and 5 it follows that

Theorem 6 Let $1 < p, q < \infty$ and $0 < \alpha < 1$. Suppose that A is the positive operator in a Banach space E. Then problem (3) is well posed in $L_p(E_{\alpha,q})$ and the coercivity inequality holds:

$$\|v''\|_{L_p(E_{\alpha,q})} + \|Av\|_{L_p(E_{\alpha,q})} \leq \frac{M(q)}{\alpha(1-\alpha)}\|f\|_{L_p(E_{\alpha,q})},$$

where $M(q)$ and M does not depend on α, p and $f(t)$.

Theorems 4-6 and theorem on the structure of the fractional spaces $E_{\alpha,q}(L_q(\mathbf{R}^n), (A^x)^{\frac{1}{2}})$ [3],[8],[20] permit us to establish the coercive inequalities in $L_p(L_2(\mathbf{R}^n))(1 < p < \infty)$ and in $L_p(W_q^{m\alpha}(\mathbf{R}^n))$ $(1 < p, q < \infty, 0 < \alpha < \frac{1}{m})$ for the solution of the nonlocal boundary-value problem on the range $\{0 \leq t \leq 1, x \in \mathbf{R}^n\}$ for elliptic equation

$$-\frac{\partial^2 u}{\partial y^2} + \sum_{|r|=2m} a_r(x)\frac{\partial^{|\tau|}u}{\partial x_1^{r_1}\ldots\partial x_n^{r_n}} + \delta u(y,x) = f(y,x), 0 \leq y \leq 1, \ x,r \in \mathbf{R}^n,$$

$$|r| = r_1 + \cdots + r_n, u(0,x) = u(1,x), u_y(0,x) = u_y(1,x), f(0,x) = f(1,x), x \in \mathbb{R}^n,$$

where $a_r(x)$ and $f(y,x)$ are given sufficiently smooth functions and $\delta > 0$ is the sufficiently large number.

REFERENCES

1. Agmon, S., *Lectures on Elliptic Boundary Value Problems*, D. Van Nostrand, Princeton:New Jersey, 1965.
2. Amann, H., *Linear and quasilinear parabolic problems*, Birkhäuser Verlag, Basel, **I** (1995). Abstract linear theory.
3. Ashyralyev, A., *Method of positive operators of investigations of the high order of accuracy difference schemes for parabolic and elliptic equations*, Doctor Sciences Thesis, Kiev, 1991. (Russian).
4. Ashyralyev, A., *On well-posedness of the nonlocal boundary value problem for elliptic equations*, Numerical Functional Analysis and Optimization (2003) **24**, no. 1-2, 1-15.
5. Ashyralyev, A., Hanalyev, A. and Sobolevskii, P. E., *Coercive solvability of nonlocal boundary value problem for parabolic equations*, Abstract and Applied Analysis 6(2001), no. 1, 53-61.
6. Ashyralyev, A. and Kendirli, B., *Well posedness of the nonlocal boundary value problem for elliptic equations*, Functional Differential Equations 9(2002), no. 1-2, 33-55.
7. Ashyralyev, A. and Sobolevskii, P.E.,*Well-Posedness of Parabolic Difference Equations*.Birkhäuser Verlag, Basel, Boston, Berlin,Operator Theory and Appl., 1994, vol.69, 349 p.
8. Ashyralyev A. and Sobolevskii P.E., *New Difference schemes for Partial Differential Equations*, Birkhäuser Verlag, Basel, Boston, Berlin,Operator Theory and Appl., 2004, vol.148,480p.
9. Da Prato, G. and Grisvard, P., *Sommes d'opérateus liéaires et équations différentielles opérationnelles*, J. Math. Pures Appl. (9)**54**(1975), no. 3, 305-387.
10. Da Prato, G. and Grisvard, P., *Équations d'évolution abstraites non linus é aires de type parabolique*, C. R. Acad. Sci. Paris Sér. A-B **283** (1976), no. 9, A709-A711.
11. Grisvard, P., *Elliptic Problems in Nonsmooth Domains,* Pitman Advanced Publishing Program: London, 1986.
12. Krein, S. G., *Linear Differential Equations in Banach space*, "Nauka", Moscow, 1966(Russian).
13. Ladas, G. E. and Lakshmikantham, V., *Differential Equations in Abstract Spaces*, Academic Press, New York, London, 1972.
14. Ladyzhenskaya, O. A. and Ural'tseva, N. N., *Linear and Quasilinear Equations of Elliptic Type*, "Nauka", Moscow, 1973. (Russian).
15. Skubachevskii A. L., *Elliptic Functional Differential Equations and Applications*, Birkhauser Verlag, Operator Theory - Advances and Applications, Vol 91, 1997.
16. Sobolevskii, P. E., *Coerciveness inequalities for abstract parabolic equations*, Dokl. Akad. Nauk SSSR **157** (1964), no. 1, 52-55. (Russian).
17. Sobolevskii, P. E., *On elliptic equations in a Banach space*, Differensialnyye Uravneniya 4(1969), no. 7, 1346–1348. (Russian).
18. Sobolevskii, P. E., *Some properties of the solutions of differential equations in fractional spaces*, in: Trudy Nauchn. -Issled. Inst. Mat. Voronezh. Gos. Univ. No. 14(1975), 68-74. (Russian).
19. Sobolevskii, P. E., *The theory of semigroups and the stability of difference schemes*, in:Operator Theory in Function Spaces (Proc. School, Novosibirsk , 1975), pp. 304-337.(Russian).
20. Triebel, H., *Interpolation Theory, Function Spaces, Differential Operators*, North-Holland, Amsterdam-New York, 1978.
21. Vishik M. I., Myshkis, A. D. and Oleinik, O. A., *Partial Differential Equations*, in: Mathematics in USSR in the Last 40 Years, 1917-1957, Vol. 1, pp. 563-599, Fizmatgiz, Moscow, 1959. (Russian).

Crossover behavior between KdV and mKdV equations in a cold plasma with negative ions

D. Grecu*, Anca Visinescu* and A.S. Cârstea*

*Department of Theoretical Physics, "Horia Hulubei" National Institute of Physics and Nuclear Engineering, Bucharest, Romania

Abstract. It is well known that the KdV equation describes the behavior of ion-acoustic waves in a cold plasma. In the presence of negative ions, if their concentration satisfies a certain condition (critical concentration) the relevant equation is the modified KdV equation. The transition between these two regimes is studied from several points of view. The multiple scales analysis is extended to higher order and the role played by the next equations in the corresponding hierarchies KdV and mKdV is discussed.
Key words. KdV equation, mKdV equation, cold plasma.
PACS. 05.45.-a.
MSC. 34G20, 47J35.

1. INTRODUCTION

The plasma physics is a big "storehouse" of nonlinear phenomena. This diversity is generated by the multitude of physical situations in which a plasma can be produced. One of the simplest cases is a cold collisionless one-dimensional plasma with isothermal electrons. It represents one of the first physical problems where the Korteweg-de Vries (KdV) equation and the solitons plays an essential role. After the seminal paper of Washini and Taniuti [7] this system was discussed by many authors. No attempt to give an exhaustive list of references will be done; only a selected list of references will be given, directly related to the present work.

We shall consider a plasma consisting of cold positive (index α) and negative (index β) ions and isothermal electrons (index e). Denoting by $Q = \frac{m_\alpha}{m_\beta}$ the ratio of ion masses the equations describing the plasma are

$$\frac{\partial n_\alpha}{\partial t} + \frac{\partial}{\partial x}(u_\alpha n_\alpha) = 0 \qquad \frac{\partial n_\beta}{\partial t} + \frac{\partial}{\partial x}(u_\beta n_\beta) = 0 \qquad \frac{\partial u_\alpha}{\partial t} + u_\alpha \frac{\partial u_\alpha}{\partial x} = -\frac{\partial \Phi}{\partial x}$$

$$\frac{\partial u_\beta}{\partial t} + u_\beta \frac{\partial u_\beta}{\partial x} = \frac{1}{Q}\frac{\partial \Phi}{\partial x} \qquad n_e = e^\Phi \qquad \frac{\partial^2 \Phi}{\partial x^2} = n_e + n_\beta - n_\alpha \qquad (1)$$

In these equations the inertial electron effects are neglected $(m_e = 0)$ and consequently the electron gas will be in equilibrium with the local potential $\Phi(x,t)$ at any time t. Also special units are used, namely: the densities are normalized to the equilibrium electron

CP729, *Global Analysis and Applied Mathematics: International Workshop on Global Analysis*,
edited by K. Taş, D. Krupka, O. Krupková, and D. Baleanu
© 2004 American Institute of Physics 0-7354-0209-4/04/$22.00

density n_0; the lengths are measured in Debye radius $\lambda_D = \left(\frac{\varepsilon_0 K_B T_e}{e_0^2 n_0}\right)^{\frac{1}{2}}$ with K_B the Boltzmann constant, e_0 the elementary charge and T_e the electron temperature; time in units ω_0^{-1} where $\omega_0 = \left(\frac{n_0 e_0^2}{\varepsilon_0 m_\alpha}\right)^{\frac{1}{2}}$ is a plasma frequency; potential in units $\frac{K_B T_e}{e_0}$; velocities (u_α, u_β) in units of sound velocity $v_s = \left(\frac{K_B T_e}{m_\alpha}\right)^{\frac{1}{2}} = \lambda_D \omega_0$. For an one-dimensional plasma the following boundary conditions will be used $(|x| \to \infty)$

$$n_e \to 1 \qquad n_\alpha \to c_\alpha \qquad n_\beta \to c_\beta \qquad \Phi \to 0 \qquad u_\alpha \to 0 \qquad u_\beta \to 0 \qquad (2)$$

where 1, c_α and c_β are the equilibrium density values for electrons, positive ions and negative ions respectively. Of course the neutrality condition $c_\alpha = 1 + c_\beta$ is assumed to be satisfied.

There are many physical situations of plasma with positive and negative ions. Here we mention only one example, a plasma with Ar positive ions and SF_6 negative ions, for which $Q = 0.274$. Experiments on this plasma are reported in [1]. It is an interesting example because the mass of the negative ions is significantly greater than of the positive ions, and a small concentration of negative ions has a major effect on the plasma behavior.

2. MULTIPLE SCALES ANALYSIS

The cumulative effects of the weak nonlinearity manifests at long time and space scales. The right way to discuss these effects is to use the asymptotic method of multiple scales [6] (also called the reductive perturbation method). Depending if we are in the non-critical region (a), or in the critical one (b) the following "slow variables" (stretched variables) are introduced

$$a) \quad \xi = \varepsilon^{\frac{1}{2}}(x - vt) \qquad \tau_3 = v\varepsilon^{\frac{3}{2}}t, \qquad \tau_5 = v\varepsilon^{\frac{5}{2}}t, \ldots$$

$$b) \quad X = \varepsilon(x - vt) \qquad T_3 = v\varepsilon^3 t, \qquad T_5 = v\varepsilon^5 t, \ldots \qquad (3)$$

where $v^2 = c_\alpha + \frac{c_\beta}{Q}$ is the phase velocity of the acoustic wave propagating in the linearized system. Also all the quantities $n_e, n_\alpha, n_\beta, u_\alpha, u_\beta, \Phi$ are expanded in power series in the small parameter ε, namely

$$n_e = 1 + \varepsilon n_e^{(1)} + \varepsilon^2 n_e^{(2)} + \ldots; \qquad n_\alpha = c_\alpha + \varepsilon n_\alpha^{(1)} + \varepsilon^2 n_\alpha^{(2)} + \ldots$$
$$n_\beta = c_\beta + \varepsilon n_\beta^{(1)} + \varepsilon^2 n_\beta^{(2)} + \ldots; \qquad u_\alpha = \varepsilon u_\alpha^{(1)} + \varepsilon^2 u_\alpha^{(2)} + \ldots \qquad (4)$$
$$u_\beta = \varepsilon u_\beta^{(1)} + \varepsilon^2 u_\beta^{(2)} + \ldots; \qquad \Phi = \varepsilon \Phi^1 + \varepsilon^2 \Phi^{(2)} + \ldots$$

The expansion (4) are in agreement with the boundary conditions (2). Inserting (3) and (4) in the system (1) in each order of ε we get a system of equations which has to be solved. In order ε^2 the elimination of second order quantities lead to a constrain on the first order ones. Looking at $\Phi^{(1)}$ it has to satisfy the KdV equation

$$\frac{\partial \Phi^{(1)}}{\partial \tau_3} + \frac{1}{2}\frac{\partial^3 \Phi^{(1)}}{\partial \xi^3} + A\Phi^{(1)}\frac{\partial \Phi^{(1)}}{\partial \xi} = 0, \quad \text{where} \quad A = \frac{1}{2}\left[\frac{3}{v^4}(c_\alpha - \frac{c_\beta}{Q^2}) - 1\right]. \quad (5)$$

The condition $A = 0$ determines the critical concentration. Writing $c_\alpha = 1+c$ and $c_\beta = c$ this is given by

$$c_0 = \frac{Q - 3 + \sqrt{9 - 6Q + 9Q^2}}{2(1+Q)}. \quad (6)$$

When $Q \gg 1$ $c_0 \to 2$, while for $Q \ll 0$ $c_0 \to \frac{1}{18}Q^2$. For the example of $Ar^+ + SF_6^-$ $c_0 = 0.042$ One sees that a small concentration of massive negative ions is enough to bring the plasma in the critical region.

In the critical region we have to use the new slow variables $X, T_3, T_5...$ defined in (3). Then in order ε^3 instead of the KdV equation (2) we get the mKdV equation [8]

$$\frac{\partial \Phi^{(1)}}{\partial T} + \frac{1}{2}\frac{\partial^3 \Phi^{(1)}}{\partial X^3} + B(\Phi^{(1)})^2\frac{\partial \Phi^{(1)}}{\partial X} = 0, \quad B = \frac{1}{4}\left[\frac{15}{v^6}(c_\alpha + \frac{c_\beta}{Q^3}) - 1\right]. \quad (7)$$

If the concentration c is near the critical concentration $c = c_0 + 2K\varepsilon$ the mKdV equation (7) is replaced by a mixed mKdV + KdV equation

$$\frac{\partial \Phi^{(1)}}{\partial T} + \frac{1}{2}\frac{\partial^3 \Phi^{(1)}}{\partial X^3} + K\left(\frac{\partial A}{\partial c}\right)_{c_0} \Phi^{(1)}\frac{\partial \Phi^{(1)}}{\partial X} + B(\Phi^{(1)})^2\frac{\partial \Phi^{(1)}}{\partial X} = 0. \quad (8)$$

All the equations (5), (7) and (8) are completely integrable, having soliton solutions.

Higher order contributions in the multiple scales analysis can be also discussed. For a plasma with only positive ions an inhomogeneous linear equation for the second order perturbed potential $\Phi^{(2)}$ is obtained [3]

$$\frac{\partial \Phi^{(2)}}{\partial \tau_3} + \frac{1}{2}\frac{\partial^3}{\partial \xi^3}\Phi^{(2)} + \frac{\partial}{\partial \xi}(\Phi^{(1)}\Phi^{(2)}) = \quad (9)$$

$$= -\frac{\partial \Phi^{(1)}}{\partial \tau_5} - \left(\frac{3}{8}\frac{\partial^5 \Phi^{(1)}}{\partial \xi^5} + \frac{5}{8}\frac{\partial \Phi^{(1)}}{\partial \xi}\frac{\partial^2 \Phi^{(1)}}{\partial \xi^2} - \frac{1}{2}\Phi^{(1)}\frac{\partial^3 \Phi^{(1)}}{\partial \xi^3}\right).$$

In the left-hand-side (lhs) of (9) we recognize the linearized KdV equation, while the source term in the right-hand-side (rhs) depends only on the first order perturbed potential $\Phi^{(1)}$. In solving it we have to eliminate secular producing terms which appears in rhs of this equation. These are members of the null space of KdV equation, and are solutions of the homogeneous linearized KdV equation (the so called "symmetries" of KdV eq.). Such a secular producing terms is the fifth order derivatives $\frac{\partial^5 \Phi^{(1)}}{\partial \xi^5}$ and it can be eliminated requiring that $\Phi^{(1)}$ satisfies the next equation in the KdV hierarchy, namely (with a convenient scaling of the variables) [4]

$$\frac{\partial \Phi^{(1)}}{\partial \tau_5} = 6\frac{\partial^5 \Phi^{(1)}}{\partial \xi^5} + 10\Phi^{(1)}\frac{\partial^3 \Phi^{(1)}}{\partial \xi} + 20\frac{\partial \Phi^{(1)}}{\partial \xi}\frac{\partial^2 \Phi^{(1)}}{\partial \xi^2} + 5(\Phi^{(1)})^2\frac{\partial \Phi^{(1)}}{\partial \xi}. \tag{10}$$

As all equations in the hierarchy have the same spectral problem, the τ_5 dependence can appear only in the initial position of the soliton solution (a linear dependence on τ_5 is obtained). Eliminating this secular terms in (9), there remains an inhomogeneous linear equation for $\Phi^{(2)}$ free of secularities. Taking for $\Phi^{(1)}$ the one-soliton solution the equation can be solved analytically (the equation can be reduced to the equation for associated Legendre functions).

The same higher order analysis can be done in the critical region. The resulting inhomogeneous equation for $\Phi^{(2)}$ is [5]

$$\frac{\partial \Phi^{(2)}}{\partial \tau_3} + \frac{1}{2}\frac{\partial^3 \Phi^{(2)}}{\partial \xi^3} + B\frac{\partial}{\partial \xi}\left[(\Phi^{(1)})^2\Phi^{(2)}\right] = \tag{11}$$

$$= \frac{2}{3}\Phi^{(1)}\frac{\partial^3 \Phi^{(1)}}{\partial \xi^3} + \frac{1}{3}\frac{\partial \Phi^{(1)}}{\partial \xi}\frac{\partial^2 \Phi^{(1)}}{\partial \xi^2} - C(\Phi^{(1)})^3\frac{\partial \Phi^{(1)}}{\partial \xi},$$

where

$$C = \frac{35}{4v^8}(c_\alpha - \frac{c_\beta}{Q^4}) - \frac{35}{6v^6}(c_\alpha + \frac{c_\beta}{Q^4}) + \frac{11}{36}. \tag{12}$$

Again in the lhs we recognize the linearized mKdV equation, but no secular terms in the rhs can be identified. Indeed we can prove easily by straightforward calculations that no combination of the form

$$\alpha\Phi^{(1)}\frac{\partial^3 \Phi^{(1)}}{\partial \xi^3} + \beta\frac{\partial \Phi^{(1)}}{\partial \xi}\frac{\partial^2 \Phi^{(1)}}{\partial \xi^2} + \gamma(\Phi^{(1)})^3\frac{\partial \Phi^{(1)}}{\partial \xi} \tag{13}$$

with constant α, β, γ can be a solution of the homogeneous linearized mKdV equation. So we conclude that equation (11) is free of secularities (these appears in the next order of approximation).

3. CROSSOVER BEHAVIOR BETWEEN MKDV AND KDV.

As mentioned in the previous section the mixed KdV-mKdV equation (3) describes the properties of the system near the critical concentration. What happens when we move away from the critical region? This can be mimed by letting the parameter $\lambda = K(\frac{\partial A}{\partial c})_{c0}$ to become greater and greater. Suppose that $\lambda \to \infty$ as ε^{-1}, $\lambda \to \frac{\lambda}{\varepsilon}$. Reminding that $T = \varepsilon^3 t = \varepsilon^{\frac{3}{2}}\tau$ $(T_3 \to T, \tau_3 \to \tau)$ and $X = \varepsilon(x - vt) = \varepsilon^{\frac{1}{2}}\xi$ we can multiply equation (8) by $\varepsilon^{\frac{3}{2}}$. It is easily seen that it transforms into $(\Phi^{(1)} \to \Phi)$

$$\frac{\partial \Phi}{\partial \tau} + \frac{1}{2}\frac{\partial^3 \Phi}{\partial \xi^3} + \lambda\Phi\frac{\partial \Phi}{\partial \xi} = -\varepsilon B\Phi^2\frac{\partial \Phi}{\partial \xi}, \tag{14}$$

which is KdV eq. with a perturbative term.

The same conclusion can be obtained using Hirota's bilinear formalism [2]. Using bilinear operators

$$D_t^n D_x^m f \cdot g = \left(\frac{\partial}{\partial t} - \frac{\partial}{\partial t'} \right)^n \left(\frac{\partial}{\partial x} - \frac{\partial}{\partial x'} \right)^m f(x,t)g(x',t')\Big|_{\substack{x=x' \\ t=t'}} \tag{15}$$

and looking for $\Phi = \frac{G}{F}$, the bilinear form of (14) is

$$(D_T + D_X^3)G \bullet F = 0; \qquad D_X^2 F \bullet F = \frac{\lambda}{3}GF + \frac{B}{3}G^2. \tag{16}$$

If in (16) we let $\lambda \to \frac{\lambda}{\varepsilon}$ and use (τ, ξ) instead of (T, X) the system writes

$$(D_\tau + D_\xi^3)G \bullet F = 0; \qquad D_\xi^2 F \bullet F = \frac{\lambda}{3}GF + \varepsilon \frac{B}{3}G^2. \tag{17}$$

If $\varepsilon \to 0$ and $\frac{G}{F} = \Phi$ we get from the second equation (17)

$$\Phi = \frac{3}{\lambda} \frac{D_\xi^2 F \bullet F}{F^2} = \frac{6}{\lambda} \partial_\xi^2 \ln F, \tag{18}$$

which introduced into (14) gives

$$(D_{\xi\tau}^2 + D_\xi^4)F \bullet F = 0. \tag{19}$$

This is the usual bilinear form of KdV equation.

4. SLOW TIME VARIATION OF EQUILIBRIUM ION DENSITIES

Now let us consider a situation in which there is a continuous transition from the non-critical to critical one. This can be realized assuming that the equilibrium ion concentrations $c_\alpha(t)$, $c_\beta(t)$ are slowly time varying. Writing $1 + c_\beta(t) = c_\alpha(t)$ the neutrality condition is always satisfied. We shall assume that the slow time variation $c(t)$ takes place on τ_3 time scale, and the simplest case is to consider a linear dependence, namely

$$c(t) = \begin{cases} c_0 - \gamma & \tau_3 < -1 \\ c_0 + \gamma \tau_3 & \tau_3 \in (-1,1) \\ c_0 + \gamma & \tau_3 > 1 \end{cases}, \tag{20}$$

where c_0 is the critical concentration. As the slow time variable τ_3 is related to the small parameter ε (we remind $\tau_3 = \varepsilon^{\frac{3}{2}}vt$) this becomes now a "physical quantity", being connected with the physical interval $I = [-\frac{t}{v}\varepsilon^{-\frac{3}{2}}, \frac{t}{v}\varepsilon^{-\frac{3}{2}}]$ in which the concentrations are varied. Outside the interval I the multiple scales analysis leads to the KdV equation (5), but with different coefficients A to the left (A_-) and to the right (A_+) of this interval.

336

For $t \in I$ the multiple scales analysis has to be slightly changed to be adapted to the situation of time dependent ion concentrations. Using the slow variables

$$x_1 = \varepsilon^{\frac{1}{2}}x, \quad \tau_1 = v\varepsilon^{\frac{1}{2}}t, \quad \tau_3 = v\varepsilon^{\frac{3}{2}}t, \quad \tau_5 = v\varepsilon^{\frac{5}{2}}t,\dots \quad \xi = x_1 - \tau_1, \tag{21}$$

it is easily shown that we have to write

$$n_\alpha^{(1)} = -\gamma\tau_1 + \bar{n}_\alpha^{(1)}, \qquad n_\beta^{(1)} = -\gamma\tau_1 + \bar{n}_\beta^{(1)} \tag{22}$$

with $\bar{n}_\alpha^{(1)}, \bar{n}_\beta^{(1)}, u_\alpha^{(1)}, u_\beta^{(1)}, \Phi^{(1)}$ functions only of $(\xi = x_1 - \tau, \tau_3\dots)$, and given by

$$u_\alpha^{(1)} = \frac{1}{v}\Phi^{(1)}; \quad \bar{n}_\alpha^{(1)} = \frac{c_\alpha}{v^2}\Phi^2; \quad u_\beta^{(1)} = -\frac{1}{Qv}\Phi^{(1)}; \quad \bar{n}_\beta^{(1)} = -\frac{c_\beta}{Qv^2}\Phi^2; \quad n_e^{(1)} = \Phi^{(1)}. \tag{23}$$

In the next order ε^2, after eliminating second order quantities $n_\alpha^{(2)}, n_\beta^{(2)}, u_\alpha^{(2)}, u_\beta^{(2)}$ and using the expressions (23) for the first order quantities, we remain with

$$2\frac{\partial\Phi^{(1)}}{\partial\tau_3} + \frac{\partial^3\Phi^{(1)}}{\partial\xi^3} + \left[\frac{3}{v^4}(c_\alpha - \frac{c_\beta}{Q^2}) - 1\right]\Phi^{(1)}\frac{\partial\Phi^{(1)}}{\partial\xi} + \tag{24}$$

$$+\frac{\partial\Phi^{(2)}}{\partial x_1} + \frac{\partial\Phi^{(2)}}{\partial\tau_1} + \frac{\gamma}{v^2}(1+\frac{1}{Q})\tau_1\frac{\partial\Phi^{(1)}}{\partial\xi} - \frac{\gamma}{2v^2}(1+\frac{1}{Q})\Phi^{(1)} = 0.$$

Taking

$$\Phi^{(2)} = \frac{\gamma}{2v^2}(1+\frac{1}{Q})\left[-\tau_1^2\frac{\partial\Phi^{(1)}}{\partial\xi} + \tau_1\Phi^{(1)}\right] + \bar{\Phi}^{(2)}(\xi), \tag{25}$$

the equation (24) simplifies and we remain only with the KdV eq. (5), but with a time dependent coefficient A. Although it has a τ_3 dependence also at denominator (through v^4) in a first approximation we can neglect it and A will be a linear function of τ_3, $A = A_0\tau_3$

$$A = A_0\tau_3, \qquad A_0 = \frac{3\gamma(1-\frac{1}{Q})}{2(1+c_0+\frac{c_0}{Q})^2}. \tag{26}$$

Formally the KdV eq. ($\tau_3 \to \tau$, $\Phi^{(1)} \to \Phi$)

$$\frac{\partial\Phi}{\partial\tau} + \frac{1}{2}\frac{\partial^3\Phi}{\partial\xi^3} + A_0\tau\Phi\frac{\partial\Phi}{\partial\xi} = 0 \tag{27}$$

through the transformation $\tau\Phi = \Psi$ is reduced to the cylindrical KdV equation (cKdV)

$$\frac{\partial\Psi}{\partial\tau} + \frac{1}{2}\frac{\partial^3\Psi}{\partial\xi^3} + A_0\Psi\frac{\partial\Phi}{\partial\xi} - \frac{\Psi}{\tau} = 0. \tag{28}$$

According as we approach the critical concentration ($\tau_3 \to 0$) the previous analysis fails down. As discussed in section 2, a right approach is to use new slow variables (3) and the mKdV eq. (7) (or (8)) emerge as the right equation to describe the system behavior.

5. CONCLUSIONS

A full discussion of a cold plasma with positive and negative ions and isothermal electrons is presented. Depending if we are far or in the proximity of the critical region the first order component of the potential satisfies a KdV, or a mKdV equation. If the plasma contains heavy negative ions the critical region is reached for a very small concentration of these ions. Going to higher order in the multiple scales analysis the second order component of the potential has to be a solution of a linear inhomogeneous equation. The homogeneous linear part is nothing but the linearized KdV or mKdV equation, depending if we are far or near the critical region. To eliminate possible secular behavior it is necessary that the first order component to be solution also of the next order equation in the corresponding hierarchies.

In the neighborhood of the critical region, the first order component of the potential verifies a mixed KdV+mKdV equation. Moving off from the critical region (we let a parameter to increase as ε^{-1}) this equation is easily transformed to the KdV eq. plus a perturbative term. The same conclusion is reached in a bilinear formalism.

A continuous transition to the critical region can be realized if the equilibrium ion concentrations are varying slowly in time. A KdV equation is found with a time depending coefficient of the nonlinear term. If this dependence is linear in time the equation is easily transformed into a cylindrical type KdV eq. Then the following picture of such a "Gedanken Experiment" can be considered. Outside the time interval in which the slow variation of the concentrations take place the system is described by two KdV equations. Inside this interval we have two cylindrical type KdV equations, separated by a mKdV eq. which works in the critical region. It would be very interesting to study the time evolution, starting with an one-soliton in the region (1).

REFERENCES

1. Cooney J.M., Gavin M.T., Longreen K.E., *Experiments on Korteweg-de Vries solitons in a positive-ion and negative-ion plasma*, Phys.Fluids, **3**, (1991), 2758-2766.
2. Hirota R., - in *"Solitons", Topics in Current Physics* **17**, p. 157-176 (eds. R.K. Bullough, P.J. Caudrey, *Springer Verlag*, 1980).
3. Ichikawa Y.H., Mitsuhasi T., Konno K., *Contribution of higher order terms in the reductive perturbation theory: I A case of weakly dispersive wave*, J. Phys. Soc. Japan, **41**, (1976), 1382-1386.
4. Kraenkel R.A., Manna M.A., Pereira J.G., *The KdV hierarchy and long water waves*, J. Math. Phys., **36**, (1995), 307-320.
5. Tagare S.G., Reddy R.V., *Effect of higher-order nonlinearity on propagation of nonlinear ion-acoustic waves in a collisionless plasma consisting of negative ions*, J. Plasma Phys., **35**, (1986), 219-237.
6. Taniuti T., Wei C.C., *Reductive perturbation method in nonlinear wave propagation*, J. Phys. Soc. Japan, **24**, (1968), 941-946.
7. Washini H., Taniuti T., *Propagation of ion-acoustic solitary waves of small amplitude*, Phys.Rev.Lett., **17**, (1966), 996-998.
8. Watanabe S., *Ion acoustic soliton in plasma with negative ion*, J. Phys. Soc. Japan, **53**, (1984), 950-956.

An Overview of Mean Field Theory in Combinatorial Optimization Problems

Suat Kasap[*] and Theodore B. Trafalis[†]

[*]Department of Industrial Engineering, Cankaya University, Ogretmenler Caddesi. No: 14, Yuzuncuyil, Ankara 06530 TURKEY
[†]School of Industrial Engineering, University of Oklahoma, 202 West Boyd, Room 124, Norman, OK 73019 USA

Abstract. In the last three decades, there has been significant interest in using mean field theory of statistical physics for combinatorial optimization. This has led to the development of powerful optimization techniques such as neural networks (NNs), simulated annealing (SA), and mean field annealing (MFA). MFA replaces the stochastic nature of SA with a set of deterministic equations named as mean field equations. The mean field equations depend on the energy function of the NNs and are solved at each temperature during the annealing process of SA. MFA advances to the optimal solution in a fundamentally different way than stochastic methods. The use of mean field techniques for the combinatorial optimization problems are reviewed in this study.
Key words. Mean Field Theory, Combinatorial Optimization, Neural Networks, Annealing.
PACS. 05.50.+q.
MSC. 90C27, 90C90.

1. INTRODUCTION

Combinatorial optimization problems are one of the most challenging problems that have received considerable attention over the last three decades. Neural networks (NNs) have constituted a new direction in solving combinatorial optimization problems since the pioneering work of Hopfield and Tank [6]. The basic idea is to map the problem onto a highly interconnected network of neurons. The new problem is to find a configuration of neurons that minimizes the network energy that corresponds to an optimal solution. Since Simulated Annealing (SA) was proposed in 1983 by Kirkpatrick et al. [9], it has been one of the most popular heuristic algorithms for the combinatorial optimization problems. SA works by reducing the chance of getting stuck in a poor local optimum by accepting bad solutions with a decreasing probability. By combining many characteristics of SA and NNs, another technique named as mean field annealing (MFA) based on the mean field theory (MFT) of statistical physics was proposed [20, 21] to solve the combinatorial optimization problems. It has been reported that MFA is much faster than SA while they have the same quality of solutions [20, 21]. An overview of MFT applications for combinatorial optimization problems are given in this paper.

This paper was organized as follows: In the next two sections the NN and SA approach

CP729, *Global Analysis and Applied Mathematics: International Workshop on Global Analysis,*
edited by K. Taş, D. Krupka, O. Krupková, and D. Baleanu
© 2004 American Institute of Physics 0-7354-0209-4/04/$22.00

to solve combinatorial problems are presented in brief respectively. In the proceeding section, the MFT from statistical physics is explained in detail. Then applications of MFT to optimization more specifically to combinatorial optimization are discussed in the rest of the paper respectively.

2. NEURAL NETWORK (NN)

To solve combinatorial optimization problems using NNs requires a mapping of the problem onto the NNs in such a way that one can identify a solution from the outputs of neurons. The first step in designing NNs for combinatorial optimization problems is to formulate an energy function that maps the problem onto NNs. The second step is to derive a dynamic equation that prescribes the motion of the activation states of the NN. The last step is to determine the architecture of the NN in terms of neurons and links between neurons with associated weights based on the dynamic equation. The objective function and constraints of the problem is mapped onto a quadratic energy function of neuron states. The energy function has the following form:

$$E(S) = -\frac{1}{2} \sum_{i=1}^{N} \sum_{j=1}^{N} C_{ij} S_i S_j - \sum_{i=1}^{N} S_i I_i, \tag{1}$$

where $S_i = \{0, 1\}$ represents the output state of neurons, C_{ij} represents the weight of the link between neurons i and j and I_i represents the input bias. The input of neurons is denoted as x_i. The dynamic equations of neurons are defined by

$$\frac{dx_i}{dt} = -x_i + \sum_{j \neq i}^{N} C_{ij} S_i + I_i, \text{ where } S_i = f_i(x_i) = sgn \left(\sum_{j=1}^{N} C_{ij} S_j + I_i \right). \tag{2}$$

3. SIMULATED ANNEALING (SA)

SA is a stochastic approach for solving optimization problems. Let one consider the minimization of $E(x)$. SA accepts all of the good solutions that decrease $E(x)$ but also allows some bad solutions which increase $E(x)$. Specifically, a bad solution is accepted under the following condition

$$e^{-\Delta E/T} < RN, \tag{3}$$

where T is the temperature, ΔE is the change in objective, and RN is a random number. The temperature T is initially high, allowing many bad solutions to be accepted and is slowly reduced like $T = \alpha T$, where $0 < \alpha < 1$, this reduction known as cooling schedule, to a value where bad solutions are almost always rejected. One can interpret the function to be minimized as the energy function of a physical system, which when cooled sufficiently, slowly converges to a state of the minimum energy. This state represents the desired solution. To guarantee convergence to a (near) global minimum, the cooling schedule must be very slow. Extensive review of applications of SA were given in [3].

4. MEAN FIELD THEORY (MFT)

NNs are similar to some models of magnetic materials namely the spin-glass models encountered in statistical physics [24]. A spin-glass [5, 14] is a set of spins (i.e. magnetic moments) whose spins S_i and S_j interact through random couplings J_{ij}. There are two models of spin-glasses; Ising-glass and Potts-glass models. In Ising-glass models, the spin S_i can only take two values $S_i = \pm 1$. A Potts-glass model [30] is a generalization of the 2-state Ising-glass model to N-states. Spins in a system interact with each other according to the following Hamiltonian:

$$H(S) = -\sum_{i=1}^{N} \sum_{j=1}^{N} J_{ij} S_i S_i \qquad (4)$$

The analogy to NNs can be realized by identifying each spin with a neuron and associating ± 1 values of spins with active or resting state of neurons. The NN input represents the field acting on a spin, and the connection or weight matrix corresponds to the random couplings for interaction between spins.

The theory of spin-glasses relies strongly on the MFT. The MFT assumes that the state of a spin S_i, depends on its local environment because of interactions with other spins. Therefore, the value of S_i is approximated by its mean value $\langle S_i \rangle$, as the name of the theory implies. Conceptually, the mean field approximation replaces the value of a spin state that occurs in an energy field by its mean value when evaluating the probability distribution of any other spins. The $\langle S_i \rangle$s over a distribution of given parameters are expressed in terms of derivatives of an effective energy with respect to auxiliary fields according to the mean field approximation as follows:

$$E_{eff}(V,U,T) = E(V) + T \left(\sum_{j=1}^{n} U_j V_j - \ln(\cosh(U_j)) \right), \qquad (5)$$

$$\frac{\partial E_{eff}(V,U,T)}{\partial U_i} = V_i - \tanh U_i = 0 \quad \text{and} \quad \frac{\partial E_{eff}(V,U,T)}{\partial V_i} = \frac{\partial E(V)}{T \partial V_i} + U_i = 0. \qquad (6)$$

The mean field approximation is an analytic manipulation that simplifies the study of a physical system at the equilibrium. The behavior of the physical system at the equilibrium can be described by a set of equations named as mean field equations. The general form of the mean field equations for the Ising model are given in equation 6. Note that the discrete variables S_i have been replaced by continuous variables V_i and U_i, and V_i is the mean value of S_i. For the NN with energy function given by equation 1, the mean field equations can be reduced to

$$V_i = \tanh \left(\frac{1}{T} \sum_{j=1}^{N} C_{ij} V_j \right), \qquad (7)$$

where for simplicity, we have set $I_i = 0$.

Each spin of the Ising model has only two states either -1 or $+1$. This makes the Ising model inconvenient to implement for combinatorial optimization problems with multi-state variables. An alternative approach, proposed by Peterson and Soderberg [21], is to

341

use the Potts model. Introducing a second index for the spin S_{ij}, the first index denotes the site i and the second index denotes the spin j within each site such that $\sum S_{ij} = 1$. Two possible values of each spin is changed to 0 or 1. Thus, for every i, S_{ij} has only one value for j and zero for the remaining values of j. Note that, by using the Potts model, some of the soft constraints of the problem are treated as hard constraints namely they hold automatically. This makes the solution space smaller than the Ising model.

The effective energy and the general form of mean field equations for the Potts model will be as follows:

$$E_{eff} = E(V) + T \left(\sum_{j=1}^{n} U_j V_j - \ln \sum_{S} e^{US} \right), \tag{8}$$

$$V_j = \frac{\sum_{S} S e^{US}}{\sum_{S} e^{US}} \quad \text{and} \quad U_j = -\frac{\partial E(V)}{T \partial V_j}. \tag{9}$$

Mostly, the mean field equations are solved iteratively. Different approaches to solve mean field equations are proposed in [25, 26, 32, 33]. The iterative approach guarantees to converge to the local minima. It converges to a global minimum in the case that the starting point is sufficiently close to a global minimum. One way of avoiding this dependency is to use Mean Field Annealing (MFA). MFA replaces the stochastic nature of SA with a set of deterministic mean field equations that need to be solved iteratively. This deterministic relaxation procedure exhibits fast convergence toward the solution of combinatorial optimization problems. The effective energy function has the general structure of $E_{eff}(V) = E(V) + TF(V)$, where T and F correspond to the temperature and entropy of the system, respectively. Notice that the effective energy $E_{eff}(V)$ is smoother than $E(S)$ because of the additional entropy term. At the high temperatures, the effective energy is dominated by the convex entropy term. Therefore the effective energy is a strictly convex, and it has a unique minimum. On the other hand, at the low temperatures, the effective energy is nearly equal to the energy of the system. MFA can be described as follows: First, the mean field equations are solved at a high T and a unique solution is obtained. Then, after slightly lowering the T, the mean field equations are solved again starting from the higher T solution. By continuing this procedure, one can get a lower T solution that corresponds to a near global minimum of the energy function.

5. COMBINATORIAL OPTIMIZATION PROBLEMS

A new method to solve combinatorial optimization problems using the MFT and NNs was introduced by Peterson and Anderson [20]. The problem was mapped onto a NN by using the Ising model and then MFA was used in order to escape from local minima. As an extension of this study, a new method based on the Potts model was presented by Peterson and Soderberg [21]. They reduced the dimension of the solution space by one by using Potts model instead of Ising model. It was stated that the solution quality and parameter insensitivity were advantages of the new method.

The following combinatorial optimization problems have been studied using MFT and NNs approaches.

The **graph partitioning (GP) problem** can be defined for given a set of N nodes with a given connectivity, partition them into K partitions each with N/K nodes such that the net connectivity (cut-size) is minimal between each set. The GP problem was studied in [13, 18, 19, 20, 21, 22, 27].

In order to map the problem onto a NN, a second index for the neuron, s_{ia}, is introduced. The first index i denotes the node $i = 1, \cdots, N$ and the second index a denotes the partition $a = 1, \cdots, K$. An energy function for this problem can be written as follows [18, 21]:

$$E(s) = -\frac{1}{2}\sum_{i=1}^{N}\sum_{j=1}^{N}T_{ij}s_is_j + \frac{\alpha}{2}\left(\sum_{i=1}^{N}s_i\right)^2 - \frac{\beta}{2}\sum_{i=1}^{N}s_i^2, \tag{10}$$

where α and β are imbalance parameters and $s_i = (s_{i1}, s_{i2}, \cdots, s_{iK})$. The first term corresponds to the cost function and second and third terms correspond to the constraints of the problem that guarantee equal partition into K subsets.

The mean field equations for this energy function by using the Potts model can be written as follows:

$$V_i = \frac{\sum_S S e^{U_i S}}{\sum_S e^{U_i S}} \quad \text{and} \quad U_i = \frac{1}{T}\left(\sum_{j=1}^{N}(T_{ij} - \alpha)V_j + \beta V_i\right). \tag{11}$$

Note that, if $K = 2$, then this problem becomes **graph bi-partitioning problem**. The energy function for this problem becomes similar to equation 10 without the last term. The mean field equations for the graph bi-partitioning problem by using the Ising model can be written as follows [18, 19, 20]:

$$V_i = \tanh\left(\sum_{j=1}^{N}(T_{ij} - \alpha)\frac{V_j}{T}\right). \tag{12}$$

Numerical studies on the GP problems were performed by comparing the MFA approach with SA in [21]. Very good quality solutions were consistently found by using the MFA approach. It was stated that the solution quality and parameter insensitivity were advantages of the method. The concept of MFA in the context of the graph partitioning problem was presented in [27]. It was indicated that the results obtained by MFA was comparable to the results obtained by SA or the Kerninghan-Lin algorithm, but the rate of convergence of MFA was much faster than that of SA.

The **travelling salesman problem (TSP)** is to find the shortest possible tour through a set of n cities and the distances, d_{ij}, visiting each one exactly once. Note that the TSP is a special case of the GP problem where $K = N$. TSP is the most studied problem of combinatorial optimization problems by using NN approaches. The TSP was studied in [6, 7, 8, 12, 13, 17, 18, 21, 29, 31].

An energy function for this problem by using the Potts model can be written as

$$E(S) = \frac{A}{2}\sum_i\sum_j\sum_a D_{ij}S_{ia}S_{j(a+1)} - \frac{B}{2}\sum_i\sum_a S_{ia}^2 + \frac{C}{2}\sum_a\left(\sum_i S_{ia}\right)^2, \quad (13)$$

where A, B, and C are the parameters, D_{ij} is the distance between cities i and j, and $a+1$ is defined Modulo N. The first term minimizes the distance, and the second and third terms ensure that each city is visited exactly once. The mean field equations for this energy function by using the Potts model can be written as follows [17, 18, 21]:

$$V_{ia} = \frac{e^{U_{ia}}}{\sum_b e^{U_{ib}}} \quad \text{and} \quad U_{ia} = \frac{1}{T}\left(-\sum_j D_{ij}(V_{j(a+1)}+V_{j(a-1)}) - \alpha\sum_j V_{ja} + \beta V_{ia}\right). \quad (14)$$

Numerical studies on the TSP problems by comparing with SA were performed in [17, 21]. Very good quality solutions were consistently found by using the MFA approach. It was shown that MFA solutions dominate clearly to SA.

The **generalized quadratic assignment problem (QAP)** represents a large class of combinatorial optimization problems arising in a variety of planning and designing concepts. The generalized QAP can be defined as minimizing a quadratic cost function for assignment of a number of M objects to N positions where $N \geq M$. The QAP was studied in [4, 13, 29].

An energy function for this problem can be written as follows [4]:

$$E = \frac{A}{2}\sum_{i=1}^N\sum_{j\neq i}\sum_{k=1}^M\sum_{l\neq k} c_{ij}d_{kl}s_{ik}s_{jl} + \frac{B}{2}\sum_{i=1}^N\sum_{k=1}^M\sum_{l\neq k} s_{ik}s_{il} + \frac{C}{2}\sum_{k=1}^M\left(1-\sum_{i=1}^N s_{ik}\right)^2, \quad (15)$$

where A, B, and C are the weight parameters. The first term corresponds to the objective function. The second term specifies the constraint that at most one position can be assigned to each object and the third term specifies the constraint that every object must be assigned to exactly one assignment. This term prevents a solution with no assignments. The mean field equations for this energy function can be written as follows:

$$V_{ik} = \tanh\left(-\frac{A}{T}\sum_{j\neq i}\sum_{l\neq k} c_{ij}d_{kl}V_{jl} - \frac{B}{T}\sum_{l\neq k}V_{il} - \frac{C}{T}\left(\sum_{j=1}^N V_{jk}-1\right)\right). \quad (16)$$

The **knapsack problem** can be defined as filling a knapsack with a subset of given N items i with associated values c_i and sizes a_{ji} such that their total values are maximized subject to a set of M size constraints. The knapsack problem was studied in [1, 4, 13, 15, 16, 19]. An energy function for this problem can be written as follows [15, 19]:

$$E(S) = -\sum_{i=1}^N c_i s_i + \alpha\sum_{j=1}^M G\left(\sum_{i=1}^N a_{ji}s_i - b_j\right), \quad (17)$$

where the first term maximizes the total values of items and the second term corresponds to the constraints. Note that G is a penalty function to ensure that the constraints are satisfied. The mean field equations for this energy function can be written as follows:

$$
V_i = \tanh\left(-\frac{c_i}{T} + \frac{\alpha}{T}\sum_{k=1}^{M}\left[G\left(\sum_{j=1}^{N}a_{kj}V_j - b_k\right)_{V_i=1} + G\left(\sum_{j=1}^{N}a_{kj}V_j - b_k\right)_{V_i=0}\right]\right).
$$
(18)

The MFT approach is compared with other approaches such as branch and bound, simulated annealing, linear programming, and greedy heuristics in [15, 19]. Results showed that, the MFT approach is very competitive as compared to other approximate methods for the hard homogeneous knapsack problems, both with respect to solution quality and time consumption.

In addition to combinatorial optimization problems that discussed above, other combinatorial optimization problems have been also studied using MFT and NNs approaches. A novel method named as the invisible hand algorithm for solving the assignment problem using MFT based on the Potts model is proposed in [10]. The MFT approach was extended to the generalized assignment problem [16]. A MFA algorithm based on the Potts glass model for the circuit partitioning problem using the net-cut model was proposed in [2]. Mathematical models combining the Ising spin-glass model and NNs to solve the maximum clique problem were proposed in [11]. A number of interesting combinatorial optimization problems such as vertex cover, maximum independent set, number partitioning, maximum matching, set cover, graph morphism, and graph coloring were studied in [13, 22].

6. CONCLUSION

An overview of mean field theory coming from statistical physics and its applications to combinatorial optimization is examined. Recently, there have been many successful studies in applying the techniques from statistical physics to many optimization and engineering problems. This has led to the development of powerful optimization techniques such as NNs, SA, and mean field annealing (MFA). MFA combines many characteristics of SA and NNs. MFA replaces the stochastic nature of SA with a set of deterministic equations named as mean field equations. The mean field equations depend on the energy function of the NNs and are solved at each temperature during the annealing process of SA. MFA advances to the optimal solution in a fundamentally different way than stochastic methods. It was stated that the solution quality and parameter insensitivity were advantages of the MFT based approach.

REFERENCES

1. Abe S., Kawakami J., and Hirasawa K., *Solving inequality constrained combinatorial optimization problems by the Hopfield neural networks*, Neural Networks, vol. **5**, (1992), 663-670.
2. Bultan T., and Aykanat C., *Circuit partitioning using mean field annealing*, Neurocomputing, vol. **8**, (1995), 171-194.
3. Dowsland K. A., *Simulated annealing*, in C.R. Reeves (Ed.), Modern Heuristic Techniques for Combinatorial Problems, John Wiley and Sons, (1993), 20-69.

4. Fang L., and Li T., *Design of competition-based neural networks for combinatorial optimization*, International Journal of Neural Systems, vol. **1**, no. 3, (1990), 221-235.
5. Fischer K. H., and Hertz J. A., *Spin Glasses*, Cambridge University Press, Cambridge, Great Britain, (1991).
6. Hopfield J. J., and Tank D. W., *"Neural" computation of decisions in optimization problems*, Biological Cybernetics, vol. **52**, (1985), 141-152.
7. Igarashi H., *A solution for combinatorial optimization problems using a two-layer random field model: Mean-field approximation*, Systems and Computers in Japan, vol. **25**, no. 8, (1994), 61-71.
8. Ishii S, and Sato M. -A., *Chaotic Potts spin model for combinatorial optimization problems*, Neural Networks, vol. **10**, no. 5, (1997), 941-963.
9. Kirkpatrick S., Gelatt C. D., and Vecchi M. P., *Optimization by simulated annealing*, Science, vol. **220**, no. 4598, (1983), 671-680.
10. Kosowsky J. J., and Yuille A. L., *The invisible hand algorithm: Solving the assignment problem with statistical physics*, Neural Networks, vol. **7**, no. 3, (1994), 477-490.
11. Lee K. -C.,and Takefuji Y., *Maximum clique problems*, in Y. Takefuji and J. Wang (Eds.), Neural Computing for Optimization and Combinatorics, World Scientific, (1996), 31-77.
12. Looi C. -K., *Neural network methods in combinatorial optimization*, Computers in Operations Research, vol. **19**, no. 3/4, (1992), 191-208.
13. Melamed I. I., *Neural networks and combinatorial optimization*, Automation and Remote Control, vol. **55**, no. 11, (1994), 1553-1584.
14. Mezard M., Parisi G., and Virasoro M. A., *Spin Glass Theory and Beyond*, World Scientific, Singapore, (1987).
15. Ohlsson M., Peterson C., and Soderberg B., *Neural networks for optimization problems with inequality constraints: The knapsack problem*, Neural Computation, vol. **5**, (1993), 331-339.
16. Ohlsson M., and Pi H., *A study of the mean field approach to knapsack problems*, Neural Networks, vol. **10**, no. 1, (1997), 263-271.
17. Peterson C., *Parallel distributed approaches to combinatorial optimization: Benchmark studies on traveling salesman problem*, Neural Computation, vol. **2**, (1990), 261-269.
18. Peterson C., *Mean field theory neural networks for feature recognition, content addressable memory and optimization*, Connection Science, vol. **3**, no. 1, (1991), 3-33.
19. Peterson C., *Solving optimization problems with mean field methods*, Physica A, vol. **200**, (1993), 570-580.
20. Peterson C., and Anderson J.R., *Neural networks and NP-complete optimization problems; a performance study on the graph bisection problem*, Complex Systems, vol. **2**, (1988), 59-89.
21. Peterson C., and Soderberg B., *A new method for mapping optimization problems onto neural networks*, International Journal of Neural Systems, vol. **1**, no. 1, (1989), 3-22.
22. Ramanujam J., and Sadayappan P., *Mapping combinatorial optimization problems onto neural networks*, Information Sciences, vol. **82**, (1995), 239-255.
23. Reeves C. R., *Modern Heuristic Techniques for Combinatorial Problems*, John Wiley and Sons, New York, USA, (1993).
24. Reichl L. E., *A Modern Course in Statistical Physics*, University of Texas Press, Texas, USA, (1980).
25. Urahama K., and Ueno S.-I., *A gradient system solution to Potts mean field equations and its electronic implementation*, International Journal of Neural Systems, vol. **4**, no. 1, (1993), 27-34.
26. Urahama K., and Yamada T., *Constrained Potts mean field systems and their electronic implementation*, International Journal of Neural Systems, vol. **5**, no. 3, (1994), 229-239.
27. Van Den Bout D. E., and Miller T. K., *Graph partitioning using annealed networks*, IEEE Transactions on Neural Networks, vol. **1**, no. 2, (1990), 192-203.
28. Van Laarhoven P. J. M., *Theoretical and Computational Aspects of Simulated Annealing*, Stichting Mathematisch Centrum, Amsterdam, Netherlands, (1988).
29. Wang J., *Deterministic neural networks for combinatorial optimization*, in O.M. Omidvar (Ed.), Progress in Neural Networks, Ablex, (1994), 319-340.
30. Wu F. Y., *The Potts model*, Reviews of Modern Physics, vol. **54**, no. 1, (1982), 235-268.
31. Yuille A. L., *Generalized Deformable models, statistical physics, and matching problems*, Neural Computation, vol. **2**, (1990), 1-24.
32. Yuille A. L., and Kosowsky J. J., *Statistical physics algorithms that converge*, in R. J. Mammone (Ed.), Artificial Neural Networksfor Speech and Vision, Chapman and Hall, (1993), 19-36.
33. Yuille A. L., and Kosowsky J. J., *Statistical physics algorithms that converge*, Neural Computation, vol. **6**, (1994), 341-356.

Differential Algebraic Equations in Primal Dual Interior Point Optimization Methods

Suat Kasap* and Theodore B. Trafalis†

*Department of Industrial Engineering, Cankaya University, Ogretmenler Caddesi, No: 14, Yuzuncuyil, Ankara 06530 TURKEY
†School of Industrial Engineering, University of Oklahoma, 202 West Boyd, Room 124, Norman, OK 73019 USA

Abstract. Primal dual Interior Point Methods (IPMs) generate points that lie in the neighborhood of the central trajectory. The key ingredient of the primal dual IPMs is the parameterization of the central trajectory. A new approach to the parameterization of the central trajectory is presented. Instead of parameterizing the central trajectory by the barrier parameter, it is parameterized by the time by describing a continuous dynamical system. Specifically, a new update rule based on the solution of an ordinary differential equation for the barrier parameter of the primal dual IPMs is presented. The resulting ordinary differential equation combined with the first order Karush-Kuhn-Tucker (KKT) conditions, which are algebraic equations, are called differential algebraic equations (DAEs). By solving DAEs, we find an optimal solution to the given problem.
Key words. Interior Point Methods, Differential Algebraic Equations, Linear Programming, Quadratic Programming, Central Trajectory.
PACS. 02.60.Pn, 02.70.-c.
MSC. 90C51, 90C05, 65L80, 65L20.

1. INTRODUCTION

Iterative algorithms generate a sequence of points that converge to the optimal solution of the problem. The points specified by the iterative scheme lie on a smooth analytical curve that is called the *central trajectory*. A central trajectory starts from an interior point of the feasible region and ends at the optimal solution of the problem. The primal dual Interior Point Methods (IPMs) generate points that lie in the neighborhood of the central trajectory. The convergence of the primal dual IPMs is achieved when the duality gap becomes close to zero and the duality gap of the primal dual IPMs depends linearly on the barrier parameter μ for the points in the central trajectory. The barrier parameter μ is a positive number that monotonically decreases at each iteration. Therefore, the updating rule for the barrier parameter becomes so critical for the convergence of the primal dual IPMs. In other words, the key ingredient of the primal dual IPMs is the parameterization of the central trajectory. In this paper, a new approach to the parameterization of the central trajectory for primal dual IPMs is presented. Instead of parameterizing the central trajectory by the barrier parameter, we parameterize it by the time. In fact, the central trajectory is described by a continuous dynamical system. Specifically, a new up-

CP729, *Global Analysis and Applied Mathematics: International Workshop on Global Analysis*,
edited by K. Taş, D. Krupka, O. Krupková, and D. Baleanu
© 2004 American Institute of Physics 0-7354-0209-4/04/$22.00

date rule based on the solution of an ordinary differential equation (ODE) for the barrier parameter of the primal dual IPMs is presented [6, 10, 11]. The resulting ordinary differential equation combined with the first order Karush-Kuhn-Tucker conditions, which are algebraic equations, are called Differential Algebraic Equations (DAEs). By solving DAEs, we follow approximately the central trajectory of the primal dual IPMs. The proposed parameterization of the central trajectory for primal dual IPMs is investigated both for linear and convex quadratic optimization problems and primal dual IPMs are modified by using the new parameterization [6]. In addition, stability analysis of the proposed parameterization of the central trajectory are also presented.

This paper was organized as follows: In the next two sections brief introduction to DAEs and the primal dual IPMs for linear and convex quadratic optimization problems are given respectively. In the proceeding section, the proposed parameterization of the central trajectory of the primal dual IPMs and its extensions to linear and convex quadratic optimization problems are presented. Then stability analysis of the proposed parameterization of the central trajectory are discussed in the rest of the paper.

2. DIFFERENTIAL ALGEBRAIC EQUATIONS

A differential equation and a nonlinear system form a Differential Algebraic Equations (DAEs). The DAEs occur frequently as an initial value problem in modelling electrical networks, the flow of incompressible fluids, mechanical systems subject to constraints, robotics, distillation process, power systems, and in many other applications [3, 5, 8]. In general, DAEs can be expressed in the following semi-explicit form

$$\frac{dx}{dt} = x' = f(x,y,t) \quad \text{and} \quad g(x,y,t) = 0. \tag{1}$$

DAEs are different from ODEs. An ODE involves integration. On the other hand, a DAE involves both integrations and differentiations. Since a DAE involves both integrations and differentiations, by applying analytical differentiations to a given system and eliminating as needed will yield an explicit ODE. *Index* is defined as the minimum number of times that all or part of the DAE system must be differentiated to get a system of ODEs. ODE systems have index zero. The DAEs in (1) have index one if g_y is nonsingular. In a sense, the index of a DAE system is a measure of its degree of singularity. The more singular a DAE system is, the more difficult to solve it numerically. Therefore, DAEs of index one are easy to solve since they can be treated as ODEs. The idea of using ODE methods for solving DAE systems directly was introduced by Gear [4]. The simplest way to solve DAEs of index 1 is to apply backward Euler's method.

3. PRIMAL DUAL INTERIOR POINT METHODS

There has been an enormous research on using Interior Point Methods (IPMs) to solve optimization problems since Karmarkar's [7] study. Primal dual IPMs, are considered the most successful techniques to solve linear optimization problems [9]. Let us consider the following primal linear optimization problem P and the dual problem D

$$P: \min \left\{ c^T x \quad \text{such that} \quad Ax = b, x \geq 0 \right\}, \tag{2}$$

$$D: \max \left\{ b^T y \quad \text{such that} \quad A^T y + z = c, z \geq 0 \right\}, \tag{3}$$

where $x, z, c \in \Re^n$, $A \in \Re^{m \times n}$, and $y, b \in \Re^m$. It is assumed that A has a full row-rank. In other words $m \leq n$ and the feasible regions of the problems P and D are not empty.

Inequality constraints of the problem can be handled by adding a logarithmic barrier term μ to the objective function of P that results to the following problem P_μ

$$P_\mu : \min \left\{ c^T x - \mu \sum_{i=1}^n \ln x_i \quad \text{such that} \quad Ax = b, x > 0 \right\}. \tag{4}$$

To solve P_μ for $\mu > 0$, corresponding Lagrangian functions can be constructed as

$$L_P(x, y, \mu) = c^T x - \mu \sum_{i=1}^n \ln x_i - y^T (Ax - b). \tag{5}$$

The KKT necessary conditions for the Lagrangian function are given as

$$\begin{aligned} Ax - b &= 0, \\ A^T y + z - c &= 0, \\ XZe - \mu e &= 0, \end{aligned} \tag{6}$$

where $X = diag(x_1, x_2, \ldots, x_n)$, $Z = diag(z_1, z_2, \ldots, z_n)$, and $e = (1, 1, \ldots, 1)^T$. The KKT necessary conditions are interpreted as primal feasibility, dual feasibility, and complementary slackness condition, respectively. The KKT necessary conditions construct a nonlinear system of $F(x, y, z) = 0$ with the following Jacobian

$$J(x^k, y^k, z^k) = \begin{bmatrix} A & 0 & 0 \\ 0 & A^T & I \\ Z & 0 & X \end{bmatrix}. \tag{7}$$

By setting $\mu > 0$, the vectors x^k, y^k, z^k that solve this system are obtained by using Newton's method. Since these vectors are dependent on the choice of the barrier parameter μ, we get a family of solutions depending on the value of μ. The *central trajectory* is defined as the set of all vectors $x^k(\mu), y^k(\mu)$ and $z^k(\mu)$, satisfying this nonlinear system. The limits of $x(\mu), y(\mu)$ and $z(\mu)$ as μ goes to 0 approach to the solution.

Given $x^0 > 0, y^0$ and $z^0 > 0$ and $\mu > 0$, the moving direction vectors dx, dy, and dz to move from the current point to a new point while satisfying the KKT necessary conditions are determined by using Newton's method as

$$\begin{aligned} dy &= (AZ^{-1}XA^T)^{-1}(AZ^{-1}Xd_D + d_P - AZ^{-1}d_w), \\ dz &= d_D - A^T dy, \\ dx &= Z^{-1}(d_w - Xdz), \end{aligned} \tag{8}$$

where $d_P = Ax^0 - b$, $d_D = A^T y^0 + z^0 - c$ and $d_w = \mu e - X^0 Z^0 e$.

Convergence of primal dual IPMs has been achieved when the duality gap becomes close to zero. Specifically, $c^T x - b^T y = \varepsilon$. Moreover, there exists a simple relation between the duality gap and the barrier parameter, μ. Specifically, $c^T x - b^T y = n\mu$. Note that this relation suggests us the following mostly used update rule for the barrier parameter μ

$$\mu^{k+1} = \sigma \frac{(x^k)^T z^k}{n} = \sigma \mu^k, \quad \text{where} 0 < \sigma < 1. \tag{9}$$

For the convex quadratic optimization problems, we have an additional quadratic term at the problem P. That results the following nonlinear KKT system

$$
\begin{aligned}
Ax - b &= 0, \\
-Qx + A^T y + z - c &= 0, \\
XZe - \mu e &= 0.
\end{aligned}
\tag{10}
$$

Moving direction vectors dx, dy, and dz for the convex quadratic case are determined by the Newton's method as

$$
\begin{aligned}
dy &= [A(Z+XQ)^{-1} X A^T]^{-1} [A(Z+XQ)^{-1} X d_D + d_P - A(Z+XQ)^{-1} d_w], \\
dx &= (Z+XQ)^{-1}(d_w - X d_D + X A^T dy), \\
dz &= X^{-1}(d_w - Z dx),
\end{aligned}
\tag{11}
$$

where $d_P = Ax^0 - b$, $d_D = -Qx^0 + A^T y^0 + z^0 - c$ and $d_w = \mu e - X^0 Z^0 e$.

4. NEW PARAMETERIZATION OF THE CENTRAL TRAJECTORY

The primal dual IPM maintains primal feasibility and dual feasibility and iterates to reduce the duality gap. The duality gap depends linearly on the barrier parameter for the points in the central trajectory. Our objective is to consider a continuous dynamical system that describes the rate of change of the barrier parameter. Rather than updating the barrier parameter by heuristic rules, we have determined the developing trajectory of the barrier parameter by using an ODE. The resulting ODE combined with KKT conditions to form a DAE system. The DAE is used to determine the central trajectory of the optimization problem. Now let us consider the primal linear optimization problem P and the corresponding problem P_μ that can also be defined as

$$\theta(\mu) = \inf \left\{ c^T x - \mu \sum_{i=1}^{n} \ln x_i \quad \text{such that} \quad Ax = b, x > 0 \right\}. \tag{12}$$

By the Theorem 9.4.3 by Bazaraa et al. [2], it is stated that the optimal solution to the problem P could be obtained by minimizing $\theta(\mu)$. In short,

$$\min \{ c^T x \quad \text{such that} \quad Ax = b, x \geq 0 \} = \inf_{\mu > 0} \theta(\mu). \tag{13}$$

Note that, (2) and (12) are strictly convex optimization problems because both the objective function is strictly convex and the constraints are convex. Consequently, for any fixed μ, $\theta(\mu)$ has a unique minimum. The minimum of $\theta(\mu)$ will be found by using the steepest descent method. Next, we need to consider μ as a function of parameter t. To find the rate of change of the barrier parameter μ, we have to move in the direction of the negative gradient of $\theta(\mu)$. Thus, $\frac{d\mu}{dt} = -\frac{d\theta(\mu)}{d\mu}$. After all, we have $\frac{d\theta(\mu)}{d\mu} = -\sum_{i=1}^{n} \ln x_i$.

Finally, we can obtain the rate of change of the barrier parameter μ as the following ODE

$$\frac{d\mu}{dt} = \sum_{i=1}^{n} \ln x_i. \tag{14}$$

The ODE (14) and algebraic equation system (6) form the following DAE system for the problem P.

$$\begin{aligned}
\frac{d\mu}{dt} &= \sum_{i=1}^{n} \ln x_i, \\
Ax - b &= 0, \\
A^T y + z &= c, \\
XZe - \mu e &= 0.
\end{aligned} \tag{15}$$

To find a solution to the DAE, we need to state the following theorem.

Theorem. *The DAE defined by (15) has index 1.*

Proof. Let us rewrite the DAE in semi-explicit form as in (1). Then by differentiating the algebraic equation with respect to t, we get

$$g_x \frac{dx}{dt} + g_y \frac{dy}{dt} + g_z \frac{dz}{dt} + g_\mu \frac{d\mu}{dt} = 0, \tag{16}$$

where $g_x = \begin{bmatrix} A \\ 0 \\ Z \end{bmatrix}$, $g_y = \begin{bmatrix} 0 \\ A^T \\ 0 \end{bmatrix}$, $g_z = \begin{bmatrix} 0 \\ I \\ X \end{bmatrix}$, $g_\mu = \begin{bmatrix} 0 \\ 0 \\ -e \end{bmatrix}$.

With some algebra, (16) becomes

$$\begin{bmatrix} g_x & g_y & g_z \end{bmatrix} \begin{bmatrix} \frac{dx}{dt} & \frac{dy}{dt} & \frac{dz}{dt} \end{bmatrix}^T = -g_\mu \frac{d\mu}{dt}. \tag{17}$$

Note that, $[g_x, g_y, g_z]$ is equal to the Jacobian of (7). Next, we need to change coordinates Let $v = (x, y, z)$ and $\frac{dv}{dt} = (\frac{dx}{dt} \frac{dy}{dt} \frac{dz}{dt})$. Then (17) becomes

$$J(v)\frac{dv}{dt} = f(v, \mu, t), \tag{18}$$

where $f(v, \mu, t) = -g_\mu \frac{d\mu}{dt}$. By the assumption of nonsingular Jacobian $J(v)$, we have the following implicit ODE

$$\frac{dv}{dt} = J^{-1}(v)f(v, \mu, t) = F(v, \mu, t). \tag{19}$$

351

By the definition of index an implicit ODE resulted after one differentiation step. This implies that the DAE defined by (15) has index 1. This concludes the proof. ∎

The algebraic equations in the DAE can be further transformed into ODEs. Differentiating (6) with respect to t, and some algebra by using the sparsity of system, the ODEs are expressed as

$$
\begin{aligned}
\frac{dx}{dt} &= \frac{d\mu}{dt}\left[I - Z^{-1}XA^T(AZ^{-1}XA^T)^{-1}A\right]Z^{-1}e, \\
\frac{dy}{dt} &= -\frac{d\mu}{dt}(AZ^{-1}XA^T)^{-1}AZ^{-1}e, \\
\frac{dz}{dt} &= \frac{d\mu}{dt}A^T(AZ^{-1}XA^T)^{-1}AZ^{-1}e.
\end{aligned}
\tag{20}
$$

By taking the initial values for x, y, and z which satisfy the algebraic equations of the DAE in the interior of the feasible region, and taking a small initial value for μ, we can solve (14) and (20).

A modified primal dual IPM by using the new parameterization of the central trajectory for linear optimization problems can be described as follows: By applying Euler's method for the DAE system (15), we obtain a nonlinear system that is to be solved by Newton's method. Given $x^0 > 0, y^0$ and $z^0 > 0$ and $\mu^0 > 0$, moving direction vectors dx, dy, dz, and $d\mu$ that move from the current point to a new point while satisfying the DAE system (15), are determined as

$$
\begin{aligned}
dx &= Z^{-1}(d - Xdz), \\
dz &= d_D - A^T dy, \\
dy &= (AZ^{-1}XA^T)^{-1}(AZ^{-1}Xd_D + d_P - AZ^{-1}d), \\
d\mu &= h\sum_{i=1}^{n}\ln x_i^0,
\end{aligned}
\tag{21}
$$

where $d_P = Ax^0 - b$, $d_D = A^T y^0 + z^0 - c$, and $d = \mu^0 e - X^0 Z^0 e + h\sum_{i=1}^{n}\ln x_i^0 e$.

The following DAE system is derived for quadratic convex optimization problems.

$$
\begin{aligned}
\frac{d\mu}{dt} &= \sum_{i=1}^{n}\ln x_i, \\
Ax - b &= 0, \\
-Qx + A^T y + z &= c, \\
XZe - \mu e &= 0.
\end{aligned}
\tag{22}
$$

Given $x^0 > 0, y^0$ and $z^0 > 0$ and $\mu^0 > 0$, moving direction vectors dx, dy, dz, and $d\mu$ for the modified primal dual IPM for quadratic convex optimization are determined as

$$
\begin{aligned}
dy &= [A(Z + XQ)^{-1}XA^T]^{-1}[A(Z + XQ)^{-1}(Xd_D - d) + d_P], \\
dx &= (Z + XQ)^{-1}[X(A^T dy - d_D) + d], \\
dz &= X^{-1}(d - Zdx),
\end{aligned}
\tag{23}
$$

$$d\mu \;=\; h\sum_{i=1}^{n}\ln x_i^{\,0},$$

where $d_P = Ax^0 - b$, $d_D = -Qx^0 + A^T y^0 + z^0 - c$, and $d = \mu^0 e - X^0 Z^0 e + h\sum_{i=1}^{n}\ln x_i^{\,0} e$.

5. STABILITY ANALYSIS

In this section, we study the stability of the dynamical system and DAE system that resulted while parameterizing the central trajectory. The stability of our dynamical system can be established by defining a function $E(x)$ and proving that $E(x)$ is a Lyapunov function. Let us consider the problem $\theta(\mu)$ in (12). Let $E(\mu) = \theta(\mu) - \theta(\mu^*)$ where μ^* minimizes $\theta(\mu)$, for all $\mu \geq 0$. Note that, $\theta(\mu)$ is strictly convex since both the objective function is strictly convex and the constraints are convex. Therefore, $E(\mu)$ is positive definite. We know that x, y, z, and μ are continuous in μ and t. Finally,

$$\frac{dE(\mu)}{dt} = \frac{d\theta(\mu)}{d\mu}\frac{d\mu}{dt} = -\left[\frac{d\theta(\mu)}{d\mu}\right]^2 < 0. \tag{24}$$

This proves that $E(\mu)$ is a Lyapunov function, which implies that $E(\mu)$ decreases monotonically until a stable state is reached in which case neither the Lyapunov function nor the state changes further. Therefore stable points are the minimizers of $\theta(\mu)$. In other words, stable points are the optimal solutions of the given problem [12].

Next, we study the stability of the DAE system that has index-1. Ascher and Petzold [1] presented that if the all of the following conditions are satisfied by the index-1 DAE system, then the DAE system is stable. Specifically, for a linear index-1 problem, if

1. it can be transformed (without differentiations) into a semi-explicit system, and from there to an ODE by eliminating the algebraic variables,
2. the transformations are all suitably well conditioned,
3. the obtained ODE problem is stable,

then the index-1 DAE problem is also stable in the usual sense. It is obvious that the DAE system has index 1 and is in a semi-explicit form. The semi-explicit DAE transformed to an ODE. The resulting ODE after this transformation is given in (19). Note that the Jacobian that defines the resulting ODE is the same as the Jacobian of the primal dual IPMs. By the assumption of nonsingularity of the Jacobian of primal dual IPMs, our transformation is well conditioned. That proves that the second condition is also satisfied. The resulting ODE is stable as we proved by defining a Lyapunov function. That concludes that all above conditions are satisfied and the DAE system is stable.

6. CONCLUSION

A new approach to the parameterization of the central trajectory for primal dual IPMs is presented. The proposed parameterization of the central trajectory is investigated both for linear optimization and convex quadratic optimization problems. Primal dual IPMs for linear and convex quadratic optimization problems are modified by using the new parameterization of central trajectory. Stability issues are also studied.

REFERENCES

1. Ascher U. M., and Petzold L. R., *Computer Methods for Ordinary Differential Equations and Differential-Algebraic Equations*, Society for Industrial and Applied Mathematics, SIAM, Philadelphia, Pennsylvania, USA, (1998).
2. Bazaraa M. S., Sherali H. D., and Shetty C. M., *Nonlinear Programming: Theory and Algorithms*, John Wiley and Sons, New York, (1993).
3. Brenan K. E., Campbell S. L., and Petzold L. R., *Numerical Solution of Initial-Value Problems in Differential-Algebraic Equations*, North-Holland, Amsterdam, Holland, (1989).
4. Gear C. W., *Simultaneous numerical solution of differential-algebraic equations*, IEEE Transactions on Circuit Theory, vol. **TC-18**, no. 1, (1971), 89-95.
5. Hindmarsh A. C., and Petzold L. R., *Algorithms and software for ordinary differential equations and differential-algebraic equations part* II: Higher *order methods and software packages*, Computers in Physics, vol. **9**, no. 2, (1995), 148-155.
6. Kasap S., *Differential-Algebraic Equations in Primal Dual Interior Point Optimization Methods: A New Approach to the Parameterization of the Central Trajectory*, Ph. D. Thesis, School of Industrial Engineering, University of Oklahoma, Norman, Oklahoma, (2003).
7. Karmarkar N., *A new polynomial-time algorithm for linear programming*, Combinatorica, vol. **4**, (1984), 373-395.
8. Lotstedt P. and Petzold L. R., *Numerical solution of nonlinear differential equations with algebraic constraints* I: *Convergence results for backward differentiation formulas*, Mathematics of Computation, vol. **46**, no. 174, (1986), 491-516.
9. Lustig I. J., Marsten R. E., and Shanno D. F., *Interior point method for linear programming*, ORSA Journal on Computing, vol. **6**, no. 1, (1994), 1-14.
10. Trafalis T. B., and Kasap S., *Differential algebraic equations in primal dual interior point methods for linear programming problems*, in C.H. Dagli, M. Akay, A.L. Buczak, O. Ersoy, B.R. Fernandez, and J. Ghosh (Eds.), Intelligent Engineering Systems Through Artificial Neural Networks, ASME Press, vol. **8**, (1998),323-328.
11. Xiong M., Wang J., and Wang P., *Differential-algebraic approach to linear programming*, Journal of Optimization Theory and Applications, vol. **114**, no. 2, (2002), 443-470.
12. Zangwill W. I., and Garcia C. B., *Pathways to Solutions, Fixed Points, and Equilibria*, Prentice-Hall, Inc., Englewoods Cliffs, New Jersey, USA, (1981).

Nambu–Poisson Dynamics and its Applications

Nugzar Makhaldiani

Laboratory of Information Technologies, Joint Institute for Nuclear Research, Dubna, Moscow Region, Russia

Abstract. After a short introduction in the Nambu-Poisson dynamics (NPD), some applications of NPD in the (in)finite-dimensional models are considered.
Key words. Nambu-Poisson dynamics; Finite dimensional integrable dynamical systems; Nonlinear Schrodinger type equations.
PACS. 02.30.Ik, 05.45.-a.
MSC. 37K10, 35Q55.

1. INTRODUCTION

The Hamiltonian mechanics (HM) is in the ground of mathematical description of the physical theories, [7]. But HM is in a sense blind, e.g., it does not make difference between two opposites: the ergodic Hamiltonian systems (with just one integral of motion) [11] and (super)integrable Hamiltonian systems (with maximal number of the integrals of motion).

By our proposal [2], Nambu's mechanics (NM) [10] is proper generalization of the HM, which makes difference between dynamical systems with different numbers of integrals of motion explicit.

In the first part of this paper we consider a system of nonlinear ordinary differential equations which in a particular case reduces to Volterra's system [16], and integrate this system using Nambu–Poisson formalism [10, 12]. In the second part of this paper we review some results of the paper [8].

2. FINITE DIMENSIONAL DYNAMICAL SYSTEM

In this section we consider the following dynamical system [2]

$$
\dot{x}_n = \gamma_n \sum_{m=1}^{p} \left(e^{x_{n+m}} - e^{x_{n-m}} \right),
$$
$$
1 \leq n \leq N,\ 1 \leq p \leq [(N-1)/2],\ 3 \leq N,
$$
$$
x_{n+N} = x_n, \tag{1}
$$

CP729, *Global Analysis and Applied Mathematics: International Workshop on Global Analysis,*
edited by K. Taş, D. Krupka, O. Krupková, and D. Baleanu
© 2004 American Institute of Physics 0-7354-0209-4/04/$22.00

where γ_n are real numbers, and $[a]$ means the integer part of a.

For $\gamma_n = 1$, $p = 1$ and $x_n = lnv_n$, the system (1) becomes Volterra's system [16]

$$\dot{v}_n = v_n(v_{n+1} - v_{n-1}).$$ (2)

Our system is connected to the Toda's lattice system [14]

$$\ddot{y}_n = e^{y_{n+1}-y_n} + e^{y_n-y_{n-1}}.$$ (3)

Indeed, if we make the following transformation

$$x_n = y_n - y_{n-1},$$ (4)

then

$$\dot{x}_n = e^{x_{n+1}} - e^{x_{n-1}}.$$ (5)

In addition, for $\gamma_n = 1$ and $p \geq 1$, the system (1) reduces to the Bogoiavlensky's lattice system [4]

$$\dot{v}_n = v_n \sum_{m=1}^{p} (v_{n+m} - v_{n-m}).$$ (6)

Finally, for $N = 3$, $p = 1$ and arbitrary γ_n, the system (1) is connected to the system of three vortexes of two-dimensional ideal hydrodynamics [1, 9].

2.1. System of three vortexes

The system of N vortexes can be described by the following system of differential equations [1]

$$\dot{z}_n = i \sum_{m \neq n}^{N} \frac{\gamma_m}{z_n^* - z_m^*}, \quad 1 \leq n \leq N,$$ (7)

where $z_n = x_n + iy_n$ are complex coordinate of the center of n-th vortex.

For $N = 3$, it is easy to verify that the quantities

$$x_1 = ln|z_2 - z_3|^2,$$ (8)
$$x_2 = ln|z_3 - z_1|^2,$$
$$x_3 = ln|z_1 - z_2|^2$$

satisfy the following system

$$\dot{x}_1 = \gamma_1(e^{x_2} - e^{x_3}),$$
$$\dot{x}_2 = \gamma_2(e^{x_3} - e^{x_1}),$$
$$\dot{x}_3 = \gamma_3(e^{x_1} - e^{x_2}),$$ (9)

after changing the time parameter as

$$dt = \frac{e^{(x_1+x_2+x_3)}}{4S} d\tau = e^{(x_1+x_2+x_3)/2} R d\tau. \tag{10}$$

Here S represents the area of the triangle with vertexes in the centers of the vortexes and R is the radius of the circle with the vortexes on it.

The system (9) has two integrals of motion

$$H_1 = \sum_{i=1}^{3} \frac{e^{x_i}}{\gamma_i}, \qquad H_2 = \sum_{i=1}^{3} \frac{x_i}{\gamma_i} \tag{11}$$

and it can be presented in the Nambu–Poisson form [9]

$$\dot{x}_i = \omega_{ijk} \frac{\partial H_1}{\partial x_j} \frac{\partial H_2}{\partial x_k} = \{x_i, H_1, H_2\} = \omega_{ijk} \frac{e^{x_j}}{\gamma_j} \frac{1}{\gamma_k}, \tag{12}$$

where

$$\omega_{ijk} = \varepsilon_{ijk}\rho, \qquad \rho = \gamma_1 \gamma_2 \gamma_3. \tag{13}$$

The Nambu–Poisson bracket of the functions A, B, C on the three-dimensional phase space is given as follows

$$\{A, B, C\} = \omega_{ijk} \frac{\partial A}{\partial x_i} \frac{\partial B}{\partial x_j} \frac{\partial C}{\partial x_k}. \tag{14}$$

From (14), the fundamental bracket has the form

$$\{x_1, x_2, x_3\} = \omega_{ijk}. \tag{15}$$

Changing the time parameter as

$$du = \rho d\tau, \tag{16}$$

then we obtain Nambu's mechanics [9]

$$\dot{x}_i = \varepsilon_{ijk} \frac{\partial H_1}{\partial x_j} \frac{\partial H_2}{\partial x_k}.$$

For N=3 degrees of freedom, we obtain two integrals of motion and we conclude that this system is superintegrable.

2.2. Four dimensional system

The next important case corresponds to $N = 4$ and $p = 1$,

$$\dot{x}_1 = \gamma_1 (e^{x_2} - e^{x_4}),$$

$$\dot{x}_2 = \gamma_2(e^{x_3} - e^{x_1}),$$
$$\dot{x}_3 = \gamma_3(e^{x_4} - e^{x_2}),$$
$$\dot{x}_4 = \gamma_4(e^{x_1} - e^{x_3}). \tag{17}$$

We have two integrals of motion

$$H_1 = \frac{e^{x_1}}{\gamma_1} + \frac{e^{x_2}}{\gamma_2} + \frac{e^{x_3}}{\gamma_3} + \frac{e^{x_4}}{\gamma_4}, \tag{18}$$

$$H_2 = \frac{x_1}{\gamma_1} + \frac{x_2}{\gamma_2} + \frac{x_3}{\gamma_3} + \frac{x_4}{\gamma_4}, \tag{19}$$

corresponding to the system (17). In order to be superintegrable, the system (17) should admit one more integral of motion, H_3. To find this integral of motion, suppose Nambu's form of the system (17) as

$$\dot{x}_n = \{x_n, H_1, H_2, H_3\} = \gamma_1 \gamma_2 \gamma_3 \gamma_4 \varepsilon_{nmkl} \frac{\partial H_1}{\partial x_m} \frac{\partial H_2}{\partial x_k} \frac{\partial H_3}{\partial x_l}. \tag{20}$$

From (20), the missing integral of motion is obtained as

$$H_3 = -\frac{1}{2} \left(\frac{x_1}{\gamma_1} - \frac{x_2}{\gamma_2} + \frac{x_3}{\gamma_3} - \frac{x_4}{\gamma_4} \right). \tag{21}$$

Having three integrals of motion, we can easily integrate the system (17). From (19) and (21), we obtain

$$x_4 = \gamma_4 \left(\frac{H_2 + 2H_3}{2} - \frac{x_2}{\gamma_2} \right),$$
$$x_3 = \gamma_3 \left(\frac{H_2 - 2H_3}{2} - \frac{x_1}{\gamma_1} \right). \tag{22}$$

One the other hand, (18) gives us

$$H_1 = \frac{e^{x_1}}{\gamma_1} + \frac{e^{x_2}}{\gamma_2} + \frac{e^{-\frac{\gamma_3}{\gamma_1} x_1}}{\gamma_3 e^{-\gamma_3(H_2/2 - H_3)}} + \frac{e^{-\frac{\gamma_4}{\gamma_2} x_2}}{\gamma_4 e^{-\gamma_4(H_2/2 + H_3)}}. \tag{23}$$

So, x_2 is an implicit function of x_1, $x_2 = n_2(x_1, H_1, H_2, H_3)$. When

$$\frac{\gamma_4}{\gamma_2} = \pm 1, \pm 2, \pm 3, -4, \tag{24}$$

the function n_2 reduces to the composition of the elementary functions.
Now, we can solve the equation,

$$\dot{x}_1 = \gamma_1(e^{x_2} - e^{x_4}) = n_1(x_1), \tag{25}$$

by one quadrature, we obtain

$$N(x_1) = \int_{x_{10}}^{x_1} \frac{dx}{n_1(x)} = t - t_0. \tag{26}$$

358

3. INFINITE DIMENSIONAL SYSTEM

As an example of the infinite dimensional Nambu-Poisson dynamics, let me consider the following extension of the Schrödinger quantum mechanics [8]

$$iV_t = \Delta V - \frac{V^2}{2}, \tag{27}$$

$$i\psi_t = -\Delta\psi + V\psi. \tag{28}$$

An interesting solution of the equation for the potential (27) is

$$V = \frac{4(4-d)}{r^2}, \tag{29}$$

where d is the dimension of the space. In the case of d=1 we have the potential of the conformal quantum mechanics.

The variational formulation of the extended quantum theory, (27,28) is given by the following Lagrangian

$$L = \left(iV_t - \Delta V + \frac{1}{2}V^2\right)\psi. \tag{30}$$

The momentum variables are

$$P_v = \frac{\partial L}{\partial V_t} = i\psi,$$

$$P_\psi = 0. \tag{31}$$

As a Hamiltonian of the Nambu-theoretic formulation we take the following integrals of motion

$$H_1 = \left(\Delta V - \frac{1}{2}V^2\right)\psi,$$

$$H_2 = P_v - i\psi,$$

$$H_3 = P_\psi. \tag{32}$$

We invent unifying vector notation, $\phi = (\phi_1, \phi_2, \phi_3, \phi_4) = (\psi, P_\psi, V, P_v)$. Then it may be verified that the equations of the extended quantum theory can be put in the following Nambu-theoretic form

$$\phi_t(x) = \{\phi(x), H_1, H_2, H_3\}$$

$$= i \int \frac{\partial(\phi(x), H_1, H_2, H_3)}{\partial(\phi_1(y), \phi_2(y), \phi_3(y), \phi_4(y))} dy, \tag{33}$$

where the bracket is defined as

$$\{A, B, C, D\} = i\varepsilon_{ijkl} \int \frac{\delta A}{\delta\phi_i(y)} \frac{\delta B}{\delta\phi_j(y)} \frac{\delta C}{\delta\phi_k(y)} \frac{\delta D}{\delta\phi_l(y)} dy. \tag{34}$$

4. CONCLUSIONS

Nambu's mechanics is a generalization of the classical Hamiltonian mechanics introduced by Yoichiro Nambu, [10]. Nambu's bracket, according to the [17], where considered by Albeggiani. In [6, 5], it was demonstrated that several Hamiltonian systems possessing dynamical symmetries can be realized in the Nambu formalism [15, 13] of generalized mechanics.

In this paper, we have considered the system (1) and investigated the integrability properties of the particular cases of the system by elementary methods using Nambu–Poisson reformulation of Hamiltonian mechanics.

For the general case, we obtained two integrals of motion for the system (1), namely

$$H_1 = \sum_{n=1}^{N} \frac{e^{x_n}}{\gamma_n}, \tag{35}$$

$$H_2 = \sum_{n=1}^{N} \frac{x_n}{\gamma_n}. \tag{36}$$

For the case $N = 2M$, we found a third integral of motion

$$H_3 = \frac{1}{2} \sum_{n=1}^{2M} \frac{(-1)^n x_n}{\gamma_n}, \tag{37}$$

but, for $N \geq 5$, we need extra integrals of motion. The integrability properties of the system (1) in the general case are under investigation [3].

ACKNOWLEDGMENTS

It is a pleasure to thank Dumitru Baleanu, Kenan Taş and their team of young scientists for various help.

REFERENCES

1. H. Aref, *Integrable, chaotic and turbulent vortex motion in two-dimensional flows*, Ann.Rev.Fluid Mech. **15** 345 (1983); A.V. Meleshko, N.N. Konstantinov, *Dynamics of vortex systems*, Kiev: Naukova Dumka, 1993.
2. D. Baleanu, N. Makhaldiani, *Nambu-Poisson Reformulation of the Finite Dimensional Dynamical Systems*, Communications of the JINR **E2-98-348** Dubna 1998, solv-int/9903002; *Roumanian J. Phys.* **44** no. 9-10,927-932, 1999.
3. D. Baleanu, N. Makhaldiani, in preparation.
4. O.I. Bogoyavlensky, *Integrable discretization of the KdV equation*, Phys.Lett. A **134** 34 (1988).
5. R. Chatterjee, *Dynamical symmetries and Nambu mechanics*, Lett.Math.Phys. **36** 117 (1996).
6. I. Cohen, *Generalization of Nambu's Mechanics*, Int. J.Theor.Phys. **12** 69 (1975).
7. L.D. Faddeev and L.A. Takhtajan, *Hamiltonian methods in the theory of solitons*, Springer, Berlin, 1987.
8. N. Makhaldiani, *New Hamiltonization of the Schrödinger Equation by Corresponding Nonlinear Equation for the Potential*, Communications of the JINR, Dubna, **E2-2000-179**, Dubna, 2000.

9. N. Makhaldiani, *The System of Three Vortexes of Two-Dimensional Ideal Hydrodynamics as a New Example of the (Integrable) Nambu-Poisson Mechanics, Communications of the JINR*, Dubna, **E2-97-407**, Dubna, 1997; solv-int/9804002.

10. Y. Nambu, *Generalized Hamiltonian mechanics, Phys.Rev.D* **7** 2405 (1973).

11. Ja.G. Sinai, *Theory of dynamical systems Part I. Ergotic theory*, Warsaw Univ. Press, 1969.

12. L.A. Ṭakhtajan, *On fondation of the generalized Nambu mechanics, Comm.Math.Phys.* **160** 295 (1994).

13. A. Teğmen, A. Verçin, *Superintegrable Systems, Multi-Hamiltonian Structures and Nambu Mechanics in an Arbitrary Dimension*, math-ph/0212070, 2002.

14. M. Toda, *Theory of Nonlinear Lattices*, Springer-Verlag, Berlin, Heidelberg, New York, 1981.

15. I. Vaisman, *A Survey on Nambu-Poisson Brackets*, math.DG/9901047, 1999.

16. V. Volterra, *Lecons sur la theorie matyhematique de la lutre pour la vie*, Cahers scientifiques YII.-Paris: Gauthier Vollars, 1931.

17. E.T. Whittaker, *A Treatise on the Analitical Dynamics*, Cambrige Univ. Press, 1961, p.337.

Relaxation phenomena in the (in)activation gates of the voltage-gated ion channels

Mahmut Özer

Department of Electrical and Electronics Engineering, Engineering Faculty, Zonguldak Karaelmas University, 67100 Zonguldak, Turkey

Abstract. We previously proposed a method for the study of the relaxation phenomena in the activation and inactivation gates of ion channels present in the excitable membranes of neurons. In order to study the relaxation phenomena, the assumption is made that the activation and inactivation gate order parameters can be treated as fluxes and forces. In the present paper, we extend the previous model as including an ensemble of gating particles, and apply it for T-type calcium channel in thalamic relay neurons. It is found that kinetic equations are characterized by two relaxation times. The kinetic coefficients are determined for its empirical model. We also determine the kinetic coefficients of linear and nonlinear thermodynamic models for the same T-type calcium channel, and compare them with the empirical ones.

Key words. Ion channels, Onsager Reciprocity theorem, Kinetic Coefficient, Relaxation.
PACS. 87.19.La, 87.16.Uv, 87.17.Aa, 05.70.Ln.
MSC. 82C35, 92B05, 92C15, 92C37.

1. INTRODUCTION

Ion channels are key molecules for cellular regulation and communication. They couple biomolecular events to electric signals. Voltage-gated ion channels open and close in a stochastic manner dependent on the transmembrane voltage. They are involved in the generation and propagation of electrical signals in the excitable cell membranes. Voltage-gated ion channels are formed by pore-like proteins whose functions are dictated by their possible conformations. They include charged regions which make their structure susceptible to the membrane potential. The voltage-dependent gating of these channels between conducting and non-conducting states is a major factor in controlling the transmembrane potential. Hodgkin and Huxley (H-H) [4] provided the first quantitative description of the voltage-dependent gating of the ion currents. In their formalism, voltage-dependent gating of the ion channel requires the movement of hypothetical gating particles able to sense the electric field across the membrane.

During the last few years, there have been enormous strides in our understanding of structure-function relationships in the ion channels. In this context, one of the most exciting recent developments in the ion channel gating is the determination of an X-ray crystal structure of a voltage-gated potassium channel [6, 7]. It was found that the voltage sensors, called paddles, are attached to the central ion-conducting pore by flexible

CP729, *Global Analysis and Applied Mathematics: International Workshop on Global Analysis*,
edited by K. Taş, D. Krupka, O. Krupková, and D. Baleanu
© 2004 American Institute of Physics 0-7354-0209-4/04/$22.00

hinges and apparently move in response to the membrane potential changes by carrying their positive charges across the membrane. There is a need to develop models that can relate the structural parameters of the channels to experimental data and thereby build a theoretical framework that can explain different sets of observations [1]. In this context, different approaches were proposed to address the mechanism of voltage sensing and gating in these channels. Yang et al [15] proposed a statistical mechanical model for ion channels in the presence of electric fields. They obtained the maximum fractions of open potassium and sodium channels by solving a self-consistent non-linear algebraic equation under a mean-field approximation with the important constraint of a static, not time-varying, electric field. Roux [14] developed a rigorous statistical mechanical formulation of the equilibrium properties of selective ion channels incorporating the influence of the membrane potential, multiple occupancy and saturation effects. In this context, we recently presented a new methodology to define the equilibrium value function of the kinetics of activation and inactivation gates based on the lowest approximation of the cluster variation method [11]. Then, we proposed a method, which combines statistical equilibrium theory and the thermodynamics of irreversible processes for the study of the relaxation phenomena in the activation and inactivation gates of ion channels [10]. We also formulated dynamics of the voltage-gated ion channels and ion channel gates by the path probability method of the nonequilibrium statistical physics [12, 13].

In the present paper, we extend the previous model [10] as including an ensemble of gating particles, and apply it for T-type calcium channel in thalamic relay neurons. It is found that kinetic equations are characterized by two relaxation times. The kinetic coefficients are determined for its empirical model. We also determine the kinetic coefficients of linear and nonlinear thermodynamic models for the same T-type calcium channel, and compare them with the empirical ones.

2. MODEL AND EQUILIBRIUM PROPERTIES

We consider a simple two-state gate system in which the conformational change consists of the movement of gating particles. At any moment, the particles are in one of two positions, 1 or 2, which are associated with closed and open states, respectively. In Eyring rate theory terms, the positions correspond to two wells in an energy profile, and there is a single energy barrier between them [3]. In order to study the dynamics of such a system, first we have to study the steady-state behavior of the system, especially to obtain the steady state equation for the gate.

In this paper, we revised the previous model [10] and included an ensemble of gating particles in ion channel gates. Given an ensemble of particles in activation gate and particles in inactivation gate in a channel in which activation and inactivation gates are independent, the internal variables will be indicated by x_1 and x_2 for the activation gate and by x_3 and x_4 for the inactivation gate, which are also called the state or point variables. x_1 and x_3 are the fractions of gating particles in position 1 with energy ε_1 and ε_3 for the activation and inactivation gates respectively, and x_2 and x_4 are the fractions of gating particles in position 2 with energy ε_2 and ε_4 for the activation and inactivation gates respectively so that $x_1 + x_2 = 1$ and $x_3 + x_4 = 1$. In our case, x_2 and x_4 correspond to m and h, respectively and x_1 and x_3 correspond to $1 - m$ and $1 - h$, respectively, where m and h show voltage-dependent probability of being open state for the activation and inactivation gates, respectively. A simple expression for the internal energy of such a system in the presence of a membrane potential is written as:

$$E = N \sum_{i=1}^{2} x_i \varepsilon_i + z_1 e_0 N x_1 V + M \sum_{i=3}^{4} x_i \varepsilon_i + z_2 e_0 M x_3 V, \tag{1}$$

where z_1 and z_2 are the number of charges on each particle in the activation and inactivation gate, respectively. e_0 is the elementary electronic charge and V is the potential difference, also called the membrane potential.

In the molecular-field approximation, the Helmholtz free energy of the system is given by

$$F = E - k_B T \ln W (m, h), \tag{2}$$

where k_B is the Boltzmann constant, T is the absolute temperature, and $W(m, h)$ is the number of configurations for given values of activation and inactivation order parameters. The second term in Eq. (2) was introduced by Ozer [10] for the activation and inactivation gates, and is given with a small modification for the present case by

$$\ln W (m, h) = -N[(1 - m) \ln (1 - m) + m \ln m] - M[(1 - h) \ln (1 - h) + h \ln h]. \tag{3}$$

By substituting Eqs. (1) and (3) in Eq. (2) we get

$$\begin{aligned} F = & \ N[\varepsilon_1(1 - m) + \varepsilon_2 m] + z_1 e_0 N(1 - m)V + M[\varepsilon_3(1 - h) + \varepsilon_4 h] + z_2 e_0 M(1 - h)V \\ & + k_B T N[(1 - m) \ln(1 - m) + m \ln m] + k_B T M[(1 - h) \ln(1 - h) + h \ln h]. \end{aligned} \tag{4}$$

The equilibrium values of the activation and inactivation order parameters are found by the conditions:

$$\frac{\partial F}{\partial m} = 0, \ \frac{\partial F}{\partial h} = 0. \tag{5}$$

One can easily find the following set of self-consistent equations from Eqs. (4) and (5) by introducing half-(in)activation voltage $V_{1/2}$:

$$m_0 = \frac{1}{1 + \exp^{-\beta z_1 e_0 (V - V_{1/2,m})}}, h_0 = \frac{1}{1 + \exp^{-\beta z_2 e_0 (V - V_{1/2,h})}}, \tag{6}$$

where $\beta = 1/k_B T$. $V_{1/2,m}$ equals to $-(\varepsilon_1 - \varepsilon_2)/z_1 e_0$ and $V_{1/2,h}$ equals to $-(\varepsilon_3 - \varepsilon_4)/z_2 e_0$. defines a voltage at which half the gates are open.

3. KINETIC COEFFICIENTS FOR THE ION CHANNEL GATES

In this section, we use the Onsager reciprocity theorem (ORT), which was derived by Onsager [8, 9], to obtain the kinetic coefficients of the activation and inactivation gates. In the thermodynamics of irreversible processes, when the system is not in the equilibrium, derivatives of the Helmholtz free energy F of the system with respect to the order parameters are not equal to zero, and can be regarded as the generalized forces which cause changes in the parameters. According to the ORT, the relation between the time rate of change of deviations α_i in the parameters from their values in equilibrium states and the generalized forces is written as

$$J_i = \frac{d\alpha_i}{dt} = \sum_{i,j} \gamma_{ij} X_j \quad (i,j = 1,2\ldots), \tag{7}$$

where J_i are thermodynamic fluxes, X_j are the generalized forces and γ_{ij} is a matrix of kinetic coefficients. α_i is also called the fluctuation. The generalized forces are obtained by differentiating the entropy production ΔS_E with respect to the deviations α_i as follows:

$$X_i = \frac{\partial(\triangle S_E)}{\partial \alpha_i} = -\sum_j \beta_{ij} \alpha_j. \tag{8}$$

The entropy is a maximum when the system is in equilibrium. Therefore, one can write the following equation for the fluctuation from this maximum due to small changes:

$$\Delta S_E = -\frac{1}{2} \sum_{i,j} \beta_{ij} \alpha_i \alpha_j, \tag{9}$$

where β_{ij} denotes the entropy production coefficients defined by

$$\beta_{ij} = -\left(\frac{\partial^2 S_E}{\partial \alpha_i \partial \alpha_j}\right)_{eq} = \frac{1}{T}\left(\frac{\partial^2 F}{\partial \alpha_i \partial \alpha_j}\right)_{eq}, \tag{10}$$

where F is the Helmholtz free energy for the system given by Eq. (4), and the subscript "eq" represents equilibrium. In our case, we have two thermodynamic quantities, m the activation and h the inactivation, and corresponding two generalized forces, X_m and X_h . Deviations in m and h from their equilibrium values are represented by $(m - m_o)$ and $(h - h_0)$ respectively. m_0 and h_0 are the equilibrium values of m and h respectively.

The entropy production for the system can be obtained by substituting Eq. (4) in Eqs. (9) and (10) as follows:

$$\Delta S_E = -\frac{1}{2}\left[A(m - m_o)^2 + 2B(m - m_o)(h - h_0) + C(h - h_0)^2\right]. \tag{11}$$

Entropy production coefficients are also obtained as follows:

$$A = \frac{1}{T}\left(\frac{\partial^2 F}{\partial m^2}\right)_{eq} = \frac{k_B N}{m_0(1 - m_0)}, \tag{12}$$

$$B = \frac{1}{T}\left(\frac{\partial^2 F}{\partial m \partial h}\right)_{eq} = 0, \tag{13}$$

$$C = \frac{1}{T}\left(\frac{\partial^2 F}{\partial h^2}\right)_{eq} = \frac{k_B M}{h_0(1 - h_0)}. \tag{14}$$

365

Generalized forces X_m and X_h are found by substituting Eq. (11) into Eq. (8) as follows:

$$X_m = \frac{\partial(\triangle S_E)}{\partial(m - m_0)} = -A(m - m_0) - B(h - h_0) = -A(m - m_0), \tag{15}$$

$$X_h = \frac{\partial(\triangle S_E)}{\partial(h - h_0)} = -B(m - m_0) - C(h - h_0) = -C(h - h_0). \tag{16}$$

The linear relations between the fluxes and the generalized forces may be written in terms of the kinetic coefficients by substituting Eqs. (15) and (16) into Eq. (7) as follows:

$$\begin{bmatrix} J_m \\ J_h \end{bmatrix} = \begin{bmatrix} \gamma_m & \gamma \\ \gamma & \gamma_h \end{bmatrix} \begin{bmatrix} X_m \\ X_h \end{bmatrix}, \tag{17}$$

where a symmetric matrix of kinetic coefficients is introduced so as to satisfy the On-sager's reciprocal relation. The off-diagonal kinetic coefficient γ couples the activation and inactivation gate fluxes in Eq. (17). In our case, the gates are independent and have no any interaction. Therefore, we take $\gamma = 0$. Then, the rate equations are obtained from the matrix Eq. (17) as follows:

$$\frac{dm}{dt} = -\gamma_m A(m - m_0), \tag{18}$$

$$\frac{dh}{dt} = -\gamma_h C(h - h_0). \tag{19}$$

Assuming a form $\exp(-t/\tau)$ for both $(m - m_0)$ and $(h - h_0)$ for Eqs. (18) and (19) one obtains the following secular equation:

$$\begin{vmatrix} (1/\tau) - \gamma_m A & 0 \\ 0 & (1/\tau) - \gamma_h C \end{vmatrix} = 0. \tag{20}$$

Two solutions of this equation are given by

$$\tau_m = \frac{1}{\gamma_m A}, \tag{21}$$

$$\tau_h = \frac{1}{\gamma_h C}. \tag{22}$$

Consequently, we determined the kinetic coefficients in terms of time constant and equilibrium values of the (in)activation gates by using Eqs. (12), (14), (21) and (22) as follows:

$$L_m = \frac{m_0(1 - m_0)}{\tau_m}, \tag{23}$$

$$L_h = \frac{h_0(1 - h_0)}{\tau_h}, \tag{24}$$

where $L_m = \gamma_m k_B N$ and $L_h = \gamma_h k_B M$. We also determined L_m and L_h in Eqs. (23) and (24) so as to satisfy the Hodgkin-Huxley mathematical formalism as follows:

$$L_m = \frac{\alpha_m \beta_m}{(\alpha_m + \beta_m)},$$ (25)

$$L_h = \frac{\alpha_h \beta_h}{(\alpha_h + \beta_h)}.$$ (26)

4. RESULTS AND DISCUSSION

We previously proposed a method, which combines statistical equilibrium theory and the thermodynamics of irreversible processes for the study of the relaxation phenomena in the activation and inactivation gates of ion channels [10]. In this study, we extended the previous model and included an ensemble of gating particles in ion channel gates. The equilibrium value function of the kinetics of activation and inactivation gates was obtained for self-consistency and given in Eq. (6). It was also founded that the kinetic activation and inactivation equations are characterized by two relaxation times, which were given in Eqs. (21) and (22), respectively. The kinetic coefficients are given by Eqs. (23) and (24) and defined in terms of the steady state activation and inactivation (m_0, h_0) and the activation and inactivation time constants (τ_m, τ_h). Since the quantities directly observable by voltage-clamp experiments are the steady state (in)activation and the (in)activation time constant, the kinetic coefficients are also convenient for fitting experimental data. Then, we determined the kinetic coefficients so as to satisfy the Hodgkin-Huxley mathematical formalism by expressing the steady state (in)activation and (in)activation time constant in terms of forward and backward rate constants, and gave them in Eqs. (25) and (26) for the activation and inactivation gates, respectively.

We considered activation and inactivation gates of T-type calcium channel in thalamic relay neurons. Huguenard and McCormick [5] developed empirical Hodgkin-Huxley style mathematical equations that describe the voltage-dependent kinetics of the gates of T-type calcium channel. We obtained the empirical activation and inactivation kinetic coefficients by substituting the steady state activation and inactivation (m_0, h_0) and the activation and inactivation time constants (τ_m, τ_h) given by Huguenard and McCormick [5] into Eqs. (23) and (24), respectively. Then, we obtained the kinetic coefficients for the linear and nonlinear thermodynamic models of the same calcium channel derived by Destexhe and Huguenard [2] by substituting their rate constant expressions into Eqs. (25) and (26), respectively. Obtained results for the activation and inactivation gates are shown in Figure 1-(a) and Figure 1-(b), respectively.

Activation and inactivation kinetic coefficients of the linear thermodynamic model largely deviated from those of the empirical model as seen in Figure 1. Despite a shift from the empirical model, activation and inactivation kinetic coefficients of the nonlinear thermodynamic model produced more consisted trajectory with those of the empirical model. Consequently, we can propose that the kinetic coefficients of activation and inactivation gates in ion channels can be used to determine how well the rate constants, derived from different models, account for experimental data.

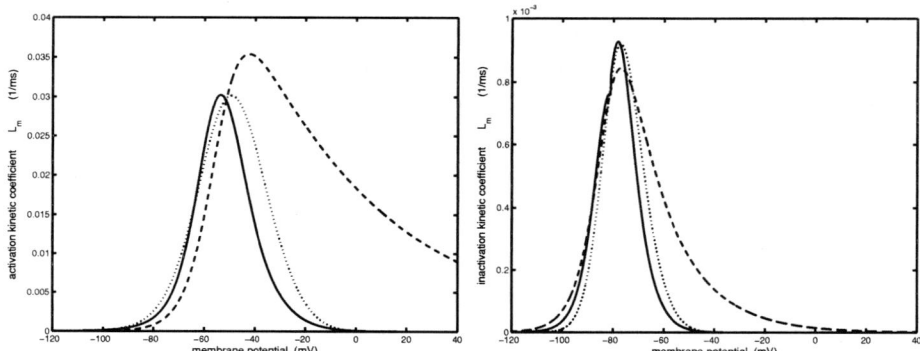

FIGURE 1. Kinetic coefficients of T-type calcium channel in thalamic relay neuron as a function of membrane potential for an empirical model (solid line), a linear thermodynamic model (dashed line) and a nonlinear thermodynamic model (dotted line). (a) Activation kinetic coefficient, (b) Inactivation kinetic coefficient.

ACKNOWLEDGMENTS

The author would like to thank Dr.Rifat Hacioglu for critical reading of the manuscript.

REFERENCES

1. Chung, S. H., Allen, T. W., Hoyles, M., and Kuyucak, S., *Permeation of ions across the potassium channel:Brownian dynamic studies*, Biophys. J., **77**, (1999), 2517-2533.
2. Destexhe, A., and Huguenard, J. R., *Nonlinear thermodynamic models of voltage-dependent currents*, J. Comput. Neurosci., **9**, (2000), 259-270.
3. Eyring, H., Lumry, R., and Woodbury, J. W., *Some applications of modern rate theory to physiological systems*, Record Chem. Progr., **10**, (1949), 100-114.
4. Hodgkin, A. L., and Huxley, A. F., *A quantitative description of membrane current and its application to conduction and excitation in nerve*, J. Physiol. Lond., **117**, (1952), 500-544.
5. Huguenard, J. R., and McCormick, D. A., *Simulations of the currents involved in rhythmic oscillations in thalamic relay neurons*, J. Neurophysiol., **68**, (1992), 1373-1383.
6. Jiang, Y., Lee, A., Chen, J., Ruta, V., Cadene, M., Chait, B. T., and MacKinnon, R., *X-ray structure of a voltage-dependent K channel*, Nature, **423**, (2003), 33-41.
7. Jiang, Y., Ruta, V., Lee, A., and MacKinnon, R., *The principle of gating charge movement in a voltage-dependent K channel*, Nature, **423**, (2003), 42-48.
8. Onsager, L., *Reciprocal relations in irreversible process.I.*, Phys. Rev., **37**, (1931), 405-426.
9. Onsager, L., *Reciprocal relations in irreversible process.II.*, Phys. Rev., **38**, (1931), 2265-2279.
10. Ozer, M., *Relaxation phenomena in the activation and inactivation gates of ionic channels*, Chinese J. Phys., **41**, (2003), 206-217.
11. Ozer, M., and Erdem, R., *A new methodology to define the equilibrium value function in the kinetics of (in)activation gates*, NeuroReport, **14**, (2003), 1071-1073.
12. Ozer, M., and Erdem, R., *Dynamics of voltage-gated ion channels in cell membranes by the path probability method*, Physica A, **331**, (2004), 51-60.
13. Ozer, M., Erdem, R., and Provaznik, I., *A new approach to define dynamics of the ion channel gates*, NeuroReport, **15**, (2004), 335-338.
14. Roux, B., *Statistical mechanical equilibrium theory of selective ion channels*, Biophys. J., **77**, (1999), 139-153.
15. Yang, Y. S., Thompson, C. J., Anderson, V., and Wood, A. W., *A statistical mechanical model of cell membrane ion channels in electric fileds: The mean-field approximation*, Physica A, **268**, (1999), 424-432.

Stability Analysis of the Steady-State Solution of a Mathematical Model in Tumor Angiogenesis

Serdal Pamuk* and Aslıhan Gürbüz[†]

*Department of Mathematics, University of Kocaeli, Ataturk Bulvari, 41300, Kocaeli - TURKEY,
spamuk@kou.edu.tr
[†]Department of Mathematics, University of Kocaeli, Ataturk Bulvari, 41300, Kocaeli - TURKEY

Abstract. The stability of the steady-state solution of endothelial cell equation in a mathematical model for tumor angiogenesis is studied. It is proven mathematically that the steady-state solution is indeed the transition probability function $\tau(c_a, f)$. Trajectories near the critical point(s) are drawn, and the biological importance of the result is expressed briefly.
Key words. Stability, steady-state solution, endothelial cell, tumor angiogenesis.
PACS. 82.39.-k.
MSC. 35K57.

1. INTRODUCTION

In this paper, we study the stability of the steady-state solution of the Endothelial Cell (EC) equation originally presented in [4]

$$\frac{\partial \eta}{\partial t} = D_\eta \frac{\partial}{\partial y}\left(\eta \frac{\partial}{\partial y}\left(\ln \frac{\eta}{\tau} \right) \right), \tag{1}$$

with the zero-flux boundary conditions

$$D_\eta \eta \frac{\partial}{\partial y} \ln\left(\frac{\eta}{\tau(c_a, f)} \right) = 0 \quad (\text{at } y = 0, 1). \tag{2}$$

Here D_η is a positive constant, the EC diffusion cofficient in the capillary, and $\eta = \eta(y,t)$ is the EC density, and τ is the so called transition probability function. We take

$$\tau = \tau(c_a, f), \tag{3}$$

where $c_a = c_a(y,t)$ is the active enzyme density and $f = f(y,t)$ is the fibronectin density $(0 < y < 1, t > 0)$. Fibronectin is a protein normally found in and around cells in various tissues in the body. A simple transition probability which reflects the influence of active enzyme and fibronectin on the motion of endothelial cells is $\tau(c_a, f) = c_a^{\gamma_1} f^{-\gamma_2}$ for positive constants $\gamma_i \ (i = 1, 2)$ [5].

CP729, *Global Analysis and Applied Mathematics: International Workshop on Global Analysis*,
edited by K. Taş, D. Krupka, O. Krupková, and D. Baleanu
© 2004 American Institute of Physics 0-7354-0209-4/04/$22.00

The biological interpretation of this choice is that endothelial cells prefer to move into the regions where c_a is large or where f is small. As in [6], we consider that there is no angiostatin supplied to the circulatory system for simplicity. Therefore, the active enzyme is the same as the total enzyme, i.e., $c_a(y,t) \equiv c(y,t)$.

We took the transition probability function as follows in [4,6]

$$\tau(c,f) = \left(\frac{a_1+c}{a_2+c} \right)^{\gamma_1} \left(\frac{b_1+f}{b_2+f} \right)^{\gamma_2}. \tag{4}$$

Here the a_i, b_i are constants such that $0 < a_1 << 1 < a_2$ and $b_1 > 1 >> b_2 > 0$. Clearly, τ is not singular for small or large values of c, f and will approximate $c^{\gamma_1} f^{-\gamma_2}$ over a considerable range of these variables [5].

2. APPROXIMATED TRANSITION PROBABILITY FUNCTION

We take the quasi-steady state enzyme and fibronectin concentrations to have the form [6]

$$c(y) = Ay^n(1-y)^n, \quad f(y) = 1 - By^n(1-y)^n, \quad 0 \le y \le 1,$$

where A and B are positive constants and $n \ge 16$. We take $\gamma_1 = \gamma_2 = 1$ in Eq.4 for simplicity. As we have said above in Chapter 2, the function τ in Eq.4 can be approximated by a function

$$\tau(c,f) = cf^{-1}.$$

Therefore, we have

$$\tau(y) = \frac{Ay^n(1-y)^n}{1 - By^n(1-y)^n} \approx Cy^n(1-y)^n = \tau^\star(y)$$

for some constant C, since $y^n(1-y)^n \ll 1$.

Figure 1 shows the transition probability function τ. The dotted line in Fig.1 is the graph of the function given by Eq.4 with the data $a_1 = 0.0001, a_2 = 2, b_1 = 10, b_2 = 0.1, A = 28 \times 10^7, B = 0.22 \times 10^9, n = 16$, and the solid line is the graph of the function $\tau^\star$ defined above with $C = 140 \times 10^7$.

Therefore, from now on, we will take $\tau^\star$ as our transition probability function, i.e, $\tau = \tau^\star$.

3. STABILITY ANALYSIS OF THE STEADY-STATE

The steady-state model obtained from Eq.(1) can be written as follows:

$$0 = \eta_{yy} - \eta_y G - \eta G_y, \tag{5}$$

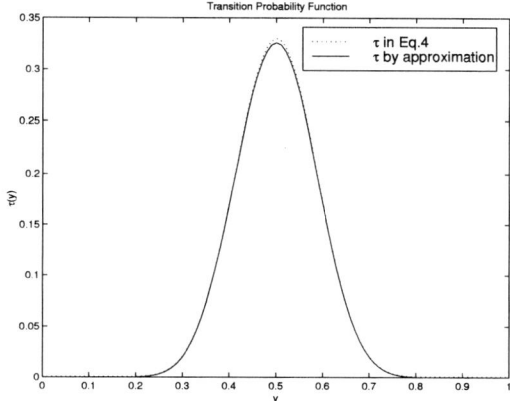

FIGURE 1. Transition Probability Function

where $G = \dfrac{\tau_y}{\tau}$. If we let $p = \eta_y$, Eq.(5) reads

$$0 = p_y - pG - \eta G_y. \tag{6}$$

Therefore, we have the following system of ordinary differential equations:

$$\eta_y = p, \tag{7}$$
$$p_y = pG + \eta G_y. \tag{8}$$

Note that one has

$$G = n\frac{1-2y}{y-y^2}, \quad G_y = -n\frac{2y^2 - 2y + 1}{(y-y^2)^2}, \quad 0 < y < 1. \tag{9}$$

Since $G_y < 0$ for $0 < y < 1$, (0,0) is the only equilibrium point of the system (7)-(8) (i.e. points satisfying $\eta_y = p_y = 0$).

The jacobien matrix $J(\eta, p)$ for the system (7)-(8) is given by

$$J(\eta, p) = \begin{bmatrix} 0 & 1 \\ G_y & G \end{bmatrix}, \tag{10}$$

and the critical point (0,0) gives rise to the same stability matrix given above, that is $J = J(0,0) = J(\eta, p)$ since the matrix does not contain the variables η and p.

If we now let $\beta = TrJ$, $\gamma = detJ$, $\delta = \beta^2 - 4\gamma = discJ$, we have $\beta = G$, $\gamma = -G_y$, $\delta = G^2 + 4G_y$. Therefore, it follows that $\beta > 0$ for $0 < y < 1/2$, $\beta < 0$ for $1/2 < y < 1$, and $\gamma > 0$ for $0 < y < 1$, which results that the critical point (0,0) is a stable node when $1/2 < y < 1$, and is an unstable node when $0 < y < 1/2$ [1,3]. Furthermore, since

$$\delta = \frac{(4n^2 - 8n)y^2 - (4n^2 - 8n)y + n^2 - 4n}{(y-y^2)^2}, \quad 0 < y < 1,$$

we have $\delta < 0$ when

$$y_1 = \frac{1}{2} - \frac{1}{2\sqrt{7}} < \frac{1}{2} - \frac{1}{\sqrt{2n-4}} < y < \frac{1}{2} + \frac{1}{\sqrt{2n-4}} < \frac{1}{2} + \frac{1}{2\sqrt{7}} = y_2, \quad n \geq 16. \text{ But,}$$

$\beta > 0$ when $y_1 < y < 1/2$, $\beta = 0$ when $y = 1/2$, and $\beta < 0$ when $1/2 < y < y_2$. Therefore, the critical point $(0,0)$ is unstable spiral when $y_1 < y < 1/2$, it is neutral center when $y = 1/2$, and is stable spiral when $1/2 < y < y_2$ [1,3].

Furthermore, Eq.(5) can be written as

$$\frac{\partial}{\partial y}(\eta_y - \eta G) = 0,$$

which results in

$$\eta_y - \eta G = 0,$$

by the boundary conditions given by Eq.(2). By solving the last equation one obtains

$$\eta = \alpha \tau(c, f), \tag{11}$$

where α is a positive constant. The result we have obtained in Eq.(11) agrees with the result obtained in [6]. Therefore, we now have

$$\eta = \beta y^{16}(1-y)^{16}, \tag{12}$$

where β is a positive constant, and the system (7)-(8) should now read

$$\eta_y = 16\beta(y-y^2)^{15}(1-2y), \tag{13}$$
$$p_y = 16\beta(y-y^2)^{14}(62y^2 - 62y + 15). \tag{14}$$

On the other hand,

$$\eta_y = \begin{cases} > 0 & \text{when } 0 < y < 1/2, \\ < 0 & \text{when } 1/2 < y < 0, \end{cases}$$

$$p_y = \begin{cases} > 0 & \text{when } 0 < y < 0.41 \ \text{or} \ 0.58 < y < 1, \\ < 0 & \text{when } 0.41 < y < 0.58. \end{cases}$$

As it is clear from the Eq.(8) that there is a linear relation between p and η for each fixed y when we set $p_y = 0$. As seen from Fig.2 and Fig.3, the equilibrium point $(0,0)$ for the system (7)-(8) is an unstable node for $y = 0.25$, whereas it is a stable node for $y = 0.75$. All we have said so far has, of course, a biological meaning since Eq.(1) is an equation for Endothelial cell migration in Tumor angiogenesis. We have shown that this simpler model, consisting of a single pde for the endothelial cell density, captures almost all of the features of the original model [4,6]. The analysis performed on the model has permitted us to focus upon the behaviour of the endothelial cells at the capillary. These cells effectively drive the capillary sprouts across the tissue towards the tumor. The analysis clearly indicates that the solution takes on different characteristics in two distinct regions of the domain, the point $y = 1/2$ marking the transition from one region to the next. The analysis also shows that in order for successful completion of angiogenesis to take place, both migration and proliferation are essential.

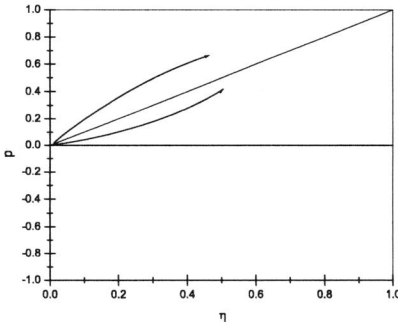

FIGURE 2. The equilibrium point $(0,0)$ is an unstable node for $y = 0.25$

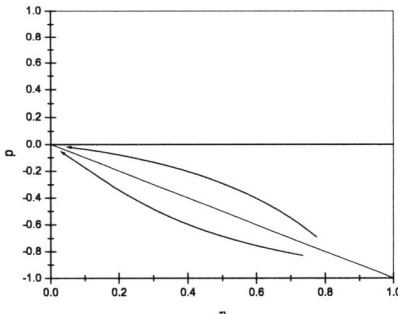

FIGURE 3. The equilibrium point $(0,0)$ is a stable node for $y = 0.75$

REFERENCES

1. W. E. Boyce and R. C. DiPrima, *Elementary Differential Equations and Boundary Value Problems*, John Wiley & Sons, Inc., USA, (1992).
2. M. A. J. Chaplin , S. M. Giles, B. D. Sleeman, R. J. Jarvis, *A mathematical Analysis of a Model for Tumour Angiogenesis*, J. Math. Biol., **33**, (1995), 744-770.
3. L. Edelstein-Keshet, *Mathematical Models in Biology*, Random House, NY, (1988).
4. H. A. Levine, S. Pamuk, B. D. Sleeman and M. Nilsen-Hamilton, *Mathematical modeling of capillary formation and development in tumor angiogenesis: Penetration into the stroma*, Bull. Math. Biol., **63**, no.5, (2001), 801-863.
5. H. A. Levine, B. D. Sleeman and M. Nilsen-Hamilton , *Mathematical modeling of the onset of capillary formation initiating angiogenesis*, J. Math. Biol., **42**, no.3, (2001), 195-238.
6. S. Pamuk, *Qualitative Analysis of a Mathematical Model for Capillary Formation in Tumor Angiogenesis*, Math. Models Methods Appl. Sci., **13**, no.1, (2003), 19-33.

The Hirota Method for Reaction-Diffusion Equations with Three Distinct Roots

Gamze Tanoğlu* and Oktay Pashaev*

*Department of Mathematics, Izmir Institute of Technology,
Gulbahce Campus, 35430, Urla, Izmir, Turkey

Abstract. The Hirota Method, with modified background is applied to construct exact analytical solutions of nonlinear reaction-diffusion equations of two types. The first equation has only nonlinear reaction part, while the second one has in addition the nonlinear transport term. For both cases, the reaction part has the form of the third order polynomial with three distinct roots. We found analytic one-soliton solutions and the relationships between three simple roots and the wave speed of the soliton. For the first case, if one of the roots is the mean value of other two roots, the soliton is static. We show that the restriction on three distinct roots to obtain moving soliton is removed in the second case by, adding nonlinear transport term to the first equation.
Key words. Soliton, Hirota method, diffusion- reaction equations, phase transition.
PACS. 02.30.jr, 05.45.-a, 11.10.Lm.
MSC. 35Q51, 35K57.

1. INTRODUCTION

There is a big variety of nonlinear phenomena which can be well described by travelling waves. They include, genetic waves in biology, vibrations in physics, phase transitions and interface phenomena in material science, and so on [1]. In the systems with degenerate ground state the kink type travelling wave implements transition between two ground state configurations. The speed of such wave is an important characteristic related with ground state values of the order parameter. According to the Landau theory of the phase transitions, the potential in effective energy has form of polynomial function of the order parameter. The simplest potential admitting two ground states and the kink soliton between these states has to be polynomial degree of four. Then, the effective force in dynamical equation for the order parameter is of the cubic form. According to the main theorem of algebra, cubic polynomial with real coefficients admits only one or three real roots. But kink type travelling wave for the real order parameter could exist only for effective force with three distinct roots.

In the present paper we consider non-linear diffusion equation,

$$U_t = U_{xx} - (U - a_1)(U - a_2)(U - a_3),\tag{1}$$

with nonlinear reaction term having three distinct roots a_1, a_2, a_3. Frequently it is called as Fisher's type equation. There are many papers dealing with solutions of this equation

CP729, *Global Analysis and Applied Mathematics: International Workshop on Global Analysis,*
edited by K. Taş, D. Krupka, O. Krupková, and D. Baleanu
© 2004 American Institute of Physics 0-7354-0209-4/04/$22.00

for the degenerate roots and with different applications of it. The main progress has been made to find travelling wave solution of the reaction-diffusion equations, by travelling wave ansatz, reducing a nonlinear PDE to a nonlinear ODE. But in ODE system, the travelling wave speed is an unknown parameter that must be fixed by the analysis and choosing special trial trajectory [1]. That is why the method seems doesn't allow extensions to more than one travelling waves, higher space dimensions and multi-component order parameter. From another site, during the last 30 years, the direct method proposed by Hirota has become a powerful tool to construct multi-soliton solutions of integrable systems [3]. This, relatively simple and algebraic rather than analytic method, allows one to avoid many analytic difficulties of more sophisticated the inverse scattering method. Moreover, it is deeply related with Plucker coordinates of Grassmanians, quantum theory of fermions, τ functions and vertex operator representation of infinite-dimensional algebras [2]. Recently, the Hirota method has been applied also to construct exact solution of some non-integrable evolution equations. The general idea of the method is first to transform nonlinear equation under consideration into bilinear equation, and then use the power series expansion to solve it. For integrable systems the series admits exact truncation for arbitrary number of solitons. While for periodic solutions it includes an infinite number of terms. To find proper bilinearization in the case of Eq. (1) with cubic nonlinearity, we propose to extend the Hirota ansatz by extracting one of the roots as a background field. Then, solution of bilinear equations gives travelling wave, implementing transition between other two roots and with velocity, depending on values of all three roots. Efficiency and simplicity of our approach is demonstrated by exact integration of extended model with additional nonlinear transport term,

$$U_t + \alpha U U_x = U_{xx} - (U - a_1)(U - a_2)(U - a_3), \tag{2}$$

where α is a constant.

The paper is organized as follows. In the next section, modified Hirota Ansatz is applied to find one soliton solution for the equation (1). Firstly, Eq. (1) is transformed into the bilinear system of differential equations, which we solve exactly using a kind of perturbational approach. We show that if the background root of our ansatz is the mean value of the other two roots, the velocity parameter is vanishing and soliton is static. In the following section, by adding the nonlinear transport term to Eq. (1) we study Eq. (2). The related, coupled Hirota bilinear form is obtained and the one soliton solution is found. We show that even in the mean value case our solution is not static. To be the case, parameter α must be fixed as some definite function of roots a_1, a_2, a_3. In conclusions we discuss some applications and possible extensions of our results to other equations.

2. REACTION- DIFFUSION EQUATION

To reduce Eq. (1) with cubic nonlinearity and three distinct roots to the bilinear form, we have to modify the standard Hirota ansatz. The solution of the problem is assumed to have a form

$$U = a_i + \frac{g}{f}, \tag{3}$$

where i=1,2,3 and g, f are real functions of x,t. For three different numbers, a_1, a_2, a_3 it provides three different solutions of the same form, but with different parameters. Thus, without loss of generality we assume that $i = 3$.

In order to obtain bilinear system of equations, all derivatives with respect to the dependent variables in Eq. (1) are expressed as

$$U_t = \frac{D_t(g \cdot f)}{f^2}, \quad U_x = \frac{D_x(g \cdot f)}{f^2}, \tag{4}$$

$$U_{xx} = \frac{D_x^2(g \cdot f)}{f^2} - \frac{g}{f}\frac{D_x^2(f \cdot f)}{f^2}, \tag{5}$$

where the Hirota derivative operator is defined as

$$D_x^n(a \cdot b) = (\frac{\partial}{\partial x_1} - \frac{\partial}{\partial x_2})^n (a(x_1)b(x_2))|_{x=x_1=x_2}. \tag{6}$$

After substituting Eqs. (4) into Eq. (1), the following expression is obtained:

$$\frac{D_t(g \cdot f)}{f^2} - \frac{D_x^2(g \cdot f)}{f^2} + \frac{g}{f}\left[\frac{D_x^2(f \cdot f)}{f^2} + \left(\frac{g}{f} - (a_1 - a_3)\right)\left(\frac{g}{f} - (a_2 - a_3)\right)\right] = 0 \tag{7}$$

Since in Eq.(3) instead of one real function U we introduced two real functions g and f, we have freedom to decouple this equation on two bilinear ones. Equating part in brackets to zero, we deduce the following bilinear system

$$(D_t - D_x^2)(g \cdot f) = 0, \tag{8}$$

$$(D_x^2 + \tilde{a}_1\tilde{a}_2)(f \cdot f) = -g^2 + (\tilde{a}_1 + \tilde{a}_2)gf, \tag{9}$$

where $\tilde{a}_1 \equiv a_1 - a_3$ and $\tilde{a}_2 \equiv a_2 - a_3$. To solve this system in the Hirota method functions f and g suppose to have the form of the perturbation series in some parameter ε

$$f = 1 + \varepsilon f_1 + \varepsilon^2 f_2 + \ldots, \tag{10}$$

$$g = g_0 + \varepsilon g_1 + \varepsilon^2 g_2 + \ldots \tag{11}$$

Substituting (10) into the system (8) and equating coefficients of the same power of ε, converts (8) into a sequence of bilinear equations of the zeroth, first, second and so on orders respectively:

$$(D_t - D_x^2)(g_0 \cdot 1) = 0, \tag{12}$$

$$(D_x^2 + \tilde{a}_1\tilde{a}_2)(1 \cdot 1) = -g_0^2 + (\tilde{a}_1 + \tilde{a}_2)g_0; \tag{13}$$

$$(D_t - D_x^2)(g_0 \cdot f_1 + g_1 \cdot 1) = 0, \tag{14}$$

$$(D_x^2 + \tilde{a}_1\tilde{a}_2)(2 \cdot f_1) = -[2g_0g_1 - (\tilde{a}_1 + \tilde{a}_2)(g_0f_1 + g_1 \cdot 1)]; \tag{15}$$

$$(D_t - D_x^2)(g_0 \cdot f_2 + g_1 \cdot f_1 + g_2 \cdot 1) = 0, \tag{16}$$

$$(D_x^2 + \tilde{a}_1\tilde{a}_2)(2 \cdot f_2 + f_1 \cdot f_1) = -2g_0g_2 - g_1^2 + (\tilde{a}_1 + \tilde{a}_2)(g_0f_2 + g_1f_1 + g_2 \cdot 1); \tag{17}$$

$$\ldots\ldots \tag{18}$$

2.1. The zeroth order solution

We assume that g_0 is a constant, then the first equation in (12) is satisfied automatically. From the second equation (12), we get

$$g_0^2 - (\tilde{a}_1 + \tilde{a}_2)g_0 + \tilde{a}_1\tilde{a}_2 = (g_0 - \tilde{a}_1)(g_0 - \tilde{a}_2) = 0, \qquad (19)$$

which is the quadratic algebraic equation for g_0 with two roots $\tilde{a}_1$ and $\tilde{a}_2$.

2.2. The first order solution

In this subsection we find the first order solutions, g_1, f_1. First, without loss of generality we assume that $g_0 = \tilde{a}_1$. Then Eqs. (14) may be rewritten as a linear system

$$\tilde{a}_1(-\partial_t - \partial_x^2)f_1 + (\partial_t - \partial_x^2)g_1 = 0, \qquad (20)$$
$$2f_{1_{xx}} + 2\tilde{a}_1\tilde{a}_2 f_1 + 2\tilde{a}_1 g_1 - (\tilde{a}_1 + \tilde{a}_2)(\tilde{a}_1 f_1 + g_1) = 0. \qquad (21)$$

The simplest nontrivial solution of this system has the form

$$g_1 = e^{\eta_1}, \; f_1 = b e^{\eta_1}, \qquad (22)$$

where $\eta_1 = k_1 x + \omega_1 t + \gamma_1$. Coefficient in front of g_1 can be absorbed by constant γ_1, it is why we can choose it equal to one. Unknown constants k_1 and ω_1 are fixed by dispersion relation. After substituting Eq. (22) into the system (20) we have the algebraic system. The nontrivial solution of this system exists only if the following dispersion relation on k_1 and ω_1 is satisfied

$$\omega_1 = k_1^2 + \tilde{a}_1(\tilde{a}_2 - \tilde{a}_1). \qquad (23)$$

The free parameter b becomes fixed as

$$b = \frac{(\tilde{a}_2 - \tilde{a}_1)}{2k_1^2 + \tilde{a}_1(\tilde{a}_2 - \tilde{a}_1)}. \qquad (24)$$

So, from above the calculations only k_1 and γ_1 are free parameters of our solution.

2.3. The second order calculation

Substituting g_1, f_1 to the system (16) and using property of bilinear Hirota operators

$$(D_t - D_x^2)(g_1 \cdot f_1) = b(D_t - D_x^2)(e^{\eta_1} \cdot e^{\eta_1}) = 0 \qquad (25)$$

for g_2, f_2 we find the following equation

$$\tilde{a}_1(-\partial_t - \partial_x^2)f_2 + (\partial_t - \partial_x^2)g_2 = 0, \qquad (26)$$

similar to Eq.(20). The simplest solutions for this equation is the trivial one $g_2 = 0$ and $f_2 = 0$. Then from Eq.(16) we find additional constraint on g_1, f_1 and b:

$$\tilde{a}_1 \tilde{a}_2 f_1^2 = -g_1^2 + (\tilde{a}_1 + \tilde{a}_2)(g_1 f_1) \tag{27}$$

Substitution of Eq. (22) into Eq. (27)results in quadratic equation for b:

$$\tilde{a}_1 \tilde{a}_2 b^2 - (\tilde{a}_1 + \tilde{a}_2)b + 1 = (\tilde{a}_1 b - 1)(\tilde{a}_2 b - 1) = 0 \tag{28}$$

The roots of this quadratic equation are

$$b_1 = \frac{1}{\tilde{a}_1}, \ b_2 = \frac{1}{\tilde{a}_2}. \tag{29}$$

If we assume that $b = \frac{1}{\tilde{a}_1}$, then after simple calculation we find that the solution of our problem becomes constant, i.e, $u = a_1$. Therefore, we assume that $b = \frac{1}{\tilde{a}_2}$. Combining this condition with Eq.(24), we find restrictions on allowed values of parameter k_1 by Eq.(31). It is easy to show that each bilinear equation, which has order greater than 2, has simple solution as $g_i = 0$ and $f_i = 0$ for $i > 2$. Therefore, we have only finite number of terms. After substituting f and g into the form of u, we obtain the exact solution of our problem as follows :

$$u = a_3 + \frac{(a_1 - a_3) + e^{\eta_1}}{(a_2 - a_3) + e^{\eta_1}}(a_2 - a_3). \tag{30}$$

This is the kink-soliton with asymptotics $u \to a_1$, for $x \to -\infty$ and $u \to a_2$, for $x \to +\infty$, at fixed time.

If in the zeroth order calculations we choose instead of $\tilde{a}_1$ the second root $g_0 = \tilde{a}_2$ and in the second order calculations (29), $b = 1/\tilde{a}_1$, then we get another solution with opposite asymptotics $u \to a_2$, for $x \to -\infty$ and $u \to a_1$, for $x \to +\infty$.

2.4. Velocity of the soliton

In this section we find the wave number parameter k_1, frequency ω_1 and the speed of the soliton v in terms of three distinct roots, a_1, a_2, a_3. By equating Eqs. (24) and (29), we get

$$k_1 = \pm \frac{1}{\sqrt{2}}(a_2 - a_1) \tag{31}$$

and from Eq.(23)

$$\omega_1 = \frac{1}{2}(a_2^2 - a_1^2) - a_3(a_2 - a_1). \tag{32}$$

The velocity v of our travelling wave is given by the equation

$$v = \left(-\frac{\omega_1}{k_1} \right) \tag{33}$$

378

or combining Eqs. (31) and (32) we obtain the following relationship between the velocity and the roots of the nonlinear part.

$$v = \pm \frac{a_1 + a_2 - 2a_3}{\sqrt{2}}. \tag{34}$$

When the root a_3 is equal to the mean value of two other roots $a_3 = (a_1 + a_2)/2$, the velocity vanishes and soliton is static. By cyclic permutation of indices $1, 2, 3$ in Eqs.(3) and (30) we find other kink solitons of our problem. But only one of them could be stable since implement transition between stable vacuum states.

3. EXTENDED REACTION-DIFFUSION EQUATION

In this section, we solve Eq. (2) by modified Hirota method introduced in section 1. We use the same ansatz for U and perturbation series expansion of f and g. After using Hirota derivatives (4), the analog form to the Equation (7) can be obtained for the Eq. (2) as well, then decoupling this equation, the following bilinear system of Eqs. can be deduced

$$(D_t + \alpha a_3 D_x - D_x^2)(g \cdot f) = 0, \tag{35}$$

$$(D_x^2 + \tilde{a}_1 \tilde{a}_2)(f \cdot f) = -g^2 + (\tilde{a}_1 + \tilde{a}_2 - \alpha D_x)(gf). \tag{36}$$

After substituting the previous ansatz for f and g, we obtain the zeroth, first and second order bilinear equations. The zeroth order solution will not change. For the first order solution, we again assume that the form of the f and g are the same as in Eq. (22). After substituting (22) into the Equations (35), we obtain the algebraic system which has nontrivial solution if and only if the following dispersion relation on parameters k_1, ω_1 and α is satisfied

$$\omega_1 = k_1^2 + \tilde{a}_1(\tilde{a}_2 - \tilde{a}_1) - (\tilde{a}_1 + a_3)\alpha k_1. \tag{37}$$

Then, parameter b becomes fixed as

$$b = \frac{(\tilde{a}_2 - \tilde{a}_1) - \alpha k_1}{2k_1^2 + \tilde{a}_1(\tilde{a}_2 - \tilde{a}_1) - \tilde{a}_1 \alpha k_1}. \tag{38}$$

If we put $\alpha = 0$ in the Eqs. (37) and (38), one can see easily show that the equations reduce to Eqs. (23) and (24), respectively.

The second order calculation will not change, and one can find that value of b is as before. Then, by equating the Eqs. (29) and (38), we find following restriction on k_1,

$$k_1 = \frac{(\tilde{a}_2 - \tilde{a}_1)(-\alpha \mp \varepsilon \sqrt{\alpha^2 + 8})}{4}, \tag{39}$$

where

$$\varepsilon = \begin{cases} 1, & a_2 > a_1 \\ -1, & a_2 < a_1. \end{cases}$$

379

Finally, the velocity of our solution can be calculated by using Eq. (33), where ω_1 and k_1 are given by Eqs. (37) and (39), respectively:

$$v = \frac{2(a_1 + a_2 - 2a_3)}{\beta} - (\frac{a_2 + a_1}{2})\alpha, \tag{40}$$

where $\beta = -\alpha \pm \varepsilon \sqrt{\alpha^2 + 8}$. From the last equation one can see that even for the mean value condition $a_3 = (a_1 + a_2)/2$ the soliton is not static. The static soliton arises for fixed value of α in terms of a_1, a_2, a_3 in Eq.(40).

4. CONCLUSION

In the present paper, the modified Hirota method is applied to find the exact analytic one soliton solution of the nonlinear reaction-diffusion equations. These equations have nonlinear reaction part which is the third order polynomial with three distinct roots. In the phase transition content, these three distinct roots correspond to the order of the system phases. From the phase plane analysis, the system has two stable and one unstable phases or one stable and two unstable phases. In the last case no stable kink soliton can exist. The solutions of equations (1) and (2) giving connection between two stable phases correspond to the one soliton solutions. If solution of the problem is written with background value of the one of the unstable phases, say a_3, then it gives connection between two stable phases, a_1 and a_2.

Finally, we showed that Hirota method is very efficient and systematic procedure to obtain exact solutions of reaction-diffusion equations with three distinct roots. We hope that the method allows one to find solutions of other extensions of the model to higher dimensions, additional terms and multi-component order parameters. These questions are under our study now.

ACKNOWLEDGMENTS

This work was supported partially by Izmir Institute of Technology grant 2002-IYTE-24 and 2002-IYTE-25.

REFERENCES

1. Alwyn Scott, Nonlinear Science, Oxford University Press, 1999
2. Alan C. Newell, Soliton and Physics, SIAM, 1985.
3. R. Hirota, Direct methods in Soliton Theory, in : "Solitons, Springer Verlag", 1980.

Group Theoretical Treatment of the Jan-Teller Systems: $T_1 \otimes (e \oplus t_2)$ Coupling

Hayriye Tütüncüler*, Ramazan Koç* and Bora Umut Türkdönmez*

*Department of Physics, Faculty of Engineering, University of Gaziantep, 27310, Gaziantep, Turkey

Abstract. A novel method is developed to construct the Jan-Teller interaction matrices. The applicability of the method is demonstrated by constructing first and second order Jan-Teller interaction matrices of the $T_1 \otimes (e \oplus t_2)$ octahedral system. The method given here is useful to obtain the Jan-Teller interaction matrices of the other systems as well as higher order interaction matrices. We also discuss the determination of the location of the stationary points on the potential energy surface, by a successive symmetry breaking of corresponding finite group into its maximal little groups.
Key words. Group theory, Coupled cluster theory, Electronic Structure, Symmetry breaking.
PACS. 31.15.Hz, 31.15.Dv, 31.70.-f.
MSC. 81R40.

1. INTRODUCTION

The Jan-Teller (JT) distortion problem is an old one, dating back over sixty years [3]. Over the years several studies have been devoted to the continuous group symmetry of linear JT systems. O'Brien [7], Pooler [8, 9], and Judd [4, 5] have made use of the special orthogonal groups to describe the case of degenerate linear coupling. According to the JT theorem any non-linear molecular system in a degenerate electronic state will be unstable and will go under distortion to form a system of lower symmetry thereby removing degeneracy. This effects plays a crucial role in explaining the structure and dynamics of solids and molecules in degenerate electronic states.

One way to construct the JT interaction matrices is to generate the continuous groups by irreducible tensor operators within a single electronic manifold provided that linear interaction matrices. Pooler, in his remarkable paper has provided a method for generating linear JT Hamiltonians with continuous group symmetries that is applicable to all real character simple phase group. In this work we provide somewhat simple method to construct JT matrices based on the invariance of a polynomial under finite group operations. In our calculations one can also obtain the part of JT interaction matrices that invariant under continuous group.

The other problem we discuss here that the determination of the location of the stationary points on potential energy surface. A convenient way to analyze the group theoretical features of JT surfaces is the epikernel principle [1]. It is obvious that the symmetry of the physical system is reduced due to the JT interaction. Epikernel principle based on the reduction of the symmetry of the physical system. We show that the same results can be obtained by breaking symmetries of the physical system into its little groups. The symmetry breaking mechanism introduced here can also be used to determine the phase transition in condensed matter physics. This open a way to analyze the superconducting phase transitions.

2. THEORY

We begin by describing the Hamiltonian that produce $D^\ell \otimes D^\ell$ surface where D^ℓ denotes the irreducible representation. The standard Hamiltonian may be written in the form

$$H = H_o + H_{JT}, \tag{1}$$

$$H_0 = -\frac{1}{2} \sum_{i=1}^{5} \frac{\partial^2}{\partial q_i^2} + \frac{1}{2} k_h \sum_{i=1}^{5} q_i^2,$$

where H_0 is the Hamiltonian of the harmonic oscillator, q_i are the reduced coordinates in which $\hbar\omega = 1$. In general, H_{JT} introduces a rotationally invariant linear coupling between the electronic state and the vibrational mode. It is known that the Hamiltonian for a linear JT coupling of the any given system is invariant under the rotational operations of $SO(3)$ group, and it will be shown that this property leads to the construction of JT interaction matrices.

Before we discuss the construction of the JT Hamiltonian, let us investigate the interrelation between the JT Hamiltonian and associated symmetry group. Totally symmetric part of direct product of an irreducible representation of a finite group, which describes the properties of JT surfaces, is written in the form of

$$\left[D^\ell \otimes D^\ell \right] = D^{\ell_1} \oplus D^{\ell_2} \oplus \cdots \oplus D^{\ell_n}, \tag{2}$$

where ℓ is the angular momentum quantum number. Decomposition of $\left[D^\ell \otimes D^\ell \right]$ implies that the JT Hamiltonian can be written in the following way

$$H_{JT} = H^{\ell_1} + H^{\ell_2} + \ldots + H^{\ell_n}, \tag{3}$$

where H^ℓ is the JT Hamiltonian and it is invariant under the symmetry operations of corresponding finite group, for the $2\ell + 1$ dimensional representation. The implication of (2) and (3) can explicitly be illustrated on an Octahedral group (O). Consider the symmetric part of the $[T \otimes T]$ coupling for Octahedral system. We recall the essential symmetry properties for a linear JT systems. The Hamiltonian of $[T \otimes T]$ coupling for octahedral system has two parts H^0, H^2 which must be separately invariant under octahedral symmetry. Vibrations subtend the configuration space, which contains all distorted configurations, may be reached by JT active coordinates. The JT Hamiltonian for this coupling can be written in the form of

TABLE 1. The representations of the $O(3)$ group with $\ell < 6$ are split by the octahedral point group.

ℓ	O	ℓ	O	ℓ	O
0	A_1	2	$E + T_2$	4	$A_1 + E + T_1 + T_2$
1	T_1	3	$A_1 + T_1 + T_2$	5	$E + 2T_1 + T_2$

$$H_{JT} = H^0 + H^2, \tag{4}$$

where H^0, and H^2 correspond to a_1 representation, $(t_2 \oplus e)$ representation respectively. The decomposition of direct product of $[T \otimes T]$ coupling in $SO(3)$ is found using the$Table$ 1 and symmetric part of the three dimensional direct product for Octahedral group can be written as

$$[T \otimes T] = a_1 \oplus (t_2 \oplus e), \tag{5}$$

where T denotes electronic level, a_1 and $(t_2 \oplus e)$ denotes vibrational levels. The symmetric part contains totally symmetric representation $H^0 = a_1$ that is trivial polaronic problem and it can be solved exactly. The H^0 vibration shifts energies but it does not cause splitting. This means that for triplet electronic states the doubly degenerate e-type vibrations and the threefold degenerate t_2-type vibrations are JT active. This is known as a $T \otimes (t_2 \oplus e)$ problem. The adiabatic potentials of all cubic systems in the subspace of $(t_2 \oplus e)$ vibrations possess the same symmetry O or T. The simplest cubic molecule, the full vibrational representation of which contains only one e and one t_2 representation, is octahedral molecule of type SF_6 (sulphur hexafluoride).

3. FINITE GROUP INVARIANCE OF THE $T \otimes (T_2 \oplus E)$ JT SYSTEM

In this section, we want to construct a polynomial function in electronic and nuclear configuration space, to get JT interaction matrices. A polynomial function which has been produced for this purpose is in the form

$$U_{\ell m}(x, q) = \sum_{i,j=1}^{2\ell+1} \sum_{k=1}^{2m+1} f_{ijk} x_i x_j q_k. \tag{6}$$

In this expression x_i and x_j correspond to electronic coordinates q_k corresponds to nuclear coordinates. The force elements or coupling coefficients should be chosen appropriately for $T \otimes (t_2 \oplus e)$. In the (6) the indices $2\ell + 1$ and $2m + 1$ stand for dimensions of electronic and active coordinates, respectively. For $T \otimes (t_2 \oplus e)$ coupling ℓ and m take value 1 and 2 respectively. The choices of ℓ and m depend on the number of electronic and nuclear coordinates of the given system, respectively.

We want to construct a polynomial function in electronic and nuclear configuration space. The invariant polynomial function for $\ell = 1$, in real basis has been computed using the three dimensional matrix (T state). The 3×3 generators transform electronic coordinates x_i $(i = 1\ldots3)$ and nuclear coordinates q_i. It can be written in the form

$$\sum_{i,j=1}^{3}\sum_{k=1}^{5} f_{ijk}x_ix_jq_k = \sum_{i,j=1}^{3}\sum_{k=1}^{5} f_{ijk}x_i'x_j'q_k',$$ (7)

where $x_i' = \sum \tau_{in}^r x_n$ and $q_i' = \sum \tau_{in}^r q_n$ are the 3×3 and 5×5 matrix elements of generators of Octahedral group, respectively. These matrices are [6]:

$$g_1 = \begin{pmatrix} 0 & 0 & 1 \\ 1 & 0 & 0 \\ 0 & 1 & 0 \end{pmatrix}, \qquad g_3 = \begin{pmatrix} 1 & 0 & 0 & 0 & 0 \\ 0 & -1 & 0 & 0 & 0 \\ 0 & 0 & -1 & 0 & 0 \\ 0 & 0 & 0 & 0 & 1 \\ 0 & 0 & 0 & -1 & 0 \end{pmatrix},$$ (8)

$$g_2 = \begin{pmatrix} 1 & 0 & 0 \\ 0 & 0 & -1 \\ 0 & 1 & 0 \end{pmatrix}, \qquad g_4 = \frac{1}{2}\begin{pmatrix} 1 & \sqrt{3} & 0 & 0 & 0 \\ -\sqrt{3} & -1 & 0 & 0 & 0 \\ 0 & 0 & 0 & 0 & 1 \\ 0 & 0 & 1 & 0 & 0 \\ 0 & 0 & 0 & 1 & 0 \end{pmatrix}.$$

The (7) is solved for f_{ijk}. In Ceulemans's paper [1], relations between polynomial coefficients f_{ijk} in (7) is expressed in terms of Clebsch Gordon series for the icosahedral point group . One of our main goal that first order JT interaction matrices can be derived by working out invariant polynomial function. The double differentiation of $U_{12}(x,q)$ with respect to electronic coordinates x_i and x_j produce linear JT interaction matrix. In general, we can write

$$\beta = \frac{\partial^2 U_{\ell m}}{\partial x_i x_j}.$$ (9)

First order linear JT interaction matrix $T \otimes (t_2 \oplus e)$ coupling as in the form

$$\beta_1 = f_t \begin{pmatrix} 0 & q_5 & q_4 \\ q_5 & 0 & q_3 \\ q_4 & q_3 & 0 \end{pmatrix} + \frac{2}{\sqrt{3}}f_e \begin{pmatrix} 2q_1 & 0 & 0 \\ 0 & -q_1 - \sqrt{3}q_2 & 0 \\ 0 & 0 & -q_1 + \sqrt{3}q_2 \end{pmatrix},$$ (10)

where f_t and f_e are coupling constants that parametrizes the strength of the Jan-Teller coupling, which is strong or weak as $f_t \gg 1$ and $f_e \gg 1$ or $f_t \ll 1$ and $f_e \ll 1$. The JT matrix is found using the finite group invariance of the Hamiltonian. In literature [2], the linear JT interaction matrix $T \otimes (t_2 \oplus e)$ coupling was found using the rotation group $SO(3)$ invariance of the Hamiltonian. In this study, we classified the JT matrix according to the 3-dimensional and 2-dimensional coordinate parameters. If $f_t = f_e$, this matrix is in agreement with the matrix given in [2]. In such away we have proved that these different approximations give the same result under the such constraints. At this point, it must be emphasized that $T \otimes (t_2 \oplus e)$ coupling in cubic symmetry is equivalent to the $T \otimes h$ coupling.

The three roots of this matrix are three energies. As we scan the five dimensional $\{q_i\}$ phase space, these three energies expand into the three adiabatic potential energy surfaces (APES), and if the potential energy of the original oscillators ($\frac{1}{2}\sum k_h q_i^2$), necessarily positive, is added, a minimum energy appears on the lowest energy appears on the lowest APES. The depth of this energy minimum below the minimum energy of the uncoupled state is called the Jan-Teller energy. It is worth noting at this point that the kinetic energy of the oscillators has not been included, and it will alter the actual energy of the ground state, which will always lie above the Jan-Teller minimum.

We can also calculate higher order JT interaction matrices. To calculate the second order JT interaction matrices for this coupling, polynomial function is in the form

$$U_{\ell m}(x,q) = \sum_{i,j=1}^{3}\sum_{k=1}^{5}\sum_{l=1}^{5} f_{ijk} x_i x_j q_k q_l. \tag{11}$$

Using the same procedure given above, second order JT matrix is found as

$$\beta_2 = f_1 A + f_1 \begin{pmatrix} 0 & q_0 & q_+ \\ q_0 & 0 & q_- \\ q_+ & q_- & 0 \end{pmatrix} + f_2 q_i^2 I, \tag{12}$$

where f_1 and f_2 are coupling constants, A is a diagonal matrix with elements $-(2q_1^2 + 2q_2^2 + 2q_3^2 - q_4^2 - 2\sqrt{3}q_1 q_5)$, $-(2q_1^2 + 2q_2^2 + q_3^2 - 2q_4^2 - 2\sqrt{3}q_1 q_5)$, $(q_1^2 + q_2^2 - 2q_3^2 - 2q_4^2 - 3q_5^2)$ respectively and the elements of second matrix are $q_- = -3q_2 q_3 - q_4(-3q_1 + \sqrt{3}q_5)$; $q_+ = -3q_2 q_4 - q_3(3q_1 + \sqrt{3}q_5)$; $q_0 = -3q_3 q_4 + 2\sqrt{3}q_2 q_5$.

4. DETERMINATION OF STATIONARY POINTS ON JT SURFACE

The second important aspect of invariant polynomial functions, in addition to, linear Jan-Teller matrices, concerns the extremum points on JT surface. Since our interest are the minimal points and the saddle points, the present section is devoted to the determining of extremum points on JT surface by breaking symmetries of octahedral group into its maximal little groups. The problem has been studied by Ceulemans by using the epikernel principle [1].

The little group is the new symmetry group of distorted molecule as a result of coupling. A real representation of any subgroup $S < G$; the degree of subduction can be computed by the relation

$$C_\ell(S) = \frac{1}{S}\sum_{p \in S} \chi_z(p), \tag{13}$$

where $\chi_z(p)$ is the character of the representation p. In a zth representation of finite group, if all subgroups $S' > S$ and $C_\ell(S') < C_\ell(S)$ then S is little group of G. The maximal little subgroups of Octahedral group have computed by using (13) and are

TABLE 2. Maximal Little groups of T_2 and E representations.

	D_2	D_3	D_4
E	$A_1 + B_1$	E	$A_1 + B_1$
T_2	$A_1 + B_2 + A_1 + B_3$	$A_1 + E$	$E + B_2$
$E \oplus T_2$	$2A_1 + B_1 + B_2 + B_3$	$A_1 + 2E$	$A_1 + B_1 + B_2 + E$

given in *Table 2*. As seen from *Table 2*, the maximal little groups predict the existence of dihedral groups D_2, D_4 and D_3 minima on the octahedral molecular cage. In the coordinate space the distortion that is the result of JT instability should conserve all maximal little group symmetry.

In our perspective the structure of Jan-Teller surfaces have been identified by the symmetry breaking of the continuous symmetry to the true finite point group of the representation space and its maximal little groups. This may be represented as follows $O \longrightarrow S'$, where S' are little groups of O group given in Table2.

4.1. Transitions Associated with $T \otimes (t_2 \oplus e)$ Coupling

The three dimensional irreducible representation of O group has three maximal little group named D_2, D_4 and D_3. In order to break symmetry of a parent group into its little groups, one should assign an appropriate q_i which can be computed by constructing set of equations such that

$$q_i = \sum_{j=1}^{5} \tau_{ij} q_j, \tag{14}$$

where τ_{ij} is the matrix elements of generator of the corresponding little group. The method of symmetry breaking predicts the existence of trigonal, tetragonal and orthorhombic turning points on JT surfaces associated with D_2, D_4 and D_3 groups. Under the given conditions; the energy eigenvalues of each little group can be computed.

4.1.1. D_2 Transition

Decomposition of three dimensional representations in D_2 group is $A_1 + B_2 + B_3$. The symmetry of the group O is broken into D_2, assigning as, $q_1 \rightarrow \frac{q_2}{\sqrt{3}}, q_2 \rightarrow q_2, q_3 \rightarrow 0, q_4 \rightarrow q_4, q_5 \rightarrow 0$ using the (14). After substituting values of q_i into JT matrix β_1, the eigenvalues of β_1 are carried out. Combinations of eigenvalues of β_1 with harmonic restoring potentials for the distortional coordinates q_2 and q_4 give JT surface energy values.

$$E(A_1) = -\frac{8}{3} f_e q_2 + \frac{1}{2} k_h (q_2^2 + q_4^2), \tag{15}$$

$$E(B_2) = \frac{1}{3}(4f_e q_2 - 3f_t q_4) + \frac{1}{2}k_h(q_2^2 + q_4^2),$$

$$E(B_3) = \frac{1}{3}(4f_e q_2 + 3f_t q_4) + \frac{1}{2}k_h(q_2^2 + q_4^2).$$

In these equations, k_h is the harmonic force constant.

4.1.2. D_3 Transition

In this case decomposition of T representation of irreducible representations of D_3 is $A_1 + E$. Following the same procedure given in previous section, symmetry of O group can be broken into D_3 by transforming coordinates in the directions; $q_1 \to 0, q_2 \to 0, q_3 \to -q_5, q_4 \to q_5, q_5 \to q_5$. The computed energy values are

$$E(A_1) = -2f_t q_5 + \frac{1}{2}k_h q_5^2, \tag{16}$$

$$E(E) = f_t q_5 + \frac{1}{2}k_h q_5^2.$$

4.1.3. D_4 Transition

In D_4 the triplet irreducible representation state reduces to $E + B_2$. Applying the same procedure as in the previous section, q_i values are found as $q_1 \to -\frac{q_2}{\sqrt{3}}, q_2 \to q_2, q_3 \to 0, q_4 \to 0, q_5 \to 0$. The corresponding energies are obtained as

$$E(B_2) = -\frac{8}{3}f_e q_2 + \frac{1}{2}k_h q_2^2, \tag{17}$$

$$E(E) = \frac{4}{3}f_e q_2 + \frac{1}{2}k_h q_2^2.$$

In general, minimal energy values for each little group can be obtained from the given energy expressions.

5. CONCLUSION

It is worth noting that the JT matrix for $T \otimes (t_2 \oplus e)$ coupling is found by using the finite group invariance. We have shown that how the symmetry breaking method is applied for the determination of the potential energies of $T \otimes (t_2 \oplus e)$ surface. In Ceulemans' paper [1], these energies were found by the method of isostationary function and potential energies of D_3 and D_5 groups were investigated for the $H \otimes (h \oplus g)$ and $H \otimes h$ state. Splitting of energy levels of octahedral system due to distortion is analyzed and amazingly interesting that, there is a proper contribution on the connection between our method and epikernel principle [1]. This method can also be used for other distorted systems.

REFERENCES

1. Ceulemans, A., and Fowler, P. W., *The Jahn–Teller instability of fivefold degenerate states in icosahedral molecules*, J. Chem. Phys., vol. **93**, (1990), 1221-1234.
2. Chancey, C. C., and O'Brien, M. C. M., *Jahn Teller Effect in C_{60} and other Icosahedral Complexes* Princeton University Press, (1997).
3. Englman, R., *The Jahn Teller Effect in Molecules and Crystals*, Wiley, London, (1972).
4. Judd, B. R., *Lie Groups and the Jahn-Teller effect*, Can. J. Phys., vol. **52**, (1974), 999-1011.
5. Judd, B. R., *Group Theory for the Jahn-Teller Effect*, Physica A, vol. **114**, (1984), 19-27.
6. Koca, M., Koç, R., and Al-Barwani, M., *Breaking $SO(3)$ into its closed subgroups by Higgs mechanism*, J. Phys. A: Math. Gen., vol. **30**, 2109-2125, (1997).
7. O'Brien, M. C. M., *Dynamic Jahn-Teller Effect in an Orbital Triplet State Coupled to Both E_g and T_2 sigma Vibrations*, Phys.Rev., vol. **187**, (1969), 407-418.
8. Pooler, D. R., *Continuous group invariances of linear Jahn-Teller systems*, J. Phys. A: Math. Gen., vol. **11**, (1978), 1045-1055.
9. Pooler, D. R., *Continuous group invariances of linear Jahn-Teller systems. II. Extension and application to icosahedral systems*, J. Phys. A: Math. Gen., vol. **11**, (1980), 1029-1042.

Modulational instability of some nonlinear continuum and discrete systems

Anca Visinescu* and D. Grecu*

*Department of Theoretical Physics, "Horia Hulubei" National Institute for Physics and Nuclear Engineering, Bucharest, Romania

Abstract. Modulational instability (also known as the Benjamin-Feir instability) of quasi-monochromatic waves propagating in dispersive and weakly nonlinear media is a general phenomenon encountered in hydrodynamics, plasma physics, condensed matter and is responsible for the generation of robust solitary waves (sometime solitons). The statistical approach is reviewed for several nonlinear systems: the nonlinear Schrödinger equation, the discrete self-trapping equation and Ablowitz-Ladik equation. An integral stability equation is deduced from a linearized kinetic equation for the two-point correlation function. This is solved for several choices of the unperturbed initial spectral function.

Key words. Modulational instability, NLS equation, Ablowitz-Ladik equation.

PACS. 05.45.-a, 63.20.Pw.

MSC. 35Q55, 34G20.

1. INTRODUCTION

Modulational instability (also known as Benjamin-Feir instability) is a general phenomenon encountered any time when a quasi-monochromatic wave is propagating in a weakly nonlinear medium [4]-[1].

It has been discussed by many authors in different physical contexts and for a list of references we shall refer to several review articles and books [5]-[1]. Two basic approaches are possible. The first is a deterministic approach (DAMI) in which small modulation around a basic plane wave solution (Stokes wave) is considered. In the second, considering the field variable as a stochastic quantity, a statistical approach is used (SAMI) [3]. In this approach a linear stability analysis is done in a kinetic equation for a 2-point correlation function, the result depending on the statistical properties of the medium. This problem of MI will be studied for several typical nonlinear equations: the nonlinear Schrödinger equation (completely integrable)

$$i\frac{\partial A}{\partial t} + \frac{\partial^2 A}{\partial x^2} + \mu |A|^2 A = 0, \tag{1}$$

a discrete NLS equation (discrete self-trapping equation)

$$i\frac{da_n}{dt} + \lambda(a_{n+1} + a_{n-1}) + \mu |a_n|^2 a_n = 0, \tag{2}$$

CP729, *Global Analysis and Applied Mathematics: International Workshop on Global Analysis,*
edited by K. Taş, D. Krupka, O. Krupková, and D. Baleanu
© 2004 American Institute of Physics 0-7354-0209-4/04/$22.00

and the Ablowitz-Ladik (completely integrable)

$$i\frac{da_n}{dt} + \lambda(a_{n+1} - 2a_n + a_{n-1}) + \mu|a_n|^2(a_{n+1} + a_{n-1}) = 0. \tag{3}$$

As DAMI is quite well known we shall present it briefly and compare the results of the continuum case (1) (NLS) with the results for the two discrete equations (2) and (3).

The Stokes wave solution of NLS is given by $A(x,t) = ae^{i(kx-\omega t)}$ with an amplitude dependent dispersion relation $\omega = k^2 - \mu|a|^2$. This is slowly modulated in amplitude $A(x,t) = a(1 + \varepsilon A_1)e^{i(kx-\omega t)}$, $\varepsilon \ll 1$ and writing $A_1 \sim e^{i(qx-\Omega t)}$ we find

$$\Omega = 2kq + iq\sqrt{2\mu|a|^2 - q^2}. \tag{4}$$

Instability ($Im\Omega > 0$) appears if $\mu > 0$, (focusing case of NLS eq.) and if $q^2 < 2\mu|a|^2$. As $|a|^2$ is a small quantity the instability region corresponds to the long wave length domain. The mechanism is well known: two side bands at the left and the right of the basic harmonic appear and their presence lead to a synchronous resonance and their mutual reinforcement.

The same analysis can be done for the discrete equations (2) and (3). The final results are

$$\Omega = 2\lambda \sin k \sin q + i4|\lambda|\cos k \sin\frac{q}{2}\sqrt{\frac{\mu}{2\lambda}|a|^2\frac{1}{\cos k} - \sin^2\frac{q}{2}} \tag{5}$$

for the discrete NLS, and

$$\Omega = 2\lambda(1 + \frac{\mu}{\lambda}|a|^2)\sin k \sin q + i2\sqrt{2}|\lambda|\cos k \sin\frac{q}{2}(1 + \frac{\mu}{\lambda}|a|^2)\sqrt{\cos q - \frac{1 - \frac{\mu}{\lambda}|a|^2}{1 + \frac{\mu}{\lambda}|a|^2}} \tag{6}$$

for Ablowitz-Ladik equation. In these relations both wave-vectors k and q are restricted to the first Brillouin zone $(-\pi, \pi)$ except the NLS case, where such a restriction doesn't exists. In both cases the instability can appear if λ and μ have the same sign. For the AL system the stability condition is $q < q_0$, $\cos q_0 = \frac{1 - \frac{\mu}{\lambda}|a|^2}{1 + \frac{\mu}{\lambda}|a|^2}$ and is independent of k. In the case of discrete NLS system, the instability region is more complex. Writing $f^2 = \frac{\mu}{2\lambda}|a|^2$ the condition is

$$\frac{f^2}{\cos k} - \sin^2\frac{q}{2} \geq 0. \tag{7}$$

It is k-dependent and for $\cos k \leq f^2$ it is satisfied for any $q \in 1 - BZ$.

2. SAMI FOR ABLOWITZ-LADIK EQUATION

As mentioned above, in SAMI the field variable is considered a stochastic quantity. It was discussed especially in hydrodynamics to study the evolution of surface water waves [3], [9], but also for the study of Landau damping of Langmuir waves [6] and the propagation of electromagnetic wave packets in nonlinear media [8].

Recently, the method has been extended to the discrete NLS equation (2) [11].

In this section we shall concentrate to the discrete completely integrable Ablowitz-Ladik system. Compared to the discrete NLS case it presents some special technical points and an exhaustive description is justified.

First, introducing the displacement operator $a_{n\pm1} = e^{\pm\frac{\partial}{\partial n}} a_n$ we write the AL eq. in the form

$$i\frac{\partial a_n}{\partial t} + 2\lambda \cosh \frac{\partial}{\partial n} a_n + 2\mu |a_n|^2 \cosh \frac{\partial}{\partial n} a_n = 0. \tag{8}$$

To find a kinetic equation for the two point correlation function $\rho(n_1,n_2) = <a_{n_1} a_{n_2}^*>$, where $< ... >$ means an average over a statistical ensemble, we write (8) for $n = n_1$ and multiply it by $a_{n_2}^*$, add to it the complex conjugated of (8) for $n = n_2$, multiplied by a_{n_1}, and finally take a statistical average. We get

$$i\frac{\partial\rho}{\partial t} + 2\lambda \left(\cosh \frac{\partial}{\partial n_1} - \cosh \frac{\partial}{\partial n_2} \right) \rho + \tag{9}$$

$$+2\mu \left[<a_{n_1} a_{n_2}^* \cosh \frac{\partial}{\partial n_1} a_{n_1} a_{n_2}^*> - <a_{n_2} a_{n_2}^* \cosh \frac{\partial}{\partial n_2} a_{n_2}^* a_{n_1}> \right] = 0,$$

a kinetic equation, which beside ρ contains 4-point correlation functions. For a Gaussian random process an exact decomposition of these 4-point correlation functions in products of 2-point correlation functions exists. We shall assume , and this is the only approximation we have, that such a factorization is acceptable also for non-Gaussian processes. Then we can write

$$<a_{n_1} a_{n_1}^* \cosh \frac{\partial}{\partial n_1} a_{n_1} a_{n_2}^*> = \bar{a}_{n_1}^2 \cosh \frac{\partial\rho}{\partial n_1} + \rho \left(\lim_{n_2 \to n_1} \cosh \frac{\partial\rho}{\partial n_1} \right),$$

$$<a_{n_2} a_{n_2}^* \cosh \frac{\partial}{\partial n_2} a_{n_2}^* a_{n_1}> = \bar{a}_{n_2}^2 \cosh \frac{\partial\rho}{\partial n_2} + \rho \left(\lim_{n_1 \to n_2} \cosh \frac{\partial\rho}{\partial n_2} \right). \tag{10}$$

Here $\bar{a}_n^2 = <a_n a_n^*>$ is the ensemble averaged mean square amplitude.

It is convenient to use new spatial coordinates, namely the center of mass coordinate $M = \frac{n_1+n_2}{2}$, and the relative coordinate $m = n_1 - n_2$ and to make a Fourier transform with respect to the relative coordinate. This is the right way to study the MI, because the initial state naturally depends only on the relative coordinate and the instability becomes to manifest by a dependence on the center of mass coordinate. With these transformations in mind, it is easy to calculate the limits appearing in (10). We get

$$\lim_{n_2 \to n_1} \cosh \frac{\partial}{\partial n_1} \rho = \cosh \frac{1}{2} \frac{\partial}{\partial M} H_c + i \sinh \frac{1}{2} \frac{\partial}{\partial M} H_s,$$

$$\lim_{n_1 \to n_2} \cosh \frac{\partial}{\partial n_2} \rho = \cosh \frac{1}{2} \frac{\partial}{\partial M} H_c - i \sinh \frac{1}{2} \frac{\partial}{\partial M} H_s, \tag{11}$$

where

$$H_{c,s}(M,t) = \frac{1}{2\pi} \int_{-\pi}^{\pi} \left(\begin{array}{c} \cos k \\ \sin k \end{array} \right) F(k,M,t) dk. \tag{12}$$

The Fourier transform of the kinetic equation (9) using (10) and (11) leads to the following equation in a mixed space-wave number representation

$$\frac{\partial F}{\partial t} + 4\lambda \sin k \sinh \frac{1}{2}\frac{\partial}{\partial M} F + 4\mu F \sinh \frac{1}{2}\frac{\partial}{\partial M}\frac{1}{2\pi}\int_{-\pi}^{\pi}\sin k' F(k',M)dk' +$$

$$+2\mu \sum_{j=0} \frac{2(-1)^j}{2j!2^{2j}}\frac{\partial^{2j}\bar{a}^2}{\partial M^{2j}}\sinh \frac{1}{2}\frac{\partial}{\partial M}\frac{\partial^{2j}}{\partial k^{2j}}(\sin k F(k,m)) + \qquad (13)$$

$$+2\mu \sum_{j=0} \frac{2(-1)^j}{(2j+1)!2^{2j+1}}\frac{\partial^{2j+1}\bar{a}^2}{\partial M^{2j+1}}\frac{\partial^{2j+1}}{\partial k^{2j+1}}(\cos k F(k,M)) +$$

$$+2\mu \sum_{j=0} \frac{2(-1)^j}{(2j+1)2^{2j+1}}\frac{\partial^{2j+1}\bar{a}^2}{\partial M^{2j+1}}(\cosh \frac{\partial}{\partial M} - 1)\frac{\partial^{2j+1}}{\partial k^{2j+1}}(\cos k F(k,M)) = 0.$$

Here $F(k,M,t)$ is the Fourier transform of $\rho(m,M,t)$, $\sum_m e^{-ikm}\rho(M+\frac{m}{2},M-\frac{m}{2},t)$.
The next step is to make a linear stability analysis of this equation. We consider

$$F(k,M,t) = f(k) + \varepsilon \mathscr{F}(k,M,t), \qquad (14)$$

where $f(k)$ is the Fourier transform of the initial condition, independent on M and t. In the same way

$$\bar{a}^2(M,t) = \bar{a}_0^2 + \varepsilon \bar{a}_1^2(M,t), \qquad (15)$$

where

$$\bar{a}_0^2 = \frac{1}{2\pi}\int_{-\pi}^{\pi}f(k)dk, \qquad \bar{a}_1^2 = \frac{1}{2\pi}\int_{-\pi}^{\pi}\mathscr{F}(k,M,t)dk. \qquad (16)$$

Keeping only first order terms in ε, we obtain

$$\frac{\partial \mathscr{F}}{\partial t} + 4(\lambda + \mu|a_0|^2)\sin k \sinh \frac{1}{2}\frac{\partial}{\partial M}\mathscr{F} + 4\mu f(k)\sinh \frac{1}{2}\frac{\partial}{\partial M}\frac{1}{2\pi}\int_{-\pi}^{\pi}\sin k' \mathscr{F}(k',M,t) +$$

$$+2\mu \sum_{j=0}^{\infty}\frac{2(-1)^j}{(2j+1)!2^{2j+1}}\frac{\partial^{2j+1}\bar{a}_1^2(M,t)}{\partial M^{2j+1}}\frac{\partial^{2j+1}}{\partial k^{2j+1}}(\cos k f(k)) = 0. \qquad (17)$$

Looking for plane wave solutions

$$\mathscr{F}(k,M,t) = g(k)e^{i(KM-\Omega t)}, \qquad (18)$$

$$\bar{a}_1^2(M,t) = e^{i(KM-\Omega t)}\frac{1}{2\pi}\int_{-\pi}^{\pi}g(k)dk = Ge^{i(KM-\Omega t)},$$

the equation (17) becomes

$$\left[-\Omega + 4(\lambda + \mu\bar{a}_0^2)\sin k \sin \frac{K}{2}\right]g + 4\mu f(k)\sin \frac{K}{2}\frac{1}{2\pi}\int_{-\pi}^{\pi}\sin k' g(k')dk' + \qquad (19)$$

$$+2\mu G \sum_{j=0}^{\infty} \frac{2}{(2j+1)!} \left(\frac{K}{2}\right)^{2j+1} \frac{\partial^{2j+1}}{\partial k^{2j+1}} (\cos k f(k)) = 0.$$

Assuming $f(k)$ an even function of k, we have

$$\frac{1}{2\pi} \int_{-\pi}^{\pi} \frac{\partial^{2j+1}}{\partial k^{2k+1}} (\cos k f(k)) = 0.$$

Then, integrating (19), we get $(H = \frac{1}{2\pi} \int_{-\pi}^{\pi} \sin k' g(k') dk'. \ )$

$$-\Omega G + 4(\lambda + 2\mu \bar{a}_0^2) \sin \frac{K}{2} H = 0, \tag{20}$$

which determines H in function of G. Considering Ω purely imaginary $(\Omega = i\Omega_i)$ and introducing the quantity $\omega = \frac{\Omega_i}{4(\lambda + \mu \bar{a}_0^2) \sin \frac{K}{2}}$, after a little algebra, the equation (19) transforms into the following integral stability equation

$$1 + i \frac{\mu}{\lambda + 2\mu \bar{a}_0^2} \omega \frac{1}{2\pi} \int_{-\pi}^{\pi} \frac{f(k)}{\sin k - i\omega} dk + \tag{21}$$

$$+ \frac{\mu}{2(\lambda + \mu \bar{a}_0^2) \sin \frac{K}{2}} \frac{1}{2\pi} \int_{-\pi}^{\pi} \frac{Q(k + \frac{K}{2}) - Q(k - \frac{K}{2})}{\sin k - i\omega} dk = 0.$$

Here $Q(k) = \cos k f(k)$ and we used the relation

$$Q(k + \frac{K}{2}) - Q(k - \frac{K}{2}) = \sum_{j=0}^{\infty} 2 \frac{(\frac{K}{2})^{2j+1}}{(2j+1)!} \frac{\partial^{2j+1}}{\partial k^{2j+1}} Q(k). \tag{22}$$

For a δ-spectrum

$$f_0(k) = 2\pi \bar{a}_0^2 \delta(k), \tag{23}$$

the calculations in (21) are straightforward. One obtains

$$1 - \frac{\mu \bar{a}_0^2}{\lambda + 2\mu \bar{a}_0^2} = \frac{\mu \bar{a}_0^2}{\lambda + \mu \bar{a}_0^2} \frac{1}{\sin^2 \frac{K}{2} + \omega^2}, \tag{24}$$

giving

$$\Omega_i = 4|\lambda + \mu \bar{a}_0^2|| \sin \frac{K}{2}| \sqrt{\bar{a}_0^2 \frac{\lambda + 2\mu \bar{a}_0^2}{(\lambda + \mu \bar{a}_0^2)^2} - \sin^2 \frac{K}{2}}. \tag{25}$$

The instability takes place $(\Omega_i > 0)$ if λ, μ have the same sign and in the long wave length region

$$\sin \frac{K}{2} < \left(\mu \bar{a}_0^2 \frac{\lambda + 2\mu \bar{a}_0^2}{(\lambda + \mu \bar{a}_0^2)^2}\right). \tag{26}$$

Using (21), more realistic initial conditions than the δ-spectrum case (23) (a δ-spectrum corresponds to an uniform distribution in space, $\rho(x) = const.$) can be worked down.

3. CONCLUSIONS

Continuing our previous investigations for DAMI in continuous and discrete nonlinear systems [11], [7], the case of Ablowitz-Ladik equation is studied. It presents some technical specific features which legitimate a full discussion of it. The final results, the integral stability equation (21) and the increment in time of the instability (25) (the imaginary part of the frequency of the modulated wave), can now be compared with the similar results obtained for the continuous NLS eq [9], [7] and the discrete NLS eq [11]. The integral stability equations are

$$1 + \frac{\mu}{2\pi\lambda K} \int_{-\infty}^{\infty} \frac{f_0(k + \frac{K}{2})f_0(k - \frac{K}{2})}{k - i\omega} dk = 0 \qquad (27)$$

for the continuous NLS case ($\omega = \frac{\Omega_i}{2\lambda K}$)

$$1 + \frac{\mu}{4pi\lambda \sin\frac{K}{2}} \int_{-\pi}^{\pi} \frac{f_0(k + \frac{K}{2})f_0(k - \frac{K}{2})}{\sin k - i\omega} dk = 0 \qquad (28)$$

for the discrete NLS equation ($\omega = \frac{\Omega_i}{4\lambda \sin\frac{K}{2}}$). The main differences between the continuous and the discrete NLS eq. (27) and (28) are the domain if integration and the presence of a trigonometric function ($\sin k$) at the denominator of the integrand in (28). The effect is a slightly more complicated way of integration in (28) than in (27).

The expressions for Ω_i for a Lorentzian initial distribution $f_0(k)$ are

$$\Omega_i = 2\lambda K \sqrt{\frac{\mu}{\lambda}\bar{a}_0^2 - \frac{K^2}{4}} - p \qquad (29)$$

for NLS equation, and

$$\Omega_i = 4\lambda \sin\frac{K}{2} \left(\sqrt{\frac{\mu}{\lambda}\bar{a}_0^2 - \sin^2\frac{K}{2}} - \alpha(K)p \right) \qquad (30)$$

for the discrete NLS case [11] (the result is obtained for $p \ll 1$; $\alpha(K)$ is a well behaved function of K). From these expressions one can see the strong influence of the statistical properties of the medium on the MI development. There is a critical value of p, p_c, and for $p > p_c$ the instability is completely suppressed. For NLS eq. from (29) we have $p_c = 2\lambda K \sqrt{\frac{\mu}{\lambda}\bar{a}_0^2 - \frac{K^2}{4}}$ which has a maximum value for $K = (2\frac{\mu}{\lambda}\bar{a}_0^2)^{\frac{1}{2}}$. Physically the instability is possible if and only if the correlations in the initial state are of long range order. For too short range of correlations the coherent growth associated with the MI is overcome. The same conclusions follows from the analysis of (30), and has to be true for the Ablowitz-Ladik system also.

REFERENCES

1. Abdulaev F. Kh., Darmanyan S.A., Garnier J., *Modulational instability of electromagnetic waves in inhomogeneous and in discrete media*, Progress in Optics, **44**, (2002), 303-365.

2. Agrawal G.P., in *"Nonlinear Fiber Optics"*, 2-nd edition Academic Press, NY, (1995).
3. Alber I.E., *The effects of randomness on the stability of twodimensional surface wavetrains*, Proc. Roy. Soc. London A, **362**, (1978), 525-546.
4. Benjamin T.B., Feir J.E., *The disintegration of wave-trains on deep water*, J.Fluid.Mech., **27**, (1967), 417-430.
5. Dodd R.K., Eilbeck J.C., Gibbon J.D., Morris H.C., in *"Solitons and Nonlinear Wave Equations"*, ch.8, Academic Press, NY, (1982).
6. Fedele R., Shukla R.K., Onorato M., Anderson D., Lisak M., *Landau damping of partially incoherent Langmuir waves*, Phys.Lett.A, **303**, (2002), 61-66.
7. Grecu D., Visinescu A., *Modulational instability in some nonlinear 1-D lattices and soliton generation*, Ann.Univ. Craiova. Physics AUC, **12**, (2002), 129-149.
8. Hall B., Lisak M., Anderson D., Fedele R., Semenov V.E., *Statistical theory for incoherent light propagation in nonlinear media*, Phys.Rev. E, **65**, (2002), 035602 R, 1-4.
9. Onorato M., Osborne A., Serio M., Fedele R., *Landau damping and coherent structures in narrow-banded 1+1 deep water gravity waves*, Phys.Rev E, **67**, *(2003), 046305, 1-6*.
10. *Stuart J.T., DiPrima R.C., The Ekhaus and Benjamin-Feir resonance mechanisms, Proc.R.Soc. Lond. A* **362**, *(1978), 27-41*.
11. *Visinescu A., Grecu D., Statistical approach of the modulational instability of the discrete self-trapping equation, Eur. Phys. J. B,* **34**, *(2003), 225-229*.

395

AUTHOR INDEX

A

Akgül, A., 309
Alzabut, J., 317
Asada, A., 71
Ashyralyev, A., 325
Avkar, T., 84

B

Balan, V., 91
Baleanu, D., 84, 99, 106
Başkal, S., 106
Bayramov, S., 309

C

Camci, U., 114
Cârstea, A. S., 332
Cortés, J., 54
Cotăescu, I. I., 124
Czudková, L., 131

D

de Fabritiis, C., 147
Defterli, Ö., 99
Dullin, H. R., 141

G

Gerdjikov, V. S., 162
Grahovski, G. G., 162
Grecu, D., 332, 389
Gürbüz, A., 369
Guseinov, G. S., 170

H

Hall, G. S., 177
Haydargil, D., 185

J

Jensen, C. V., 193

K

Kasap, S., 339, 347
Koç, R., 185, 381
Kruglikov, B., 39
Krupka, D., 3
Krupková, O., 19

L

Léandre, R., 200
Lychagin, V., 39

M

Makhaldiani, N., 355
Manno, G., 207
Matsumoto, K., 218
Matsuyama, Y., 225
Matveev, V. S., 141
Mestdag, T., 232
Musilová, J., 131
Muslih, S. I., 240

N

Niculescu, C. P., 248

Q

Özer, M., 362

P

Pamuk, S., 369
Pashaev, O., 374
Pekşen, Ö., 257

397

S

Saunders, D. J., 262
Schneider, B., 274
Šeděnková, J., 281
Shapiro, M., 274
Smetanová, D., 289
Swaczyna, M., 297

T

Tanoğlu, G., 374
Trafalis, T. B., 339, 347
Türkdönmez, B. U., 381

Tütüncüler, H., 381

V

Visinescu, A., 332, 389
Visinescu, M., 124

Z

Zafer, A., 317